Applications and q-Extensions of Hypergeometric Functions

CONTEMPORARY MATHEMATICS

819

Applications and q-Extensions of Hypergeometric Functions

Minisymposium on All Things Hypergeometric, q-series and Generalizations
16th International Symposium on Orthogonal Polynomials,
Special Functions and Applications (OPSFA-16)
June 13–17, 2022
Centre de Recherches Mathématiques, Montréal, Québec, Canada

AMS Special Session on Hypergeometric Functions and q-series
2022 AMS Fall Western Sectional Meeting
October 22–23, 2022
University of Utah, Salt Lake City, Utah

AMS Special Session on Hypergeometric Functions, q-series
and Generalizations
2023 AMS Spring Eastern Virtual Sectional Meeting
April 1–2, 2023

Howard S. Cohl
Roberto S. Costas-Santos
Robert S. Maier
Editors

AMS AMERICAN MATHEMATICAL SOCIETY

2020 *Mathematics Subject Classification.* Primary 33C05, 33C20,33C45, 33C90, 33D15, 33E05, 14H52, 81Q05.

Library of Congress Cataloging-in-Publication Data

Names: Cohl, Howard S., 1968– editor. | Costas-Santos, Roberto S., 1974– editor. | Maier, Robert S., 1968– editor. | Minisymposium on All Things Hypergeometric, q-series and Generalizations, International Symposium on Orthogonal Polynomials, Special Functions and Applications (OPSFA-16) (16th : 2022 : Montréal, Québec). | AMS Special Session on Hypergeometric Functions and q-series (2022 : Salt Lake City, Utah). | Virtual AMS Special Session on Hypergeometric Functions, q-series and Generalizations (2023 : online).

Title: Applications and q-extensions of hypergeometric functions / Minisymposium on All Things Hypergeometric, q-series and Generalizations, 16th International Symposium on Orthogonal Polynomials, Special Functions and Applications (OPSFA-16), June 13–17, 2022, Centre de Recherches Mathématiques, Montréal, Québec, Canada ; AMS Special Session on Hypergeometric Functions and q-series, October 22–23, 2022, 2022 AMS Fall Western Sectional Meeting, University of Utah, Salt Lake City, Utah, USA; AMS Special Session on Hypergeometric Functions, q-series and Generalizations, April 1–2, 2023, AMS Spring Eastern Virtual Sectional Meeting; Howard S. Cohl, Roberto S. Costas-Santos, Robert S. Maier, editors.

Description: Providence, Rhode Island : American Mathematical Society, [2025] | Series: Contemporary mathematics, 0271-4132 ; volume 819 | Includes bibliographical references.

Identifiers: LCCN 2024039092 | ISBN 9781470476540 (paperback) | ISBN 9781470479398 (ebook)

Subjects: LCSH: Q-series–Congresses. | Hypergeometric functions–Congresses. | AMS: Special functions – Hypergeometric functions – Classical hypergeometric functions. | Special functions – Hypergeometric functions – Generalized hypergeometric series. | Special functions – Hypergeometric functions – Orthogonal polynomials and functions of hypergeometric type (Jacobi, Laguerre, Hermite, Askey scheme, etc.). | Special functions – Hypergeometric functions – Applications. | Special functions – Basic hypergeometric functions – Basic hypergeometric functions in one variable. | Special functions – Other special functions – Elliptic functions and integrals. | Algebraic geometry – Curves – Elliptic curves. | Quantum theory – General mathematical topics and methods in quantum theory – Closed and approximate solutions to the Schrödinger, Dirac, Klein-Gordon and other equations of quantum mechanics.

Classification: LCC QA295 .A68 2025 | DDC 515/.5–dc23/eng/20250207

LC record available at https://lccn.loc.gov/2024039092

Contemporary Mathematics ISSN: 0271-4132 (print); ISSN: 1098-3627 (online)

DOI: https://doi.org/10.1090/conm/819

∞ The paper used in this book is acid-free and falls within the guidelines
established to ensure permanence and durability.
Visit the AMS home page at `https://www.ams.org/`

10 9 8 7 6 5 4 3 2 1 30 29 28 27 26 25

Dedicated to Richard Askey, who noticed and expounded on the utility and functionality of hypergeometric functions.

Contents

Preface

Hypergeometric functions, and the hypergeometric series by which they can be defined, are parametric mathematical objects of much interest in classical analysis. 'The' hypergeometric function is the classical function now denoted by ${}_2F_1(z)$, which was investigated by Gauss and is now named after him, though it was previously studied by Euler. It satisfies a second-order differential equation with respect to z, and the coefficients of its defining power series satisfy a two-term recurrence. Beginning with Kummer, ${}_2F_1$ was generalized to ${}_rF_s$. All these functions satisfy contiguous function relations in which parameters are displaced by integers [**8**], and they have many interesting transformation properties. The nicest generalized hypergeometric functions may be the non-confluent ones, for which $r = s + 1$.

These classical functions are prominent among the many special functions that arise in mathematical physics and elsewhere. They also appear repeatedly in the theory of orthogonal polynomials. For instance, Jacobi polynomials are simply terminating hypergeometric series of the ${}_2F_1$ (Gauss) type, and the three-term recurrences that these orthogonal polynomials satisfy can be viewed as specializations of certain contiguous function relations satisfied by the parametric function ${}_2F_1$. Jacobi polynomials include Gegenbauer and Legendre polynomials, as specializations. The classical hypergeometric functions ${}_rF_s$ have q-extensions, the so-called *basic* hypergeometric functions, which satisfy q-difference equations and are objects of intense interest in q-calculus. (The classical functions are obtained by suitable $q \to 1$ limits.) The interested reader will find a wealth of related information in the monographs of Andrews, Askey, and Roy [**1**], Gasper and Rahman [**4**], and Ismail [**7**].

This is the second volume of a two-volume collection of recent research results related to hypergeometric functions, which come with full bibliographies and should serve as entry points to the current literature. The first volume is "Classical Hypergeometric Functions and Generalizations" [**2**]. The present volume is divided into two parts. The papers in the first part focus on applications of classical hypergeometric functions and some of their extensions, in mathematical physics and elsewhere. The functions appearing include complete and incomplete elliptic integrals, and the applications include the explicit solution of the stationary Schrödinger equation with various potentials. The papers in the second part deal with basic hypergeometric functions and their properties, which may be similar to or richer than the properties of their classic counterparts. For instance, the Askey scheme of hypergeometric orthogonal polynomials has a natural "q-Askey" generalization. The second part also includes an important $q \to -1$ limit of basic hypergeometric orthogonal polynomials.

The first part contains five papers of varying lengths. The first paper (by J. Dereziński, C. Gaß and B. Ruba) is *Generalized integrals of Macdonald and Gegenbauer functions.* The second paper (by L. Ellis, I. Ellis, C. Koutschan and S. K. Suslov) is *On potentials integrated by the Nikiforov–Uvarov method.* The third paper (by Jeongsu Kyeong, I. Mitrea and K. Ott) is *On the Mellin symbol of singular integral operators associated with the biharmonic equation in infinite sectors.* The fourth paper (by the late N. Lebovitz) is *Gravitational potentials of ellipsoidal masses.* The fifth paper (by I. Sitiraju) is *Spherical distributions on the de Sitter space and their spectral singularities.*

The second, 'basic' part of this volume treats such topics as q-Appell functions; very-well-poised basic hypergeometric functions; q-continued fractions; basic hypergeometric orthogonal polynomials; limits of basic hypergeometric orthogonal polynomials; Bailey chains; product representations; and basic and elliptic extensions of classical identities. It features seven papers. The first paper (by G. E. Andrews) is *The surprising first q-Appell function.* The second paper (by G. Bhatnagar) is *Ramanujan's q-continued fractions.* The third paper (by H. S. Cohl and R. S. Costas-Santos) is *Orthogonality of the big* -1 *Jacobi polynomials for non-standard parameters.* The fourth paper (by N. Hoshi, M. Katori, T. H. Koornwinder and M. J. Schlosser) is *On an identity of Chaundy and Bullard. III. Basic and elliptic extensions.* The fifth paper (by T. Huber, J. Mc Laughlin, and Dongxi Ye) is *Generalizations of some q-product identities of Ramanujan and others*, which focuses on finding new basic hypergeometric expansions for some q-products and series that Ramanujan expressed in terms of Lambert series. The sixth paper (by L. Verde-Star) is *Discrete orthogonality of the polynomial sequences in the q-Askey scheme.* The seventh paper (by Jianan Xu, Ying Zhang and Xinrong Ma) is *A new three-term relation for an indefinite sum of a very-well-poised basic hypergeometric series*, which closely studies Bailey's very-well-poised ${}_6\psi_6$ summation formula.

It should be noted that the paper in the first part by Norman Lebovitz (1935–2022) was graciously provided by his family and the Department of Mathematics at the University of Chicago, after he passed away. Because Norman is no longer available to comment on or to correct possible errors and misprints in his work, we are indebted to several individuals who provided significant commentary and reports on this beautiful piece of work, namely Joel Tohline, Toshio Fukushima, George Dassios, and Hans Volkmer. It was pointed out by the last two reviewers that although the paper contains very decent references to the history of the gravitational potential of ellipsoids, one should also cite other historically important references such as Rodrigues [**9**], Heine [**5**, pp. 31–173], and Hobson [**6**]. In Tohline's review he eloquently remarks on the importance of the Lebovitz manuscript: [*F*]*rom my perspective, the strength of this manuscript lies in its practical utility; it is a contribution by an extraordinarily talented applied mathematician to his colleagues in the astrophysics community, providing a firm analytic foundation to those of us who are striving to model the structure and evolution of nontrivial self-gravitating fluid flows. This is a task that we would not relish tackling on our own, given that Lebovitz himself has acknowledged that a particular verification "is rather arduous"; and that another result has been obtained "somewhat tediously." With this work in hand as a guide, we can relax our previous focus on incompressible fluids and begin to draw upon the analytically known "Ferrers potentials" to test the tools* *that we have built to numerically examine compressible (i.e., non-uniform density) triaxial flows.*

This two-volume collection follows in the tradition of an earlier volume in the AMS Contemporary Mathematics series, "Special Functions and Orthogonal Polynomials," edited by D. Dominici and R. S. Maier [**3**]. The contributions to that volume by the late Prof. Diego Dominici are warmly remembered.

Howard S. Cohl
Roberto S. Costas-Santos
Robert S. Maier

Bibliography

[1] G. E. Andrews, R. Askey, and R. Roy, *Special functions*, Encyclopedia of Mathematics and its Applications, vol. 71, Cambridge University Press, Cambridge, 1999, DOI 10.1017/CBO9781107325937. MR1688958

[2] H. S. Cohl, R. S. Costas-Santos, and R. S. Maier (eds.), *Classical hypergeometric functions and generalizations*, Contemporary Mathematics, vol. 818, American Mathematical Society, Providence, RI, 2025. Minisymposium on All Things Hypergeometric, q-series and Generalizations—16th International Symposium on Orthogonal Polynomials, Special Functions and Applications (OPSFA-16), June 13–17, 2022, Centre de Recherches Mathématiques, Montréal, Québec, Canada; AMS Special Session on Hypergeometric Functions and q-series—2022 AMS Fall Western Sectional Meeting, October 22–23, 2022, University of Utah, Salt Lake City, Utah; AMS Special Session on Hypergeometric Functions, q-series and Generalizations—2023 AMS Spring Eastern Virtual Sectional Meeting, April 1–2, 2023; Edited by Howard S. Cohl, Roberto S. Costas-Santos and Robert S. Maier, DOI 10.1090/conm/818. MR4901156

[3] D. Dominici and R. S. Maier (eds.), *Special functions and orthogonal polynomials*, Contemporary Mathematics, vol. 471, American Mathematical Society, Providence, RI, 2008. Papers from the Special Session held at the University of Arizona, Tucson, AZ, April 21–22, 2007; Edited by Diego Dominici and Robert S. Maier, DOI 10.1090/conm/471. MR2367444

[4] G. Gasper and M. Rahman, *Basic hypergeometric series*, 2nd ed., Encyclopedia of Mathematics and its Applications, vol. 96, Cambridge University Press, Cambridge, 2004. With a foreword by Richard Askey, DOI 10.1017/CBO9780511526251. MR2128719

[5] E. Heine, *Handbuch der Kugelfunctionen, Theorie und Anwendungen (volume 2)*. Druck und Verlag von G. Reimer, Berlin, 1881.

[6] E. W. Hobson, *The theory of spherical and ellipsoidal harmonics*, Chelsea Publishing Co., New York, 1955. MR64922

[7] M. E. H. Ismail, *Classical and quantum orthogonal polynomials in one variable*, Encyclopedia of Mathematics and its Applications, vol. 98, Cambridge University Press, Cambridge, 2005. With two chapters by Walter Van Assche; With a foreword by Richard A. Askey, DOI 10.1017/CBO9781107325982. MR2191786

[8] E. D. Rainville, *Special functions*, The Macmillan Company, New York, 1960. MR107725

[9] O. Rodrigues, Mémoire sur l'attraction des spheroïdes. *Corresp. sur l'école polyt.*, III:361–385, 1815.

Applications

Contemporary Mathematics
Volume **819**, 2025
https://doi.org/10.1090/conm/819/16387

Generalized integrals of Macdonald and Gegenbauer functions

Jan Dereziński, Christian Gaß, and Błażej Ruba

Abstract. We compute bilinear integrals involving Macdonald and Gegenbauer functions. These integrals are convergent only for a limited range of parameters. However, when one uses *generalized integrals*, they can be computed essentially without restricting the parameters. The generalized integral is a linear functional extending the standard integral to a certain class of functions involving finitely many homogeneous non-integrable terms at the endpoints of the interval. For generic values of parameters, generalized bilinear integrals of Macdonald and Gegenbauer functions can be obtained by analytic continuation from the region in which the integrals are convergent. In the case of integer parameters we obtain expressions with explicit additional terms related to an *anomaly*, namely the failure of the generalized integral to be scaling invariant.

1. Introduction

Consider a *Sturm–Liouville operator*

$$\mathcal{C} := -\rho(r)^{-1}\big(\partial_r p(r)\partial_r + q(r)\big) \tag{1.1}$$

acting on functions on an interval $]a,b[$. $\mathcal{C}$ is formally symmetric for the bilinear scalar product with the *density* ρ:

$$\langle f|g\rangle := \int_a^b f(r)g(r)\rho(r)\,\mathrm{d}r. \tag{1.2}$$

Let f be an eigenfunction of $\mathcal{C}$, that is, $\mathcal{C}f = Ef$. In important applications one needs to know the value of the scalar product of f with itself:

$$\langle f|f\rangle = \int_a^b f(r)^2\rho(r)\,\mathrm{d}r. \tag{1.3}$$

2020 *Mathematics Subject Classification.* Primary 33C05, 33C10, 47A52.

The work of J. D. and C. G. was supported by the National Science Center of Poland under the grant UMO-2019/35/B/ST1/01651. J. D. would like to thank Howard Cohl for an inspiring discussion at an early stage of this work. We thank Michał Wrochna for pointing out the references **[13, 18]**.

There exists a simple method, which often allows us to evaluate (1.3). If f_i are eigenfunctions corresponding to two different eigenvalues E_i, then

$$\int_a^b f_1(r)f_2(r)\rho(r)\,\mathrm{d}r = \frac{\mathcal{W}(f_1,f_2,b)-\mathcal{W}(f_1,f_2,a)}{E_1-E_2}, \tag{1.4}$$

$$\text{where}\quad \mathcal{W}(f_1,f_2,r) := p(r)\big(f_1(r)f_2'(r)-f_1'(r)f_2(r)\big). \tag{1.5}$$

(1.4) is sometimes called *Green's identity*, or the *integrated Lagrange identity*. The rhs of (1.4) can often be easily evaluated. Typically this is possible if the endpoints a, b are singular points of the corresponding differential equation. We will then say that $[a, b]$ is a *natural interval* for the operator $\mathcal{C}$. (1.4) is undefined for $f = f_1 = f_2$, but one can often use a limiting procedure to derive (1.3) from (1.4). We will call integrals of the form (1.3) or (1.4) *bilinear integrals*.

In this paper we consider two families of Sturm–Liouville operators: the *Bessel operator* and the *Gegenbauer operator*:

$$\mathcal{B}_\alpha := -\partial_r^2 - \frac{1}{r}\partial_r + \frac{\alpha^2}{r^2}, \tag{1.6}$$

$$\mathcal{G}_\alpha := -(1-w^2)\partial_w^2 + 2(1+\alpha)w\partial_w. \tag{1.7}$$

$\mathcal{B}_\alpha$ and $\mathcal{G}_\alpha$ are very common in applications, e.g., in mathematical physics. Separation of variables in the Laplace equation on $\mathbb{R}^d$ leads to the Bessel operator with $\alpha = \frac{d-2}{2}$, while separation of variables in the Laplace equation on the sphere $\mathbb{S}^d$ and on the hyperbolic space $\mathbb{H}^d$ lead to the Gegenbauer operator.

The well-known *modified Bessel equation* is the eigenequation of (1.6) with the eigenvalue -1. The even better-known *(standard) Bessel equation* is its eigenequation for the eigenvalue 1. $[0, \infty[$ is a natural interval for $\mathcal{B}_\alpha$, and the density is $\rho(r) = 2r$. *Macdonald functions* are eigenfunctions of $\mathcal{B}_\alpha$ decaying fast at infinity.

The *Gegenbauer equation* is the eigenequation of (1.7) with an eigenvalue that we parametrize as $\lambda^2 - \big(\alpha + \frac{1}{2}\big)^2$. The natural intervals for $\mathcal{G}_\alpha$ are $[-1, 1]$ with $\rho(w) = (1-w^2)^\alpha$, and $[1, \infty[$ with $\rho(w) = (w^2-1)^\alpha$.

The main goal of our paper is to describe bilinear integrals involving Macdonald and Gegenbauer functions. These integrals are convergent only for a limited range of parameters. Therefore, we introduce a concept of the *generalized integral*, which allows us to compute (1.3) and (1.4) for essentially all parameters.

The generalized integral is a linear functional on a certain class of functions on the halfline. This class consists of functions integrable on $[1, \infty[$ whose restriction to $[0, 1]$ is a linear combination of an integrable function and homogeneous functions. On integrable functions, the generalized and the usual (Lebesgue) integral coincide. The generalized integral of non-integrable homogeneous terms is defined as follows:

$$\mathrm{gen}\!\int_0^1 r^\lambda\,\mathrm{d}r = \begin{cases} \frac{1}{\lambda+1} & \text{for } \lambda \neq -1, \\ 0 & \text{for } \lambda = -1. \end{cases} \tag{1.8}$$

Note that the case $\lambda = -1$ is troublesome.

We say that the generalized integral is *anomalous* if at least one of coefficients at negative integer powers of r is nonzero. In the anomalous case the generalized integral has more complicated properties. This case often appears in applications and is especially interesting.

The behavior of Macdonald and Gegenbauer functions near the endpoints of integration crucially depends on the parameter α. This parameter determines the index of the singular point of the Bessel/Gegenbauer equation. The bilinear integrals of Macdonald and Gegenbauer functions are convergent in the standard sense if and only if $|\mathrm{Re}(\alpha)| < 1$. For all other α they are well-defined only in the generalized sense. For $\alpha \in \mathbb{Z}\backslash\{0\}$, these integrals are anomalous.

Let us describe the main application of bilinear integrals (1.3) of Macdonald and Gegenbauer functions that we have in mind **[8]**. Consider a Schrödinger operator on $\mathbb{R}^d$, $\mathbb{S}^d$ or $\mathbb{H}^d$ with a potential confined to a very small region. One often approximates its Green function (that is, the integral kernel of its resolvent) by an expression derived by assuming that the potential is supported at a point. This expression involves a bilinear integral of Macdonald functions for $\mathbb{R}^d$ and of Gegenbauer functions for $\mathbb{H}^d$ and $\mathbb{S}^d$. This is a usual convergent integral in dimensions $d = 1, 2, 3$, which corresponds to $\alpha = -\frac{1}{2}, 0, \frac{1}{2}$. In these dimensions one obtains a 1-parameter family of self-adjoint realizations of the Laplacian perturbed by a point-like potential. The situation is different in dimensions $d \geqslant 4$. The usual bilinear integral is divergent and point interactions do not lead to self-adjoint realizations of the Laplacian.

We expect **[8, 9]** that in all dimensions Green functions of the Laplacian with perturbations supported in a small region are asymptotic to certain *renormalized Green functions*, which depend on a finite number of parameters. If $d \geqslant 4$, except for the trivial (unperturbed) one, these renormalized Green functions are not integral kernels of bounded operators—hence the Laplacian perturbed by a point potential cannot be interpreted as a self-adjoint operator. The parameters of nontrivial renormalized Green functions form naturally an affine space. The generalized integral leads to a distinuished renormalized Green function—it provides a reference point in the affine space of these parameters.

Note that there is a considerable difference between even and odd dimensions. In even dimension bilinear integrals are anomalous, while in odd dimensions they are not. This corresponds to logarithmic terms in even dimensions, which are absent in odd dimensions.

The case $d = 4$, corresponding to $\alpha = 1$, is of particular interest. It is the borderline case: the bilinear integrals are well defined in the usual sense for smaller α, but for $\alpha = 1$ one has to use their generalized version. Besides, they are anomalous.

The above analysis is closely related to renormalization in Quantum Field Theory, especially based on the dimensional regularization. After the Wick rotation, the Minkowski space becomes the Euclidean space $\mathbb{R}^4$, and the d'Alembertian becomes the Laplacian. As described in **[4]**, the configuration space version of dimensional regularization in Quantum Field Theory can be carried out by allowing the order α of Bessel functions describing the propagators to be complex, so that certain integrals are well-defined in the usual sense. Then one has to subtract the terms which diverge if α approaches the physical value and take the limit. Thus renormalization in Quantum Field Theory can be interpreted as an application of the (anomalous) generalized integral.

1.1. Remarks about literature. The generalized integral is closely related to the standard theory of extensions of *homogeneous distributions*, as described by Hörmander **[17]**, and earlier by Hadamard **[15, 16]**, Riesz **[25]**, and Gel'fand **[13]**.

Various kinds of generalized integrals appear in the literature. One can divide them into two kinds: those involving non-integrability at a finite point and those for which the problem comes from ∞. In our paper we use only the former setting:

we assume that the integrand is integrable close to infinity, however close to the finite endpoint at the left-hand side of the interval it may involve non-integrable homogeneous terms. In this setting definitions equivalent to our generalized integrable were given by Hadamard [**15**, **16**] and Riesz [**25**]. A recent exposition of this topic can be found in a book by Paycha [**23**, Chapter 1].

A very general discussion of the generalized integral is contained in [**18**]. The generalized integral is defined there in terms of the Mellin transform, and one can account for sums of almost homogeneous singularities at 0 and ∞ simultaneously.

Let us mention that various forms of the generalized integrals where the integrand is not integrable near ∞ are probably more common in the literature. Starting from Chapter 2, the book [**23**] is devoted mostly to this setting. The noncommutative version of the generalized integral for polyhomogeneous pseudodifferential operators in the non-anomalous case is sometimes called the *Vishik–Kontsevich trace.* In the anomalous case it is related to the *Wodzicki residue.*

The Gegenbauer equation is essentially a special case of the hypergeometric equation, where we put the finite singular points at $-1, 1$ and we assume the symmetry with respect to $w \to -w$. Its particular solutions for special values of parameters are the well-known *Gegenbauer polynomials.* The Gegenbauer equation is equivalent through a simple transformation to the *associated Legendre equation.* In the literature, *associated Legendre functions* are much more common than Gegenbauer functions. However, the use of Gegenbauer functions with the conventions that we introduce, rather than associated Legendre functions, leads to a simplification of various identities, and therefore seems to be preferable.

The application of bilinear integrals to Green's functions of the Laplacian with point interactions will be described in a separate paper [**8**]. It is a classic result that Green's functions without point interactions can be expressed in terms of Macdonald functions for $\mathbb{R}^d$ and Gegenbauer (or associated Legendre) functions for $\mathbb{H}^d$ and $\mathbb{S}^d$ (cf., e.g., [**5**, **26**]). Green's functions with point potentials in dimensions 1,2,3, at least in the flat case, are also well-known [**1**–**3**, **19**, **24**]. Green's functions with point potentials in all dimensions, and also for $\mathbb{S}^d$ and $\mathbb{H}^d$, will be described in our subsequent paper [**8**].

1.2. Outline of the paper.

The main goal of the present article is to pedagogically describe elements of the theory the generalized integral and of Bessel and Gegenbauer equation needed in [**8**].

In Section 2, we define the generalized integral and study its properties. We describe its behavior under a change of variables. We describe a method of computing generalized integrals which resembles the *dimensional regularization* in Quantum Field Theory, and actually can be traced back to an earlier work of M. Riesz.

In Section 3 we recall elements of the theory of the Bessel equation. The main new result is the computation of generalized bilinear integrals of Macdonald functions using the definitions and methods of Section 2.

Section 4 is devoted to the Gegenbauer equation. We introduce two kinds of Gegenbauer functions: $\mathbf{S}_{\alpha,\lambda}$ with a simple behavior near 1 and $\mathbf{Z}_{\alpha,\lambda}$ with a simple behavior near $+\infty$. We compute bilinear integrals of $\mathbf{S}_{\alpha,\lambda}$ and $\mathbf{Z}_{\alpha,\lambda}$, both usual and generalized. We discuss various kinds of integral representations of Gegenbauer functions. These representations are then used to show that Gegenbauer functions are asymptotic to Macdonald functions. We also prove that, if we choose the variables correctly, then bilinear integrals of Gegenbauer functions are asymptotic to bilinear integrals of Macdonald functions.

For obvious reasons, Section 4 about the Gegenbauer equation is more complex than Section 3. However, both sections have to a large extent a parallel structure.

This paper has two appendices. In Appendix A, we list several useful properties of special functions related to the Gamma function, which are needed in the main part of the paper. Appendix B contains an overview of conventions for functions related to Gegenbauer functions.

2. Generalized integral

2.1. Definition of the generalized integral. In this section we introduce the *generalized integral*, which extends the usual integral to a certain class of non-integrable functions. We restrict ourselves to functions which fail to be integrable near the (finite) left endpoint of the integration interval. Our definition essentially coincides with similar concepts introduced in [**23**, Chapter 1], and goes back to the works of Riesz and Hadamard.

One should mention that it is also natural to define the generalized integral for functions not integrable close to ∞. This is described, e.g., in [**23**] starting from Chapter 2, and is actually more common in the literature. In our work this will not be considered.

DEFINITION 2.1. Let $a \in \mathbb{R}$. We say that a function f on $]a,\infty[$ is *integrable in the generalized sense* if it is integrable on $]a+1,\infty[$ and there exists a finite set $\Omega \subset \mathbb{C}$ and complex coefficients $(f_k)_{k\in\Omega}$ such that

$$f - \sum_{k\in\Omega} f_k (r-a)^k \tag{2.1}$$

is integrable on $]a, a+1[$. We define

$$\begin{aligned} &\operatorname{gen}\int_a^\infty f(r)\,\mathrm{d}r \\ &:= \sum_{k\in\Omega\setminus\{-1\}} \frac{f_k}{k+1} + \int_a^{a+1} \Big(f(r) - \sum_{k\in\Omega} f_k (r-a)^k\Big)\,\mathrm{d}r + \int_{a+1}^\infty f(r)\,\mathrm{d}r. \end{aligned} \tag{2.2}$$

We note that the set $\{k \in \Omega \,|\, \mathrm{Re}(k) \leqslant -1\}$ and the corresponding f_k are uniquely determined by f. It is convenient to allow $k \in \Omega$ with $\mathrm{Re}(k) > -1$. The generalized integral of f does not depend on the choice of Ω.

The generalized integral extends the standard integral:

$$\operatorname{gen}\int_a^\infty f(r)\,\mathrm{d}r = \int_a^\infty f(r)\,\mathrm{d}r \quad \text{for } f \in L^1[a,\infty[. \tag{2.3}$$

It is translation invariant in the following sense:

$$\operatorname{gen}\int_a^\infty f(r)\,\mathrm{d}r = \operatorname{gen}\int_{a-\alpha}^\infty f(u+\alpha)\,\mathrm{d}u, \quad \alpha \in \mathbb{R}. \tag{2.4}$$

In particular, we can without loss of generality assume $a = 0$.

If f is defined on an interval $]0,b[$ with $b > 0$, we define $\operatorname{gen}\int_0^b f(r)\,\mathrm{d}r$ as the generalized integral of the extension of f to $[0,\infty[$ by zero.

PROPOSITION 2.2. *If f satisfies Def. 2.1 on $[0,\infty[$, then*

$$\operatorname{gen}\int_0^\infty f(r)\,\mathrm{d}r = \lim_{\delta\searrow 0}\left(\int_\delta^\infty f(r)\,\mathrm{d}r + \sum_{k\in\Omega\setminus\{-1\}}\frac{f_k}{k+1}\delta^{k+1} + f_{-1}\ln(\delta)\right). \tag{2.5}$$

PROOF. For any $0<\delta\leqslant 1$ we have

$$\begin{aligned}\operatorname{gen}\int_0^\infty f(r)\,\mathrm{d}r &= \int_\delta^\infty f(r)\,\mathrm{d}r + \sum_{k\in\Omega\setminus\{-1\}}\frac{f_k}{k+1}\delta^{k+1} + f_{-1}\ln(\delta)\\ &\quad+\int_0^\delta\Big(f(r)-\sum_{k\in\Omega}f_k r^k\Big)\,\mathrm{d}r.\end{aligned} \tag{2.6}$$

The last term of (2.6) converges to 0 as $\delta\to 0$. □

2.2. Change of variables. The standard formula for a change of variables in an integral is not true for the generalized integral, but one may write down the correction terms explicitly.

PROPOSITION 2.3. *Suppose that $g\colon [0,\infty[\to[0,\infty[$ is a bijection, smooth down to zero, such that $g(0)=0$ and $g'(0)\neq 0$. If f is integrable in the generalized sense, then the same is true for $(f\circ g)g'$ and we have*

$$\begin{aligned}&\operatorname{gen}\int_0^\infty f(g(u))g'(u)\,\mathrm{d}u - \operatorname{gen}\int_0^\infty f(r)\,\mathrm{d}r\\ &\quad= -f_{-1}\ln g'(0) + \sum_{l=2}^\infty \frac{f_{-l}}{(l-1)(l-1)!}\,\frac{\mathrm{d}^{l-1}}{\mathrm{d}u^{l-1}}\left(\frac{u}{g(u)}\right)^{l-1}\Bigg|_{u=0},\end{aligned} \tag{2.7}$$

where we put $f_k=0$ for $k\notin\Omega$. In particular for $g(u)=\alpha u$, $\alpha>0$:

$$\operatorname{gen}\int_0^\infty f(\alpha u)\alpha\,\mathrm{d}u = \operatorname{gen}\int_0^\infty f(r)\,\mathrm{d}r - f_{-1}\ln\alpha. \tag{2.8}$$

PROOF. The first statement easily follows from the Taylor expansion of g around zero. Let $\Delta_f(g)$ be the left-hand side of (2.7). We have

$$\Delta_f(g) = \int_{g(1)}^1 f(r)\,\mathrm{d}r + \operatorname{gen}\int_0^1 \big(f(g(r))g'(r)-f(r)\big)\,\mathrm{d}r, \tag{2.9}$$

where we used the standard change of variables formula for the integral over $[1,\infty[$ and $\int_{g(1)}^1 - \int_1^\infty = \int_{g(1)}^1$. Now decompose $f(r)=F(r)+\sum_{k\in\Omega}f_k r^k$. It is easy to check that F gives no contribution in (2.9), so

$$\Delta_f(g) = \sum_{k\in\Omega} f_k\left(\int_{g(1)}^1 r^k\,\mathrm{d}r + \operatorname{gen}\int_0^1\big(g(r)^k g'(r) - r^k\big)\,\mathrm{d}r\right). \tag{2.10}$$

We will analyse this expression term by term. First consider $k\neq -1$. Then

$$\begin{aligned}&\int_{g(1)}^1 r^k\,\mathrm{d}r + \operatorname{gen}\int_0^1\big(g(r)^k g'(r)-r^k\big)\,\mathrm{d}r\\ &\quad= -\frac{g(1)^{k+1}}{k+1} + \frac{1}{k+1}\operatorname{gen}\int_0^1\frac{\mathrm{d}}{\mathrm{d}r}g(r)^{k+1}\,\mathrm{d}r.\end{aligned} \tag{2.11}$$

We have a Taylor expansion of the form

$$g(r)^{k+1} = r^{k+1}\left(\sum_{j=0}^{N} a_j r^j + R(r)\right), \tag{2.12}$$

where $R(r) = \mathcal{O}(r^{N+1})$ and N is such that $\mathrm{Re}(N+k+2) > 0$. Then

$$\begin{aligned}\operatorname{gen}\int_0^1 \frac{\mathrm{d}}{\mathrm{d}r} g(r)^{k+1}\,\mathrm{d}r &= \sum_{j=0}^{N} (k+j+1) a_j \operatorname{gen}\int_0^1 r^{k+j}\,\mathrm{d}r + R(1)\\ &= \sum_{\substack{j=0\\ k+j\neq -1}}^{N} a_j + R(1) = g(1)^{k+1} - a_{-k-1},\end{aligned} \tag{2.13}$$

where we put $a_{-k-1} = 0$ if $-k-1 \notin \mathbb{N}$. Therefore, (2.11) vanishes for $-k-1 \notin \mathbb{N}$. If $k = -l$ with $l \in \{2, 3, \dots\}$, then (2.11) is equal to

$$-\frac{1}{-l+1} a_{l-1} = \frac{1}{(l-1)(l-1)!}\frac{\mathrm{d}^{l-1}}{\mathrm{d}r^{l-1}}\left(\frac{r}{g(r)}\right)^{l-1}\Bigg|_{r=0}. \tag{2.14}$$

To handle the case $k = -1$, we write $g(r) = g'(0) r\,\mathrm{e}^{R(r)}$ with $R(0) = 0$ and compute

$$\operatorname{gen}\int_0^1 \frac{g'(r)}{g(r)}\,\mathrm{d}r = R(1) = \ln g(1) - \ln g'(0). \tag{2.15}$$

□

Formula (2.8) shows that the generalized integral is invariant under scaling if and only if $f_{-1} = 0$, and invariant under a large class of a change of variables if $f_k = 0$ for every negative integer k.

DEFINITION 2.4. The generalized integral (2.2) is called *anomalous* if there exists $n = 1, 2, \dots$ such that $f_{-n} \neq 0$.

Remarkably, the generalized integral is always invariant under changes of variables given by power functions.

PROPOSITION 2.5. *Let f be integrable in the generalized sense. Then for $\alpha > 0$*

$$\operatorname{gen}\int_0^\infty f(r)\,\mathrm{d}r = \operatorname{gen}\int_0^\infty f(u^\alpha)\,\alpha u^{\alpha-1}\,\mathrm{d}u. \tag{2.16}$$

PROOF. The formula is true for integrable f, so by linearity it is sufficient to verify it for $f(r) = r^k \mathbb{1}_{[0,1]}(r)$. This is an elementary calculation. □

2.3. Dimensional regularization. In this subsection we describe a method to compute generalized integrals by analytic continuation. It is closely related to dimensional regularization used in QFT.

Let F be a holomorphic function on $U \setminus \{z_0\}$, where $z_0 \in U$ and U is open. Then one has a Laurent expansion

$$F(z) = \sum_{j\in\mathbb{Z}} F_j (z-z_0)^j \tag{2.17}$$

convergent for $z \neq z_0$ sufficiently close to z_0. Coefficients $\operatorname{res} F(z_0) := F_{-1}$ and $\operatorname{fp}_{z\to z_0} F(z) := F_0$ are called the residue and the finite part of F at z_0.

Let $N \in \mathbb{N}$ and let $f\colon\ ]0,\infty[\times\{\alpha \in \mathbb{C} \mid \mathrm{Re}(\alpha) > -N-1\} \to \mathbb{C}$ be a function such that $f(r,\cdot)$ is holomorphic for each r, $\|f(\cdot,\alpha)\|_{L^1[1,\infty[}$ is bounded locally uniformly in α, and there exist holomorphic functions $f_0, \dots, f_N$ of α such that the $L^1]0,1]$ norm of $f(r,\alpha) - \sum_{n=0}^N r^{\alpha+n} f_n(\alpha)$ is bounded locally uniformly in α. Then $f(\cdot,\alpha)$ is integrable in the generalized sense, and for $-\alpha \notin \{1,\dots,N\}$ one has

$$\begin{aligned}&\operatorname{gen}\int_0^\infty f(r,\alpha)\,\mathrm{d}r \\ &= \sum_{n=0}^N \frac{f_n(\alpha)}{\alpha+n+1} + \int_0^1 \Big(f(r,\alpha) - \sum_{n=0}^N r^{\alpha+n} f_n(\alpha)\Big)\,\mathrm{d}r + \int_1^\infty f(r,\alpha)\,\mathrm{d}r.\end{aligned} \tag{2.18}$$

By Morera's theorem, the right-hand side is, away from the poles at $-1,\dots,-N$, a holomorphic function of α. Therefore, to obtain (2.18) in the non-anomalous case it is enough to compute (2.18) in the region where the usual integral is convergent and continue analytically.

Let $m \in \{1,\dots,N\}$. The right-hand side of (2.18) has a simple pole at $\alpha = -m$ with residue $f_{m-1}(-m)$ (possibly zero). Its finite part is

$$\operatorname*{fp}_{\alpha\to-m} \operatorname{gen}\int_0^\infty f(r,\alpha)\,\mathrm{d}r = \lim_{\alpha\to-m}\left(\operatorname{gen}\int_0^\infty f(r,\alpha)\,\mathrm{d}r - \frac{f_{m-1}(-m)}{\alpha+m}\right). \tag{2.19}$$

One may be tempted to think that the finite part is equal to the generalized integral of $f(\cdot,-m)$, but in fact one has to subtract also a certain finite term.

PROPOSITION 2.6. *The generalized integral at* $\alpha = -m$ *is given by*

$$\operatorname{gen}\int_0^\infty f(r,-m)\,\mathrm{d}r = \operatorname*{fp}_{\alpha\to-m} \operatorname{gen}\int f(r,\alpha)\,\mathrm{d}r - f'_{m-1}(-m). \tag{2.20}$$

PROOF. By definition, the generalized integral of $f(\cdot,-m)$ is given by (2.18) evaluated at $-m$, with the term $n = m-1$ in the summation omitted. Therefore,

$$\operatorname{gen}\int_0^\infty f(r,-m)\,\mathrm{d}r = \lim_{\alpha\to-m}\left(\operatorname{gen}\int_0^\infty f(r,\alpha)\,\mathrm{d}r - \frac{f_{m-1}(\alpha)}{\alpha+m}\right). \tag{2.21}$$

The formula (2.20) immediately follows. □

The following facts deserve emphasis.

- The term $f'_{m-1}(-m)$ in (2.20) may be nonzero even if the generalized integral of $f(r,\alpha)$ has a removable singularity at $\alpha = -m$.
- The finite part for $\alpha \to -m$ of the analytic continuation of $\int_0^\infty f(r,\alpha)\,\mathrm{d}r$ is not uniquely determined by the function $f(r,-m)$, in contrast to the generalized integral of $f(r,-m)$.

2.4. Examples. Let us give a few examples of generalized integrals.

EXAMPLE 2.7.

$$\operatorname{gen}\int_0^1 \alpha r^{-1+\alpha}\,\mathrm{d}r = \begin{cases} 1 & \text{for } \alpha \neq 0; \\ 0 & \text{for } \alpha = 0. \end{cases} \tag{2.22}$$

Therefore, the limit of generalized integrals of $f(r,\alpha) = \alpha r^{-1+\alpha}\mathbb{1}_{[0,1]}(r)$ for $\alpha \to 0$ is not the generalized integral of $f(r,0)$. The finite part for $\alpha \to 0$ (in this case, the limit) is nonzero even though $f(r,0) = 0$.

EXAMPLE 2.8 (Gamma integral).

$$\operatorname{gen}\int_0^\infty \mathrm{e}^{-r} r^{-1+\alpha}\,\mathrm{d}r = \begin{cases} \Gamma(\alpha), & \alpha \notin -\mathbb{N}_0; \\ \frac{(-1)^m}{m!}\psi(m+1), & \alpha = -m \in -\mathbb{N}_0. \end{cases}$$

EXAMPLE 2.9 (Beta integrals). Assume that $\mathrm{Re}(v) > 0$. Then

$$\operatorname{gen}\int_0^1 r^{-1+u}(1-r)^{v-1}\,\mathrm{d}r = \begin{cases} \frac{\Gamma(u)\Gamma(v)}{\Gamma(u+v)}, & u \notin -\mathbb{N}_0; \\ \frac{(-1)^m (v-m)_m \left(\psi(m+1)-\psi(v-m)\right)}{m!}, & u = -m \in -\mathbb{N}_0. \end{cases}$$

$$\operatorname{gen}\int_0^\infty r^{-1+u}(1+r)^{v-1}\,\mathrm{d}r = \begin{cases} \frac{\Gamma(u)\Gamma(1-u-v)}{\Gamma(1-v)}, & u \notin -\mathbb{N}_0; \\ \frac{(-1)^m (1-v)_m \left(\psi(m+1)-\psi(1+m-v)\right)}{m!}, & u = -m \in -\mathbb{N}_0. \end{cases}$$

3. Bessel equation

3.1. Modified Bessel equation. Here is the *modified Bessel equation*:

$$\left(\partial_r^2 + \frac{1}{r}\partial_r - \frac{\alpha^2}{r^2} - 1\right) v(r) = 0. \tag{3.1}$$

There are two standard solutions of (3.1). The first is the *modified Bessel function*, which can be defined by the power series

$$I_\alpha(r) = \sum_{n=0}^\infty \frac{\left(\frac{r}{2}\right)^{2n+\alpha}}{n!\,\Gamma(\alpha+n+1)}.$$

The second solution is the *Macdonald function*, which for $\mathrm{Re}(r) > 0$ and all α can be defined by the absolutely convergent integral

$$K_\alpha(r) := \frac{1}{2}\int_0^\infty \exp\left(-\frac{r}{2}(s+s^{-1})\right) s^{\alpha-1}\,\mathrm{d}s. \tag{3.2}$$

We have the identities

$$K_{-\alpha}(r) = K_\alpha(r) = \frac{\pi}{2\sin\pi\alpha}(I_{-\alpha}(r) - I_\alpha(r)), \tag{3.3}$$

$$I_\alpha(r) = \frac{1}{\mathrm{i}\pi}(K_\alpha(\mathrm{e}^{-\mathrm{i}\pi} r) - \mathrm{e}^{\mathrm{i}\pi\alpha} K_\alpha(r)), \tag{3.4}$$

the asymptotics for $|\arg r| < \pi - \epsilon$, $\epsilon > 0$,

$$\lim_{|r|\to\infty} \left(\sqrt{\frac{\pi}{2r}}\,\mathrm{e}^{-r}\right)^{-1} K_\alpha(r) = 1, \tag{3.5}$$

and the recurrence relations

$$\left(\frac{1}{r}\partial_r\right)^n r^{\pm\alpha} I_\alpha(r) = r^{\pm\alpha-n} I_{\alpha\mp n}(r), \tag{3.6}$$

$$\left(-\frac{1}{r}\partial_r\right)^n r^{\pm\alpha} K_\alpha(r) = r^{\pm\alpha-n} K_{\alpha\mp n}(r). \tag{3.7}$$

We note also the inequality **[22]**

$$\frac{K_\alpha(r)}{K_\alpha(R)} < \mathrm{e}^{R-r}\left(\frac{R}{r}\right)^\alpha, \qquad \alpha > \frac{1}{2},\ 0 < r < R. \tag{3.8}$$

To see (3.8), first note that (3.2) can be transformed into

$$K_\alpha(r) := \frac{1}{2}\int_1^\infty \exp\Big(-\frac{r}{2}(s+s^{-1})\Big)(s^\alpha+s^{-\alpha})s^{-1}\,\mathrm{d}s, \tag{3.9}$$

which easily implies

$$K_\alpha(r) < K_{\alpha'}(r), \quad 0 \leqslant \alpha < \alpha'. \tag{3.10}$$

By (3.7)

$$\frac{K_\alpha'(r)}{K_\alpha(r)} = -\frac{K_{\alpha-1}(r)}{K_\alpha(r)} - \frac{\alpha}{r}. \tag{3.11}$$

Hence, for $\alpha \geqslant \frac{1}{2}$, using (3.10) we obtain

$$\frac{K_\alpha'(r)}{K_\alpha(r)} > -1 - \frac{\alpha}{r}. \tag{3.12}$$

Integrating (3.12) from r to R we obtain (3.8).

3.2. Degenerate case. The case $\alpha \in \mathbb{Z}$ is called degenerate, because then the standard solutions characterized by asymptotics at 1 coincide. Using for (3.14) the de l'Hôpital rule, one finds for $\alpha \in \{0, 1, 2, \dots\}$

$$I_{\pm\alpha}(r) = \sum_{n=0}^\infty \frac{\left(\frac{r}{2}\right)^{2n+\alpha}}{n!(n+\alpha)!}, \tag{3.13}$$

$$\begin{aligned} K_{\pm\alpha}(r) = {} & \frac{1}{2}\sum_{k=0}^{\alpha-1} \frac{(-1)^k(\alpha-k-1)!}{k!}\Big(\frac{r}{2}\Big)^{2k-\alpha} \\ & + \frac{(-1)^\alpha}{2}\sum_{j=0}^\infty \frac{\big(H_j + H_{\alpha+j} - 2\gamma_{\mathrm{E}} - 2\ln(\frac{r}{2})\big)}{j!(\alpha+j)!}\Big(\frac{r}{2}\Big)^{2j+\alpha}. \end{aligned} \tag{3.14}$$

3.3. Half-integer case. In the half-integer case the Bessel equation is solved by elementary functions. More precisely, for $k = 0, 1, 2, \dots,$ we have

$$I_{\frac{1}{2}+k}(r) = \Big(\frac{2}{\pi}\Big)^{\frac{1}{2}} r^{\frac{1}{2}+k}\Big(\frac{1}{r}\partial_r\Big)^k \frac{\sinh r}{r}, \tag{3.15}$$

$$I_{-\frac{1}{2}-k}(r) = \Big(\frac{2}{\pi}\Big)^{\frac{1}{2}} r^{\frac{1}{2}+k}\Big(\frac{1}{r}\partial_r\Big)^k \frac{\cosh r}{r}, \tag{3.16}$$

$$K_{\pm(\frac{1}{2}+k)}(r) = \Big(\frac{\pi}{2}\Big)^{\frac{1}{2}} r^{\frac{1}{2}+k}\Big(-\frac{1}{r}\partial_r\Big)^k \frac{\mathrm{e}^{-r}}{r}. \tag{3.17}$$

3.4. Bilinear integrals. First let us describe some integral identities satisfied by Bessel functions involving the usual integral. For $\mathrm{Re}(a+b) > 0$ and $|\mathrm{Re}(\alpha)| < 1$ we have

$$\int_0^\infty K_\alpha(ar)K_\alpha(br)2r\,\mathrm{d}r = \frac{\pi}{\sin(\pi\alpha)}\frac{(a/b)^\alpha - (b/a)^\alpha}{a^2-b^2}, \qquad \alpha \neq 0, \tag{3.18}$$

$$\int_0^\infty K_0(ar)K_0(br)2r\,\mathrm{d}r = \frac{2\ln\frac{a}{b}}{a^2-b^2}. \tag{3.19}$$

If $b = a$ with $\mathrm{Re}(a) > 0$, the above integrals reduce to

$$\int_0^\infty K_\alpha(ar)^2 2r\,\mathrm{d}r = \frac{\pi\alpha}{a^2\sin(\pi\alpha)}, \qquad \alpha \neq 0, \tag{3.20}$$

$$\int_0^\infty K_0(ar)^2 2r\,\mathrm{d}r = \frac{1}{a^2}, \qquad \mathrm{Re}(a) > 0. \tag{3.21}$$

Formula (3.18) is well-known [**14**, Chapter 6.521]. It can be obtained from the identity

$$\begin{aligned}&(a^2-b^2)K_\alpha(ar)K_\alpha(br)\\&= (\partial_r^2 + r^{-1}\partial_r)K_\alpha(ar)K_\alpha(br) - K_\alpha(ar)(\partial_r^2 + r^{-1}\partial_r)K_\alpha(br),\end{aligned}$$

which follows directly from Bessel's equation and integration by parts. The remaining formulas are then obtained by applying the de l'Hôpital rule.

If one replaces the integration in (3.18)–(3.21) by generalized integration, it makes sense for every α. If we treat r^2 as the variable of integration, these generalized integrals are non-anomalous for $\alpha \in \mathbb{Z}\setminus\{0\}$, so expressions stated in (3.18),(3.20) for $\alpha \neq 0$ are true for all $\alpha \notin \mathbb{Z}$. Below we calculate the anomalous integrals for $\alpha \in \mathbb{Z}$.

PROPOSITION 3.1. *Let $\alpha \in \mathbb{Z}$ and let $\mathrm{Re}(a+b) > 0$. If $a \neq b$, then*

$$\begin{aligned}\mathrm{gen}\!\int_0^\infty K_\alpha(ar)K_\alpha(br)2r\,\mathrm{d}r &= (-1)^\alpha 2\frac{\left(\frac{a}{b}\right)^\alpha \ln\left(\frac{a}{2}\right) - \left(\frac{b}{a}\right)^\alpha \ln\left(\frac{b}{2}\right)}{a^2-b^2}\\&\quad - \frac{(-1)^\alpha}{ab}\sum_{k=0}^{|\alpha|-1}\left(\frac{a}{b}\right)^{2k-|\alpha|+1}\big(\psi(1+k)+\psi(|\alpha|-k)\big).\end{aligned} \tag{3.22}$$

In the case $a = b$, one has

$$\mathrm{gen}\!\int_0^\infty K_\alpha(br)^2 2r\,\mathrm{d}r = \frac{(-1)^\alpha}{b^2}\Big(1 + |\alpha|\ln\left(\tfrac{b^2}{4}\right) + 2|\alpha|\big(1-\psi(1+|\alpha|)\big)\Big). \tag{3.23}$$

PROOF. Both sides of (3.22) are invariant with respect to the sign flip of α, so it is enough to consider $\alpha = -m$ with $m \in \mathbb{N}$. We will use dimensional regularization.

We change integration variables to r^2 (using Proposition 2.5). Let

$$f(r,\alpha) := \left(\frac{ab}{4}\right)^{-\alpha} K_\alpha(ar)K_\alpha(br). \tag{3.24}$$

The generalized integral of $f(\cdot,\alpha)$ is non-anomalous for non-integer α, so from (3.18)

$$\mathrm{gen}\!\int_0^\infty f(r,\alpha)\,\mathrm{d}r^2 = \frac{\pi}{\sin\pi\alpha}\,\frac{\left(\frac{b}{2}\right)^{-2\alpha} - \left(\frac{a}{2}\right)^{-2\alpha}}{a^2-b^2}. \tag{3.25}$$

Equation (3.25) has a simple pole at $\alpha = -m$, with residue and finite part given by

$$\lim_{\alpha\to-m}(\alpha+m)\,\mathrm{gen}\!\int_0^\infty f(r,\alpha)\,\mathrm{d}r^2 = (-1)^m\frac{\left(\frac{b}{2}\right)^{2m} - \left(\frac{a}{2}\right)^{2m}}{a^2-b^2}, \tag{3.26}$$

$$\operatorname*{fp}_{\alpha\to-m}\,\mathrm{gen}\!\int_0^\infty f(r,\alpha)\,\mathrm{d}r^2 = (-1)^m 2\frac{\left(\frac{a}{2}\right)^{2m}\ln\left(\frac{a}{2}\right) - \left(\frac{b}{2}\right)^{2m}\ln\left(\frac{b}{2}\right)}{a^2-b^2}. \tag{3.27}$$

By (3.3), all terms of $f(r,\alpha)$ singular for non-integer α with $\mathrm{Re}(\alpha)<0$ are in:

(3.28)
$$\Big(\frac{ab}{4}\Big)^{-\alpha}\frac{\pi^2}{4\sin^2\pi\alpha}I_\alpha(ar)I_\alpha(br)$$
$$=\frac14\Bigg(\sum_{k=0}^\infty\frac{\big(\frac a2\big)^{2k}r^{2k+\alpha}(-1)^k\Gamma(-\alpha-k)}{k!}\Bigg)\Bigg(\sum_{j=0}^\infty\frac{\big(\frac b2\big)^{2j}r^{2j+\alpha}(-1)^j\Gamma(-\alpha-j)}{j!}\Bigg).$$

The coefficient of (3.28) at $(r^2)^{\alpha+m-1}$ is

$$f_{m-1}(\alpha)=\frac{(-1)^{m-1}}{4}\sum_{k=0}^{m-1}\frac{\big(\frac a2\big)^{2k}\big(\frac b2\big)^{2m-2-2k}\Gamma(-\alpha-k)\Gamma(-\alpha-m+1+k)}{k!(m-1-k)!}.\tag{3.29}$$

We calculate its derivative:

$$f'_{m-1}(-m)=\frac{(-1)^{m-1}}{4}\sum_{k=0}^{m-1}\Big(\frac a2\Big)^{2k}\Big(\frac b2\Big)^{2m-2-2k}\big(-\psi(1+k)-\psi(m-k)\big).\tag{3.30}$$

Formula (3.22) follows from (2.20). It is easy to see that the generalized integral of $K_\alpha(ar)K_\alpha(br)2r$ is a holomorphic function of b, even in the anomalous case. Therefore, (3.23) may be obtained from (3.22) using the de l'Hôpital rule and identity (A.24). □

3.5. Poisson-type integral representations. There exist two basic kinds of integral representations of solutions of the Bessel equation. Representations similar to or derived from (3.2) are sometimes called *Bessel–Schläfli-type.* There exists also another family of integral representations, sometimes called *Poisson-type*:

$$I_\alpha(r)=\frac{1}{\sqrt\pi\Gamma\big(\alpha+\frac12\big)}\Big(\frac r2\Big)^\alpha\int_{-1}^1(1-t^2)^{\alpha-\frac12}\,\mathrm e^{\pm rt}\,\mathrm dt,\quad \mathrm{Re}(\alpha)>-\frac12,\tag{3.31}$$

$$K_\alpha(r)=\frac{\sqrt\pi\,\big(\frac r2\big)^\alpha}{\Gamma(\alpha+\frac12)}\int_1^\infty\mathrm e^{-sr}(s^2-1)^{\alpha-\frac12}\,\mathrm ds,\quad \mathrm{Re}(\alpha)>-\frac12,\ \mathrm{Re}(r)>0,\tag{3.32}$$

$$K_\alpha(r)=\frac{\Gamma(\alpha+\frac12)}{2\sqrt\pi}\Big(\frac r2\Big)^{-\alpha}\int_{-\infty}^\infty\mathrm e^{-\mathrm i sr}(s^2+1)^{-\alpha-\frac12}\,\mathrm ds,\quad \mathrm{Re}(\alpha)>0,\ r>0.\tag{3.33}$$

3.6. Standard Bessel equation. The (*standard*) *Bessel equation* is obtained by setting $r\to\pm\mathrm i r$ in the modified one:

$$\Big(\partial_r^2+\frac1r\partial_r-\frac{\alpha^2}{r^2}+1\Big)v(r)=0.\tag{3.34}$$

We regard the modified Bessel equation as more basic than the standard Bessel equation because its standard solutions holomorphic on $\mathbb C\backslash]-\infty,0]$ have simpler properties than those of the standard Bessel equation. The latter is useful mostly with r restricted to $\mathbb R_+$.

We have several kinds of standard solutions of (3.34). The most important is the *Bessel function*, defined as

$$J_\alpha(r)=\mathrm e^{\pm\mathrm i\pi\frac\alpha2}\,I_\alpha(\mp\mathrm i r).\tag{3.35}$$

The two *Hankel functions* also solve (3.34):

$$H_\alpha^\pm(r)=\frac2\pi\,\mathrm e^{\mp\mathrm i\frac\pi2(\alpha+1)}\,K_\alpha(\mp\mathrm i r).\tag{3.36}$$

We have an integral representation

$$H_\alpha^\pm(r) = \pm\frac{1}{\pi\,\mathrm{i}}\,\mathrm{e}^{\mp\mathrm{i}\,\frac{\pi}{2}\alpha}\int_0^\infty \exp\Big(\pm\mathrm{i}\,\frac{r}{2}(s+s^{-1})\Big)\,s^{\alpha-1}\,\mathrm{d}s,\quad \pm\mathrm{Im}(r)>0. \tag{3.37}$$

REMARK 3.2. In the literature the usual notation for Hankel functions is

$$H_\alpha^{(1)}(r) = H_\alpha^+(r),\quad H_\alpha^{(2)}(r) = H_\alpha^-(r). \tag{3.38}$$

Note the identities

$$J_\alpha(r) = \frac{1}{2}\left(H_\alpha^+(r) + H_\alpha^-(r)\right), \tag{3.39}$$

$$J_{-\alpha}(r) = \frac{1}{2}\left(\mathrm{e}^{\alpha\pi\,\mathrm{i}}\,H_\alpha^+(r) + \mathrm{e}^{-\alpha\pi\,\mathrm{i}}\,H_\alpha^-(r)\right), \tag{3.40}$$

$$\mathrm{e}^{\mp\alpha\pi\,\mathrm{i}}\,H_{-\alpha}^\pm(r) = H_\alpha^\pm(r) = \pm\frac{\mathrm{i}\,\mathrm{e}^{\mp\alpha\pi\,\mathrm{i}}\,J_\alpha(r) - \mathrm{i}\,J_{-\alpha}(r)}{\sin\alpha\pi}. \tag{3.41}$$

We have recurrence relations

$$\left(\pm\frac{1}{r}\partial_r\right)^n r^{\pm\alpha}J_\alpha(r) = r^{\pm\alpha-n}J_{\alpha\mp n}(r). \tag{3.42}$$

In the above recurrence relations one may replace $J_\alpha(r)$ with $H_\alpha^\pm(r)$.

3.7. Special values of parameters. For $\alpha = 0, 1, 2, \ldots$, we have

$$(-1)^\alpha J_{-\alpha}(r) = J_\alpha(r) = \sum_{n=0}^\infty \frac{(-1)^n\left(\frac{r}{2}\right)^{2n+\alpha}}{n!(n+\alpha)!}, \tag{3.43}$$

$$\begin{aligned}(-1)^\alpha H_{-\alpha}^\pm(r) = H_\alpha^\pm(r) &= \mp\frac{\mathrm{i}}{\pi}\sum_{k=0}^{\alpha-1}\left(\frac{r}{2}\right)^{2k-\alpha}\frac{(\alpha-k-1)!}{k!}\\ &\mp\frac{\mathrm{i}}{\pi}\sum_{j=0}^\infty \frac{(-1)^j\left(H_j + H_{\alpha+j} - 2\gamma_{\mathrm{E}} \pm \mathrm{i}\,\pi - 2\ln(\frac{r}{2})\right)}{j!(\alpha+j)!}\left(\frac{r}{2}\right)^{2j+\alpha}.\end{aligned} \tag{3.44}$$

For $k = 0, 1, 2, \ldots$ we have

$$J_{\frac{1}{2}+k}(r) = \left(\frac{2}{\pi}\right)^{\frac{1}{2}} r^{\frac{1}{2}+k}\left(-\frac{1}{r}\partial_r\right)^k\frac{\sin r}{r}, \tag{3.45}$$

$$J_{-\frac{1}{2}-k}(r) = \left(\frac{2}{\pi}\right)^{\frac{1}{2}} r^{\frac{1}{2}+k}\left(\frac{1}{r}\partial_r\right)^k\frac{\cos r}{r}, \tag{3.46}$$

$$H_{-\frac{1}{2}-k}^\pm(r) = \pm\mathrm{i}(-1)^k H_{\frac{1}{2}+k}^\pm(r) = \left(\frac{2}{\pi}\right)^{\frac{1}{2}} r^{\frac{1}{2}+k}\left(\frac{1}{r}\partial_r\right)^k\frac{\mathrm{e}^{\pm\mathrm{i}\,r}}{r}. \tag{3.47}$$

4. The Gegenbauer equation

4.1. Gegenbauer functions. Here is the *Gegenbauer equation*:

$$\left((1-w^2)\partial_w^2 - 2(1+\alpha)w\partial_w + \lambda^2 - \left(\alpha+\frac{1}{2}\right)^2\right)f(w) = 0. \tag{4.1}$$

Its solutions can be expressed in terms of the Gauss hypergeometric function $F(a, b; c; z)$. We will often use this function with the so-called *Olver's normalization*

$$\mathbf{F}(a,b;c;z) := \frac{F(a,b;c;z)}{\Gamma(c)} = \sum_{n=0}^\infty \frac{(a)_n(b)_n z^n}{\Gamma(c+n)n!}. \tag{4.2}$$

The defining series converges only in the unit disc, but $\mathbf{F}(a,b;c;z)$ extends to a holomorphic function on $\mathbb{C}\backslash[1,\infty[$ (as well as on a universal cover of $\mathbb{C}\backslash\{0,1\}$, but we will not use the latter point of view here).

In what follows complex power functions should be interpreted as their principal branches (holomorphic on $\mathbb{C}\backslash]-\infty,0]$). For example $w\mapsto(1-w)^\alpha$ is holomorphic away from $[1,\infty[$. In addition, we will frequently use the notation

$$(w^2-1)^\alpha_\bullet := (w-1)^\alpha(w+1)^\alpha. \tag{4.3}$$

The function $(w^2-1)^\alpha_\bullet$ is holomorphic on $\mathbb{C}\backslash]-\infty,1]$, whereas $(w^2-1)^\alpha$ is holomorphic on $\mathbb{C}\backslash([-1,1]\cup\mathrm{i}\,\mathbb{R})$. One has $(w^2-1)^\alpha_\bullet=(w^2-1)^\alpha$ only for $\mathrm{Re}(w)>0$. However, $(1-w^2)^\alpha=(1-w)^\alpha(1+w)^\alpha$ for all $w\notin]-\infty,-1]\cup[1,\infty[$.

Standard solutions of the Gegenbauer equations are characterized by simple behavior at one of the three singular points $1,-1,\infty$. Due to the $w\mapsto -w$ symmetry of the equation (4.1), solutions of the second type are obtained from solutions of the first type by negating the argument. Therefore we consider four functions, corresponding to two behaviors at 1 and two behaviors at ∞. All of them are holomorphic on $\mathbb{C}\backslash]-\infty,1]$.

- The solution characterized by asymptotics ~ 1 at 1:

$$S_{\alpha,\pm\lambda}(w) := F\Big(\frac12+\alpha+\lambda,\frac12+\alpha-\lambda;\alpha+1;\frac{1-w}{2}\Big) \tag{4.4}$$

$$= \Big(\frac{2}{w+1}\Big)^\alpha F\Big(\frac12+\lambda,\frac12-\lambda;\alpha+1;\frac{1-w}{2}\Big). \tag{4.5}$$

  $S_{\alpha,\lambda}$ is distinguished among the four solutions introduced here by the fact that it is holomorphic on $\mathbb{C}\backslash]-\infty,-1]$ rather than only on $\mathbb{C}\backslash]-\infty,1]$. The equality of (4.4) and (4.5) is a special case of the well-known *Euler identity*. On the right half-plane we have an alternative expression obtained by a quadratic transformation:

$$S_{\alpha,\lambda}(w)=F\Big(\frac14+\frac{\alpha}{2}+\frac{\lambda}{2},\frac14+\frac{\alpha}{2}-\frac{\lambda}{2};\alpha+1;1-w^2\Big),\qquad \mathrm{Re}(w)>0. \tag{4.6}$$

- The solution $\frac{2^{2\alpha}}{(w^2-1)^\alpha_\bullet}S_{-\alpha,\lambda}(w)$ is characterized by asymptotics $\sim\frac{2^\alpha}{(w-1)^\alpha}$ at 1.
- The solution characterized by asymptotics $\sim w^{-\frac12-\alpha-\lambda}$ at ∞:

$$\begin{aligned} Z_{\alpha,\lambda}(w) &= (w\pm1)^{-\frac12-\alpha-\lambda}F\Big(\frac12+\lambda,\frac12+\lambda+\alpha;1+2\lambda;\frac{2}{1\pm w}\Big)\\ &= w^{-\frac12-\alpha-\lambda}F\Big(\frac14+\frac{\alpha}{2}+\frac{\lambda}{2},\frac34+\frac{\alpha}{2}+\frac{\lambda}{2};1+\lambda;\frac{1}{w^2}\Big).\end{aligned} \tag{4.7}$$

- The solution $Z_{\alpha,-\lambda}(w)$ is characterized by asymptotics $\sim w^{-\frac12-\alpha+\lambda}$ at ∞.

All these four functions can be expressed in terms of $S_{\alpha,\lambda}$, but for typographical reasons it is convenient to introduce also $Z_{\alpha,\lambda}$. We will often use Olver's normalization:

$$\mathbf{S}_{\alpha,\lambda}(w):=\frac{1}{\Gamma(\alpha+1)}S_{\alpha,\lambda}(w),\qquad \mathbf{Z}_{\alpha,\lambda}(w):=\frac{1}{\Gamma(\lambda+1)}Z_{\alpha,\lambda}(w). \tag{4.8}$$

We note the identities

$$\mathbf{S}_{\alpha,\lambda}(w)=\mathbf{S}_{\alpha,-\lambda}(w),\quad \mathbf{Z}_{\alpha,\lambda}(w)=\frac{\mathbf{Z}_{-\alpha,\lambda}(w)}{(w^2-1)^\alpha_\bullet} \tag{4.9}$$

as well as the slightly more subtle *Whipple transformations*:

$$\mathbf{Z}_{\alpha,\lambda}(w) := (w^2-1)_\bullet^{-\frac{1}{4}-\frac{\alpha}{2}-\frac{\lambda}{2}}\mathbf{S}_{\lambda,\alpha}\left(\frac{w}{(w^2-1)_\bullet^{\frac{1}{2}}}\right), \tag{4.10}$$

$$\mathbf{S}_{\alpha,\lambda}(w) := (w^2-1)_\bullet^{-\frac{1}{4}-\frac{\alpha}{2}-\frac{\lambda}{2}}\mathbf{Z}_{\lambda,\alpha}\left(\frac{w}{(w^2-1)_\bullet^{\frac{1}{2}}}\right), \qquad \operatorname{Re}(w)>0. \tag{4.11}$$

Note that $w \mapsto \frac{w}{(w^2-1)_\bullet^{\frac{1}{2}}}$, defined on $\mathbb{C}\setminus[-1,1]$, is a holomorphic double cover of $\{z \in \mathbb{C}\setminus[-1,1]\,|\,\operatorname{Re}(z)>0\}$. One has $f(-w) = f(w)$ for $w \notin [-1,1]$ and $f(f(w)) = \operatorname{sgn}(\operatorname{Re}(w))w$ for $w \notin [-1,1]\cup \mathrm{i}\,\mathbb{R}$. The reason why (4.11) (but not (4.10)) holds only on the right half-plane is that $\mathbf{Z}_{\lambda,\alpha}(w)$ has a branch cut on $[-1,1]$, implying that the right-hand side of (4.11) is discontinuous on $\mathrm{i}\,\mathbb{R}$.

Here are the connection formulas:

$$\begin{aligned}&\mathbf{S}_{\alpha,\lambda}(-w)\\ &= -\frac{\cos(\pi\lambda)}{\sin(\pi\alpha)}\mathbf{S}_{\alpha,\lambda}(w) + \frac{2^{2\alpha}\pi}{\sin(\pi\alpha)\Gamma(\frac{1}{2}+\alpha+\lambda)\Gamma(\frac{1}{2}+\alpha-\lambda)}\frac{\mathbf{S}_{-\alpha,-\lambda}(w)}{(1-w^2)^{\alpha}},\end{aligned} \tag{4.12}$$

$$\begin{aligned}&\mathbf{Z}_{\alpha,\lambda}(w)\\ &= -\frac{2^{\lambda-\alpha-\frac{1}{2}}\sqrt{\pi}}{\sin(\pi\alpha)\Gamma(\frac{1}{2}-\alpha+\lambda)}\mathbf{S}_{\alpha,\lambda}(w) + \frac{2^{\lambda+\alpha-\frac{1}{2}}\sqrt{\pi}}{\sin(\pi\alpha)\Gamma(\frac{1}{2}+\alpha+\lambda)}\frac{\mathbf{S}_{-\alpha,-\lambda}(w)}{(w^2-1)_\bullet^{\frac{\alpha}{2}}}.\end{aligned} \tag{4.13}$$

Using the connection formulas (4.12) and (4.13) (and for $\alpha \in \mathbb{Z}$ in addition the de l'Hôpital rule, see also (4.23) and (4.24)), we derive the asymptotics of the Gegenbauer functions near $w=1$ for $\operatorname{Re}(\alpha)>0$:

$$\mathbf{S}_{\alpha,\lambda}(-w) \simeq \frac{2^{\alpha}\Gamma(\alpha)}{\Gamma(\frac{1}{2}+\alpha+\lambda)\Gamma(\frac{1}{2}+\alpha-\lambda)(1-w)^{\alpha}}, \tag{4.14}$$

$$\mathbf{Z}_{\alpha,\lambda}(w) \simeq \frac{2^{-\frac{1}{2}+\lambda}\Gamma(\alpha)}{\sqrt{\pi}\Gamma(\frac{1}{2}+\alpha+\lambda)(w-1)^{\alpha}}. \tag{4.15}$$

4.2. Recurrence relations. The Gegenbauer functions satisfy various recurrence relations, whose proof is straightforward:

$$\begin{aligned}
\partial_w \mathbf{S}_{\alpha,\lambda}(w) &= -\frac{1}{2}\Big(\Big(\frac{1}{2}+\alpha\Big)^2 - \lambda^2\Big)\mathbf{S}_{\alpha+1,\lambda}(w),\\
\big((1-w^2)\partial_w - 2\alpha w\big)\,\mathbf{S}_{\alpha,\lambda}(w) &= -2\mathbf{S}_{\alpha-1,\lambda}(w),\\
\Big((1-w^2)\partial_w - \Big(\frac{1}{2}+\alpha+\lambda\Big)w\Big)\,\mathbf{S}_{\alpha,\lambda}(w) &= -\Big(\frac{1}{2}+\alpha+\lambda\Big)\mathbf{S}_{\alpha,\lambda+1}(w),\\
\Big((1-w^2)\partial_w - \Big(\frac{1}{2}+\alpha-\lambda\Big)w\Big)\,\mathbf{S}_{\alpha,\lambda}(w) &= -\Big(\frac{1}{2}+\alpha-\lambda\Big)\mathbf{S}_{\alpha,\lambda-1}(w);
\end{aligned}$$

and

$$\partial_w \mathbf{Z}_{\alpha,\lambda}(w) = -\Big(\frac{1}{2}+\alpha+\lambda\Big)\mathbf{Z}_{\alpha+1,\lambda}(w),$$
$$\big((1-w^2)\partial_w - 2\alpha w\big)\,\mathbf{Z}_{\alpha,\lambda}(w) = \Big(\frac{1}{2}-\alpha+\lambda\Big)\mathbf{Z}_{\alpha-1,\lambda}(w),$$

$$\Big((1-w^2)\partial_w - \Big(\frac{1}{2}+\alpha+\lambda\Big)w\Big)\,\mathbf{Z}_{\alpha,\lambda}(w) = -\frac{1}{2}\Big(\Big(\frac{1}{2}+\lambda\Big)^2-\alpha^2\Big)\mathbf{Z}_{\alpha,\lambda+1}(w),$$
$$\Big((1-w^2)\partial_w - \Big(\frac{1}{2}+\alpha-\lambda\Big)w\Big)\,\mathbf{Z}_{\alpha,\lambda}(w) = 2\mathbf{Z}_{\alpha,\lambda-1}(w).$$

REMARK 4.1. It might be interesting to note that the above recurrence relations correspond to short roots of the Lie algebra $so(5)$, as explained in [**6, 7**].

4.3. Gegenbauer polynomials. In the literature one can find two kinds of polynomials related to the functions $S_{\alpha,\lambda}$. The *Jacobi polynomials* with $\alpha=\beta$ are defined as

$$\begin{aligned} P_n^{\alpha,\alpha}(w) &:= \frac{(\alpha+1)_n}{n!}F\Big(-n, n+2\alpha+1; 1+\alpha; \frac{1-w}{2}\Big) \\ &= \frac{\Gamma(\alpha+1+n)}{n!}\mathbf{S}_{\alpha,\frac{1}{2}+\alpha+n}(w). \end{aligned} \tag{4.16}$$

However, one usually prefers the *Gegenbauer polynomials*

$$\begin{aligned} C_n^{\alpha+\frac{1}{2}}(w) &:= \frac{(2\alpha+1)_n}{(\alpha+1)_n}P_n^{\alpha,\alpha}(w) \\ &= \frac{(2\alpha+1)_n}{n!}F\Big(-n, n+2\alpha+1; 1+\alpha; \frac{1-w}{2}\Big). \end{aligned} \tag{4.17}$$

Note that

$$P_n^{\alpha,\alpha}(-w) = (-1)^n P_n^{\alpha,\alpha}(w), \quad C_n^{\frac{1}{2}+\alpha}(-w) = (-1)^n C_n^{\frac{1}{2}+\alpha}(w). \tag{4.18}$$

We have the well-known special cases: the *Legendre polynomials*

$$P_n(w) = P_n^{0,0}(w) = \mathbf{S}_{0,\frac{1}{2}+n}(w); \tag{4.19}$$

and the *Chebyshev polynomials of the first and second kind*:

$$\begin{aligned} T_n(w) &= \frac{n!}{(\frac{1}{2})_n}P_n^{-\frac{1}{2},-\frac{1}{2}}(w) = \sqrt{\pi}\mathbf{S}_{-\frac{1}{2},n}(w) \\ &= \frac{1}{2}\Big(\big(w+\mathrm{i}\sqrt{1-w^2}\big)^n + \big(w-\mathrm{i}\sqrt{1-w^2}\big)^n\Big), \end{aligned} \tag{4.20}$$

$$\begin{aligned} U_n(w) &= \frac{n!}{(\frac{3}{2})_n}P_n^{\frac{1}{2},\frac{1}{2}}(w) = C_n^1(w) = \frac{n+1}{2}\sqrt{\pi}\mathbf{S}_{\frac{1}{2},n+1}(w) \\ &= \frac{1}{2\mathrm{i}\sqrt{1-w^2}}\Big(\big(w+\mathrm{i}\sqrt{1-w^2}\big)^{n+1} - \big(w-\mathrm{i}\sqrt{1-w^2}\big)^{n+1}\Big). \end{aligned} \tag{4.21}$$

4.4. Degenerate case. The case $\alpha\in\mathbb{Z}$ is called degenerate, because then the standard solutions characterized by asymptotics at 1 become linearly dependent. More precisely, for $\alpha\in\mathbb{N}$ we have

$$\mathbf{S}_{\alpha,\lambda}(w) = \frac{2^{2\alpha}}{(\frac{1}{2}+\lambda)_\alpha(\frac{1}{2}-\lambda)_\alpha(1-w^2)^\alpha}\mathbf{S}_{-\alpha,\pm\lambda}(w). \tag{4.22}$$

The functions $\mathbf{S}_{\alpha,\lambda}(-w)$ and $\mathbf{Z}_{\alpha,\lambda}(w)$ with $\alpha \in \mathbb{N}$ have the following expansions in the disc $|w-1|<2$, involving a logarithmic singularity near 1:

$$\begin{aligned}
&\Gamma(\tfrac12+\lambda+\alpha)\Gamma(\tfrac12-\lambda+\alpha)\mathbf{S}_{\alpha,\lambda}(-w)\\
&=\Big(\frac{1-w}{2}\Big)^{-\alpha}\sum_{k=0}^{\alpha-1}\frac{(\frac12+\lambda-k)_{2k}(\alpha-k-1)!}{k!}\Big(\frac{1-w}{2}\Big)^{k}\\
&\quad+\sum_{j=0}^{\infty}\frac{(\frac12+\lambda-\alpha-j)_{2\alpha+2j}}{j!(j+\alpha)!}\Big(\frac{w-1}{2}\Big)^{j}\Big(\psi(1+\alpha+j)+\psi(1+j)\\
&\quad-\psi(\tfrac12+\lambda+\alpha+j)-\psi(\tfrac12-\lambda+\alpha+j)-\ln\Big(\frac{1-w}{2}\Big)\Big),
\end{aligned}\tag{4.23}$$

and

$$\begin{aligned}
&\sqrt{2\pi}\,(-1)^{\alpha}2^{\alpha-\lambda}\Gamma(\tfrac12+\lambda+\alpha)\mathbf{Z}_{\alpha,\lambda}(w)\\
&=\Big(\frac{1-w}{2}\Big)^{-\alpha}\sum_{k=0}^{\alpha-1}\frac{(\frac12+\lambda-k)_{2k}(\alpha-k-1)!}{k!}\Big(\frac{1-w}{2}\Big)^{k}\\
&\quad+\sum_{j=0}^{\infty}\frac{(\frac12+\lambda-\alpha-j)_{2\alpha+2j}}{j!(j+\alpha)!}\Big(\frac{w-1}{2}\Big)^{j}\Big(\psi(1+\alpha+j)+\psi(1+j)\\
&\quad-\psi(\tfrac12+\lambda+\alpha+j)-\psi(\tfrac12+\lambda-\alpha-j)-\ln\Big(\frac{w-1}{2}\Big)\Big).
\end{aligned}\tag{4.24}$$

Let us sketch a proof of (4.24). Using the connection formula (4.13) we can write

$$\begin{aligned}
&\sqrt{2\pi}\,2^{\alpha-\lambda}\Gamma(\tfrac12+\lambda+\alpha)\mathbf{Z}_{\alpha,\lambda}(w)\\
&=\frac{\pi}{\sin(\pi\alpha)}\bigg(-\frac{\Gamma(\frac12+\alpha+\lambda)}{\Gamma(\frac12-\alpha+\lambda)}\mathbf{F}\Big(\tfrac12+\alpha+\lambda,\tfrac12+\alpha-\lambda;\alpha+1;\frac{1-w}{2}\Big)\\
&\quad+\frac{2^{2\alpha}}{(w^2-1)^{\frac{\alpha}{2}}}\Big(\frac{2}{w+1}\Big)^{-\alpha}\mathbf{F}\Big(\tfrac12+\lambda,\tfrac12-\lambda;-\alpha+1;\frac{1-w}{2}\Big)\bigg)\\
&=\frac{\pi}{\sin(\pi\alpha)}\bigg(-\sum_{j=0}^{\infty}\frac{(-1)^j\Gamma(\frac12+\alpha+\lambda+j)}{\Gamma(\frac12-\alpha+\lambda-j)\Gamma(\alpha+1+j)j!}\Big(\frac{1-w}{2}\Big)^{j}\\
&\quad+\Big(\frac{2}{w-1}\Big)^{\alpha}\sum_{k=0}^{\infty}\frac{(\frac12+\lambda)_k(\frac12-\lambda)_k}{\Gamma(-\alpha+1+k)k!}\Big(\frac{1-w}{2}\Big)^{k}\bigg).
\end{aligned}$$

Then we apply the de l'Hôpital rule. We proceed similarly for (4.23).

4.5. Half-integer case. If α is a half-integer, then the Gegenbauer functions can be expressed in terms of elementary functions. They are particularly simple for $\alpha=-\frac12$ and $\alpha=\frac12$:

$$\sqrt{\pi}\mathbf{S}_{-\frac12,\lambda}(w)=S_{-\frac12,\lambda}(w)=\frac{\big(w+\mathrm{i}\sqrt{1-w^2}\big)^{\lambda}+\big(w-\mathrm{i}\sqrt{1-w^2}\big)^{\lambda}}{2},\tag{4.25}$$

$$\frac{\sqrt{\pi}}{2}\mathbf{S}_{\frac12,\lambda}(w)=S_{\frac12,\lambda}(w)=\frac{\big(w+\mathrm{i}\sqrt{1-w^2}\big)^{\lambda}-\big(w-\mathrm{i}\sqrt{1-w^2}\big)^{\lambda}}{2\lambda\mathrm{i}\sqrt{1-w^2}}\tag{4.26}$$

and

$$\Gamma(1+\lambda)\mathbf{Z}_{-\frac{1}{2},\lambda}(w) = Z_{-\frac{1}{2},\lambda}(w) = 2^{\lambda}\big((w^2-1)_{\bullet}^{\frac{1}{2}} + w\big)^{-\lambda}, \tag{4.27}$$

$$\Gamma(1+\lambda)\mathbf{Z}_{\frac{1}{2},\lambda}(w) = Z_{\frac{1}{2},\lambda}(w) = 2^{\lambda}\frac{\big((w^2-1)_{\bullet}^{\frac{1}{2}} + w\big)^{-\lambda}}{(w^2-1)_{\bullet}^{\frac{1}{2}}}. \tag{4.28}$$

Using the recurrence relations we can find expressions for Gegenbauer functions with $\alpha \in \frac{1}{2}+\mathbb{Z}$. More precisely, for $n = 0, 1, \dots,$ we have

$$\mathbf{S}_{-\frac{1}{2}-n,\lambda}(w) = \frac{(1-w^2)^{\frac{1}{2}+n}}{2\sqrt{\pi}(-2)^n}\partial_w^n \frac{\big(w + \mathrm{i}\sqrt{1-w^2}\big)^{\lambda} + \big(w - \mathrm{i}\sqrt{1-w^2}\big)^{\lambda}}{\sqrt{1-w^2}}, \tag{4.29}$$

$$\mathbf{S}_{\frac{1}{2}+n,\lambda}(w) = \frac{2^n}{\mathrm{i}\sqrt{\pi}(\lambda-n)_{2n+1}}\partial_w^n \frac{\big(w + \mathrm{i}\sqrt{1-w^2}\big)^{\lambda} - \big(w - \mathrm{i}\sqrt{1-w^2}\big)^{\lambda}}{\sqrt{1-w^2}} \tag{4.30}$$

and

$$\mathbf{Z}_{-\frac{1}{2}-n,\lambda}(w) = \frac{(-1)^n 2^{\lambda}(w^2-1)_{\bullet}^{\frac{1}{2}+n}}{\Gamma(1+\lambda+n)}\partial_w^n \frac{\big((w^2-1)_{\bullet}^{\frac{1}{2}} + w\big)^{-\lambda}}{(w^2-1)_{\bullet}^{\frac{1}{2}}}, \tag{4.31}$$

$$\mathbf{Z}_{\frac{1}{2}+n,\lambda}(w) = \frac{(-1)^n 2^{\lambda}}{\Gamma(1+\lambda+n)}\partial_w^n \frac{\big((w^2-1)_{\bullet}^{\frac{1}{2}} + w\big)^{-\lambda}}{(w^2-1)_{\bullet}^{\frac{1}{2}}}. \tag{4.32}$$

4.6. Integral representations. In this subsection we describe the four basic types of integral representations of the Gegenbauer functions $\mathbf{S}_{\alpha,\lambda}$ and $\mathbf{Z}_{\alpha,\lambda}$. In order to understand their form it is useful to start with recalling some basic facts about the hypergeometric equation.

The hypergeometric equation has 6 standard solutions, which are traditionally presented in *Kummer's table.* To represent these solutions in a form as symmetric as possible, instead of the usual parameters a, b, c it is convenient to use the parameters $\alpha = c-1$, $\beta = a+b-c$, $\mu = a-b$. Various elements of Kummer's table correspond to permutations of α, β, μ and switching their signs, as explained, e.g., in **[6, 7]**.

The hypergeometric function $\mathbf{F}(a, b; c; z)$ is of course one of the standard solutions of the hypergeometric equation. For $\mathrm{Re}(1+\alpha) > |\mathrm{Re}(\beta \mp \mu)|$ it satisfies the following pair of identities known already to Euler (see, e.g., **[6, 7]** and **[20**, Eq. (15.6.1)**]**):

$$\begin{aligned}&\mathbf{F}\Big(\frac{1+\alpha+\beta+\mu}{2}, \frac{1+\alpha+\beta-\mu}{2}; \alpha+1; z\Big)\\ &= \frac{1}{\Gamma(\frac{1+\alpha+\beta\mp\mu}{2})\Gamma(\frac{1+\alpha-\beta\pm\mu}{2})}\int_1^{\infty} t^{\frac{-1-\alpha+\beta\pm\mu}{2}}(t-1)^{\frac{-1+\alpha-\beta\pm\mu}{2}}(t-z)^{\frac{-1-\alpha-\beta\mp\mu}{2}}\,\mathrm{d}t.\end{aligned} \tag{4.33}$$

This may be verified for $|z| < 1$ by Taylor expanding $t-z$ around $z = 0$ and integrating term by term. It then follows that the identity is true everywhere because both left and right-hand side are holomorphic on $\mathbb{C}\setminus[1,\infty[$. Note that the change of the sign of μ does not affect the left-hand side of (4.33), but yields two distinct integral representations on the right-hand side.

By choosing intervals that join pairs of singular points $0, 1, \infty, z$ of the integrand of (4.33), one obtains integral representations of all 6 standard solutions (see, e.g., **[6, 7]**). Let us quote the pair of such representations of another standard solution, the first for $\mathrm{Re}(1+\mu) > |\mathrm{Re}(\alpha-\beta)|$, the second for $\mathrm{Re}(1+\mu) > |\mathrm{Re}(\alpha+\beta)|$:

$$z^{\frac{-1-\alpha-\beta-\mu}{2}}\mathbf{F}\Big(\frac{1+\mu+\beta+\alpha}{2},\frac{1+\mu+\beta-\alpha}{2};\mu+1;z^{-1}\Big)$$

$$(4.34)\qquad = \frac{1}{\Gamma(\frac{1+\mu+\beta-\alpha}{2})\Gamma(\frac{1+\mu-\beta+\alpha}{2})}\int_0^1 t^{\frac{-1-\alpha+\beta+\mu}{2}}(1-t)^{\frac{-1+\alpha-\beta+\mu}{2}}(z-t)^{\frac{-1-\alpha-\beta-\mu}{2}}\,\mathrm{d}t,$$

$$(4.35)\qquad = \frac{1}{\Gamma(\frac{1+\mu+\beta+\alpha}{2})\Gamma(\frac{1+\mu-\beta-\alpha}{2})}\int_z^\infty t^{\frac{-1-\alpha+\beta-\mu}{2}}(t-1)^{\frac{-1+\alpha-\beta-\mu}{2}}(t-z)^{\frac{-1-\alpha-\beta+\mu}{2}}\,\mathrm{d}t.$$

To obtain representations of the Gegenbauer functions we need to set $\alpha=\beta$, $\mu=2\lambda$, $t=\frac{s+1}{2}$ in the above formulas. By setting $z=\frac{1-w}{2}$ in (4.33), we obtain the following pair of representations valid for $\frac{1}{2} > \mp\mathrm{Re}(\lambda) > -\frac{1}{2}-\mathrm{Re}(\alpha)$:

$$(4.36)\qquad \mathbf{S}_{\alpha,\lambda}(w) = \frac{2^{\frac{1}{2}+\alpha\mp\lambda}}{\Gamma(\frac{1}{2}+\alpha\mp\lambda)\Gamma(\frac{1}{2}\pm\lambda)}\int_1^\infty (s^2-1)^{-\frac{1}{2}\pm\lambda}(s+w)^{-\frac{1}{2}-\alpha\mp\lambda}\,\mathrm{d}s.$$

Setting $z=\frac{1+w}{2}$ in (4.34) we obtain the identity valid for $\mathrm{Re}(\lambda)+\frac{1}{2}>0$:

$$(4.37)\qquad \mathbf{Z}_{\alpha,\lambda}(w) = \frac{1}{\sqrt{\pi}\Gamma(\frac{1}{2}+\lambda)}\int_{-1}^1 (1-s^2)^{-\frac{1}{2}+\lambda}(w-s)^{-\frac{1}{2}-\alpha-\lambda}\,\mathrm{d}s,$$

Setting $z=\frac{1+w}{2}$ in (4.35) yields the formula valid for $\mathrm{Re}(\lambda)+\frac{1}{2} > |\mathrm{Re}(\alpha)|$:

$$(4.38)\qquad \mathbf{Z}_{\alpha,\lambda}(w) = \frac{2^{2\lambda}\Gamma(\lambda+\frac{1}{2})}{\sqrt{\pi}\Gamma(\frac{1}{2}+\lambda-\alpha)\Gamma(\frac{1}{2}+\lambda+\alpha)}\int_w^\infty (s^2-1)_\bullet^{-\frac{1}{2}-\lambda}(s-w)^{-\frac{1}{2}-\alpha+\lambda}\,\mathrm{d}s.$$

Derivation of (4.37) and (4.38) uses the duplication formula (A.8).

Inserting $w=\frac{z}{(z^2-1)_\bullet^{\frac{1}{2}}}$ and using the Whipple transformations (4.10) and (4.11), we derive another family of integral representations. The first follows by inserting $s=\frac{t-z}{(z^2-1)_\bullet^{\frac{1}{2}}}$ into (4.36) and is valid for $\mathrm{Re}(\lambda)+\frac{1}{2} > -\mathrm{Re}(\alpha) > -\frac{1}{2}$:

$$(4.39)\qquad \mathbf{Z}_{\alpha,\lambda}(z) = \frac{2^{\frac{1}{2}+\alpha+\lambda}}{\Gamma(\frac{1}{2}+\alpha+\lambda)\Gamma(\frac{1}{2}-\alpha)}\int_{(z^2-1)_\bullet^{\frac{1}{2}}+z}^{\infty} (t^2-2tz+1)^{-\alpha-\frac{1}{2}}t^{-\frac{1}{2}+\alpha-\lambda}\,\mathrm{d}t.$$

To avoid having to discuss integration contour deformations involved in the change of variables, it is convenient to check this first for $z\in[1,\infty[$ and then use analyticity.

The second representation follows by inserting $s=\frac{-t+z}{(z^2-1)_\bullet^{\frac{1}{2}}}$ into (4.37) and holds for $\mathrm{Re}(\alpha)+\frac{1}{2}>0$:

$$(4.40)\qquad \mathbf{S}_{\alpha,\lambda}(z) = -\frac{\mathrm{i}(1-z^2)^{-\alpha}}{\sqrt{\pi}\Gamma(\frac{1}{2}+\alpha)}\int_{z-\mathrm{i}\sqrt{1-z^2}}^{z+\mathrm{i}\sqrt{1-z^2}} (t^2-2tz+1)_\bullet^{\alpha-\frac{1}{2}}t^{-\frac{1}{2}-\alpha\pm\lambda}\,\mathrm{d}t,$$

where $(t^2-2tz+1)_\bullet^\mu = (t-z-\mathrm{i}\sqrt{1-z^2})^\mu(t-z+\mathrm{i}\sqrt{1-z^2})^\mu$, and the integration contour is chosen so that it passes zero from the right without encircling it and does not cross the horizontal half-lines $z\pm\mathrm{i}\sqrt{1-z^2}-\mathbb{R}_+$. Such contours varying smoothly with z on $\mathbb{C}\setminus(]-\infty,-1]\cup[1,\infty[)$ exist because on this region

$\mathrm{Im}(z - \mathrm{i}\sqrt{1-z^2}) < 0$ and $\mathrm{Im}(z + \mathrm{i}\sqrt{1-z^2}) > 0$. Since $\mathbf{S}_{\alpha,\lambda}(z)$ is continuous on $[1,\infty[$, the right-hand side of (4.40) has the same limits as z approaches an element of $[1,\infty[$ from above or below. This is not obvious directly from (4.40). The next representation follows by inserting $s = \frac{t+z}{(z^2-1)^{\frac{1}{2}}}$ into (4.38), and is valid for $\mathrm{Re}(\alpha) + \frac{1}{2} > |\mathrm{Re}(\lambda)|$:

$$\text{(4.41)}\quad \mathbf{S}_{\alpha,\lambda}(z) = \frac{2^{2\alpha}\Gamma(\alpha+\frac{1}{2})}{\sqrt{\pi}\Gamma(\frac{1}{2}+\alpha-\lambda)\Gamma(\frac{1}{2}+\alpha+\lambda)} \int_0^\infty (t^2+2tz+1)^{-\alpha-\frac{1}{2}} t^{-\frac{1}{2}+\alpha-\lambda}\,\mathrm{d}t.$$

It is convenient to make an exponential change of variables in the representations above. Substituting $z = \cosh\theta$, with $\mathrm{Re}(\theta) > 0$, $|\mathrm{Im}(\theta)| < \pi$, and $t = \mathrm{e}^{\phi}$ in (4.39), we obtain (with the same restrictions on α, λ):

$$\text{(4.42)}\quad \mathbf{Z}_{\alpha,\lambda}(\cosh\theta) = \frac{2^{\lambda}}{\Gamma(\frac{1}{2}+\alpha+\lambda)\Gamma(\frac{1}{2}-\alpha)} \int_\theta^\infty \big(\cosh\phi - \cosh\theta\big)^{-\alpha-\frac{1}{2}} \mathrm{e}^{-\lambda\phi}\,\mathrm{d}\phi.$$

Under the same conditions on θ, combining (4.9) and (4.42) gives

(4.43)

$$\mathbf{Z}_{\alpha,\lambda}(\cosh\theta) = \frac{2^{\lambda}}{\Gamma(\frac{1}{2}-\alpha+\lambda)\Gamma(\frac{1}{2}+\alpha)(\sinh\theta)^{2\alpha}} \int_\theta^\infty (\cosh\phi - \cosh\theta)^{\alpha-\frac{1}{2}} \mathrm{e}^{-\lambda\phi}\,\mathrm{d}\phi.$$

Substituting $t = \mathrm{e}^{\mathrm{i}\,\phi}$ in (4.40) we obtain for $0 < \mathrm{Re}(\theta) < \pi$

$$\text{(4.44)}\qquad \mathbf{S}_{\alpha,\lambda}(\cos\theta) = \frac{2^{\alpha-\frac{1}{2}}}{\sqrt{\pi}\Gamma(\frac{1}{2}+\alpha)(\sin\theta)^{2\alpha}} \int_{-\theta}^{\theta} (\cos\phi - \cos\theta)^{\alpha-\frac{1}{2}} \mathrm{e}^{\pm\mathrm{i}\,\lambda\phi}\,\mathrm{d}\phi.$$

Substituting $t = \mathrm{e}^{\phi}$, in (4.41) we get for $|\mathrm{Re}(\theta)| < \pi$

(4.45)

$$\mathbf{S}_{\alpha,\lambda}(\cos\theta) = \frac{2^{\alpha-\frac{1}{2}}\Gamma(\alpha+\frac{1}{2})}{\sqrt{\pi}\Gamma(\frac{1}{2}+\alpha-\lambda)\Gamma(\frac{1}{2}+\alpha+\lambda)} \int_{-\infty}^{\infty} (\cosh\phi + \cos\theta)^{-\alpha-\frac{1}{2}} \mathrm{e}^{-\lambda\phi}\,\mathrm{d}\phi.$$

4.7. Asymptotics. From the integral representations of the Gegenbauer functions, we can deduce the following asymptotics.

THEOREM 4.2. *Let $\alpha \geqslant -\frac{1}{2}$ and $\pi > \delta > 0$ be fixed. Then we have the following estimates:*

(1) *Uniformly for $\theta \in [0, \pi-\delta]$ and $\beta \to \infty$,*

$$\text{(4.46)}\qquad \frac{(\sin\theta)^{\alpha+\frac{1}{2}}}{2^{\alpha}\theta^{\alpha+\frac{1}{2}}} \mathbf{S}_{\alpha,\pm\mathrm{i}\,\beta}(\cos\theta) = (\theta\beta)^{-\alpha} I_\alpha(\theta\beta)\Big(1 + \theta O(\beta^{-1})\Big).$$

(2) *Uniformly for $\theta \in [0, \pi - \delta]$ and $\beta \to \infty$,*

$$\frac{\pi\,\mathrm{e}^{-\pi\beta}(\sin\theta)^{\alpha+\frac{1}{2}}}{2^{\alpha}\theta^{\alpha+\frac{1}{2}}}\mathbf{S}_{\alpha,\pm\mathrm{i}\beta}(-\cos\theta) = (\theta\beta)^{-\alpha}K_{\alpha}(\beta\theta)\big(1+\mathcal{O}(\beta^{-1})\big). \tag{4.47}$$

(3) *Uniformly for $\theta \geqslant 0$ and $\lambda \to \infty$,*

$$\frac{\sqrt{\pi}\Gamma(\frac{1}{2}-\alpha+\lambda)(\sinh\theta)^{\alpha+\frac{1}{2}}}{2^{\lambda+\frac{1}{2}}\theta^{\alpha+\frac{1}{2}}}\mathbf{Z}_{\alpha,\lambda}(\cosh\theta) = (\lambda\theta)^{-\alpha}K_{\alpha}(\lambda\theta)\big(1+\mathcal{O}(\lambda^{-1})\big). \tag{4.48}$$

REMARK 4.3. Similar statements about associated Legendre functions can for example be found in [**5**, **12**]. See also [**21**, Chapter 12, §§12,13] for detailed asymptotics of Legendre functions including also non-leading terms.

In the proof of Theorem 4.2, we will use the functions

$$\operatorname{sinc}\theta := \frac{\sin\theta}{\theta}, \quad \operatorname{sinhc}\theta := \frac{\sinh\theta}{\theta}. \tag{4.49}$$

We note that $\operatorname{sinc} 0 = \operatorname{sinhc} 0 = 1$, sinc is decreasing on $[0, \pi]$, sinhc and all its derivatives are increasing on $[0, \infty[$, and for $\theta \in [0, \pi]$ we have

$$0 \leqslant \operatorname{sinc}\theta \leqslant 1 - \frac{\theta}{\pi}, \qquad -\frac{1}{3}\theta \leqslant \operatorname{sinc}'\theta \leqslant 0. \tag{4.50}$$

PROOF. We note first that the limiting case $\alpha = -\frac{1}{2}$ can be proved explicitly by using the representations of modified Bessel functions and Gegenbauer functions in the half-integer case, given in Sections 3.3 and 4.5, respectively. For the following, let us thus assume $\alpha > -\frac{1}{2}$. In the proof, c will be the notation for various constants independent of θ, s.

Let us prove (4.46). The integral representation (3.31) implies

$$(\theta\beta)^{-\alpha}I_{\alpha}(\theta\beta) = \frac{1}{\sqrt{2\pi}\Gamma(\alpha+\frac{1}{2})}\int_{-1}^{1}\Big(\frac{1-s^2}{2}\Big)^{\alpha-\frac{1}{2}}\mathrm{e}^{-\theta\beta s}\,\mathrm{d}s. \tag{4.51}$$

We use the integral representation (4.44), the substitution $\phi = \theta s$ and the identity $\cos(s\theta) - \cos\theta = 2\sin\big(\theta\frac{1-s}{2}\big)\sin\big(\theta\frac{1+s}{2}\big)$ to find

$$\begin{aligned} &\frac{(\sin\theta)^{\alpha+\frac{1}{2}}}{2^{\alpha}\theta^{\alpha+\frac{1}{2}}}\mathbf{S}_{\alpha,\pm\mathrm{i}\beta}(\cos\theta) \\ &= \frac{1}{\sqrt{2\pi}\Gamma(\frac{1}{2}+\alpha)}\int_{-1}^{1}\left(\frac{2\sin\big(\theta\frac{1-s}{2}\big)\sin\big(\theta\frac{1+s}{2}\big)}{\theta\sin\theta}\right)^{\alpha-\frac{1}{2}}\mathrm{e}^{-\theta\beta s}\,\mathrm{d}s. \end{aligned} \tag{4.52}$$

Subtracting (4.51) from (4.52), we obtain

$$\begin{aligned} &\frac{(\sin\theta)^{\alpha+\frac{1}{2}}}{2^{\alpha}\theta^{\alpha+\frac{1}{2}}}\mathbf{S}_{\alpha,\pm\mathrm{i}\beta}(\cos\theta) - (\theta\beta)^{-\alpha}I_{\alpha}(\theta\beta) \\ &= \frac{1}{\sqrt{2\pi}\Gamma(\alpha+\frac{1}{2})}\int_{-1}^{1}\Big(\frac{1-s^2}{2}\Big)^{\alpha-\frac{1}{2}}\left(\Big(\frac{\operatorname{sinc}\big(\theta\frac{1-s}{2}\big)\operatorname{sinc}\big(\theta\frac{1+s}{2}\big)}{\operatorname{sinc}\theta}\Big)^{\alpha-\frac{1}{2}}-1\right)\mathrm{e}^{-s\beta\theta}\,\mathrm{d}s. \end{aligned} \tag{4.53}$$

Next,

$$\begin{aligned} &\frac{\operatorname{sinc}\left(\theta\frac{1-s}{2}\right)\operatorname{sinc}\left(\theta\frac{1+s}{2}\right)}{\operatorname{sinc}\theta}-1 \\ &=\Big(\operatorname{sinc}\frac{(1-|s|)\theta}{2}-1\Big)\frac{\operatorname{sinc}\frac{(1+|s|)\theta}{2}}{\operatorname{sinc}\theta}+\frac{\operatorname{sinc}\frac{(1+|s|)\theta}{2}-\operatorname{sinc}\theta}{\operatorname{sinc}\theta}=:I+II. \end{aligned} \tag{4.54}$$

Now

$$|I|\leqslant c\frac{\theta^2(1-|s|)^2}{\operatorname{sinc}\theta},\quad |II|\leqslant c\frac{\theta^2(1-|s|)}{\operatorname{sinc}\theta},\quad \text{hence}\quad |(4.54)|\leqslant c\frac{\theta^2(1-|s|)}{\operatorname{sinc}\theta}.$$

II is estimated as follows:

$$\Big|\operatorname{sinc}\frac{(1+|s|)\theta}{2}-\operatorname{sinc}\theta\Big|\leqslant\Big|\frac{(1+|s|)}{2}\theta-\theta\Big|\sup_{\phi\in[0,\theta]}|\operatorname{sinc}'\phi|\leqslant c\theta^2(1-|s|). \tag{4.55}$$

Hence, using

$$|x^\kappa-1|=\Big|\kappa\int_1^x y^{\kappa-1}\,\mathrm{d}y\Big|\leqslant|\kappa||x-1|\max(x^{\kappa-1},1) \tag{4.56}$$

and $\frac{1}{\operatorname{sinc}(c)}\leqslant c$, $\frac{\operatorname{sinc}\left(\theta\frac{1-s}{2}\right)\operatorname{sinc}\left(\theta\frac{s+1}{2}\right)}{\operatorname{sinc}\theta}\geqslant c>0$ on $[0,\pi-\delta]$, we obtain

$$\left|\Big(\frac{\operatorname{sinc}\left(\theta\frac{1-s}{2}\right)\operatorname{sinc}\left(\theta\frac{s+1}{2}\right)}{\operatorname{sinc}\theta}\Big)^{\alpha-\frac12}-1\right|\leqslant c\theta^2(1-|s|). \tag{4.57}$$

Thus, applying (4.57) and then integrating by parts, we obtain

$$\begin{aligned} |(4.53)|&\leqslant c\theta^2\int_{-1}^1\Big(\frac{1-s^2}{2}\Big)^{\alpha-\frac12}(1-|s|)\,\mathrm{e}^{s\beta\theta}\,\mathrm{d}s \\ &=\frac{c\theta}{\beta}\int_{-1}^1\Big(\frac{1-s^2}{2}\Big)^{\alpha-\frac12}\frac{2\alpha s+\operatorname{sgn}(s)}{1+|s|}\,\mathrm{e}^{s\beta\theta}\,\mathrm{d}s \\ &\leqslant\frac{c\theta}{\beta}(\beta\theta)^{-\alpha}I_\alpha(\beta\theta). \end{aligned} \tag{4.58}$$

This proves (4.46).

We next prove (4.47). The integral representation (3.32) gives

$$(\theta\beta)^{-\alpha}K_\alpha(\theta\beta)=\frac{\sqrt{\pi}}{\sqrt{2}\Gamma(\frac12+\alpha)}\int_1^\infty\Big(\frac{s^2-1}{2}\Big)^{\alpha-\frac12}\mathrm{e}^{-\theta\beta s}\,\mathrm{d}s. \tag{4.59}$$

The integral representation (4.44), the substitution $\phi=\theta s-\pi$ and the identity $\cos\theta-\cos(s\theta)=2\sin\left(\theta\frac{s-1}{2}\right)\sin\left(\theta\frac{s+1}{2}\right)$ yield

$$\begin{aligned} &\frac{\pi\,\mathrm{e}^{-\pi\beta}(\sin\theta)^{\alpha+\frac12}}{2^\alpha\theta^{\alpha+\frac12}}\mathbf{S}_{\alpha,\pm\mathrm{i}\beta}(-\cos\theta) \\ &=\frac{\sqrt{\pi}}{\sqrt{2}\Gamma(\frac12+\alpha)}\int_1^{\frac{2\pi}{\theta}-1}\left(\frac{2\sin\left(\theta\frac{s-1}{2}\right)\sin\left(\theta\frac{1+s}{2}\right)}{\theta\sin\theta}\right)^{\alpha-\frac12}\mathrm{e}^{-\theta\beta s}\,\mathrm{d}s. \end{aligned} \tag{4.60}$$

Subtracting (4.59) from (4.60) we obtain
(4.61)

$$\begin{aligned}
&\frac{\pi(\sin\theta)^{\alpha+\frac{1}{2}}\,\mathrm{e}^{-\beta\pi}}{2^{\alpha}\theta^{\alpha+\frac{1}{2}}}\mathbf{S}_{\alpha,\pm\mathrm{i}\beta}(-\cos\theta)-(\beta\theta)^{-\alpha}K_{\alpha}(\beta\theta)\\
&=\frac{\sqrt{\pi}}{\sqrt{2}\Gamma(\alpha+\frac{1}{2})}\int_{1}^{\frac{2\pi}{\theta}-1}\Big(\frac{s^2-1}{2}\Big)^{\alpha-\frac{1}{2}}\bigg(\Big(\frac{\operatorname{sinc}\theta\frac{s-1}{2}\operatorname{sinc}\theta\frac{s+1}{2}}{\operatorname{sinc}\theta}\Big)^{\alpha-\frac{1}{2}}-1\bigg)\,\mathrm{e}^{-s\beta\theta}\,\mathrm{d}s\\
&\quad-\frac{\sqrt{\pi}}{\sqrt{2}\Gamma(\alpha+\frac{1}{2})}\int_{\frac{2\pi}{\theta}-1}^{\infty}\Big(\frac{s^2-1}{2}\Big)^{\alpha-\frac{1}{2}}\mathrm{e}^{-s\beta\theta}\,\mathrm{d}s\;=:\;\frac{\sqrt{\pi}}{\sqrt{2}\Gamma(\alpha+\frac{1}{2})}\int_{1}^{\infty}F(s)\,\mathrm{d}s.
\end{aligned}$$

Arguing as in the proof of (4.55), we obtain

$$\left|\frac{\operatorname{sinc}\left(\theta\frac{s-1}{2}\right)\operatorname{sinc}\left(\theta\frac{s+1}{2}\right)}{\operatorname{sinc}\theta}-1\right|\leqslant c\theta(s-1). \tag{4.62}$$

The power of θ is worse than in (4.57), because for $|s|>1$ we can no longer bound II by a term proportional to θ^2.

Unfortunately, due to the zero of $\operatorname{sinc}\theta\frac{s+1}{2}$ at $s=\frac{2\pi}{\theta}-1$, the remaining part of the argument must be split into three cases.

First we consider $\alpha\geqslant\frac{3}{2}$. Using (4.56) we obtain

$$\left|\Big(\frac{\operatorname{sinc}\left(\theta\frac{s-1}{2}\right)\operatorname{sinc}\left(\theta\frac{s+1}{2}\right)}{\operatorname{sinc}\theta}\Big)^{\alpha-\frac{1}{2}}-1\right|\leqslant c\theta(s-1). \tag{4.63}$$

Therefore,

$$\Big|\int_{1}^{\frac{2\pi}{\theta}-1}F(s)\,\mathrm{d}s\Big|\leqslant c\theta\int_{1}^{\frac{2\pi}{\theta}-1}\Big(\frac{s^2-1}{2}\Big)^{\alpha-\frac{1}{2}}(s-1)\,\mathrm{e}^{-s\beta\theta}\,\mathrm{d}s. \tag{4.64}$$

Next, using $\theta(s-1)\geqslant 2\pi-2\theta$ we obtain

$$\Big|\int_{\frac{2\pi}{\theta}-1}^{\infty}F(s)\,\mathrm{d}s\Big|\leqslant\theta(2\pi-2\theta)^{-1}\int_{\frac{2\pi}{\theta}-1}^{\infty}\Big(\frac{s^2-1}{2}\Big)^{\alpha-\frac{1}{2}}(s-1)\,\mathrm{e}^{-s\beta\theta}\,\mathrm{d}s \tag{4.65}$$

Then we sum up (4.64) and (4.65). We choose the new c to be the maximum of the old c and $(2\delta)^{-1}$. Next we integrate by parts:

$$\begin{aligned}
\Big|\int_{1}^{\infty}F(s)\,\mathrm{d}s\Big|&\leqslant c\theta\int_{1}^{\infty}\Big(\frac{s^2-1}{2}\Big)^{\alpha-\frac{1}{2}}(s-1)\,\mathrm{e}^{-s\beta\theta}\,\mathrm{d}s\\
&=\frac{c}{\beta}\int_{1}^{\infty}\Big(\frac{s^2-1}{2}\Big)^{\alpha-\frac{1}{2}}\frac{2\alpha s+1}{1+s}\,\mathrm{e}^{-s\beta\theta}\,\mathrm{d}s\\
&\leqslant\frac{c}{\beta}(\beta\theta)^{-\alpha}K_{\alpha}(\beta\theta).
\end{aligned} \tag{4.66}$$

Next we consider $\frac{1}{2}\leqslant\alpha\leqslant\frac{3}{2}$. Choose $\sigma\in]\pi,\pi+\delta[$. For $1\leqslant s\leqslant\frac{\sigma}{\theta}$, we have

$$\pi-\theta\frac{s+1}{2}\geqslant\frac{\pi-\sigma+\delta}{2}>0, \tag{4.67}$$

so for such s the argument of $\operatorname{sinc}\left(\theta\frac{s+1}{2}\right)$ is separated from the location of the zero by a constant independent of θ and s. Hence we have a bound

$$\frac{\operatorname{sinc}\left(\theta\frac{s-1}{2}\right)\operatorname{sinc}\left(\theta\frac{s+1}{2}\right)}{\operatorname{sinc}\theta}\geqslant c>0. \tag{4.68}$$

Therefore, we can apply (4.56) and obtain (4.63) on this interval. Thus

$$\Big|\int_1^{\frac{\sigma}{\theta}} F(s)\,\mathrm{d}s\Big| \leqslant c\theta \int_1^{\frac{\sigma}{\theta}} \Big(\frac{s^2-1}{2}\Big)^{\alpha-\frac12}(s-1)\,\mathrm{e}^{-s\beta\theta}\,\mathrm{d}s. \tag{4.69}$$

On $\frac{\sigma}{\theta} \leqslant s \leqslant \frac{2\pi}{\theta}-1$ we have $\delta < \sigma-\theta \leqslant \theta(s-1)$. We estimate the left-hand side of (4.63) by a constant and get

$$\Big|\int_{\frac{\sigma}{\theta}}^{\frac{2\pi}{\theta}-1} F(s)\,\mathrm{d}s\Big| \leqslant c\frac{\theta}{\delta} \int_{\frac{\sigma}{\theta}}^{\frac{2\pi}{\theta}-1} \Big(\frac{s^2-1}{2}\Big)^{\alpha-\frac12}(s-1)\,\mathrm{e}^{-s\beta\theta}\,\mathrm{d}s. \tag{4.70}$$

Now combine (4.69), (4.70), and (4.65) and argue as in (4.66).

Finally, consider $-\frac12 < \alpha \leqslant \frac12$. We need to consider only the interval $\frac{\sigma}{\theta} \leqslant s \leqslant \frac{2\pi}{\theta}-1$. Contributions of the remaining part of the integration region are $\mathcal{O}(\beta^{-1})$ by previous arguments.

First,

$$s^2-1 \geqslant \frac{\sigma^2-\theta^2}{\theta^2} \geqslant \frac{c}{\theta^2} \quad\Rightarrow\quad \Big(\frac{s^2-1}{2}\Big)^{\alpha-\frac12} \leqslant c\theta^{1-2\alpha}. \tag{4.71}$$

Moreover, using (4.50) we have

$$\frac{1}{\sin\theta} \geqslant c, \quad \operatorname{sinc}\big(\theta\tfrac{s\pm1}{2}\big) \geqslant c\theta\big(\tfrac{2\pi}{\theta}-s\mp1\big), \tag{4.72}$$

which implies

$$\begin{aligned}&\Big(\frac{\operatorname{sinc}\big(\theta\frac{s-1}{2}\big)\operatorname{sinc}\big(\theta\frac{s+1}{2}\big)}{\operatorname{sinc}\theta}\Big)^{\alpha-\frac12}-1 \leqslant \Big(\frac{\operatorname{sinc}\big(\theta\frac{s-1}{2}\big)\operatorname{sinc}\big(\theta\frac{s+1}{2}\big)}{\operatorname{sinc}\theta}\Big)^{\alpha-\frac12}\\ &\leqslant c\theta^{2\alpha-1}\Big(\big(\tfrac{2\pi}{\theta}-s\big)^2-1\Big)^{\alpha-\frac12} = c\theta^{2\alpha-1}(t^2-1)^{\alpha-\frac12},\end{aligned} \tag{4.73}$$

where we set $t=\frac{2\pi}{\theta}-s$. Then we estimate (very roughly)

$$\begin{aligned}\Big|\int_{\frac{\sigma}{\theta}}^{\frac{2\pi}{\theta}-1} F(s)\,\mathrm{d}s\Big| &\leqslant c\int_1^{\frac{2\pi-\sigma}{\theta}} (t^2-1)^{\alpha-\frac12}\,\mathrm{e}^{-\beta2\pi+\beta\theta t}\,\mathrm{d}t\\ &\leqslant c\int_1^{\frac{2\pi-\sigma}{\theta}} (t^2-1)^{\alpha-\frac12}\,\mathrm{e}^{\beta2(\pi-\sigma)-\beta\theta t}\,\mathrm{d}t\\ &\leqslant c\mathrm{e}^{\beta2(\pi-\sigma)}(\beta\theta)^{-\alpha}K_\alpha(\beta\theta) = (\beta\theta)^{-\alpha}K_\alpha(\beta\theta)\mathcal{O}(\beta^{-\infty}).\end{aligned} \tag{4.74}$$

Finally, let us prove (4.48). Substituting $\phi = s\theta$ in (4.43) and using the identity $\cosh(s\theta)-\cosh(\theta) = 2\sinh\big(\theta\frac{s-1}{2}\big)\sinh\big(\theta\frac{s+1}{2}\big)$, we find

$$\begin{aligned}&\frac{\sqrt{\pi}\Gamma(\frac12-\alpha+\lambda)(\sinh\theta)^{\alpha+\frac12}}{2^{\lambda+\frac12}\theta^{\alpha+\frac12}}\mathbf{Z}_{\alpha,\lambda}(\cosh\theta)\\ &= \frac{\sqrt{\pi}}{\sqrt{2}\Gamma(\frac12+\alpha)}\int_1^\infty \left(\frac{2\sinh\big(\theta\frac{s-1}{2}\big)\sinh\big(\theta\frac{s+1}{2}\big)}{\theta\sinh\theta}\right)^{\alpha-\frac12}\mathrm{e}^{-\lambda\theta s}\,\mathrm{d}s.\end{aligned} \tag{4.75}$$

Subtracting (4.59) with β replaced with λ from (4.75), we obtain

$$\begin{aligned}&\frac{\sqrt{\pi}\Gamma(-\alpha+\lambda+\frac12)(\sinh\theta)^{\alpha+\frac12}}{2^{\lambda+\frac12}\theta^{\alpha+\frac12}}\mathbf{Z}_{\alpha,\lambda}(\cosh\theta)-(\lambda\theta)^{-\alpha}K_\alpha(\lambda\theta)\\ &= c\int_1^\infty \Big(\frac{s^2-1}{2}\Big)^{\alpha-\frac12}\left(\Big(\frac{\operatorname{sinhc}\big(\theta\frac{s-1}{2}\big)\operatorname{sinhc}\big(\theta\frac{s+1}{2}\big)}{\operatorname{sinhc}\theta}\Big)^{\alpha-\frac12}-1\right)\mathrm{e}^{-s\lambda\theta}\,\mathrm{d}s.\end{aligned} \tag{4.76}$$

Function $s \mapsto \frac{\operatorname{sinhc}\left(\theta\frac{s-1}{2}\right)\operatorname{sinhc}\left(\theta\frac{s+1}{2}\right)}{\operatorname{sinhc}(\theta)}$ is monotonically increasing and equals 1 at $s = 1$. Arguing as in the proof of (4.57) we obtain

$$\left|\frac{\operatorname{sinhc}\left(\theta\frac{s-1}{2}\right)\operatorname{sinhc}\left(\theta\frac{s+1}{2}\right)}{\operatorname{sinhc}\theta} - 1\right| \leqslant c\theta(s-1)\,\mathrm{e}^{\theta(s-1)}. \tag{4.77}$$

Next we note that

$$\begin{aligned}\frac{\operatorname{sinhc}\left(\theta\frac{s-1}{2}\right)\operatorname{sinhc}\left(\theta\frac{s+1}{2}\right)}{\operatorname{sinhc}\theta} &= \frac{2}{\theta(s^2-1)}(\mathrm{e}^{\theta(s-1)}-1)\frac{1-\mathrm{e}^{-\theta(s+1)}}{1-\mathrm{e}^{-2\theta}}\\ &\leqslant \frac{2\,\mathrm{e}^{\theta(s-1)}}{s+1}\frac{1-\mathrm{e}^{-\theta(s+1)}}{1-\mathrm{e}^{-2\theta}} \leqslant \mathrm{e}^{\theta(s-1)},\end{aligned} \tag{4.78}$$

where in the first inequality we used $\mathrm{e}^{x}-1 \leqslant x\,\mathrm{e}^{x}$, and in the second one we optimized $\frac{1-\mathrm{e}^{-\theta(s+1)}}{1-\mathrm{e}^{-2\theta}}$ with respect to θ. Invoking (4.56) again, we get

$$\left|\left(\frac{\operatorname{sinhc}\theta\frac{s-1}{2}\operatorname{sinhc}\theta\frac{s+1}{2}}{\operatorname{sinhc}\theta}\right)^{\alpha-\frac{1}{2}} - 1\right| \leqslant c\theta(s-1)\,\mathrm{e}^{\theta(s-1)\rho(\alpha)}, \tag{4.79}$$

where $\rho(\alpha) = \max\{1, \alpha - \frac{1}{2}\}$. Then, integrating by parts, we obtain

$$\begin{aligned}|(4.76)| &\leqslant c\theta\int_1^\infty\left(\frac{s^2-1}{2}\right)^{\alpha-\frac{1}{2}}(s-1)\,\mathrm{e}^{-s\lambda\theta+(s-1)\theta\rho(\alpha)}\,\mathrm{d}s\\ &= \frac{c}{\lambda-\rho(\alpha)}\,\mathrm{e}^{-\theta\rho(\alpha)}\int_1^\infty\left(\frac{s^2-1}{2}\right)^{\alpha-\frac{1}{2}}\frac{2\alpha s+1}{1+s}\,\mathrm{e}^{-s\theta(\lambda-\rho(\alpha))}\,\mathrm{d}s\\ &\leqslant \frac{c\,\mathrm{e}^{-\theta\rho(\alpha)}}{\lambda-\rho(\alpha)}(\theta(\lambda-\rho(\alpha)))^{-\alpha}K_\alpha(\theta(\lambda-\rho(\alpha)))\\ &\leqslant \frac{c}{\lambda-\rho(\alpha)}\left(\frac{\lambda}{\lambda-\rho(\alpha)}\right)^{2\alpha}(\lambda\theta)^{-\alpha}K_\alpha(\lambda\theta),\end{aligned} \tag{4.80}$$

where in the last step we used (3.8). □

4.8. Bilinear integrals. In this subsection we compute certain integrals involving products of two Gegenbauer functions. They are analogs of the integrals for products of Macdonald functions considered in Subsect. 3.4. We will use the integration variable $2w$, because this will facilitate comparison of the integrals for Gegenbauer functions with those for Macdonald functions. Indeed, if we consider the spherical, resp. hyperbolic interpretation of these identities, for small r, we have $2(w \mp 1) \approx r^2$. Therefore, $\mathrm{d}2w$ corresponds to $2r\,\mathrm{d}r = \mathrm{d}r^2$. This is especially important in the case of anomalous generalized integrals in Theorem 4.10, where without a good choice of variables we would not have obtained a correct limit.

THEOREM 4.4. *For* $|\mathrm{Re}(\alpha)| < 1$, *the following identities hold:*

$$\begin{aligned}(4.81)\quad &\int_{-2}^{2} \mathbf{S}_{\alpha,\mathrm{i}\,\beta_1}(w)\mathbf{S}_{\alpha,\mathrm{i}\,\beta_2}(w)(1-w^2)^{\alpha}\,\mathrm{d}2w \\ &= \frac{2^{2\alpha+2}}{(\beta_1^2-\beta_2^2)\sin\pi\alpha}\Big(\frac{\cosh(\pi\beta_1)}{\Gamma(\frac12+\alpha-\mathrm{i}\,\beta_2)\Gamma(\frac12+\alpha+\mathrm{i}\,\beta_2)} - (\beta_1\leftrightarrow\beta_2)\Big),\end{aligned}$$

$$\begin{aligned}(4.82)\quad &\int_{-2}^{2} \mathbf{S}_{0,\mathrm{i}\,\beta_1}(w)\mathbf{S}_{0,\mathrm{i}\,\beta_2}(w)\,\mathrm{d}2w \\ &= \frac{4\cosh(\pi\beta_1)\cosh(\pi\beta_2)\Big(\psi(\frac12-\mathrm{i}\,\beta_1)+\psi(\frac12+\mathrm{i}\,\beta_1)-(\beta_1\leftrightarrow\beta_2)\Big)}{\pi^2(\beta_1^2-\beta_2^2)},\end{aligned}$$

$$\begin{aligned}(4.83)\quad &\int_{-2}^{2} \mathbf{S}_{\alpha,\mathrm{i}\,\beta}(w)^2(1-w^2)^{\alpha}\,\mathrm{d}2w = \frac{2^{2\alpha+1}\,\mathrm{i}\cosh(\pi\beta)}{\beta\sin\pi\alpha\,\Gamma(\frac12+\alpha-\mathrm{i}\,\beta)\Gamma(\frac12+\alpha+\mathrm{i}\,\beta)} \\ &\quad\times\Big(\psi(\tfrac12+\alpha+\mathrm{i}\,\beta)-\psi(\tfrac12+\alpha-\mathrm{i}\,\beta)+\psi(\tfrac12-\mathrm{i}\,\beta)-\psi(\tfrac12+\mathrm{i}\,\beta)\Big),\end{aligned}$$

$$(4.84)\quad \int_{-2}^{2} \mathbf{S}_{0,\mathrm{i}\,\beta}(w)^2\,\mathrm{d}2w = \frac{2\,\mathrm{i}\cosh^2(\pi\beta)\Big(\psi'(\frac12+\mathrm{i}\,\beta)-\psi'(\frac12-\mathrm{i}\,\beta)\Big)}{\beta\pi^2},$$

and

$$(4.85)\quad \int_{-2}^{2} \mathbf{S}_{\alpha,0}(w)^2(1-w^2)^{\alpha}\,\mathrm{d}2w = \frac{2^{2\alpha+1}\left(\pi^2-2\psi'(\frac12+\alpha)\right)}{\sin(\pi\alpha)\Gamma(\frac12+\alpha)^2},$$

$$(4.86)\quad \int_{-2}^{2} \mathbf{S}_{0,0}(w)^2\,\mathrm{d}2w = -\frac{4\psi''(\frac12)}{\pi^2}.$$

PROOF. The Gegenbauer equation implies

$$\begin{aligned}&(\beta_1^2-\beta_2^2)\int_{-1}^{1} \mathbf{S}_{\alpha,\mathrm{i}\,\beta_1}(w)\mathbf{S}_{\alpha,\mathrm{i}\,\beta_2}(w)(1-w^2)^{\alpha}\,\mathrm{d}w \\ &= \int_{-1}^{1} \mathbf{S}_{\alpha,\mathrm{i}\,\beta_1}(w)\partial_w(1-w^2)^{\alpha+1}\partial_w\mathbf{S}_{\alpha,\mathrm{i}\,\beta_2}(w)\,\mathrm{d}w \\ &\quad - \int_{-1}^{1}\Big(\partial_w(1-w^2)^{\alpha+1}\partial_w\mathbf{S}_{\alpha,\mathrm{i}\,\beta_1}(w)\Big)\mathbf{S}_{\alpha,\mathrm{i}\,\beta_2}(w)\,\mathrm{d}w \\ &= \lim_{w\searrow-1}\Big((1-w^2)^{\alpha+1}\Big(\mathbf{S}_{\alpha,\mathrm{i}\,\beta_1}(w)\partial_w\mathbf{S}_{\alpha,\mathrm{i}\,\beta_2}(w)-\big(\partial_w\mathbf{S}_{\alpha,\mathrm{i}\,\beta_1}(w)\big)\mathbf{S}_{\alpha,\mathrm{i}\,\beta_2}(w)\Big)\Big).\end{aligned}$$

Then we use the connection formula (4.12) and obtain (4.81). The remaining identities follow by the de l'Hôpital rule. □

Integrals in Theorem 4.4 are defined as generalized integrals for $\mathrm{Re}(\alpha) > -1$ (beyond this region one would need to use a generalized integral which also takes care of non-integrability at the right endpoint). If $\alpha \notin \mathbb{N}$, these generalized integrals are non-anomalous and hence formulas from Theorem 4.4 remain true. Below we compute anomalous generalized integrals for $\alpha \in \mathbb{N}$.

THEOREM 4.5. *Let $\alpha \in \mathbb{N}$. Then*

(4.87)
$$
\begin{aligned}
&(-\tfrac{1}{4})^{\alpha+1}\Big(\prod_{i=1}^{2}\prod_{\pm}\Gamma(\tfrac{1}{2}+\alpha\pm\mathrm{i}\,\beta_i)\Big)\,\mathrm{gen}\!\int_{-2}^{2}\mathbf{S}_{\alpha,\mathrm{i}\,\beta_1}(w)\mathbf{S}_{\alpha,\mathrm{i}\,\beta_2}(w)(1-w^2)^{\alpha}\,\mathrm{d}2w\\
&=\frac{1}{\beta_1^2-\beta_2^2}\bigg(\big(\tfrac{1}{2}+\mathrm{i}\,\beta_1\big)_{\alpha}\big(\tfrac{1}{2}-\mathrm{i}\,\beta_1\big)_{\alpha}\Big(\ln 4-\sum_{\pm}\psi\big(\tfrac{1}{2}+\alpha\pm\mathrm{i}\,\beta_1\big)\Big)-(\beta_1\leftrightarrow\beta_2)\bigg)\\
&\quad+\sum_{k=0}^{\alpha-1}\bigg(\prod_{\pm}\frac{(\frac{1}{2}\pm\mathrm{i}\,\beta_2)_{\alpha}(\frac{1}{2}\pm\mathrm{i}\,\beta_1)_k}{(\frac{1}{2}\pm\mathrm{i}\,\beta_2)_{k+1}}\bigg)\\
&\quad\times\Big(\psi(\alpha-k)+\psi(1+k)-\sum_{\pm}H_{\alpha-1-k}\big(\tfrac{1}{2}-\alpha\pm\mathrm{i}\,\beta_2\big)\Big),
\end{aligned}
$$

(4.88)
$$
\begin{aligned}
&(-\tfrac{1}{4})^{\alpha+1}\frac{\pi}{\cosh\pi\beta}\Big(\prod_{\pm}\Gamma(\tfrac{1}{2}+\alpha\pm\mathrm{i}\,\beta)\Big)\,\mathrm{gen}\!\int_{-2}^{2}\mathbf{S}_{\alpha,\mathrm{i}\,\beta}(w)^2(1-w^2)^{\alpha}\,\mathrm{d}2w\\
&=\sum_{k=0}^{\alpha-1}\frac{1}{(k+\frac{1}{2})^2+\beta^2}\Big(\ln 4+\psi(\alpha-k)+\psi(k+1)-\sum_{\pm}\psi(-\tfrac{1}{2}-k\pm\mathrm{i}\,\beta)\Big)\\
&\quad-\frac{\mathrm{i}}{2\beta}\big(\psi'(\tfrac{1}{2}+\alpha+\mathrm{i}\,\beta)-\psi'(\tfrac{1}{2}+\alpha-\mathrm{i}\,\beta)\big),
\end{aligned}
$$

and

$$
\begin{aligned}
&(-\tfrac{1}{4})^{\alpha+1}\pi\Gamma(\tfrac{1}{2}+\alpha)^2\,\mathrm{gen}\!\int_{-2}^{2}\mathbf{S}_{\alpha,0}(w)^2(1-w^2)^{\alpha}\,\mathrm{d}2w\\
&=\psi''(\tfrac{1}{2}+\alpha)+\sum_{k=0}^{\alpha-1}\frac{\ln 4+\psi(\alpha-k)+\psi(k+1)-2\psi(-\frac{1}{2}-k)}{(k+\frac{1}{2})^2}.
\end{aligned}\tag{4.89}
$$

REMARK 4.6. The right-hand side of (4.87) is not manifestly symmetric in β_1, β_2, although the left-hand side is. This is a nontrivial sum identity which follows from the equality of (4.4) and (4.5). The same is true for (4.109) below.

PROOF. Let us first compute (4.87). We will use the dimensional regularization. Thus α at first takes arbitrary complex values with $\mathrm{Re}(\alpha) > -1$. We define

$$
f(\alpha,w):=2^{-4\alpha}\left[\prod_{i=1}^{2}\prod_{\pm}\Gamma(\tfrac{1}{2}+\alpha\pm\mathrm{i}\,\beta_i)\right]\mathbf{S}_{\alpha,\mathrm{i}\,\beta_1}(w)\mathbf{S}_{\alpha,\mathrm{i}\,\beta_2}(w)(1-w^2)^{\alpha}.\tag{4.90}
$$

In terms of $f(\alpha,w)$, (4.81) becomes (for $|\mathrm{Re}(\alpha)|<1$)

(4.91)
$$
\int_{-2}^{2}f(\alpha,w)\,\mathrm{d}2w=\frac{2^{-2\alpha+2}\pi}{(\beta_1^2-\beta_2^2)\sin\pi\alpha}\bigg(\prod_{\pm}\frac{\Gamma(\frac{1}{2}+\alpha\pm\mathrm{i}\,\beta_1)}{\Gamma(\frac{1}{2}\pm\mathrm{i}\,\beta_1)}-(\beta_1\leftrightarrow\beta_2)\bigg).
$$

Let $m \in \mathbb{N}_0$. The residue and the finite part of (4.91) at $\alpha = m$ are

$$
\mathrm{res}(m)=\frac{(-1)^m2^{-2m+2}}{\beta_1^2-\beta_2^2}\Big(\big(\tfrac{1}{2}-\mathrm{i}\,\beta_1\big)_m\big(\tfrac{1}{2}+\mathrm{i}\,\beta_1\big)_m-(\beta_1\leftrightarrow\beta_2)\Big)\tag{4.92}
$$

and
(4.93)

$$(-1)^m 2^{2m-2}(\beta_1^2-\beta_2^2)\ \underset{\alpha\to m}{\mathrm{fp}}\int_{-2}^{2} f(\alpha,w)\,\mathrm{d}2w$$
$$=\big(\tfrac12-\mathrm{i}\,\beta_1\big)_m\big(\tfrac12+\mathrm{i}\,\beta_1\big)_m\Big(-\ln(4)+\sum_{\pm}\psi\big(\tfrac12+m\pm\mathrm{i}\,\beta_1\big)\Big)-(\beta_1\leftrightarrow\beta_2).$$

All negative powers of $w+1$ in $f(\alpha,w)$ are contained in

$$f^{\mathrm{sing}}(\alpha,w):=\frac{2^{-2\alpha}\pi^2\big(\frac{1-w}{2}\big)^{-\alpha}\big(\frac{1+w}{2}\big)^{-\alpha}\mathbf{S}_{-\alpha,\mathrm{i}\,\beta_1}(-w)\mathbf{S}_{-\alpha,\mathrm{i}\,\beta_2}(-w)}{\sin^2\pi\alpha}.$$

In order to avoid having to expand $(1-w)^\alpha$ in terms of $1+w$, we choose to expand one Gegenbauer function using (4.4) and the other one using (4.5):

$$(2(1+w))^\alpha f^{\mathrm{sing}}(\alpha,w)=\frac{\pi^2}{\sin^2\pi\alpha}\mathbf{F}\Big(\frac12+\mathrm{i}\,\beta_1,\frac12-\mathrm{i}\,\beta_1,-\alpha+1,\frac{1+w}{2}\Big)$$
$$\times\,\mathbf{F}\Big(\frac12-\alpha+\mathrm{i}\,\beta_2,\frac12-\alpha-\mathrm{i}\,\beta_2,-\alpha+1,\frac{1+w}{2}\Big)$$
(4.94)
$$=\Bigg(\sum_{k=0}^{\infty}\frac{(-1)^k(\frac12+\mathrm{i}\,\beta_1)_k(\frac12-\mathrm{i}\,\beta_1)_k\Gamma(\alpha-k)}{k!}\Big(\frac{1+w}{2}\Big)^k\Bigg)$$
$$\times\Bigg(\sum_{j=0}^{\infty}\frac{(-1)^j(\frac12-\alpha+\mathrm{i}\,\beta_2)_j(\frac12-\alpha-\mathrm{i}\,\beta_2)_j\Gamma(\alpha-j)}{j!}\Big(\frac{1+w}{2}\Big)^j\Bigg).$$

The coefficient at $\big(2(1+w)\big)^{-\alpha+m-1}$ in $f^{\mathrm{sing}}(\alpha,w)$ is

$$f_{m-1}(\alpha)=\sum_{k=0}^{m-1}\frac{\Gamma(\alpha-k)\Gamma(\alpha-m+1+k)}{k!(m-1-k)!(-4)^{m-1}}\prod_{\pm}\big(\tfrac12\pm\mathrm{i}\,\beta_1\big)_k\big(\tfrac12-\alpha\pm\mathrm{i}\,\beta_2\big)_{m-1-k}.$$

We evaluate the derivative

(4.95)
$$f'_{m-1}(m)=(-\tfrac14)^{m-1}\sum_{k=0}^{m-1}\prod_{\pm}\big(\tfrac12\pm\mathrm{i}\,\beta_1\big)_k\big(\tfrac12-m\pm\mathrm{i}\,\beta_2\big)_{m-1-k}$$
$$\times\Bigg(\psi(m-k)+\psi(1+k)-\sum_{\pm}H_{m-1-k}\Big(\frac12-m\pm\mathrm{i}\,\beta_2\Big)\Bigg).$$

Then we use

$$\big(\tfrac12-m+\mathrm{i}\,\beta_2\big)_{m-1-k}\big(\tfrac12-m-\mathrm{i}\,\beta_2\big)_{m-1-k}=\frac{(\frac12+\mathrm{i}\,\beta_2)_m(\frac12-\mathrm{i}\,\beta_2)_m}{(\frac12+\mathrm{i}\,\beta_2)_{k+1}(\frac12-\mathrm{i}\,\beta_2)_{k+1}}.\tag{4.96}$$

Since the singular exponent is $-\alpha+m-1$, we have to *add* $f'_{m-1}(m)$ to the finite part of the bilinear integral.[1] Thus, we obtain the generalized integral (4.87) for positive integers by multiplying

$$\mathrm{gen}\int_{-2}^{2} f(m,w)\,\mathrm{d}2w=\underset{\alpha\to m}{\mathrm{fp}}\int_{-2}^{2} f(\alpha,w)\,\mathrm{d}2w+f'_{m-1}(m)\tag{4.97}$$

[1] For $\mathrm{Re}(\alpha)>0$, the derivative term has to be added to the finite part, not subtracted:
$$-\frac{f_{m-1}(\alpha)}{m-\alpha}=-\frac{f_{m-1}(m)+(\alpha-m)f'_{m-1}(m)}{m-\alpha}+\mathcal{O}(\alpha-m)=\frac{\mathrm{res}(m)}{\alpha-m}+f'_{m-1}(m)+\mathcal{O}(\alpha-m).$$

with

$$\frac{2^{4m}}{\Gamma(\frac{1}{2}+m+\mathrm{i}\beta_1)\Gamma(\frac{1}{2}+m-\mathrm{i}\beta_1)\Gamma(\frac{1}{2}+m+\mathrm{i}\beta_2)\Gamma(\frac{1}{2}+m-\mathrm{i}\beta_2)}. \tag{4.98}$$

Let us now prove (4.88). Since the exponents of the non-integrable terms do not depend on β, it suffices to take the limit $\beta_1, \beta_2 \to \beta$ of (4.87). The multiplicative factor in front of the left-hand side becomes

$$(-\tfrac{1}{4})^{\alpha+1}\prod_{\pm}\Gamma(\tfrac{1}{2}+\alpha\pm\mathrm{i}\beta)^2 = (-\tfrac{1}{4})^{\alpha+1}\frac{\pi}{\cosh\pi\beta}\prod_{\pm}\Gamma(\tfrac{1}{2}+\alpha\pm\mathrm{i}\beta)(\tfrac{1}{2}\pm\mathrm{i}\beta)_\alpha. \tag{4.99}$$

Applying the de l'Hôpital rule to the second line of (4.87) then yields

$$\begin{aligned}&\frac{\mathrm{i}(\frac{1}{2}+\mathrm{i}\beta)_\alpha(\frac{1}{2}-\mathrm{i}\beta)_\alpha}{2\beta}\Big(\big(\ln 4-\sum_{\pm}\psi(\tfrac{1}{2}+\alpha\pm\mathrm{i}\beta)\big)\big(\sum_{\pm}\pm H_\alpha(\tfrac{1}{2}\pm\mathrm{i}\beta)\big)\\&\qquad-\sum_{\pm}\pm\psi'(\tfrac{1}{2}+\alpha\pm\mathrm{i}\beta)\Big)\end{aligned} \tag{4.100}$$

The third and fourth line of (4.87) can be converted to the following form:

$$\begin{aligned}&(\tfrac{1}{2}+\mathrm{i}\beta)_\alpha(\tfrac{1}{2}-\mathrm{i}\beta)_\alpha\sum_{k=0}^{\alpha-1}\frac{1}{\left(\frac{1}{2}+k\right)^2+\beta^2}\\&\quad\times\Big(\psi(\alpha-k)+\psi(k+1)-\sum_{\pm}H_{\alpha-1-k}(\tfrac{1}{2}-\alpha\pm\mathrm{i}\beta)\Big).\end{aligned} \tag{4.101}$$

Next we use the identities

$$\sum_{k=0}^{\alpha-1}\frac{1}{(\frac{1}{2}+k)^2+\beta^2}=\frac{\mathrm{i}}{2\beta}\Big(H_\alpha(\tfrac{1}{2}+\mathrm{i}\beta)-H_\alpha(\tfrac{1}{2}-\mathrm{i}\beta)\Big), \tag{4.102}$$

$$\psi(\tfrac{1}{2}+\alpha+\mathrm{i}\beta)+\psi(\tfrac{1}{2}+\alpha-\mathrm{i}\beta)=\psi(\tfrac{1}{2}-\alpha+\mathrm{i}\beta)+\psi(\tfrac{1}{2}-\alpha-\mathrm{i}\beta), \tag{4.103}$$

$$H_{\alpha-1-k}(\tfrac{1}{2}-\alpha\pm\mathrm{i}\beta)=\psi(-\tfrac{1}{2}-k\pm\mathrm{i}\beta)-\psi(\tfrac{1}{2}-\alpha\pm\mathrm{i}\beta), \tag{4.104}$$

to combine (4.101) and (4.100). This proves (4.88).

Similarly, applying de l'Hôpital to (4.88) for $\beta\to 0$ yields (4.89). □

THEOREM 4.7. *Let* $|\mathrm{Re}(\alpha)|<1$ *and* $\mathrm{Re}(\lambda)>0$. *Then*

$$\begin{aligned}&\int_2^\infty \mathbf{Z}_{\alpha,\lambda_1}(w)\mathbf{Z}_{\alpha,\lambda_2}(w)(w^2-1)^\alpha\,\mathrm{d}2w\\&\quad=\frac{2^{\lambda_1+\lambda_2+1}}{(\lambda_1^2-\lambda_2^2)\sin\pi\alpha}\Big(\frac{1}{\Gamma(\frac{1}{2}-\alpha+\lambda_1)\Gamma(\frac{1}{2}+\alpha+\lambda_2)}-(\lambda_1\leftrightarrow\lambda_2)\Big),\end{aligned} \tag{4.105}$$

$$\int_2^\infty \mathbf{Z}_{0,\lambda_1}(w)\mathbf{Z}_{0,\lambda_2}(w)\,\mathrm{d}2w=\frac{2^{\lambda_1+\lambda_2+2}(\psi(\frac{1}{2}+\lambda_1)-\psi(\frac{1}{2}+\lambda_2))}{\pi(\lambda_1^2-\lambda_2^2)\Gamma(\frac{1}{2}+\lambda_1)\Gamma(\frac{1}{2}+\lambda_2)}, \tag{4.106}$$

$$\int_2^\infty \mathbf{Z}_{\alpha,\lambda}(w)^2(w^2-1)^\alpha\,\mathrm{d}2w=\frac{2^{2\lambda}(\psi(\frac{1}{2}+\alpha+\lambda)-\psi(\frac{1}{2}-\alpha+\lambda))}{\lambda\sin\pi\alpha\Gamma(\frac{1}{2}-\alpha+\lambda)\Gamma(\frac{1}{2}+\alpha+\lambda)}, \tag{4.107}$$

$$\int_2^\infty \mathbf{Z}_{0,\lambda}(w)^2\,\mathrm{d}2w=\frac{2^{2\lambda+1}\psi'(\frac{1}{2}+\lambda)}{\pi\lambda\Gamma(\frac{1}{2}+\lambda)^2}. \tag{4.108}$$

PROOF.

$$
\begin{aligned}
(\lambda_1^2-\lambda_2^2)&\int_1^\infty \mathbf{Z}_{\alpha,\lambda_1}(w)\mathbf{Z}_{\alpha,\lambda_2}(w)(w^2-1)^\alpha\,\mathrm{d}w\\
&=\int_1^\infty \mathbf{Z}_{\alpha,\lambda_1}(w)\partial_w(w^2-1)^{\alpha+1}\partial_w\mathbf{Z}_{\alpha,\lambda_2}(w)\,\mathrm{d}w\\
&\quad-\int_1^\infty \Big(\partial_w(w^2-1)^{\alpha+1}\partial_w\mathbf{Z}_{\alpha,\lambda_1}(w)\Big)\mathbf{Z}_{\alpha,\lambda_2}(w)\,\mathrm{d}w\\
&=\lim_{w\searrow 1}\Big(\mathbf{Z}_{\alpha,\lambda_1}(w)(w^2-1)^{\alpha+1}\partial_w\mathbf{Z}_{\alpha,\lambda_2}(w)-(w^2-1)^{\alpha+1}\big(\partial_w\mathbf{Z}_{\alpha,\lambda_1}(w)\big)\mathbf{Z}_{\alpha,\lambda_2}(w)\Big).
\end{aligned}
$$

Then we use the connection formula (4.13) to decompose the **Z** functions into the **S** functions. The remaining formulas follow by the rule of de l'Hôpital. □

REMARK 4.8. The integrand $\mathbf{Z}_{\alpha,\lambda}(w)^2(w^2-1)^\alpha$ has asymptotics $\sim w^{-1-2\lambda}$ for $w\to\infty$ and therefore the case $\mathrm{Re}(\lambda)\leqslant 0$ escapes our considerations.

As before, the above integral formulas remain valid in the non-anomalous case $\alpha\in\mathbb{C}\setminus\mathbb{Z}$ if the integral is replaced by the generalized integral. We compute the anomalous generalized integrals for $\alpha\in\mathbb{Z}$.

THEOREM 4.9. *Let $\alpha\in\mathbb{Z}$. For $\mathrm{Re}(\lambda_i)>0$, we have*

(4.109)
$$
\begin{aligned}
&\frac{\pi}{2^{\lambda_1+\lambda_2+1}}\Big(\prod_{i=1}^2\Gamma\big(\tfrac12+|\alpha|+\lambda_i\big)\Big)\,\mathrm{gen}\!\int_2^\infty \mathbf{Z}_{\alpha,\lambda_1}(w)\mathbf{Z}_{\alpha,\lambda_2}(w)(w^2-1)^\alpha\,\mathrm{d}2w\\
&\quad=\left(\frac{(\frac12+\lambda_1)_{|\alpha|}(\frac12-\lambda_1)_{|\alpha|}}{\lambda_1^2-\lambda_2^2}\Big(-\ln(4)+\sum_\pm\psi\big(\tfrac12\pm\alpha+\lambda_1\big)\Big)+(\lambda_1\leftrightarrow\lambda_2)\right)\\
&\qquad-\big(\tfrac12+\lambda_2\big)_{|\alpha|}\big(\tfrac12-\lambda_2\big)_{|\alpha|}\sum_{k=0}^{|\alpha|-1}\frac{(\frac12+\lambda_1)_k(\frac12-\lambda_1)_k}{(\frac12+\lambda_2)_{k+1}(\frac12-\lambda_2)_{k+1}}\\
&\qquad\times\Big(-\psi(|\alpha|-k)-\psi(1+k)+\sum_\pm H_{|\alpha|-1-k}\big(\tfrac12-|\alpha|\pm\lambda_2\big)\Big).
\end{aligned}
$$

Moreover, if $\mathrm{Re}(\lambda)>0$,

$$
\begin{aligned}
&\frac{(-1)^\alpha\pi}{2^{2\lambda+1}}\Big(\prod_\pm\Gamma\big(\tfrac12\pm\alpha+\lambda\big)\Big)\,\mathrm{gen}\!\int_2^\infty \mathbf{Z}_{\alpha,\lambda}(w)^2(w^2-1)^\alpha\,\mathrm{d}2w\\
&\quad=\frac{1}{2\lambda}\sum_\pm\Big(\psi'\big(\tfrac12\pm\alpha+\lambda\big)\mp H_{|\alpha|}\big(\tfrac12\pm\lambda\big)\ln(4)\Big)\\
&\qquad+\sum_{k=0}^{|\alpha|-1}\frac{\psi(\frac32+k+\lambda)+\psi(-\frac12-k+\lambda)-\psi(|\alpha|-k)-\psi(1+k)}{\lambda^2-\big(\frac12+k\big)^2}.
\end{aligned}
\tag{4.110}
$$

PROOF. The proof works similarly to the proof with the anomalous generalized **S**-integrals. For $\alpha\in\mathbb{C}$ we set

$$
f(w,\alpha)=\frac{\Gamma(\frac12-\alpha+\lambda_1)\Gamma(\frac12-\alpha+\lambda_2)}{2^{\lambda_1+\lambda_2-2\alpha-1}}\mathbf{Z}_{\alpha,\lambda_1}(w)\mathbf{Z}_{\alpha,\lambda_2}(w)(w^2-1)^\alpha.
\tag{4.111}
$$

Then for $\alpha \notin \mathbb{Z}$,

$$\operatorname{gen}\int_2^\infty f(w,\alpha)\,\mathrm{d}2w = \frac{2^{2\alpha+2}}{(\lambda_1^2-\lambda_2^2)\sin\pi\alpha}\left(\frac{\Gamma(\frac12-\alpha+\lambda_2)}{\Gamma(\frac12+\alpha+\lambda_2)} - (\lambda_1 \leftrightarrow \lambda_2)\right). \tag{4.112}$$

Let $m \in \mathbb{Z}$. The residue and the finite part of (4.112) at $\alpha = m$ are

$$\operatorname{res}(m) = \frac{(-1)^m 2^{2m+2}}{\pi(\lambda_1^2-\lambda_2^2)}\left(\frac{\Gamma(\frac12-m+\lambda_2)}{\Gamma(\frac12+m+\lambda_2)} - (\lambda_1 \leftrightarrow \lambda_2)\right) \tag{4.113}$$

and

(4.114)

$$\begin{aligned}\operatorname{fp}\int_2^\infty f(w,\alpha)\,\mathrm{d}2w &= \frac{(-1)^m 2^{2m+2}}{\pi(\lambda_1^2-\lambda_2^2)}\\ &\times\left(\frac{\Gamma(\frac12-m+\lambda_2)}{\Gamma(\frac12+m+\lambda_2)}\big(\ln(4)-\psi(\tfrac12-m+\lambda_2)-\psi(\tfrac12+m+\lambda_2)\big) - (\lambda_1 \leftrightarrow \lambda_2)\right).\end{aligned}$$

For $\operatorname{Re}(\alpha) \leqslant 0$, the singular part of $f(\alpha,w)$ is contained in

$$\begin{aligned} f^{\text{sing}}(\alpha,w) &= \frac{\pi}{\sin^2\pi\alpha}\mathbf{S}_{\alpha,\lambda_1}(w)\mathbf{S}_{\alpha,\lambda_2}(w)(w^2-1)^\alpha\\ &= \frac{2^\alpha\pi(w-1)^\alpha}{\sin^2\pi\alpha}\mathbf{F}\Big(\frac12+\lambda_1,\frac12-\lambda_1,\alpha+1,\frac{1-w}{2}\Big)\\ &\times\mathbf{F}\Big(\frac12+\alpha+\lambda_2,\frac12+\alpha-\lambda_2,\alpha+1,\frac{1-w}{2}\Big)\\ &= \frac{2^\alpha(w-1)^\alpha}{\pi}\left(\sum_{k=0}^\infty\frac{(\frac12+\lambda_1)_k(\frac12-\lambda_1)_k\Gamma(-\alpha-k)}{k!}\Big(\frac{w-1}{2}\Big)^k\right)\\ &\times\left(\sum_{j=0}^\infty\frac{(\frac12+\alpha+\lambda_2)_j(\frac12+\alpha-\lambda_2)_j\Gamma(-\alpha-j)}{j!}\Big(\frac{w-1}{2}\Big)^j\right).\end{aligned} \tag{4.115}$$

Let m be a positive integer. The coefficient of (4.115) at $\big(2(w-1)\big)^{\alpha+m-1}$ is

(4.116)

$$\begin{aligned} &f_{m-1}(\alpha)\\ &= \frac{2^{-2m+2}}{\pi}\sum_{k=0}^{m-1}\frac{(\frac12+\lambda_1)_k(\frac12-\lambda_1)_k(\frac12+\alpha+\lambda_2)_{m-1-k}(\frac12+\alpha-\lambda_2)_{m-1-k}}{k!(m-1-k)!}\\ &\times\Gamma(-\alpha-k)\Gamma(-\alpha-m+1+k).\end{aligned}$$

Now

$$\begin{aligned}\frac{\mathrm{d}}{\mathrm{d}\alpha}f_{m-1}(\alpha)\Big|_{\alpha=-m} &= \frac{2^{-2m+2}}{\pi}\\ &\times\sum_{k=0}^{m-1}(\tfrac12+\lambda_1)_k(\tfrac12-\lambda_1)_k(\tfrac12-m+\lambda_2)_{m-1-k}(\tfrac12-m-\lambda_2)_{m-1-k}\\ &\times\Big(H_{m-1-k}(\tfrac12-m+\lambda_2)+H_{m-1-k}(\tfrac12-m-\lambda_2)-\psi(m-k)-\psi(1+k)\Big).\end{aligned}$$

Then we use

$$\big(\tfrac12-m+\lambda_2\big)_{m-1-k}\big(\tfrac12-m-\lambda_2\big)_{m-1-k} = \frac{(\frac12+\lambda_2)_m(\frac12-\lambda_2)_m}{(\frac12+\lambda_2)_{k+1}(\frac12-\lambda_2)_{k+1}}. \tag{4.117}$$

Thus we find the integral at negative integers:

$$\begin{aligned}
&\operatorname{gen}\int_2^\infty \mathbf{Z}_{-m,\lambda_1}(w)\mathbf{Z}_{-m,\lambda_2}(w)(w^2-1)^{-m}\,\mathrm{d}2w \\
&= \frac{2^{\lambda_1+\lambda_2-2m+1}}{\Gamma(\frac12+m+\lambda_1)\Gamma(\frac12+m+\lambda_2)}\operatorname{gen}\int_2^\infty f(-m,w)\,\mathrm{d}2w \\
&= \frac{2^{\lambda_1+\lambda_2-2m+1}}{\Gamma(\frac12+m+\lambda_1)\Gamma(\frac12+m+\lambda_2)}\left(\operatorname{fp}\int_2^\infty f(-m,w)\,\mathrm{d}2w - \frac{\mathrm{d}}{\mathrm{d}\alpha}f_{m-1}(\alpha)\Big|_{\alpha=-m}\right).
\end{aligned}$$

This proves (4.109) for negative integers. But (4.109) is invariant with respect to the flip of the sign of α, because

$$\mathbf{Z}_{-m,\lambda_1}(w)\mathbf{Z}_{-m,\lambda_2}(w)(w^2-1)^{-m} = \mathbf{Z}_{m,\lambda_1}(w)\mathbf{Z}_{m,\lambda_2}(w)(w^2-1)^{m}. \tag{4.118}$$

Thus (4.109) has been proven.

To prove (4.110), we set $\lambda=\lambda_1=\lambda_2$ in (4.109), using the de l'Hôpital rule where needed:

(4.119)

$$\begin{aligned}
&\operatorname{gen}\int_2^\infty \mathbf{Z}_{\alpha,\lambda}(w)\mathbf{Z}_{\alpha,\lambda}(w)(w^2-1)^{\alpha}\,\mathrm{d}2w = \frac{(-1)^\alpha 2^{2\lambda+1}}{\pi\Gamma(\frac12-\alpha+\lambda)\Gamma(\frac12+\alpha+\lambda)} \\
&\times\Bigg(\frac{\psi'(\frac12-\alpha+\lambda)+\psi'(\frac12+\alpha+\lambda)}{2\lambda} \\
&+\frac{(\psi(\frac12-\alpha+\lambda)+\psi(\frac12+\alpha+\lambda)-\ln(4))(-H_{|\alpha|}(\frac12-\lambda)+H_{|\alpha|}(\frac12+\lambda))}{2\lambda} \\
&+\sum_{k=0}^{|\alpha|-1}\frac{H_{|\alpha|-1-k}(\frac12-|\alpha|-\lambda)+H_{|\alpha|-1-k}(\frac12-|\alpha|+\lambda)-\psi(|\alpha|-k)-\psi(1+k)}{\lambda^2-\left(\frac12+k\right)^2}\Bigg).
\end{aligned}$$

Then we use

$$\sum_{k=0}^{|\alpha|-1}\frac{1}{\lambda^2-\left(\frac12+k\right)^2} = \frac{H_{|\alpha|}(\frac12+\lambda)-H_{|\alpha|}(\frac12-\lambda)}{2\lambda} \tag{4.120}$$

and

$$\psi(\tfrac12-|\alpha|+\lambda)+H_{|\alpha|-1-k}(\tfrac12-|\alpha|+\lambda)=\psi(-\tfrac12-k+\lambda), \tag{4.121}$$

$$\begin{aligned}
&\psi(\tfrac12+|\alpha|+\lambda)+H_{|\alpha|-1-k}(\tfrac12-|\alpha|-\lambda) \\
&=\psi(\tfrac12+|\alpha|+\lambda)-H_{|\alpha|-1-k}(\tfrac32+k+\lambda)=\psi(\tfrac32+k+\lambda). \qquad \square
\end{aligned} \tag{4.122}$$

4.9. Asymptotics of bilinear integrals. The bilinear (generalized) integrals for Gegenbauer functions converge to the corresponding integrals for Macdonald functions consistently with (4.47) and (4.48). This is relatively straightforward in the non-anomalous case. It is also true in the anomalous case, because we have chosen the right variables. Otherwise there would be a discrepancy.

THEOREM 4.10. *For $\beta\to\infty$ with fixed α satisfying* $\mathrm{Re}(\alpha)>-1$ *we have*

$$\begin{aligned}
&\frac{\pi^2\mathrm{e}^{-2\pi\beta}\beta^{2\alpha}}{2^{2\alpha}}\operatorname{gen}\int_{-2}^2 \mathbf{S}_{\alpha,\mathrm{i}\beta}(w)^2(1-w^2)^\alpha\,\mathrm{d}2w \\
&=\left(1+\mathcal{O}(\tfrac1\beta)\right)\operatorname{gen}\int_0^\infty K_\alpha(\beta r)^2 2r\,\mathrm{d}r.
\end{aligned} \tag{4.123}$$

For $\lambda \to \infty$ and all $\alpha \in \mathbb{C}$ we have

$$\begin{aligned}\frac{\pi\Gamma\big(\frac12+\alpha+\lambda\big)^2}{2^{2\lambda+1}\lambda^{2\alpha}}\,\mathrm{gen}\!\int_2^\infty \mathbf{Z}_{\alpha,\lambda}(w)^2(w^2-1)^\alpha\,\mathrm{d}2w \\ = \Big(1+\mathcal{O}\big(\tfrac{1}{\lambda}\big)\Big)\,\mathrm{gen}\!\int_0^\infty K_\alpha(\lambda r)^2 2r\,\mathrm{d}r.\end{aligned} \tag{4.124}$$

PROOF. Let us first consider the non-anomalous case, that is $\alpha \in \mathbb{C}\setminus\mathbb{Z}$. For the $\mathbf{S}$-functions, we have

$$\begin{aligned}&\mathrm{LHS}(4.123)\\ &= \frac{2\pi^2\beta^{2\alpha}}{\mathrm{e}^{2\pi\beta}}\frac{\mathrm{i}\cosh(\pi\beta)\big(\psi\big(\frac12+\alpha+\mathrm{i}\beta\big)-\psi\big(\frac12+\alpha-\mathrm{i}\beta\big)+\psi\big(\frac12-\mathrm{i}\beta\big)-\psi\big(\frac12+\mathrm{i}\beta\big)\big)}{\beta\sin\pi\alpha\,\Gamma\big(\frac12+\alpha-\mathrm{i}\beta\big)\Gamma\big(\frac12+\alpha+\mathrm{i}\beta\big)}\\ &= \frac{\pi\alpha}{\beta^2\sin\pi\alpha}\Big(1+\mathcal{O}\big(\tfrac{1}{\beta}\big)\Big) = \mathrm{RHS}(4.123),\end{aligned}$$

where we used

$$\begin{aligned}2\mathrm{e}^{-\pi\beta}\cosh(\pi\beta) &= 1+\mathcal{O}\big(\tfrac{1}{\beta}\big),\\ \Gamma\big(\tfrac12+\alpha-\mathrm{i}\beta\big)\Gamma\big(\tfrac12+\alpha+\mathrm{i}\beta\big) &= 2\pi\beta^{2\alpha}\,\mathrm{e}^{-\beta\pi}\big(1+\mathcal{O}\big(\tfrac{1}{\beta}\big)\big),\\ \psi\big(\tfrac12+\alpha+\mathrm{i}\beta\big)-\psi\big(\tfrac12+\alpha-\mathrm{i}\beta\big)+\psi\big(\tfrac12-\mathrm{i}\beta\big)-\psi\big(\tfrac12+\mathrm{i}\beta\big) &= -\frac{2\,\mathrm{i}\,\alpha}{\beta}+\mathcal{O}\big(\tfrac{1}{\beta^2}\big).\end{aligned}$$

For the $\mathbf{Z}$-functions, we have

$$\begin{aligned}&\mathrm{LHS}(4.124)\\ &= \frac{\pi\Gamma\big(\frac12+\alpha+\lambda\big)\big(\psi\big(\frac12+\alpha+\lambda\big)-\psi\big(\frac12-\alpha+\lambda\big)\big)}{2\lambda^{2\alpha+1}\sin\pi\alpha\,\Gamma\big(\frac12-\alpha+\lambda\big)}\\ &= \frac{\pi\alpha}{\lambda^2\sin\pi\alpha}\Big(1+\mathcal{O}\big(\tfrac{1}{\lambda}\big)\Big) = \mathrm{RHS}(4.124)\end{aligned}$$

where we used

$$\frac{\Gamma\big(\frac12+\alpha+\lambda\big)}{\Gamma\big(\frac12-\alpha+\lambda\big)} = \lambda^{2\alpha}\big(1+\mathcal{O}\big(\tfrac{1}{\lambda}\big)\big),$$

$$\psi\big(\tfrac12+\alpha+\lambda\big)-\psi\big(\tfrac12-\alpha+\lambda\big) = \frac{2\alpha}{\lambda}+\mathcal{O}\big(\tfrac{1}{\lambda^2}\big).$$

In the anomalous case, that is $\alpha \in \mathbb{N}$, we have

$$\begin{aligned}&\mathrm{LHS}(4.123)\\ &= \frac{(-1)^\alpha 4\cosh(\pi\beta)\,\mathrm{e}^{-2\pi\beta}\,\beta^{2\alpha}}{\pi\Gamma\big(\frac12+\alpha+\mathrm{i}\beta\big)\Gamma\big(\frac12+\alpha-\mathrm{i}\beta\big)}\Bigg(\frac{\mathrm{i}}{2\beta}\Big(\psi'\big(\tfrac12+\alpha+\mathrm{i}\beta\big)-\psi'\big(\tfrac12+\alpha-\mathrm{i}\beta\big)\Big)\\ &\quad+\sum_{k=0}^{|\alpha|-1}\frac{\psi\big(-\frac12-k+\mathrm{i}\beta\big)+\psi\big(-\frac12-k-\mathrm{i}\beta\big)-\psi(|\alpha|-k)-\psi(1+k)-\ln 4}{\big(\frac12+k\big)^2+\beta^2}\Bigg)\\ &= (-1)^\alpha\Big(\frac{1}{\beta^2}-\frac{|\alpha|}{\beta^2}\ln(4)+\sum_{k=0}^{|\alpha|-1}\frac{(2\ln(\beta)-\psi(|\alpha|-k)-\psi(1+k))}{\beta^2}\Big)\Big(1+O\big(\tfrac{1}{\beta}\big)\Big)\\ &= \mathrm{RHS}(4.123),\end{aligned}$$

where in the last equality we used (A.24). Similarly, for **Z** functions

$$\begin{aligned}
&\mathrm{LHS}(4.124)\\
&= \frac{(-1)^\alpha \Gamma(\frac12+\alpha+\lambda)}{\Gamma(\frac12-\alpha+\lambda)\lambda^{2\alpha}}\Bigg(\frac{\psi'(\frac12-\alpha+\lambda)+\psi'(\frac12+\alpha+\lambda)}{2\lambda}\\
&\quad+\frac{H_{|\alpha|}(\frac12-\lambda)-H_{|\alpha|}(\frac12+\lambda)}{2\lambda}\ln(4)\\
&\quad+\sum_{k=0}^{|\alpha|-1}\frac{\psi(\frac32+k+\lambda)+\psi(-\frac12-k+\lambda)-\psi(|\alpha|-k)-\psi(1+k)}{\lambda^2-\big(\frac12+k\big)^2}\Bigg)\\
&=(-1)^\alpha\Big(\frac{1}{\lambda^2}-\frac{|\alpha|}{\lambda^2}\ln(4)+\sum_{k=0}^{|\alpha|-1}\frac{\big(2\ln(\lambda)-\psi(|\alpha|-k)-\psi(1+k)\big)}{\lambda^2}\Big)\Big(1+O(\tfrac{1}{\lambda})\Big)\\
&=\mathrm{RHS}(4.124). \qquad \square
\end{aligned}$$

Appendix A. About the Gamma function

The *Gamma function* and its logarithmic derivative, the *digamma function*, play a central role in the theory of special functions. In this section we collect properties of these functions, which we will often use in our paper. We will also describe the properties of the *Pochhammer symbol* and *harmonic numbers*, which are also closely related to the Gamma and digamma functions.

A.1. Definitions.

Recall the definition of the digamma function,

$$\psi(z) := \partial_z \ln\Gamma(z) = \frac{\partial_z\Gamma(z)}{\Gamma(z)}. \tag{A.1}$$

It is also useful to introduce the Pochhammer symbol,

$$(z)_k := \frac{\Gamma(z+k)}{\Gamma(z)} = \begin{cases} (z)(z+1)\cdots(z+k-1), & k \geqslant 0,\\ \frac{1}{(z+k)(z+k+1)\cdots(z-1)}, & k\leqslant 0;\end{cases} \tag{A.2}$$

(A.3) the shifted kth harmonic number $\quad H_k(z) := \frac{1}{z}+\cdots+\frac{1}{z+k-1},$

(A.4) and the kth harmonic number $\quad H_k := H_k(1) = \frac{1}{1}+\cdots+\frac{1}{k}.$

A.2. Basic properties.

$$\Gamma(z+1) = z\Gamma(z), \tag{A.5}$$

$$\frac{1}{\Gamma(1-z)\Gamma(z)} = \frac{\sin\pi z}{\pi}, \tag{A.6}$$

$$\frac{1}{\Gamma(\frac12+z)\Gamma(\frac12-z)} = \frac{\cos\pi z}{\pi}, \tag{A.7}$$

$$\frac{\Gamma(2\lambda+1)}{\Gamma(\frac12+\lambda)\Gamma(1+\lambda)} = \frac{2^{2\lambda}}{\sqrt{\pi}}. \tag{A.8}$$

$\frac{1}{\Gamma(z)}$ is an entire function with zeros at $0,-1,-2,\dots$ and derivative

$$\partial_z\frac{1}{\Gamma(z)} = -\frac{\psi(z)}{\Gamma(z)}. \tag{A.9}$$

The digamma function satisfies the following functional relations [**14**, Chapter 8.365]:

$$\psi(1+z) = \psi(z) + \frac{1}{z}, \tag{A.10}$$

$$\psi(z) - \psi(1-z) = -\pi\cot(\pi z), \tag{A.11}$$

$$\psi(\tfrac{1}{2}+z) - \psi(\tfrac{1}{2}-z) = \pi\tan(\pi z), \tag{A.12}$$

$$2\psi(2z) = 2\ln 2 + \psi(z) + \psi(z+\tfrac{1}{2}), \tag{A.13}$$

$$\psi(z+k) = \psi(z) + H_k(z), \tag{A.14}$$

$$\psi(1+k) = -\gamma_{\mathrm{E}} + H_k. \tag{A.15}$$

The following relations among the harmonic numbers and the Pochhammer symbol are immediate from there definitions:

$$H_{k+n}(z) = H_n(z) + H_k(z+n), \tag{A.16}$$

$$H_k(z) = -H_k(1-z-k), \tag{A.17}$$

$$(z)_k(z+k)_n = (z)_{k+n}, \tag{A.18}$$

$$(z)_k = (-1)^k(1-k-z)_k, \tag{A.19}$$

$$(z)_{-k} = \frac{1}{(z-k)_k} = \frac{(-1)^k}{(1-z)_k}, \tag{A.20}$$

$$(\tfrac{1}{2}+z)_k(\tfrac{1}{2}-z)_k = (-1)^k(\tfrac{1}{2}+z-k)_{2k}, \tag{A.21}$$

$$\partial_z(z)_k = H_k(z)(z)_k. \tag{A.22}$$

The Pochammer symbol is also useful to expand powers around a different center:

$$(1-z)^{-a} = \sum_{k=0}^{\infty}\frac{(a)_k}{k!}z^k, \quad |z|<1. \tag{A.23}$$

The following identities follow by induction:

$$\sum_{k=0}^{m-1}\psi(1+k) = m\big(\psi(1+m)-1\big), \tag{A.24}$$

$$\sum_{k=0}^{m-1}H_k = m\big(H_m-1\big). \tag{A.25}$$

A.3. Special values. We have [**14**, Chapter 8.366]

$$\psi(1) = -\gamma_{\mathrm{E}}, \qquad \psi(\tfrac{1}{2}) = -\gamma_{\mathrm{E}} - 2\ln 2, \qquad \psi'(\tfrac{1}{2}) = \frac{\pi^2}{2} \tag{A.26}$$

with the Euler–Mascheroni constant γ_{E}. Moreover, we have

$$(1/2)_k = \frac{\Gamma(k+\frac{1}{2})}{\sqrt{\pi}} = \frac{(2k)!}{2^{2k}k!} = \frac{(2k-1)!}{2^{2k-1}(k-1)!} \tag{A.27}$$

and

$$\partial_z\frac{1}{\Gamma(z)}\bigg|_{z=-n} = (-1)^n n!, \quad n=0,1,2,\ldots. \tag{A.28}$$

A.4. Asymptotics for large arguments. Let $\epsilon > 0$. Then for $|\arg(z)| < \pi - \epsilon$, as $|z| \to \infty$, we have

$$\Gamma(z) = z^{z-\frac{1}{2}} \mathrm{e}^{-z} \sqrt{2\pi}\Big(1 + \mathcal{O}\big(\tfrac{1}{z}\big)\Big), \tag{A.29}$$

$$\ln \Gamma(z) = z \ln z - z + \frac{1}{2} \ln \frac{2\pi}{z} + \mathcal{O}\big(\tfrac{1}{z}\big), \tag{A.30}$$

$$\psi(z) = \ln(z) - \frac{1}{2z} + \mathcal{O}\big(\tfrac{1}{z^2}\big), \tag{A.31}$$

$$\psi'(z) = \frac{1}{z} + \frac{1}{2z^2} + \mathcal{O}\big(\tfrac{1}{z^3}\big). \tag{A.32}$$

(A.29) is the famous Stirling formula. (A.30) is obviously equivalent to (A.29). Formally, (A.31) and (A.32) follow from (A.30) by differentiation. However, rigorously, we are not allowed to differentiate asymptotic series, and therefore (A.31) and (A.32) require an independent proof. We can use for this the so-called 2nd Binet's identity **[27]** and its derivatives:

$$\ln \Gamma(z) = \Big(z - \frac{1}{2}\Big) \ln z - z + \frac{1}{2} \ln 2\pi + 2 \int_0^\infty \frac{\arctan \frac{t}{z}}{\mathrm{e}^{2\pi t} - 1} \mathrm{d}t, \tag{A.33}$$

$$\psi(z) = \ln z - \frac{1}{2z} - 2 \int_0^\infty \frac{t \, \mathrm{d}t}{(z^2 + t^2)(\mathrm{e}^{2\pi t} - 1)}, \tag{A.34}$$

$$\psi'(z) = \frac{1}{z} + \frac{1}{2z^2} + 4 \int_0^\infty \frac{z t \, \mathrm{d}t}{(z^2 + t^2)^2(\mathrm{e}^{2\pi t} - 1)}. \tag{A.35}$$

Let us show, e.g., that (A.33) implies (A.30). Let $0 < 2\theta < \pi$. We have $|\arctan u| \leqslant c|u|$ for $|\arg u| < \theta$. We set $t = \mathrm{e}^{\mathrm{i}\theta} s$. Then we use

$$\left| \frac{\arctan \frac{t}{z}}{\mathrm{e}^{2\pi t} - 1} \right| \leqslant \frac{c}{|z|} \frac{1}{\mathrm{e}^{2\pi s \cos \theta}}. \tag{A.36}$$

One can present the asymptotics (A.30), (A.31) and (A.32) in a different equivalent form. Let us fix α and let $|\arg \lambda| < \pi - \epsilon$. Then as $|\lambda| \to \infty$, we have

$$\ln \Gamma(\tfrac{1}{2} + \alpha + \lambda) = (\alpha + \lambda) \ln \lambda - \lambda + \frac{1}{2} \ln(2\pi) + \mathcal{O}\big(\tfrac{1}{\lambda}\big), \tag{A.37}$$

$$\psi(\tfrac{1}{2} + \alpha + \lambda) = \ln(\lambda) + \frac{\alpha}{\lambda} + \mathcal{O}\big(\tfrac{1}{\lambda^2}\big), \tag{A.38}$$

$$\psi'(\tfrac{1}{2} + \alpha + \lambda) = \frac{1}{\lambda} - \frac{\alpha}{\lambda^2} + \mathcal{O}\big(\tfrac{1}{\lambda^3}\big). \tag{A.39}$$

Here is a consequence of the Stirling formula **[5**, Lemma 2.1]:

$$\frac{\Gamma(a + \lambda)}{\Gamma(b + \lambda)} = \lambda^{a-b}\Big(1 + \mathcal{O}\big(\tfrac{1}{\lambda}\big)\Big). \tag{A.40}$$

Here is another consequence, where $\beta > 0$ and $\beta \to \infty$:

$$\Gamma(\alpha + \mathrm{i}\beta)\Gamma(\alpha - \mathrm{i}\beta) = 2\pi \beta^{2\alpha - 1} \mathrm{e}^{-\pi\beta} \Big(1 + \mathcal{O}\big(\tfrac{1}{\beta}\big)\Big). \tag{A.41}$$

Appendix B. Other conventions

This appendix is included to facilitate the comparison of our formulas with those contained in a part of the literature.

B.1. Associated Legendre equation. In the literature (see for example [**5**, **20**, **21**]), instead of the Gegenbauer equation (4.1), one often considers the *associated Legendre equation* given by the operator

$$\begin{aligned}(B.1)\quad &(z^2-1)^{\frac{\alpha}{2}}\Big((1-z^2)\partial_z^2-2(1+\alpha)z\partial_z+\lambda^2-\Big(\alpha+\frac{1}{2}\Big)^2\Big)(z^2-1)^{-\frac{\alpha}{2}}\\ &=\partial_z(1-z^2)\partial_z-\frac{\alpha^2}{1-z^2}+\lambda^2-\frac{1}{4}.\end{aligned}$$

In the standard literature the parameter λ is shifted by $\frac{1}{2}$. Thus, with $\mu=\alpha$ and $\nu=\lambda-\frac{1}{2}$, (B.1) becomes

$$(B.2)\qquad (1-z^2)\partial_z^2-2z\partial_z-\frac{\mu^2}{1-z^2}+\nu(\nu+1).$$

Certain functions annihilated by (B.2) are called *associated Legendre functions.* In this subsection we show how to convert the functions $\mathbf{S}$ and $\mathbf{Z}$ into associated Legendre functions. The material of this subsection will not be used elsewhere in this paper.

The *associated Legendre functions of the first kind* are defined as

$$\begin{aligned}\mathbf{P}_\nu^\mu(z)&=\Big(\frac{z+1}{z-1}\Big)^{\frac{\mu}{2}}\mathbf{F}\Big(-\nu,\nu+1;1-\mu;\frac{1-z}{2}\Big)\\ (B.3)\quad&=\frac{2^\mu}{(z^2-1)^{\frac{\mu}{2}}}\mathbf{F}\Big(1-\mu+\nu,-\mu-\nu;1-\mu;\frac{1-z}{2}\Big)\\ &=\frac{2^\mu}{(z^2-1)^{\frac{\mu}{2}}}\mathbf{S}_{-\mu,\nu+\frac{1}{2}}(z).\end{aligned}$$

The functions $\mathbf{P}_\nu^\mu$ are well adapted to the halfline $[0,\infty[$. On the interval $]-1,1[$ one prefers to use *Ferrer's associated Legendre functions of the first kind*

$$\begin{aligned}\mathrm{P}_\nu^\mu(z)&=\Big(\frac{z+1}{1-z}\Big)^{\frac{\mu}{2}}\mathbf{F}\Big(-\nu,\nu+1;1-\mu;\frac{1-z}{2}\Big)\\ (B.4)\quad&=\frac{2^\mu}{(1-z^2)^{\frac{\mu}{2}}}\mathbf{F}\Big(1-\mu+\nu,-\mu-\nu;1-\mu;\frac{1-z}{2}\Big)\\ &=\frac{2^\mu}{(1-z^2)^{\frac{\mu}{2}}}\mathbf{S}_{-\mu,\nu+\frac{1}{2}}(z).\end{aligned}$$

The *associated Legendre functions of the second kind* are given by

$$\begin{aligned}\mathbf{Q}_\nu^\mu(z)&=\mathrm{e}^{\mu\mathrm{i}\pi}\sqrt{\pi}\Gamma(\mu+\nu+1)\frac{(z^2-1)^{\frac{\mu}{2}}}{2^{\nu+1}z^{\nu+\mu+1}}\mathbf{F}\Big(\frac{\nu+\mu+2}{2},\frac{\nu+\mu+1}{2};\nu+\frac{3}{2};z^{-2}\Big)\\ (B.5)\quad&=\frac{\mathrm{e}^{\mu\mathrm{i}\pi}\sqrt{\pi}\Gamma(\mu+\nu+1)(z^2-1)^{\frac{\mu}{2}}}{2^{\nu+1}(1+z)^{\nu+\mu+1}\Gamma(\nu+\frac{3}{2})}F\Big(\nu+1,\nu+\mu+1;2\nu+2;\frac{2}{z+1}\Big)\\ &=\mathrm{e}^{\mu\mathrm{i}\pi}\sqrt{\pi}\Gamma(\mu+\nu+1)\frac{(z^2-1)^{\frac{\mu}{2}}}{2^{\nu+1}}\mathbf{Z}_{\mu,\nu+\frac{1}{2}}(z).\end{aligned}$$

B.2. Other conventions for Gegenbauer functions. One sometimes finds other conventions for the Gegenbauer functions in the literature [**10, 11, 20**]:

$$C_\lambda^\alpha(z) = \frac{\Gamma(\lambda+2\alpha)}{\Gamma(1+\lambda)\Gamma(2\alpha)} F\big(-\lambda, \lambda+2\alpha; \alpha+\tfrac{1}{2}; \tfrac{1-z}{2}\big), \tag{B.6}$$

$$D_\lambda^\alpha(z) = \frac{\mathrm{e}^{\mathrm{i}\,\pi\alpha}\,\Gamma(\lambda+2\alpha)(2z)^{-\lambda-2\alpha}}{\Gamma(\alpha)\Gamma(1+\alpha+\lambda)} F\big(\tfrac{1}{2}\lambda+\alpha, \tfrac{1}{2}+\tfrac{1}{2}\lambda+\alpha; 1+\alpha+\lambda; z^{-2}\big), \tag{B.7}$$

where the second identity is valid for $|z|>1$. Comparing these definitions with (4.4), (4.7), (B.3) and (B.5), we find

$$\begin{aligned} \frac{\Gamma(\alpha)\Gamma(1+\lambda)}{\sqrt{\pi}\Gamma(\lambda+2\alpha)} C_\lambda^\alpha(z) &= 2^{\frac{1}{2}-\alpha}(z^2-1)^{\frac{1}{4}-\frac{\alpha}{2}}\mathbf{P}^{\frac{1}{2}-\alpha}_{-\frac{1}{2}+\alpha+\lambda}(z) \\ &= 2^{1-2\alpha}\mathbf{S}_{\alpha-\frac{1}{2},\alpha+\lambda}(z), \end{aligned} \tag{B.8}$$

$$\begin{aligned} \frac{\Gamma(\alpha)\Gamma(1+\alpha+\lambda)}{\Gamma(\lambda+2\alpha)} D_\lambda^\alpha(z) &= \frac{2^{\frac{1}{2}-\alpha}\,\mathrm{e}^{\mathrm{i}\,\frac{\pi}{2}}(z^2-1)^{\frac{1}{4}-\frac{\alpha}{2}}}{\sqrt{\pi}\Gamma\big(\frac{1}{2}+2\alpha+\lambda\big)}\mathbf{Q}^{\alpha-\frac{1}{2}}_{-\frac{1}{2}+\alpha+\lambda}(z) \\ &= \mathrm{e}^{\mathrm{i}\,\pi\alpha}\,2^{-\lambda-2\alpha}Z_{\alpha-\frac{1}{2},\alpha+\lambda}(z). \end{aligned} \tag{B.9}$$

References

[1] S. Albeverio, F. Gesztesy, R. Høegh-Krohn, and H. Holden, *Solvable models in quantum mechanics*, 2nd ed., AMS Chelsea Publishing, Providence, RI, 2005. With an appendix by Pavel Exner, DOI 10.1090/chel/350. MR2105735

[2] S. Albeverio and P. Kurasov, *Singular perturbations of differential operators: Solvable Schrödinger type operators*, London Mathematical Society Lecture Note Series, vol. 271, Cambridge University Press, Cambridge, 2000, DOI 10.1017/CBO9780511758904. MR1752110

[3] F. A. Berezin and L. D. Faddeev, *Remark on the Schrödinger equation with singular potential* (Russian), Dokl. Akad. Nauk SSSR **137** (1961), 1011–1014. MR129309

[4] C. G. Bollini and J. J. Giambiagi, *Dimensional regularization in configuration space*, Phys. Rev. D 53(10), 1996.

[5] H. S. Cohl, T. H. Dang, and T. M. Dunster, *Fundamental solutions and Gegenbauer expansions of Helmholtz operators in Riemannian spaces of constant curvature*, SIGMA Symmetry Integrability Geom. Methods Appl. **14** (2018), Paper No. 136, 45, DOI 10.3842/SIGMA.2018.136. MR3894535

[6] J. Dereziński, *Hypergeometric type functions and their symmetries*, Ann. Henri Poincaré **15** (2014), no. 8, 1569–1653, DOI 10.1007/s00023-013-0282-4. MR3232581

[7] J. Dereziński, *Group-theoretical origin of symmetries of hypergeometric class equations and functions*, in Complex differential and difference equations, Proceedings of the school and conference held at Będlewo, Poland, September 2–15, 2018, Filipuk, G., Lastra, A.; Michalik, S.; Takei, Y., Żołądek, H., Eds.; De Gruyter Proceedings in Mathematics, Berlin, 2020.

[8] J. Dereziński, B. Ruba, and C. Gaß, *Point potentials on Euclidean space, hyperbolic space and sphere in any dimension*, Ann. Henri Poincaré (2024), DOI 10.1007/s00023-024-01496-1.

[9] J. Dereziński, B. Ruba, and C. Gaß, In preparation.

[10] L. Durand, P. M. Fishbane, and L. M. Simmons Jr., *Expansion formulas and addition theorems for Gegenbauer functions*, J. Mathematical Phys. **17** (1976), no. 11, 1933–1948, DOI 10.1063/1.522831. MR422726

[11] L. Durand, *Complex asymptotics in λ for the Gegenbauer functions $C_\lambda^\alpha(z)$ and $D_\lambda^\alpha(z)$ with $z\in(-1,1)$*, Symmetry 2019, 11, 1465.

[12] L. Durand, *Asymptotic Bessel-function expansions for Legendre and Jacobi functions*, J. Math. Phys. 60, 013501, 2019.

[13] I. M. Gel'fand and G. E. Shilov, *Generalized functions. Vol. I: Properties and operations*, Academic Press, New York-London, 1964. Translated by Eugene Saletan. MR166596

[14] I. S. Gradshteyn and I. M. Ryzhik, *Table of Integrals, Series, and Products*, translated by Scripta Technica, Inc. (7th edn.), Academic Press, 2007.

[15] J. Hadamard, *Lectures on Cauchy's Problem in Linear Partial Differential Equations*, Dover Phoenix editions, Dover Publications, New York, 1923.

[16] J. Hadamard, *Le Problème de Cauchy et les Équations aux Dérivées Partielles Linéaires Hyperboliques* (in French), Paris, Hermann & Cie, 1932.

[17] L. Hörmander, *The Analysis of Linear Partial Differential Operators I*, Springer, Berlin, 2nd edition, 1990.

[18] M. Lesch, *Differential Operators of Fuchs Type, Conical Singularities, and Asymptotic Methods*, Vieweg+Teubner Verlag, 1997.

[19] E. Lieb, R. Seiringer, J. P Solovej, and J. Yngvason, *The Mathematics of the Bose Gas and its Condensation*, Birkhäuser Verlag, Basel, Boston, Berlin 2005.

[20] F. W. J. Olver, A. B. Olde Daalhuis, D. W. Lozier, B. I. Schneider, R. F. Boisvert, C. W. Clark, B. R. Miller, B. V. Saunders, H. S. Cohl, and M. A. McClain (Editors), *NIST Digital Library of Mathematical Functions*, Release 1.1.6 of 2022-06-30, available at http://dlmf.nist.gov/.

[21] F. W. J. Olver, *Asymptotics and Special Functions*, Academic Press, New York, 1974.

[22] A. Laforgia, *Bounds for modified Bessel functions*, J. Comput. Appl. Math. **34** (1991), no. 3, 263–267, DOI 10.1016/0377-0427(91)90087-Z. MR1102583

[23] S. Paycha, *Regularised integrals, sums and traces: An analytic point of view*, University Lecture Series, vol. 59, American Mathematical Society, Providence, RI, 2012, DOI 10.1090/ulect/059. MR2987296

[24] Reed, M. and Simon, B.: *Methods of Modern Mathematical Physics, II. Fourier Analysis, Self-Adjointness*, Academic Press, London, 1975.

[25] M. Riesz, *L'intégrale de Riemann-Liouville et le problème de Cauchy* (French), Acta Math. **81** (1949), 1–223, DOI 10.1007/BF02395016. MR30102

[26] R. Szmytkowski, *Closed forms of the Green's function and the generalized Green's function for the Helmholtz operator on the N-dimensional unit sphere*, J. Phys. A **40** (2007), no. 5, 995–1009, DOI 10.1088/1751-8113/40/5/009. MR2307727

[27] E. T. Whittaker and G. N. Watson, *A course of modern analysis: An introduction to the general theory of infinite processes and of analytic functions; with an account of the principal transcendental functions*, Cambridge Mathematical Library, Cambridge University Press, Cambridge, 1996. Reprint of the fourth (1927) edition, DOI 10.1017/CBO9780511608759. MR1424469

Department of Mathematical Methods in Physics, Faculty of Physics, University of Warsaw, Pasteura 5, 02-093 Warszawa, Poland
Email address: jan.derezinski@fuw.edu.pl

Department of Mathematical Methods in Physics, Faculty of Physics, University of Warsaw, Pasteura 5, 02-093 Warszawa, Poland
Email address: christian.gass@fuw.edu.pl

Department of Mathematics, University of Copenhagen, Universitetsparken 5, DK-2100 Copenhagen Ø, Denmark
Email address: btr@math.ku.dk

Contemporary Mathematics
Volume **819**, 2025
https://doi.org/10.1090/conm/819/16388

On potentials integrated by the Nikiforov–Uvarov method

Lina Ellis, Ikumi Ellis, Christoph Koutschan, and Sergei K. Suslov

ABSTRACT. We discuss basic potentials of nonrelativistic and relativistic quantum mechanics that can be integrated in the Nikiforov–Uvarov paradigm with the aid of a computer algebra system. This approach may help readers to study modern analytical methods of quantum physics.

> Building on ideas of DE BROGLIE and EINSTEIN, I tried to show that the ordinary differential equations of mechanics, which attempt to define the co-ordinates of a mechanical system as functions of the time, are no longer applicable for "small" systems; instead there must be introduced a certain *partial* differential equation, which defines a variable ψ ("wave function") as a function of the co-ordinates and the time.
>
> *Erwin Schrödinger* [**36**]

1. Introduction

The discovery of the relativistic and nonrelativistic Schrödinger equations [**33**–**38**] is discussed in [**3**] (see also the references therein).

Invented about a century ago, the stationary Schrödinger equation turns out to have an enormously wide range of applications, from the quantum theory of atoms and molecules to solid state physics, quantum crystals, superfluidity, and superconductivity.

Finding the energy levels and the corresponding normalized wave functions of various systems is one of the basic problems of quantum physics. Only in a few elementary cases are the exact solutions known. They are usually investigated by different techniques. Nonetheless, those completely integrable problems are important in the creation of mathematical models for complex quantum systems. Moreover, they may provide a useful testing ground for verification of numerical methods. A true story of the calculation of the energy levels for two-electron atoms [**29**] is presented in [**20**].

In this review article, we assemble analytical solutions for a range of potentials in nonrelativistic and relativistic quantum mechanics that are available in the literature. Data for most of the potentials that can be studied, in a unified way, by the so-called Nikiforov–Uvarov method [**27**] are collected, independently verified, and

2020 *Mathematics Subject Classification.* Primary 81Q05; Secondary 33C45.
C. K. was supported by the Austrian Science Fund (FWF): 10.55776/I6130.

completed with the help of the Mathematica computer algebra system. Only bound states are discussed. In a more traditional approach, for each of those problems one has to identify and factor out the singularities of the corresponding square-integrable wave functions and find the remaining terminating power series expansions or use algebraic methods (see, for example, [**4**,**6**,**7**,**9**,**10**,**12**,**17**,**18**,**23**,**39**,**40**]). As a result, each such problem must be treated separately, which is not suitable for a unified computer algebra approach.

One should admit that, despite our successful unification and creation of elaborate Mathematica codes, each of the collected cases, with the exception of a few trivial ones, has required a separate consideration with many specific details that are far from a formal application of the Nikiforov–Uvarov method, as one might have expected at first glance. We hope that the interested reader, already confronted or about to be confronted with all those quantization problems in study, research or teaching, will appreciate our diligent work. Applications of hypergeometric functions are important for all mathematicians to know.

An interesting new case of the modified Hulthén potential and its rotation correction is treated, which can be useful in the study of vibration–rotation spectra of diatomic molecules; this is a work in progress. In addition, our analysis has revealed incomplete information and typos for some valuable potentials discussed in the classical sources. The complementary Mathematica notebook makes it possible to fix those typos immediately. In general, despite a long history of applications of computer algebra methods in quantum physics, we were unable to find suitable techniques in the available literature, and as a result, had to create most of our Mathematica codes from scratch, of course by using our preliminary experience and enthusiasm in solving quantum physics and similar problems with the aid of computers.

This article is organized as follows. In the next section, we introduce the basics of the Nikiforov–Uvarov approach, and then successively apply it in the following Sections 3–19 to the main problems of introductory quantum mechanics, such as harmonic oscillators, Bessel functions, Coulomb problems, Pöschl–Teller potential holes, Kratzer's molecular potential, Hulthén potentials, and Morse potentials. All calculations are verified in the complementary Mathematica notebook, which can serve both educational and research purposes. Appendices A and B respectively contain, for the reader's convenience, data on classical orthogonal polynomials and a useful integral evaluation, in order to make our presentation as self-contained as possible. Appendix C describes the Mathematica notebook.

This review is written for those who study quantum mechanics and would like to see more details than are in the classical textbooks by utilizing the advanced computer algebra system, Mathematica. It is motivated by an introductory course in mathematics of quantum mechanics which one of the authors (SKS) has been teaching at Arizona State University for more than two decades.

2. Summary of the Nikiforov–Uvarov approach

The generalized equation of hypergeometric type

$$u'' + \frac{\widetilde{\tau}(x)}{\sigma(x)}u' + \frac{\widetilde{\sigma}(x)}{\sigma^2(x)}u = 0 \tag{2.1}$$

(the coefficient functions $\sigma(x)$, $\widetilde{\sigma}(x)$ are polynomials of degrees at most two and $\widetilde{\tau}(x)$ is a polynomial of degree at most one), by the substitution

$$u = \varphi(x) y, \tag{2.2}$$

can be reduced to the form

$$\sigma(x) y'' + \tau(x) y' + \lambda y = 0 \tag{2.3}$$

if

$$\frac{\varphi'}{\varphi}(x) = \frac{\pi(x)}{\sigma(x)}, \qquad \pi(x) = \frac{1}{2}\left(\tau(x) - \widetilde{\tau}(x)\right), \tag{2.4}$$

$$k = \lambda - \pi'(x), \tag{2.5}$$

and also,

$$\pi(x) = \frac{\sigma' - \widetilde{\tau}}{2} \pm \sqrt{\left(\frac{\sigma' - \widetilde{\tau}}{2}\right)^2 - \widetilde{\sigma} + k\sigma} \tag{2.6}$$

is a linear function. (One uses the choice of the constant k to complete the square under the radical sign; see [**27**] and our argument below for more details.)

In Nikiforov–Uvarov's method, the energy levels are obtained from the quantization rule

$$\lambda + n\tau' + \frac{1}{2} n(n-1)\sigma'' = 0 \qquad (n = 0, 1, 2, \dots) \tag{2.7}$$

and the corresponding square-integrable solutions are classical orthogonal polynomials, up to a factor. They can be found by a *Rodrigues-type formula* [**27**]:

$$y_n(x) = \frac{B_n}{\rho(x)} \left(\sigma^n(x)\rho(x)\right)^{(n)}, \qquad (\sigma\rho)' = \tau\rho, \tag{2.8}$$

where B_n is a constant (see also [**42**] and Table 19).

Let us try to reduce the differential equation (2.1) to its simplest form by the substitution $u = \varphi(x)\, y$, with the help of some special choice of function $\varphi(x)$.

Substituting $u = \varphi(x)\, y$ in (2.1) one gets

$$y'' + \left(\frac{\widetilde{\tau}}{\sigma} + 2\frac{\varphi'}{\varphi}\right) y' + \left(\frac{\widetilde{\sigma}}{\sigma^2} + \frac{\widetilde{\tau}}{\sigma}\frac{\varphi'}{\varphi} + \frac{\varphi''}{\varphi}\right) y = 0. \tag{2.9}$$

Equation (2.9) should not be more complicated than the original equation (2.1). Thus, it is natural to require that the coefficient in front of y' has the form $\tau(x)/\sigma(x)$, where $\tau(x)$ is a polynomial of degree at most one. This implies the first-order differential equation

$$\frac{\varphi'}{\varphi}(x) = \frac{\pi(x)}{\sigma(x)} \tag{2.10}$$

for the function $\varphi(x)$, where

$$\pi(x) = \frac{1}{2}\left(\tau(x) - \widetilde{\tau}(x)\right) \tag{2.11}$$

is a polynomial of degree at most one. As a result, equation (2.9) takes the form

$$y'' + \frac{\tau(x)}{\sigma(x)}\, y' + \frac{\overline{\sigma}(x)}{\sigma^2(x)}\, y = 0, \tag{2.12}$$

where

$$\overline{\sigma}(x) = \widetilde{\sigma}(x) + \pi^2(x) + \pi(x)\left[\widetilde{\tau}(x) - \sigma'(x)\right] + \pi'(x)\sigma(x). \tag{2.13}$$

The functions $\tau(x)$ and $\overline{\sigma}(x)$ are polynomials of degrees at most one and two in x, respectively. Therefore, equation (2.12) is an equation of the same type as our original equation (2.1).

By making a special choice of the polynomial $\pi(x)$ we can reduce (2.12) to its simplest form, in which

$$\overline{\sigma}(x) = \lambda\, \sigma(x), \tag{2.14}$$

where λ is some constant. Then, equation (2.12) takes the form (2.3). We call equation (2.3) a *differential equation of hypergeometric type* and its solutions *functions of hypergeometric type.* In this context, it is natural to call equation (2.1) a *generalized differential equation of hypergeometric type* [**27**].

The condition (2.14) can be rewritten as

$$\pi^2 + \left(\widetilde{\tau} - \sigma'\right)\, \pi + \widetilde{\sigma} - k\sigma = 0, \tag{2.15}$$

where

$$k = \lambda - \pi'(x) \tag{2.16}$$

is a constant. Assuming that this constant is known, we can find $\pi(x)$ as a solution (2.6) of the quadratic equation (2.15). But $\pi(x)$ is a polynomial, therefore the second degree polynomial

$$p(x) = \left(\frac{\sigma'(x) - \widetilde{\tau}(x)}{2}\right)^2 - \widetilde{\sigma}(x) + k\sigma(x) \tag{2.17}$$

under the radical in (2.6) should be the square of a linear function, and the discriminant of $p(x)$ should be zero. This condition gives an equation for the constant k, which is, generally, a quadratic equation. Given k as a solution of this equation, we find $\pi(x)$ by the quadratic formula (2.6), then $\tau(x)$ and λ by (2.11) and (2.16). Finally, we find the function $\varphi(x)$ as a solution of (2.10). It is clear that the reduction of equation (2.1) to the simplest form (2.3) can be accomplished in a few different ways, in accordance with different choices of the constant k and the sign in the equation (2.6) for $\pi(x)$.

A closed form for the constant k can be obtained as follows [**3**]. Let

$$p(x) = \left(\frac{\sigma' - \widetilde{\tau}}{2}\right)^2 - \widetilde{\sigma} + k\sigma = q(x) + k\sigma(x), \tag{2.18}$$

where

$$q(x) = \left(\frac{\sigma' - \widetilde{\tau}}{2}\right)^2 - \widetilde{\sigma}. \tag{2.19}$$

Completing the square, one gets

$$p(x) = \frac{p''}{2}\left(x + \frac{p'(0)}{p''}\right)^2 - \frac{\left(p'(0)\right)^2 - 2p''p(0)}{2p''}, \tag{2.20}$$

where the last term must be eliminated:

$$\left(p'(0)\right)^2 - 2p''p(0) = 0. \tag{2.21}$$

Therefore,

$$(q'(0) + k\sigma'(0))^2 - 2(q'' + k\sigma'')(q(0) + k\sigma(0)) = 0, \tag{2.22}$$

which results in the quadratic equation

$$ak^2 + 2bk + c = 0, \tag{2.23}$$

where

$$a = (\sigma'(0))^2 - 2\sigma''\sigma(0), \tag{2.24}$$

$$b = q'(0)\sigma'(0) - \sigma''q(0) - \sigma(0)q'', \tag{2.25}$$

$$c = (q'(0))^2 - 2q''q(0). \tag{2.26}$$

Solutions are

$$k_0 = -\frac{c}{2b} \qquad \text{if } a = 0, \tag{2.27}$$

and

$$k_{1,2} = \frac{-b \pm \sqrt{d}}{a} \qquad \text{if } a \neq 0, \tag{2.28}$$

where
(2.29)

$$d = b^2 - ac = (\sigma(0)q'' - \sigma''q(0))^2 - 2(\sigma'(0)q'' - \sigma''q'(0))(\sigma(0)q'(0) - \sigma'(0)q(0)).$$

As a result,

$$p(x) = \left[\frac{p''}{2}\left(x + \frac{p'(0)}{p''}\right)^2\right]_{k=k_{0,1,2}}, \tag{2.30}$$

which permits evaluating the linear function $\pi(x)$ in the Nikiforov–Uvarov technique.

Examples of this, for most integrable cases that are available in the literature, are presented in Sections 3–19 and the included Tables 1–18 below. All these analytical procedures are implemented in our complementary Mathematica notebook (see Appendix C).

It must be added that "quantizations" of some solutions of (2.3) require generalized power series expansions, given by

$$y(x) = (x-a)^\alpha \sum_{m=0}^{\infty} \prod_{k=0}^{m-1} \frac{(\lambda - \lambda_{\alpha+k})(a-x)}{\tau_{\alpha+k}(a)(\alpha+k+1)}, \tag{2.31}$$

where

$$\tau_\mu(x) = \tau(x) + \mu\sigma'(x), \qquad \lambda_\mu = -\mu\tau' - \mu(\mu-1)\sigma''/2, \tag{2.32}$$

provided that $\sigma(a) = 0$ and $\alpha\tau_{\alpha-1}(a) = 0$ (in particular when $\alpha = 0$). More details can be found in [**42**].

TABLE 1. Details for the case of a stationary Schrödinger equation with the harmonic potential $U(x) = \frac{1}{2}m\omega^2x^2$, treated in §3.

$\sigma(\xi)$	1
$\widetilde{\sigma}(\xi)$	$2\varepsilon - \xi^2$
$\widetilde{\tau}(\xi)$	0
k	2ε
$\pi(\xi)$	$\pm\xi$
$\tau(\xi) = \widetilde{\tau} + 2\pi$	-2ξ
$\lambda = k + \pi'$	$2\varepsilon - 1$
$\varphi(\xi)$	$e^{-\xi^2/2}$
$\rho(\xi)$	$e^{-\xi^2}$
$y_n(\xi)$	$C_nH_n(\xi)$
C_n^2	$\dfrac{1}{\sqrt{\pi}\,2^n n!}$

3. The harmonic oscillator

Let us consider the one-dimensional stationary Schrödinger equation for the harmonic oscillator,

$$-\frac{\hbar^2}{2m}\frac{\mathrm{d}^2\psi}{\mathrm{d}x^2} + \frac{1}{2}m\omega^2x^2\psi = E\psi, \tag{3.1}$$

with the real-valued wave function satisfying

$$\int_{-\infty}^{\infty} \psi^2(x)\,\mathrm{d}x = 1. \tag{3.2}$$

Introducing dimensionless variables

$$\psi(x) = u(\xi), \qquad x = \xi\sqrt{\frac{\hbar}{m\omega}}, \qquad E = \hbar\omega\varepsilon, \tag{3.3}$$

one gets

$$u'' + \left(2\varepsilon - \xi^2\right)u = 0, \tag{3.4}$$

in which the coefficient functions are $\sigma(\xi) = 1$, $\widetilde{\tau}(\xi) = 0$, and $\widetilde{\sigma}(\xi) = 2\varepsilon - \xi^2$. Therefore,

$$\pi(\xi) = \pm\sqrt{k - 2\varepsilon + \xi^2} = \pm\xi, \qquad k = 2\varepsilon. \tag{3.5}$$

We choose $\pi = -\xi$, which gives a negative derivative for

$$\tau(\xi) = \widetilde{\tau}(\xi) + 2\pi(\xi) = -2\xi. \tag{3.6}$$

Then

$$\frac{\varphi'}{\varphi}(\xi) = \frac{\pi(\xi)}{\sigma(\xi)} = -\xi, \qquad \varphi(\xi) = e^{-\xi^2/2}, \tag{3.7}$$

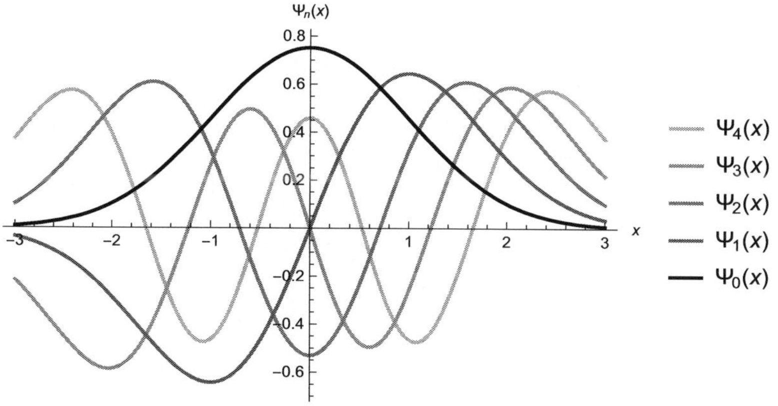

FIGURE 1. "The first five proper vibrations of the Planck oscillator according to undulatory mechanics. Outside the region $-3 \leqslant x \leqslant 3$, represented here, all five functions approach the x-axis in a monotonic fashion." — generated by Mathematica following the original Schrödinger article [**36**].

and $\lambda = 2\varepsilon - 1$, $\rho(\xi) = e^{-\xi^2}$. The energy levels are $\varepsilon = \varepsilon_n = n + 1/2$ with $n = 0, 1, 2, \ldots$, from (2.7). The eigenfunctions

$$y_n(\xi) = B_n e^{\xi^2} \frac{\mathrm{d}^n}{\mathrm{d}\xi^n} \left(e^{-\xi^2}\right) \tag{3.8}$$

are, up to a normalization, the Hermite polynomials (Table 19).

As a result, the orthonormal wave functions are given by [**23, 34, 36**]

$$\psi(x) = \left(\frac{m\omega}{\pi\hbar}\right)^{1/4} \frac{1}{\sqrt{2^n n!}} \exp\left(-\frac{m\omega}{2\hbar}x^2\right) H_n\left(x\sqrt{\frac{m\omega}{\hbar}}\right), \tag{3.9}$$

corresponding to the discrete energy levels

$$E_n = \hbar\omega\left(n + \frac{1}{2}\right) \qquad (n = 0, 1, 2, \ldots) \tag{3.10}$$

in cgs (Gaussian) units. Detailed results are presented in Table 1; see also Figure 1 for graphs of the first five wave functions. (More general, "missing" solutions of the time-dependent Schrödinger equation are discussed in [**19, 21, 24**].)

4. Bessel functions

Let us also mention some solutions of the *Bessel equation*:

$$z^2 u'' + z u' + \left(z^2 - \nu^2\right) u = 0. \tag{4.1}$$

With the aid of the subtitution $u = \varphi(z)\, y$ when $\varphi(z) = z^\nu e^{iz}$ this equation can be reduced to the hypergeometric form

$$z y'' + (2iz + 2\nu + 1) y' + i(2\nu + 1) y = 0 \tag{4.2}$$

(Table 2), and one can obtain *Poisson integral representations* for the Bessel functions of the first kind, $J_\nu(z)$, and the Hankel functions of the first and second kind,

TABLE 2. Details for the case of the Bessel equation, treated in §4.

$\sigma(z)$	z
$\widetilde{\sigma}(z)$	$z^2 - \nu^2$
$\widetilde{\tau}(z)$	1
k	$\pm 2i\nu$
$\pi(z)$	$\pm\nu \pm iz$
$\tau(z) = \widetilde{\tau} + 2\pi$	$1 + 2\nu + 2iz$
$\lambda = k + \pi'$	$i(2\nu + 1)$
$\varphi(z)$	$z^{\pm\nu} e^{\pm iz}$
$\rho(z)$	$z^{2\nu} e^{2iz}$

TABLE 3. Details for the case of the equation for spherical harmonics, treated in §5.

$\sigma(\xi)$	$1 - \xi^2, \qquad \xi = \cos\theta, \quad 0 \leqslant \theta \leqslant \pi$
$\widetilde{\sigma}(\xi)$	$\mu(1 - \xi^2) - m^2$
$\widetilde{\tau}(\xi)$	-2ξ
k	$\mu - m^2$
$\pi(\xi)$	$-\lvert m\rvert\xi$
$\tau(\xi) = \widetilde{\tau} + 2\pi$	$-2\left(\lvert m\rvert + 1\right)\xi$
$\lambda = k + \pi'$	$\mu - \lvert m\rvert\left(\lvert m\rvert + 1\right)$
$\varphi(\xi)$	$(1 - \xi^2)^{\lvert m\rvert/2} = (\sin\theta)^{\lvert m\rvert}$
$\rho(\xi)$	$(1 - \xi^2)^{\lvert m\rvert}$
$y_n(\xi)$	$N_{l\,m}\, P_{l-\lvert m\rvert}^{(\lvert m\rvert,\,\lvert m\rvert)}(\cos\theta)\,, \quad n = l - \lvert m\rvert$
$N_{l\,m}$	$\dfrac{1}{2^{\lvert m\rvert}\, l!}\sqrt{\dfrac{2l+1}{2}\,(l-m)!(l+m)!}$

$H_\nu^{(1)}(z)$ and $H_\nu^{(2)}(z)$:

$$J_\nu(z) = \frac{(z/2)^\nu}{\sqrt{\pi}\,\Gamma(\nu + 1/2)} \int_{-1}^{1} \left(1 - t^2\right)^{\nu - 1/2} \cos(zt)\, dt, \tag{4.3}$$

$$H_\nu^{(1,2)}(z) = \sqrt{\frac{2}{\pi z}}\, \frac{e^{\pm i\left(z - \frac{\pi}{2}\nu - \frac{\pi}{4}\right)}}{\Gamma(\nu + 1/2)} \int_0^\infty e^{-t} t^{\nu - 1/2} \left(1 \pm \frac{it}{2z}\right)^{\nu - 1/2} dt, \tag{4.4}$$

where $\Re\nu > -1/2$. It is then possible to deduce from these integral representations all the remaining properties of these functions. (For details, see **[27]** and **[42]** or **[44]** and **[45]**).

5. A central field: Spherical harmonics

The stationary Schrödinger equation in a central field with the potential energy $U(r)$ is given by

$$\Delta\psi + \frac{2m}{\hbar^2}\left(E - U(r)\right)\psi = 0. \tag{5.1}$$

The Laplace operator in the spherical coordinates r, θ, φ has the form [**26, 27**]

$$\Delta = \Delta_r + \frac{1}{r^2}\Delta_\omega \tag{5.2}$$

with

$$\Delta_r = \frac{1}{r^2}\frac{\partial}{\partial r}\left(r^2\frac{\partial}{\partial r}\right), \qquad \Delta_\omega = \frac{1}{\sin\theta}\frac{\partial}{\partial\theta}\left(\sin\theta\frac{\partial}{\partial\theta}\right) + \frac{1}{\sin^2\theta}\frac{\partial^2}{\partial\varphi^2}, \tag{5.3}$$

and separation of variables $\psi = R(r)Y(\theta,\varphi)$ gives

$$\Delta_\omega Y(\theta,\varphi) + \mu Y(\theta,\varphi) = 0, \tag{5.4}$$

$$\frac{1}{r^2}\frac{\mathrm{d}}{\mathrm{d}r}\left(r^2\frac{\mathrm{d}R(r)}{\mathrm{d}r}\right) + \left(\frac{2m}{\hbar^2}\left(E - U(r)\right) - \frac{\mu}{r^2}\right)R(r) = 0. \tag{5.5}$$

Bounded single-valued solutions of equation (5.4) on the sphere S^2 exist only when $\mu = l(l+1)$ with $l = 0, 1, 2, \ldots$. They are the spherical harmonics $Y = Y_{lm}(\theta,\varphi)$.

Looking for solutions in the form $Y = e^{im\varphi}f(\theta)$ with $m = 0, \pm 1, \pm 2, \ldots$, one gets

$$\frac{1}{\sin\theta}\frac{\mathrm{d}}{\mathrm{d}\theta}\left(\sin\theta\frac{\mathrm{d}f}{\mathrm{d}\theta}\right) - \frac{m^2}{\sin^2\theta}f + \mu f = 0. \tag{5.6}$$

The change of variables $\xi = \cos\theta$ and $F(\xi) = f(\theta)$ results in the generalized equation of hypergeometric type

$$\left(1-\xi^2\right)F'' - 2\xi F' + \left(\mu - \frac{m^2}{1-\xi^2}\right)F = 0, \tag{5.7}$$

with $\sigma(\xi) = 1-\xi^2$, $\widetilde{\tau}(\xi) = -2\xi$ and $\widetilde{\sigma}(\xi) = \mu\left(1-\xi^2\right) - m^2$, which can be reduced to a simpler form by the standard substitution $F = \varphi\, y$:

$$\left(1-\xi^2\right)y'' - 2\left(|m|+1\right)\xi\, y' + \left(\mu - |m|\left(|m|+1\right)\right)y = 0. \tag{5.8}$$

Indeed, by (2.6),

$$\pi(\xi) = \pm\sqrt{(\mu-k)\xi^2 + k + m^2 - \mu}, \tag{5.9}$$

or

$$\pi(\xi) = \begin{cases} \pm|m|, & k = \mu, \\ \pm|m|\xi, & k = \mu - m^2, \end{cases} \tag{5.10}$$

where we should choose the case when the linear function $\tau(\xi) = \widetilde{\tau}(\xi) + 2\pi(\xi)$ will have a negative derivative and a zero on the interval $(-1, 1)$.

Then

$$\frac{\varphi'}{\varphi}(\xi) = \frac{-|m|\xi}{1-\xi^2}, \qquad \ln\varphi = -|m|\int\frac{\xi\,\mathrm{d}\xi}{1-\xi^2} = \frac{|m|}{2}\ln\left(1-\xi^2\right), \tag{5.11}$$

and up to a constant factor,

$$\varphi(\xi) = \left(1-\xi^2\right)^{|m|/2} = \left(\sin\theta\right)^{|m|} \tag{5.12}$$

(see our complementary Mathematica notebook and Table 3 for further details of the calculations).

The final result is given by

(5.13)
$$Y_{l\,m}(\theta,\varphi)=A_m\,\frac{e^{im\varphi}}{2^{|m|}\,l!}\sqrt{\frac{2l+1}{4\pi}\,(l-m)!(l+m)!}\ (\sin\theta)^{|m|}\ P_{l-|m|}^{(|m|,|m|)}(\cos\theta).$$

Here, $P_n^{(\alpha,\alpha)}(\xi)$ are the Jacobi polynomials (Table 19); $A_m=(-1)^m$, $m\geqslant 0$ and $A_m=1$, $m<0$, or $A_m=(-1)^{(m+|m|)/2}$. (See **[26, 27, 43]** for additional details.)

Solutions of the Laplace equation $\Delta u=0$ in spherical coordinates have the form

$$u=C_1r^lY_{lm}(\theta,\varphi)+C_2r^{-l-1}Y_{lm}(\theta,\varphi), \tag{5.14}$$

where C_1 and C_2 are constants. Here, $r^lY_{lm}(\theta,\varphi)$ are homogeneous polynomials in x, y, z **[43]**:

(5.15)
$$r^lY_{lm}(\theta,\varphi)=\sqrt{\frac{2l+1}{4\pi}(l+m)!(l-m)!}\sum_{p,q,r}\frac{1}{p!q!r!}\left(\frac{x+iy}{2}\right)^p\left(\frac{x-iy}{2}\right)^q z^r,$$

where p, q, r are positive integers subject to the conditions $p+q+r=l$ and $p-q=m$. (See the Mathematica notebook for direct verifications.)

6. The nonrelativistic Coulomb problem

Due to the identity

$$\frac{1}{r^2}\frac{\mathrm{d}}{\mathrm{d}r}\left(r^2\frac{\mathrm{d}R}{\mathrm{d}r}\right)=\frac{1}{r}\frac{\mathrm{d}^2}{\mathrm{d}r^2}(rR), \tag{6.1}$$

the substitution $F(r)=rR(r)$ into (5.5) results in the standard radial equation

$$F''+\left[\frac{2m_e}{\hbar^2}(E-U(r))-\frac{l(l+1)}{r^2}\right]F=0,\qquad U(r)=-\frac{Ze^2}{r} \tag{6.2}$$

for the nonrelativistic Coulomb problem in spherical coordinates. In dimensionless units,

$$F(r)=u(x),\quad x=\frac{r}{a_0},\quad \varepsilon_0=\frac{E}{E_0}\quad\left(a_0=\frac{\hbar^2}{m_ee^2},\quad E_0=\frac{e^2}{a_0}\right), \tag{6.3}$$

and the radial equation is a generalized equation of hypergeometric type,

$$u''+\left[2\left(\varepsilon_0+\frac{Z}{x}\right)-\frac{l(l+1)}{x^2}\right]u=0, \tag{6.4}$$

in which the coefficient functions are

$$\sigma(x)=x,\qquad \widetilde{\tau}(x)=0,\qquad \widetilde{\sigma}(x)=2\varepsilon_0x^2+2Zx-l(l+1). \tag{6.5}$$

Therefore, one can utilize Nikiforov and Uvarov's approach to determine the corresponding wave functions and discrete energy levels.

We transform (6.4) to an equation of hypergeometric type (2.3). The linear function $\pi(x)$ takes the form

$$\pi(x)=\frac{1}{2}\pm\sqrt{\frac{1}{4}-2\varepsilon_0x^2-2x+l(l+1)+kx}, \tag{6.6}$$

TABLE 4. Details for the case of a Schrödinger equation with the Coulomb potential $U(r) = -Ze^2/r$, treated in §6. Dimensionless quantities: $r = a_0 x$, $a_0 = \hbar^2/(m_e e^2) \simeq 0.5 \cdot 10^{-8}$ cm, $E_0 = e^2/a_0$, $R(r) = F(x) = u(x)/x$.

$\sigma(x)$	x
$\widetilde{\sigma}(x)$	$2\varepsilon_0\, x^2 + 2Z\, x - l(l+1), \quad \varepsilon_0 = E/E_0$
$\widetilde{\tau}(x)$	0
k	$2Z - (2l+1)\sqrt{-2\varepsilon_0}$
$\pi(x)$	$l + 1 - \sqrt{-2\varepsilon_0}\, x$
$\tau(x) = \widetilde{\tau} + 2\pi$	$2\left(l + 1 - \sqrt{-2\varepsilon_0}\, x\right)$
$\lambda = k + \pi'$	$2\left(Z - (l+1)\sqrt{-2\varepsilon_0}\right)$
$\varphi(x)$	$x^{l+1} e^{-x\sqrt{-2\varepsilon_0}}$
$\rho(x)$	$x^{2l+1} e^{-(2Z\, x)/n}, \quad x = r/a_0$
$y_{n_r}(x)$	$C_{n_r} L_{n_r}^{2l+1}(\eta), \quad \eta = \dfrac{2Z}{n} x = \dfrac{2Z}{n}\left(\dfrac{r}{a_0}\right)$
$C_{n_r}^2$	$\dfrac{Z}{a_0 n^2} \dfrac{(n-l-1)!}{(n+l)!}, \quad n_r = n - l - 1$

or

$$\pi(x) = \frac{1}{2} \pm \begin{cases} \sqrt{-2\varepsilon_0}\, x + l + 1/2, & k = 2Z + (2l+1)\sqrt{-2\varepsilon_0} \\ \sqrt{-2\varepsilon_0}\, x - l - 1/2, & k = 2Z - (2l+1)\sqrt{-2\varepsilon_0}, \end{cases} \tag{6.7}$$

where we should choose the case when the linear function $\tau(x) = \widetilde{\tau}(x) + 2\pi(x)$ will have a negative derivative and a zero on $(0, +\infty)$:

$$\tau(x) = 2\left(l + 1 - x\sqrt{-2\varepsilon_0}\right).$$

This choice corresponds to

$$\pi(x) = l + 1 - x\sqrt{-2\varepsilon_0}, \qquad \varphi(x) = x^{l+1} \exp\left(-x\sqrt{-2\varepsilon_0}\right)$$

and

$$\lambda = k + \pi' = 2\left[Z - (l+1)\sqrt{-2\varepsilon_0}\,\right].$$

The energy values are given by (2.7):

$$\varepsilon_0 = \frac{E}{E_0} = -\frac{Z^2}{2\left(n_r + l + 1\right)^2}, \qquad E_0 = \frac{e^2}{a_0}. \tag{6.8}$$

Here, $n = n_r + l + 1$ is known as the principal quantum number.

In order to obtain a Rodrigues-type formula, one computes

$$\frac{\rho'}{\rho} = \frac{\tau - \sigma'}{\sigma} = \frac{2l+1}{x} - \frac{2Z}{n},$$

or

$$\rho(x) = x^{2l+1} \exp\left(-\frac{2Z}{n} x\right), \qquad x = \frac{r}{a_0}.$$

Therefore,

$$y_{n_r}(x) = \frac{B_{n_r}}{x^{2l+1}e^{-\eta}} \frac{\mathrm{d}^{n_r}}{\mathrm{d}x^{n_r}} \left(x^{n_r+2l+1}e^{-\eta}\right) = L_{n_r}^{2l+1}(\eta), \tag{6.9}$$

where

$$\eta = \frac{2Z}{n}x = \frac{2Z}{n}\left(\frac{r}{a_0}\right) = 2x\sqrt{-2\varepsilon_0},$$

and up to a constant factor,

$$F(r) = rR(r) = C_{nl}\,\eta^{l+1}e^{-\eta/2}L_{n_r}^{2l+1}(\eta). \tag{6.10}$$

In view of the normalization condition

$$1 = \int_0^\infty F^2\,\mathrm{d}r = C_{nl}^2\left(\frac{na_0}{2Z}\right)\int_0^\infty \eta^{2l+2}e^{-\eta}\left[L_{n_r}^{2l+1}(\eta)\right]^2\,\mathrm{d}\eta,$$

the three-term recurrence relation

$$\eta L_n^\alpha = -(n+1)L_{n+1}^\alpha + (\alpha+2n+1)L_n^\alpha - (\alpha+n)L_{n-1}^\alpha,$$

and the orthogonality property of the Laguerre polynomials (Table 19), one gets

$$C_{nl}^2 = \frac{Z}{a_0 n^2}\frac{(n-l-1)!}{(n+l)!}. \tag{6.11}$$

Details appear in Table 4, and additional details can be found in [**27**] and [**42**]. (See also Appendices A and B.)

As a result, the nonrelativistic Coulomb wave functions obtained by the method of separation of variables in spherical coordinates (see above) are

$$\psi = \psi_{nlm}(\mathbf{r}) = R_{nl}(r)\,Y_{lm}(\theta,\varphi), \tag{6.12}$$

where $Y_{lm}(\theta,\varphi)$ are the spherical harmonics, the radial functions $R_{nl}(r)$ are given in terms of the Laguerre polynomials (Table 19) [**4, 10, 23, 27, 39**]:

$$R(r) = R_{nl}(r) = \frac{2}{n^2}\left(\frac{Z}{a_0}\right)^{3/2}\sqrt{\frac{(n-l-1)!}{(n+l)!}}\,e^{-\eta/2}\eta^l\,L_{n-l-1}^{2l+1}(\eta) \tag{6.13}$$

with

$$\eta = \frac{2Z}{n}\left(\frac{r}{a_0}\right), \qquad a_0 = \frac{\hbar^2}{m_e e^2}, \tag{6.14}$$

and the normalization is

$$\int_0^\infty R_{nl}^2(r)r^2\,\mathrm{d}r = 1. \tag{6.15}$$

Here $n = 1, 2, 3, \ldots$ is the principal quantum number of the hydrogen-like atom in the nonrelativistic Schrödinger theory; $l = 0, 1, \ldots, n-1$ and $m = -l, -l+1, \ldots, l-1, l$ are the quantum numbers of the angular momentum and its projection on the z-axis, respectively. The corresponding discrete energy levels in cgs (Gaussian) units are given by Bohr's formula [**33**]:

$$E = E_n = -\frac{m_e Z^2 e^4}{2\hbar^2 n^2}, \tag{6.16}$$

where $n = 1, 2, 3, \ldots$ is the principal quantum number; they do not depend on the quantum number of orbital angular momentum l.

TABLE 5. Details for the case of a relativistic Schrödinger equation with the Coulomb potential $U(r) = -Ze^2/r$, treated in §7. Dimensionless quantities: $\varepsilon = E/(mc^2)$, $x = \beta r = (mc/\hbar)r$, $\mu = Ze^2/(\hbar c)$, $R(r) = F(x) = u(x)/x$.

$\sigma(x)$	x
$\widetilde{\sigma}(x)$	$\left(\varepsilon^2-1\right)\, x^2 + 2\mu\varepsilon\, x + \mu^2 - l(l+1)$
$\widetilde{\tau}(x)$	0
k	$2\mu\varepsilon - (2\nu+1)\sqrt{1-\varepsilon^2}, \quad \nu = -\dfrac{1}{2} + \sqrt{\left(l+\dfrac{1}{2}\right)^2 - \mu^2}$
$\pi(x)$	$\nu + 1 - a\, x, \qquad a = \sqrt{1-\varepsilon^2}$
$\tau(x) = \widetilde{\tau} + 2\pi$	$2\left(\nu+1-a\, x\right), \quad \tau' < 0$
$\lambda = k + \pi'$	$2\left(\mu\varepsilon - (\nu+1)a\right)$
$\varphi(x)$	$x^{\nu+1}e^{-a\ x}$
$\rho(x)$	$x^{2\nu+1}e^{-2a\ x}$
$y_{n_r}(x)$	$C_{n_r} L_{n_r}^{2\nu+1}(\xi), \quad \xi = 2ax = 2a\beta r = 2\sqrt{1-\varepsilon^2}\dfrac{mc}{\hbar}r$
C_{n_r}	$2(a\beta)^{3/2}(2a)^{\nu}\sqrt{\dfrac{n_r!}{(\nu+n_r+1)\Gamma(2\nu+n_r+2)}}$

REMARK. In the original Nikiforov–Uvarov approach, the variable coefficients $\sigma(x)$ and $\tau(x)$ in (2.3) should not depend on the eigenvalue λ. Here, we obtain

$$xy'' + 2\left(l+1-x\sqrt{-2\varepsilon_0}\right)y' + 2\left(Z-(l+1)\sqrt{-2\varepsilon_0}\right)y = 0. \tag{6.17}$$

Nonetheless, the change of variables $y(x) = Y(\eta)$ with $\eta = 2x\sqrt{-2\varepsilon_0}$ results in

$$\eta Y'' + (2l+2-\eta)\, Y' + \left(\frac{Z}{\sqrt{-2\varepsilon_0}} - l - 1\right) Y = 0. \tag{6.18}$$

Thus the Nikiforov–Uvarov method can be applied, and the uniqueness of square-integrable solutions holds.

In a similar fashion, one can consider the solution of the Kepler problem in so-called parabolic coordinates, which are important in the theory of the Stark effect [**23**, **35**].

7. The relativistic Schrödinger equation

The stationary relativistic Schrödinger equation has the form [**3**, **7**, **17**, **39**]:

$$\left(E + \frac{Ze^2}{r}\right)^2 \chi = \left(-\hbar^2c^2\Delta + m^2c^4\right)\chi. \tag{7.1}$$

We separate the variables in spherical coordinates, $\chi(r,\theta,\varphi) = R(r)Y_{lm}(\theta,\varphi)$, where $Y_{lm}(\theta,\varphi)$ are spherical harmonics with familiar properties [**43**]. As a result,

$$\frac{1}{r^2}\frac{\mathrm{d}}{\mathrm{d}r}\left(r^2\frac{\mathrm{d}R}{\mathrm{d}r}\right) + \left[\frac{\left(E+Ze^2/r\right)^2 - m^2c^4}{\hbar^2c^2} - \frac{l(l+1)}{r^2}\right]R = 0 \qquad (l = 0, 1, 2, \dots). \tag{7.2}$$

With dimensionless quantities

$$\varepsilon = \frac{E}{mc^2}, \qquad x = \beta r = \frac{mc}{\hbar} r, \qquad \mu = \frac{Ze^2}{\hbar c}, \tag{7.3}$$

for the new radial function

$$R(r) = F(x) = \frac{u(x)}{x} \tag{7.4}$$

one obtains

$$\frac{1}{x^2} \frac{\mathrm{d}}{\mathrm{d}x} \left(x^2 \frac{\mathrm{d}F}{\mathrm{d}x} \right) + \left[\left(\varepsilon + \frac{\mu}{x} \right)^2 - 1 - \frac{l(l+1)}{x^2} \right] F = 0, \tag{7.5}$$

and given the identity $(x^2 F')' = x(xF)''$,

$$u'' + \left[\left(\varepsilon + \frac{\mu}{x} \right)^2 - 1 - \frac{l(l+1)}{x^2} \right] u = 0. \tag{7.6}$$

This is a generalized equation of hypergeometric type (2.1), with coefficient functions

$$\sigma(x) = x, \qquad \widetilde{\tau}(x) \equiv 0, \qquad \widetilde{\sigma}(x) = \left(\varepsilon^2 - 1 \right) x^2 + 2\mu\varepsilon x + \mu^2 - l(l+1). \tag{7.7}$$

The normalization condition takes the form

$$\int_0^\infty R^2(r) r^2 \, \mathrm{d}r = 1, \quad \text{or} \quad \int_0^\infty u^2(x) \, \mathrm{d}x = \beta^3, \quad \beta = \frac{mc}{\hbar}. \tag{7.8}$$

Here, $u = \varphi y$. For further computational details, see our complementary Mathematica notebook, as well as Refs. [**3**] and [**27**]. Final results are presented in Table 5.

In particular, one gets Schrödinger's fine structure formula (for a charged spin-zero particle in a Coulomb field):

$$E = E_{n_r} = \frac{mc^2}{\sqrt{1 + \left(\dfrac{\mu}{n_r + \nu + 1} \right)^2}} \qquad (n = n_r = 0, 1, 2, \dots). \tag{7.9}$$

Here,

$$\mu = \frac{Ze^2}{\hbar c}, \qquad \nu = -\frac{1}{2} + \sqrt{\left(l + \frac{1}{2} \right)^2 - \mu^2}. \tag{7.10}$$

The corresponding eigenfunctions are given by the Rodrigues-type formula

$$y_n(x) = \frac{B_n}{x^{2\nu+1} e^{-2ax}} \frac{\mathrm{d}^n}{\mathrm{d}x^n} \left(x^{n+2\nu+1} e^{-2ax} \right) = C_n L_n^{2\nu+1}(2ax). \tag{7.11}$$

Up to a constant factor, they are Laguerre polynomials (Table 19). The normalization condition (7.8) becomes

$$\begin{aligned} \beta^3 &= \int_0^\infty u^2(x) \, \mathrm{d}x = C_n^2 \int_0^\infty \left[\varphi(x) L_n^{2\nu+1}(2ax) \right]^2 \mathrm{d}x \\ &= \frac{C_n^2}{(2a)^{2\nu+3}} \int_0^\infty e^{-\xi} \, \xi^{2\nu+2} \left(L_n^{2\nu+1}(\xi) \right)^2 \mathrm{d}\xi, \quad \xi = 2ax. \end{aligned} \tag{7.12}$$

The corresponding integral is given (see [**41, 42**] and Appendix B) by

$$I_1 = J_{nn1}^{\alpha\alpha} = \int_0^\infty e^{-x} x^{\alpha+1} \left(L_n^\alpha(x) \right)^2 \mathrm{d}x = (\alpha + 2n + 1) \frac{\Gamma(\alpha + n + 1)}{n!}. \tag{7.13}$$

As a result,

$$C_n = 2(a\beta)^{3/2}(2a)^{\nu}\sqrt{\frac{n!}{\Gamma(2\nu+n+2)}}. \tag{7.14}$$

The normalized radial eigenfunctions, corresponding to the relativistic energy levels (7.9), are explicitly given by

$$R(r) = R_{n_r}(r) = 2(a\beta)^{3/2}\sqrt{\frac{n_r!}{\Gamma(2\nu+n_r+2)}}\,\xi^{\nu}e^{-\xi/2}\,L_{n_r}^{2\nu+1}(\xi), \tag{7.15}$$

where

$$\xi = 2ax = 2a\beta r = 2\sqrt{1-\varepsilon^2}\,\frac{mc}{\hbar}\,r. \tag{7.16}$$

(More details can be found in [**3**, **7**, **39**].)

Let us analyse the nonrelativistic limit of Schrödinger's fine structure formula (7.9)–(7.10):

$$\begin{aligned}\varepsilon_{\text{Schrödinger}} = \frac{E_{n_r,l}}{mc^2} &= \frac{1}{\sqrt{1+\dfrac{\mu^2}{\left(n_r+\frac{1}{2}+\sqrt{\left(l+\frac{1}{2}\right)^2-\mu^2}\right)^2}}}\\ &= 1-\frac{\mu^2}{2n^2}-\frac{\mu^4}{2n^4}\left(\frac{n}{l+1/2}-\frac{3}{4}\right)+\mathrm{O}(\mu^6), \qquad \mu\to 0,\end{aligned} \tag{7.17}$$

which can be derived by a direct Taylor expansion and/or verified by a computer algebra system (see our complementary Mathematica notebook). Here, $n = n_r + l + 1$ is the corresponding nonrelativistic principal quantum number. The first term in this expansion is simply the rest mass energy $E_0 = mc^2$ of the charged spin-zero particle, the second term coincides with the energy eigenvalue in the nonrelativistic Schrödinger theory, and the third term gives the so-called fine structure of the energy levels, which removes the degeneracy between states of the same n and different l.

Once again, the equation

$$xy'' + 2(\nu+1-ax)y' + 2\left(\mu\varepsilon-(\nu+1)a\right)y = 0 \tag{7.18}$$

by a change of variables $y(x) = Y(\xi)$ with $\xi = 2ax$ can be transformed into the required form:

$$\xi Y'' + (2\nu+2-\xi)Y'\left(\frac{\mu\varepsilon}{\sqrt{1-\varepsilon^2}}-\nu-1\right)Y = 0. \tag{7.19}$$

Therefore, the set of square-integrable solutions above is unique.

8. The relativistic Coulomb problem: The Dirac equation

8.1. A system of radial equations. The radial Dirac equations are derived in Refs. [**27**, **41**, **42**] by separation of variables in spherical coordinates (see also [**4**, **7**, **10**, **12**, **17**]). Then the radial functions $F(r)$ and $G(r)$ satisfy a system of two first-order ordinary differential equations:

$$\frac{dF}{dr} + \frac{1+\kappa}{r} F = \frac{mc^2 + E - U(r)}{\hbar c} G, \tag{8.1}$$

$$\frac{dG}{dr} + \frac{1-\kappa}{r} G = \frac{mc^2 - E + U(r)}{\hbar c} F, \tag{8.2}$$

where $\kappa = \kappa_\pm = \pm(j + 1/2) = \pm 1, \pm 2, \pm 3, \dots$. For the relativistic Coulomb problem, when $U = -Ze^2/r$, we introduce the dimensionless quantities

$$\varepsilon = \frac{E}{mc^2}, \qquad x = \beta r = \frac{mc}{\hbar} r, \qquad \mu = \frac{Ze^2}{\hbar c} \tag{8.3}$$

and change the variable in the radial functions by

$$f(x) = F(r), \qquad g(x) = G(r). \tag{8.4}$$

The Dirac radial system becomes

$$\frac{df}{dx} + \frac{1+\kappa}{x} f = \left(1 + \varepsilon + \frac{\mu}{x}\right) g, \tag{8.5}$$

$$\frac{dg}{dx} + \frac{1-\kappa}{x} g = \left(1 - \varepsilon - \frac{\mu}{x}\right) f. \tag{8.6}$$

(One can show later that in the nonrelativistic limit, $c \to \infty$, the following estimate holds: $|f(x)| \gg |g(x)|$; see, for example, Refs. [**27**, **41**, **42**] for more details.)

8.2. Decoupling of the radial system. We follow [**27**] with somewhat different details. Let us rewrite the system (8.5)–(8.6) in matrix form. If

$$u = \begin{pmatrix} u_1 \\ u_2 \end{pmatrix} = \begin{pmatrix} xf(x) \\ xg(x) \end{pmatrix}, \qquad u' = \begin{pmatrix} u_1' \\ u_2' \end{pmatrix}. \tag{8.7}$$

Then

$$u' = Au, \tag{8.8}$$

where

$$A = \begin{pmatrix} a_{11} & a_{12} \\ a_{21} & a_{22} \end{pmatrix} = \begin{pmatrix} -\dfrac{\kappa}{x} & 1 + \varepsilon + \dfrac{\mu}{x} \\ 1 - \varepsilon - \dfrac{\mu}{x} & \dfrac{\kappa}{x} \end{pmatrix}. \tag{8.9}$$

To find $u_1(x)$, we eliminate $u_2(x)$ from the system (8.8), obtaining a second-order differential equation

$$\begin{aligned} u_1'' &- \left(a_{11} + a_{22} + \frac{a_{12}'}{a_{12}}\right) u_1' \\ &+ \left(a_{11}a_{22} - a_{12}a_{21} - a_{11}' + \frac{a_{12}'}{a_{12}} a_{11}\right) u_1 = 0. \end{aligned} \tag{8.10}$$

Similarly, eliminating $u_1(x)$, one gets an equation for $u_2(x)$:

$$\begin{aligned} u_2'' &- \left(a_{11} + a_{22} + \frac{a_{21}'}{a_{21}}\right) u_2' \\ &+ \left(a_{11}a_{22} - a_{12}a_{21} - a_{22}' + \frac{a_{21}'}{a_{21}} a_{22}\right) u_2 = 0. \end{aligned} \tag{8.11}$$

The components of the matrix A have the generic form

$$a_{ik} = b_{ik} + c_{ik}/x, \tag{8.12}$$

where b_{ik} and c_{ik} are constants. Equations (8.10) and (8.11) are not generalized equations of hypergeometric type (2.1). Indeed,

$$\frac{a'_{12}}{a_{12}} = -\frac{c_{12}}{c_{12}x + b_{12}x^2},$$

and the coefficients of $u'_1(x)$ and $u_1(x)$ in (8.10) are

$$a_{11} + a_{22} + \frac{a'_{12}}{a_{12}} = \frac{p_1(x)}{x} - \frac{c_{12}}{c_{12}x + b_{12}x^2},$$

$$a_{11}a_{22} - a_{12}a_{21} - a'_{11} + \frac{a'_{12}}{a_{12}}a_{11} = \frac{p_2(x)}{x^2} - \frac{c_{12}(c_{11} + b_{11}x)}{(c_{12} + b_{12}x)x^2},$$

where $p_1(x)$ and $p_2(x)$ are polynomials of degrees at most one and two, respectively (see our complementary Mathematica notebook for their explicit forms). Equation (8.10) will become a generalized equation of hypergeometric type (2.1) with $\sigma(x) = x$ if either $b_{12} = 0$ or $c_{12} = 0$.

8.3. A similarity transformation. The following consideration helps. By a linear transformation

$$\begin{pmatrix} v_1 \\ v_2 \end{pmatrix} = C \begin{pmatrix} u_1 \\ u_2 \end{pmatrix} \tag{8.13}$$

with a nonsingular matrix C that is independent of x, we transform the original system (8.8) to the similar one

$$v' = \widetilde{A}v, \tag{8.14}$$

where

$$v = \begin{pmatrix} v_1 \\ v_2 \end{pmatrix}, \qquad \widetilde{A} = CAC^{-1} = \begin{pmatrix} \widetilde{a}_{11} & \widetilde{a}_{12} \\ \widetilde{a}_{21} & \widetilde{a}_{22} \end{pmatrix}.$$

The new coefficients $\widetilde{a}_{ik}$ are linear combinations of the original ones, a_{ik}. Hence they have the similar form

$$\widetilde{a}_{ik} = \widetilde{b}_{ik} + \widetilde{c}_{ik}/x, \tag{8.15}$$

where $\widetilde{b}_{ik}$ and $\widetilde{c}_{ik}$ are constants.

The equations for $v_1(x)$ and $v_2(x)$ are similar to (8.10) and (8.11):

$$\begin{aligned} v''_1 &- \left(\widetilde{a}_{11} + \widetilde{a}_{22} + \frac{\widetilde{a}'_{12}}{\widetilde{a}_{12}}\right) v'_1 \\ &+ \left(\widetilde{a}_{11}\widetilde{a}_{22} - \widetilde{a}_{12}\widetilde{a}_{21} - \widetilde{a}'_{11} + \frac{\widetilde{a}'_{12}}{\widetilde{a}_{12}}\,\widetilde{a}_{11}\right) v_1 = 0, \end{aligned} \tag{8.16}$$

$$\begin{aligned} v''_2 &- \left(\widetilde{a}_{11} + \widetilde{a}_{22} + \frac{\widetilde{a}'_{21}}{\widetilde{a}_{21}}\right) v'_2 \\ &+ \left(\widetilde{a}_{11}\widetilde{a}_{22} - \widetilde{a}_{12}\widetilde{a}_{21} - \widetilde{a}'_{22} + \frac{\widetilde{a}'_{21}}{\widetilde{a}_{21}}\,\widetilde{a}_{22}\right) v_2 = 0. \end{aligned} \tag{8.17}$$

The calculation of the coefficients in (8.16) and (8.17) is facilitated by a similarity of the matrices A and $\widetilde{A}$:

$$\widetilde{a}_{11}+\widetilde{a}_{22}=a_{11}+a_{22},\qquad \widetilde{a}_{11}\widetilde{a}_{22}-\widetilde{a}_{12}\widetilde{a}_{21}=a_{11}a_{22}-a_{12}a_{21}.$$

By a previous consideration, in order for (8.16) to be an equation of hypergeometric type, it is sufficient to choose either $\widetilde{b}_{12}=0$ or $\widetilde{c}_{12}=0$. Similarly for (8.17): either $\widetilde{b}_{21}=0$ or $\widetilde{c}_{21}=0$. These conditions impose certain restrictions on our choice of the transformation matrix C. Let

$$C=\begin{pmatrix}\alpha & \beta\\ \gamma & \delta\end{pmatrix}. \tag{8.18}$$

Then

$$C^{-1}=\frac{1}{\Delta}\begin{pmatrix}\delta & -\beta\\ -\gamma & \alpha\end{pmatrix},\qquad \Delta=\det C=\alpha\delta-\beta\gamma,$$

and

(8.19)

$$\begin{aligned}\widetilde{A}&=CAC^{-1}\\ &=\frac{1}{\Delta}\begin{pmatrix}a_{11}\alpha\delta-a_{12}\alpha\gamma+a_{21}\beta\delta-a_{22}\beta\gamma & a_{12}\alpha^2-a_{21}\beta^2+(a_{22}-a_{11})\alpha\beta\\ a_{21}\delta^2-a_{12}\gamma^2+(a_{11}-a_{22})\gamma\delta & a_{12}\alpha\gamma-a_{11}\beta\gamma+a_{22}\alpha\delta-a_{21}\beta\delta\end{pmatrix}.\end{aligned}$$

(Here, we have corrected typos in Eqs. (3.74) of [**42**]; see also [**27**] and our complementary Mathematica notebook.) For the Dirac system (8.8)–(8.9):

$$a_{11}=-\frac{\kappa}{x},\qquad a_{12}=1+\varepsilon+\frac{\mu}{x},$$
$$a_{21}=1-\varepsilon-\frac{\mu}{x},\qquad a_{22}=\frac{\kappa}{x},$$

and

$$\Delta\,\widetilde{a}_{12}=\alpha^2-\beta^2+\left(\alpha^2+\beta^2\right)\varepsilon+\frac{\left(\alpha^2+\beta^2\right)\mu+2\alpha\beta\kappa}{x}, \tag{8.20}$$

$$\Delta\,\widetilde{a}_{21}=\delta^2-\gamma^2-\left(\delta^2+\gamma^2\right)\varepsilon-\frac{\left(\delta^2+\gamma^2\right)\mu+2\gamma\delta\kappa}{x}. \tag{8.21}$$

The	condition	$\widetilde{b}_{12}=0$	yields	$(1+\varepsilon)\,\alpha^2-(1-\varepsilon)\,\beta^2=0,$
"	"	$\widetilde{c}_{12}=0$	"	$\left(\alpha^2+\beta^2\right)\mu+2\alpha\beta\kappa=0,$
"	"	$\widetilde{b}_{21}=0$	"	$(1+\varepsilon)\,\gamma^2-(1-\varepsilon)\,\delta^2=0,$
"	"	$\widetilde{c}_{21}=0$	"	$\left(\delta^2+\gamma^2\right)\mu+2\gamma\delta\kappa=0.$

We see that there are several possibilities for choosing the elements α, β, γ, δ of the transition matrix C. All quantum mechanics textbooks use the original one, namely $\widetilde{b}_{12}=0$ and $\widetilde{b}_{21}=0$, due to Darwin [**8**] and Gordon [**16**]. Nikiforov and Uvarov [**27**] take another path: they choose $\widetilde{c}_{12}=0$ and $\widetilde{c}_{21}=0$, and show that it is more convenient for taking the nonrelativistic limit $c \to \infty$. These conditions are satisfied if

$$C = \begin{pmatrix} \mu & \nu - \kappa \\ \nu - \kappa & \mu \end{pmatrix}, \tag{8.22}$$

where $\nu = \sqrt{\kappa^2 - \mu^2}$, and we finally arrive at a system of first-order equations for $v_1(x)$ and $v_2(x)$:

$$v_1' = \left(\frac{\varepsilon\mu}{\nu} - \frac{\nu}{x}\right) v_1 + \left(1 + \frac{\varepsilon\kappa}{\nu}\right) v_2, \tag{8.23}$$

$$v_2' = \left(1 - \frac{\varepsilon\kappa}{\nu}\right) v_1 + \left(\frac{\nu}{x} - \frac{\varepsilon\mu}{\nu}\right) v_2, \tag{8.24}$$

where

$$\operatorname{Tr} \widetilde{A} = \widetilde{a}_{11} + \widetilde{a}_{22} = 0, \quad \det \widetilde{A} = \varepsilon^2 - 1 + \frac{2\varepsilon\mu}{x} - \frac{\nu^2}{x^2}, \quad \nu^2 = \kappa^2 - \mu^2; \tag{8.25}$$

which is simpler than the original choice in [**27**]. The corresponding second-order differential equations (8.16)–(8.17) become

$$v_1'' + \frac{\left(\varepsilon^2 - 1\right) x^2 + 2\varepsilon\mu x - \nu(\nu + 1)}{x^2} v_1 = 0, \tag{8.26}$$

$$v_2'' + \frac{\left(\varepsilon^2 - 1\right) x^2 + 2\varepsilon\mu x - \nu(\nu - 1)}{x^2} v_2 = 0. \tag{8.27}$$

They are generalized equations of hypergeometric type (2.1) of the simplest form $\widetilde{\tau} = 0$, thus resembling the one-dimensional Schrödinger equation; the second equation can be obtained from the first one by replacing $\nu \to -\nu$ (see also Eqs. (3.81)–(3.82) in Ref. [**42**]).

8.4. Nikiforov–Uvarov paradigm. All details of the calculations are presented in Table 6 (see also our complementary notebook and Refs. [**27, 41, 42**] for more details). The corresponding energy levels $\varepsilon = \varepsilon_n$ are determined by

$$\varepsilon\mu = a(\nu + n + 1), \tag{8.28}$$

and the eigenfunctions are given by the Rodrigues–type formula

$$y_n(x) = \frac{C_n}{\rho(x)} \left(\sigma^n(x)\rho(x)\right)^{(n)} = C_n\, x^{-2\nu-1} e^{2ax} \frac{\mathrm{d}^n}{\mathrm{d}x^n} \left(x^{2\nu+n+1} e^{-2ax}\right). \tag{8.29}$$

These functions are, up to certain constants, Laguerre polynomials $L_n^{2\nu+1}(\xi)$ (Table 19), with $\xi = 2ax$. The corresponding eigenfunctions have the form

$$v_1(x) = \begin{cases} 0, & n = 0, \\ A_n \xi^{\nu+1} e^{-\xi/2} L_{n-1}^{2\nu+1}(\xi), & n = 1, 2, 3, \ldots. \end{cases} \tag{8.30}$$

They are square-integrable functions on $(0, \infty)$. Their counterparts are

$$v_2(x) = B_n \xi^{\nu} e^{-\xi/2} L_n^{2\nu-1}(\xi), \qquad n = 0, 1, 2, \ldots. \tag{8.31}$$

It is easily seen that the solution $\varepsilon = -\nu/\kappa$ is included in this formula when $n = 0$.

TABLE 6. Details for the case of a Dirac equation with the Coulomb potential $U(r) = -Ze^2/r$, treated in §8. Dimensionless quantities: $\varepsilon = E/(mc^2)$, $x = \beta r = (mc/\hbar)r$, $\mu = Ze^2/(\hbar c)$, $F(r) = f(x) = u_1(x)/x$, $G(r) = g(x) = u_2(x)/x$.

$\sigma(x)$	x
$\widetilde{\sigma}(x)$	$(\varepsilon^2-1)x^2 + 2\mu\varepsilon\, x - \nu(\nu+1); \nu = \sqrt{\kappa^2-\mu^2}, \kappa = \pm\left(j+\dfrac{1}{2}\right)$
$\widetilde{\tau}(x)$	0
k	$2\mu\varepsilon \pm \sqrt{1-\varepsilon^2}(2\nu+1)$
$\pi(x)$	$\dfrac{1}{2} \pm \left(\sqrt{1-\varepsilon^2}\, x \pm \left(\nu + \dfrac{1}{2}\right)\right)$
$\tau(x) = \widetilde{\tau} + 2\pi$	$2(\nu+1-a\, x), \quad a = \sqrt{1-\varepsilon^2}; \quad \tau' < 0$
$\lambda = k + \pi'$	$2\left(\mu\varepsilon - (\nu+1)a\right)$
$\varphi(x)$	$x^{\nu+1}e^{-ax}$
$\rho(x)$	$x^{2\nu+1}e^{-2ax}$
$y_n(x)$	$A_n L_n^{2\nu+1}(\xi); \ \xi = 2ax = 2a\beta r = 2\sqrt{1-\varepsilon^2}\,\dfrac{mc}{\hbar}\, r$ $n = n_r = 0, 1, \dots$
$A_n = \dfrac{a}{\kappa\varepsilon - \nu} B_n$	$B_n = a\beta^{3/2}\sqrt{\dfrac{(\kappa-\nu)(\kappa\varepsilon-\nu)\, n!}{\mu\Gamma(2\nu+n)}}$

As a result,

$$xf(x) = \frac{B_n}{2\nu(\kappa-\nu)}\xi^\nu e^{-\xi/2}\left(f_1\xi L_{n-1}^{2\nu+1}(\xi) + f_2 L_n^{2\nu-1}(\xi)\right), \tag{8.32}$$

$$xg(x) = \frac{B_n}{2\nu(\kappa-\nu)}\xi^\nu e^{-\xi/2}\left(g_1\xi L_{n-1}^{2\nu+1}(\xi) + g_2 L_n^{2\nu-1}(\xi)\right), \tag{8.33}$$

where

$$f_1 = \frac{a\mu}{\varepsilon\kappa - \nu}, \quad f_2 = \kappa - \nu, \quad g_1 = \frac{a(\kappa-\nu)}{\varepsilon\kappa - \nu}, \quad g_2 = \mu. \tag{8.34}$$

(These formulas remain valid for $n = 0$; in this case the terms containing $L_{-1}^{2\nu+1}(\xi)$ must be taken to be zero.) Thus we derive a representation for the radial functions up to the constant B_n in terms of Laguerre polynomials (Table 19). The normalization condition

$$\int_{\mathbb{R}^3} \psi^\dagger\, \psi\, \mathrm{dv} = \int_0^\infty r^2\left(F^2(r) + G^2(r)\right)\mathrm{d}r = 1 \tag{8.35}$$

gives the value of this constant as follows [**27**]:

$$B_n = a\beta^{3/2}\sqrt{\frac{(\kappa-\nu)(\varepsilon\kappa-\nu)n!}{\mu\Gamma(n+2\nu)}}. \tag{8.36}$$

(This is verified in Section 5.4 of Ref. [**42**]. Observe that Eq. (8.36) applies when $n = 0$.)

8.5. Summary: Wave functions and energy levels. The end results, namely, the complete wave functions and the corresponding discrete energy levels, are given by Eqs. (3.11)–(3.17) of Ref. [**42**]. The WKB, or semiclassical, approximation for the Dirac equation with Coulomb potential is discussed in [**3**].

The relativistic energy levels of an electron in the central Coulomb field are given by

$$E = E_{n_r, j} = \frac{mc^2}{\sqrt{1 + \mu^2/(n_r + \nu)^2}}, \qquad \mu = \frac{Ze^2}{\hbar c} \qquad (n_r = 0, 1, 2, \ldots). \tag{8.37}$$

In Dirac's theory,

$$\nu = \nu_{\text{Dirac}} = \sqrt{(j + 1/2)^2 - \mu^2}, \tag{8.38}$$

where $j = 1/2, 3/2, 5/2, \ldots$ is the total angular momentum including the spin of the relativistic electron. More details on the solution of this problem, including the nonrelativistic limit, can be found in [**27, 41, 42**] (following Nikiforov–Uvarov's paradigm), or in classical sources [**4, 7, 8, 12, 16, 39**].

In Dirac's theory of the relativistic electron, the corresponding limit has the form [**4, 7, 39, 42**]:

$$\varepsilon_{\text{Dirac}} = \frac{E_{n_r, j}}{mc^2} = 1 - \frac{\mu^2}{2n^2} - \frac{\mu^4}{2n^4}\left(\frac{n}{j + 1/2} - \frac{3}{4}\right) + \mathrm{O}(\mu^6), \quad \mu \to 0, \tag{8.39}$$

where $n = n_r + j + 1/2$ is the principal quantum number of the nonrelativistic hydrogenlike atom. Once again, the first term in this expansion is the rest mass energy of the relativistic electron, the second term coincides with the energy eigenvalue in the nonrelativistic Schrödinger theory, and the third term gives the so-called fine structure of the energy levels: the correction obtained for the energy in the Pauli approximation, which includes the interaction of the spin of the electron with its orbital angular momentum. (See our complementary Mathematica notebook for a computer algebra proof.)

9. A model of the 3D-confinement potential

Looking for solutions of the Schrödinger equation (5.1) in spherical coordinates,

$$\psi = \frac{1}{r} R(r) Y_{lm}(\theta, \varphi), \tag{9.1}$$

with the model central field potential

$$U(r) = V_0 \left(\frac{r}{a} - \frac{a}{r}\right)^2, \tag{9.2}$$

one gets a radial equation of the form

$$R'' + \frac{2m}{\hbar^2}\left[(E + 2V_0) - V_0\left(\frac{a^2}{r^2} + \frac{r^2}{a^2}\right) - \frac{\hbar^2 l(l+1)}{2mr^2}\right] R = 0 \quad (l = 0, 1, 2, \ldots). \tag{9.3}$$

This is not a generalized equation of hypergeometric type, and therefore cannot be treated right away by the Nikiforov–Uvarov method. However, by using the substitution

$$R(r) = u(\xi), \qquad \xi = \alpha r^2, \qquad \alpha^2 = \frac{2mV_0}{\hbar^2 a^2}, \tag{9.4}$$

TABLE 7. Details for the case of a model with the $3D$-confinement potential $U(r)$ given in (9.2), treated in §9. Dimensionless quantities: $\xi = \alpha r^2$, $\alpha = \sqrt{2mV_0}/(\hbar a)$, $R(r) = u(\xi)$.

$\sigma(\xi)$	ξ
$\widetilde{\sigma}(\xi)$	$\frac{1}{4}\left[\frac{2m}{\alpha\hbar^2}(E+2V_0)\xi - \alpha^2 a^4 - l(l+1) - \xi^2\right]$
$\widetilde{\tau}(\xi)$	$1/2$
k	$\frac{1}{2}\left[\frac{m}{\alpha\hbar^2}(E+2V_0) \pm \sqrt{\alpha^2 a^4 + (l+1/2)^2}\right]$
$\pi(\xi)$	$\frac{1}{4} \pm \frac{\xi \pm \sqrt{\alpha^2 a^4 + (l+1/2)^2}}{2}$
$\tau(\xi) = \widetilde{\tau} + 2\pi$	$1 + \beta - \xi, \quad \beta = \sqrt{\frac{2mV_0 a^2}{\hbar^2} + \left(l + \frac{1}{2}\right)^2}$
$\lambda = k + \pi'$	$\frac{1}{2}\left[\frac{m}{\alpha\hbar^2}(E+2V_0) - \sqrt{\alpha^2 a^4 + (l+1/2)^2}\right] - \frac{1}{2}$
$\varphi(\xi)$	$\xi^{\nu/2}\exp(-\xi/2), \ \nu = \beta + \frac{1}{2} = \frac{1}{2} + \sqrt{\alpha^2 a^4 + \left(l+\frac{1}{2}\right)^2}$
$\rho(\xi)$	$\xi^\beta \exp(-\xi)$
$y_n(\xi)$	$C_n L_n^\beta(\xi), \quad \xi = \alpha r^2, \quad n = 0, 1, \ldots$
C_n^2	$\frac{2n!\sqrt{\alpha}}{\Gamma(\beta+n+1)}, \quad \alpha = \frac{\sqrt{2mV_0}}{\hbar a}$

we obtain equation (2.1) with the coefficient functions

$$\sigma(\xi) = \xi, \quad \widetilde{\tau}(\xi) = \frac{1}{2}, \quad \widetilde{\sigma}(\xi) = \frac{1}{4}\left[\frac{2m}{\alpha\hbar^2}(E+2V_0)\xi - \alpha^2 a^4 - l(l+1) - \xi^2\right]. \tag{9.5}$$

In the Nikiforov–Uvarov method, the energy levels and the corresponding radial wave functions are obtained by (2.7) and (2.8). As a result, they are given by

$$E_{n,l} = \hbar\sqrt{\frac{8V_0}{ma^2}}\left[n + \frac{1}{2} + \frac{1}{2}\left(\sqrt{\frac{2mV_0 a^2}{\hbar^2} + \left(l+\frac{1}{2}\right)^2} - \sqrt{\frac{2mV_0 a^2}{\hbar^2}}\right)\right] \tag{9.6}$$

and

$$R(r) = R_{n,l}(r) = C_n \xi^{(\beta-1/2)/2}\exp(-\xi/2) L_n^\beta(\xi), \tag{9.7}$$

provided

$$\int_0^\infty R^2(r)\,\mathrm{d}r = 1. \tag{9.8}$$

Here

$$C_n^2 = \frac{2n!\sqrt{\alpha}}{\Gamma(\beta+n+1)}, \qquad \alpha = \frac{\sqrt{2mV_0}}{\hbar a}, \qquad \beta = \sqrt{\frac{2mV_0 a^2}{\hbar^2} + \left(l+\frac{1}{2}\right)^2}. \tag{9.9}$$

TABLE 8. Details for the case of the $3D$-spherical harmonic oscillator potential $U(x) = \frac{1}{2}m\omega^2 r^2$, treated in §10. Substitution: $\xi = \mu r^2$, $\mu = m\omega/\hbar$, and $R(r) = u(\xi)$.

$\sigma(\xi)$	ξ
$\widetilde{\sigma}(\xi)$	$\frac{1}{4}\left[2\varepsilon\,\xi - \xi^2 - l(l+1)\right], \quad \varepsilon = \frac{E}{\hbar\omega}$
$\widetilde{\tau}(\xi)$	$\frac{1}{2}$
k	$-\frac{1}{2}\left(l + \frac{1}{2} - \varepsilon\right)$
$\pi(\xi)$	$\frac{1}{2}(l + 1 - \xi)$
$\tau(\xi) = \widetilde{\tau} + 2\pi$	$l + \frac{3}{2} - \xi$
$\lambda = k + \pi'$	$-\frac{1}{2}\left(l + \frac{3}{2} - \varepsilon\right)$
$\varphi(\xi)$	$\xi^{(l+1)/2}\exp\left(-\xi/2\right)$
$\rho(\xi)$	$\xi^{l+1/2}\exp\left(-\xi\right)$
$y_n(\xi)$	$C_n L_n^{l+1/2}(\xi), \quad n = 0, 1, \ldots$
C_n^2	$\frac{2n!\sqrt{\mu}}{\Gamma(l+n+3/2)}, \quad \mu = \frac{m\omega}{\hbar}$

Details of the calculations are presented in Table 7. (The case $l = 0$ corresponds to a one-dimensional problem from [**14**].)

In this case, the spectrum is linear, as for the harmonic oscillator. There is no continuous spectrum, which resembles the confinement property in quantum chromodynamics.

Energy levels (9.6) can be expanded as follows:

$$E_{n,l} = \hbar\omega\left(n + \frac{1}{2}\right) + \frac{\hbar^2}{2ma^2}\left(l + \frac{1}{2}\right)^2 - \frac{\hbar^4}{4aV_0(2m)^{3/2}}\left(l + \frac{1}{2}\right)^4 + \ldots, \qquad \frac{(l+1/2)^2}{2maV_0} \ll 1. \tag{9.10}$$

Here

$$\omega = \sqrt{\frac{8V_0}{ma^2}}\,, \tag{9.11}$$

and the first term represents the familiar harmonic oscillations, the second one is the rotational energy, and the last term gives its correction in the nonsymmetric well.

10. The 3D-spherical oscillator

Looking for solutions of the Schrödinger equation (5.1) with harmonic potential

$$U(r) = \frac{1}{2} m\omega^2 r^2 \tag{10.1}$$

in spherical coordinates as in (9.1), one gets the following radial equation:

$$R'' + \left[\frac{2mE}{\hbar^2} - \frac{m^2\omega^2}{\hbar^2} r^2 - \frac{l(l+1)}{r^2} \right] R = 0 \qquad (l = 0, 1, 2, \dots). \tag{10.2}$$

Using the abbreviations [**10**]

$$\frac{2mE}{\hbar^2} = \kappa^2, \qquad \frac{m\omega}{\hbar} = \mu, \qquad \frac{\kappa^2}{2\mu} = \frac{E}{\hbar\omega} = \varepsilon, \tag{10.3}$$

one can rewrite the radial equation in the standard form

$$\frac{\mathrm{d}^2 R}{\mathrm{d}r^2} + \left[\kappa^2 - \mu^2 r^2 - \frac{l(l+1)}{r^2} \right] R = 0. \tag{10.4}$$

The substitution $R(r) = u(\xi)$ with $\xi = \mu r^2$ results in a generalized equation of hypergeometric type with the coefficient functions

$$\sigma(\xi) = \xi, \quad \widetilde{\tau}(\xi) = \frac{1}{2}, \quad \widetilde{\sigma}(\xi) = \frac{1}{4} \left[\frac{\kappa^2}{\mu} \xi - \xi^2 - l(l+1) \right]. \tag{10.5}$$

Therefore,

$$2k - \varepsilon = \pm \left(l + \frac{1}{2} \right), \qquad \pi(\xi) = \frac{1}{4} \pm \frac{1}{2} \left[\xi \pm \left(l + \frac{1}{2} \right) \right]. \tag{10.6}$$

Further details of the calculations are presented in Table 8 and in the corresponding Mathematica file.

As a result, the energy levels are given by

$$E_n = \hbar\omega \left(2n + l + \frac{3}{2} \right), \qquad n = 0, 1, 2, \dots \tag{10.7}$$

and the corresponding radial wave functions are related to the Laguerre polynomials (Table 19):

$$R_n(r) = C_n \xi^{l+1} \exp\left(-\frac{\xi}{2} \right) L_n^{l+1/2}(\xi), \quad n = 0, 1, 2, \dots. \tag{10.8}$$

Here,

$$\int_0^\infty R_n^2(r)\, \mathrm{d}r = 1. \tag{10.9}$$

The extension to the case of n dimensions is discussed in [**26**].

11. The Pöschl–Teller potential hole

Let us consider the one-dimensional stationary Schrödinger equation

$$-\frac{\hbar^2}{2m} \frac{\mathrm{d}^2\psi}{\mathrm{d}x^2} + U(x)\psi = E\psi, \tag{11.1}$$

where

$$U(x) = \frac{1}{2} V_0 \left[\frac{a(a-1)}{\sin^2(\alpha x)} + \frac{b(b-1)}{\cos^2(\alpha x)} \right], \qquad V_0 = \frac{\hbar^2 \alpha^2}{m}, \tag{11.2}$$

TABLE 9. Details for the case of a Schrödinger equation with the Pöschl–Teller potential hole $U(x)$ given in (11.2), treated in §11. Substitution: $\xi = \sin^2(\alpha x)$, $\psi(x) = u(\xi)$, $0 < x < \pi/(2\alpha)$.

$\sigma(\xi)$	$\xi(1-\xi) = \xi - \xi^2$
$\widetilde{\sigma}(\xi)$	$-\frac{1}{4}\left[\frac{c^2}{\alpha^2}\xi^2 - \left(\frac{c^2}{\alpha^2} + (a-b)(a+b-1)\right)\xi + a(a-1)\right], c^2 = \frac{2mE}{\hbar^2}$
$\widetilde{\tau}(\xi)$	$\frac{1}{2} - \xi$
k	$\frac{1}{4}\left[\frac{c^2}{\alpha^2} - (a+b)(a+b-2)\right]$
$\pi(\xi)$	$\frac{1}{2}\left[a - (a+b)\,\xi\right]$
$\tau(\xi) = \widetilde{\tau} + 2\pi$	$(a+1/2) - (a+b+1)\xi$
$\lambda = k + \pi'$	$\frac{1}{4}\left[\frac{c^2}{\alpha^2} - (a+b)^2\right]$
$\varphi(\xi)$	$\xi^{a/2}(1-\xi)^{b/2} = \sin^a(\alpha x)\cos^b(\alpha x)$
$\rho(\xi)$	$\xi^{a-1/2}(1-\xi)^{b-1/2}$
$y_n(\xi)$	$C_n P_n^{(a-1/2,\, b-1/2)}\left((1-\xi)/2)\right) = C_n P_n^{(a-1/2,\, b-1/2)}\left(\cos(2\alpha x)\right)$
C_n^2	$\frac{2\alpha n!(a+b+2n)\Gamma(a+b+n)}{\Gamma(a+n+1/2)\Gamma(b+n+1/2)} \quad (n = 0, 1, 2, \ldots)$

with real-valued parameters $a > 1$, $b > 1$, in the finite region $0 < x < \pi/(2\alpha)$ bounded by the singularities of $U(x)$ (see [**10, 30, 32**] for original references and applications). Here, we are looking for orthonormal real-valued wave functions satisfying

$$\int_0^{\pi/(2\alpha)} \psi^2(x)\,\mathrm{d}x = 1. \tag{11.3}$$

Introducing new quantities

$$\psi(x) = u(\xi), \qquad \xi = \sin^2(\alpha x), \quad 1 - \xi = \cos^2(\alpha x), \tag{11.4}$$

one gets the following generalized equation of hypergeometric type:

$$\begin{aligned} &\xi(1-\xi)u'' + \left(\frac{1}{2} - \xi\right)u' \\ &+ \frac{1}{4}\left[\frac{c^2}{\alpha^2} - \frac{a(a-1)}{\xi} - \frac{b(b-1)}{1-\xi}\right]u = 0, \qquad c^2 = \frac{2mE}{\hbar^2}. \end{aligned} \tag{11.5}$$

Here, the coefficient functions are

$$\begin{aligned} &\sigma(\xi) = \xi(1-\xi), \qquad \widetilde{\tau}(\xi) = \left(\frac{1}{2} - \xi\right), \\ &\widetilde{\sigma}(\xi) = \frac{1}{4}\left[\frac{c^2}{\alpha^2}\xi(1-\xi) - a(a-1)(1-\xi) - b(b-1)\xi\right], \end{aligned} \tag{11.6}$$

and the boundary conditions take the form $u(0) = u(1) = 0$.

Therefore,

$$p(\xi) = \frac{1}{4}\left[(\kappa+1)\xi^2 - (\kappa+1+(a-b)(a+b-1))\xi + a(a-1) + \frac{1}{4}\right], \tag{11.7}$$
$$\kappa = \frac{c^2}{\alpha^2} - 4k.$$

Equation (2.21) takes the form

$$[\kappa+1+(a-b)(a+b-1)]^2 = (\kappa+1)\left[4a(a-1)+1\right] \tag{11.8}$$

and there are two solutions:

$$\kappa_1 = (a+b)(a+b-2), \qquad \kappa_2 = -(b-a-1)(b-a+1). \tag{11.9}$$

If one chooses

$$\frac{c^2}{\alpha^2} - 4k = (a+b)(a+b-2), \tag{11.10}$$

then

$$\pi(\xi) = \frac{1}{2}\left(\frac{1}{2} - \xi\right) - \left(\frac{a+b-1}{2}\xi - \frac{(2a-1)}{4}\right) = \frac{a}{2} - \frac{a+b}{2}\xi \tag{11.11}$$

and

$$\tau(\xi) = a + \frac{1}{2} - (a+b+1)\xi, \qquad \lambda = \frac{1}{4}\left[\frac{c^2}{\alpha^2} - (a+b)^2\right]. \tag{11.12}$$

Further details of the calculations are presented in Table 9 (see also the corresponding Mathematica file).

As a result, the energy levels are given by (2.7):

$$E_n = \frac{1}{2}V_0(a+b+2n)^2, \quad n = 0, 1, 2, \ldots \tag{11.13}$$

and the corresponding wave functions are related to the Jacobi polynomials:

$$\psi_n(x) = C_n \sin^a(\alpha x)\cos^b(\alpha x)\, P_n^{(a-1/2,\, b-1/2)}(\cos(2\alpha x)) \tag{11.14}$$

with C_n a normalization constant. Indeed, by the Rodrigues-type formula (2.8),

$$y_n = \frac{B_n}{\xi^{a-1/2}(1-\xi)^{b-1/2}} \frac{d^n}{d\xi^n}\left[\xi^{n+a-1/2}(1-\xi)^{n+b-1/2}\right], \tag{11.15}$$

and with the aid of the substitution $\eta = 1 - 2\xi = \cos(2\alpha x)$ one gets

(11.16)
$$\begin{aligned} y_n(\eta) &= \frac{(-1)^n}{2^n}\frac{B_n}{(1-\eta)^{a-1/2}(1+\eta)^{b-1/2}}\frac{d^n}{d\eta^n}\left[(1-\eta)^{n+a-1/2}(1+\eta)^{n+b-1/2}\right] \\ &= C_n\, P_n^{(a-1/2,\, b-1/2)}(\eta). \end{aligned}$$

TABLE 10. Details for the case of a Schrödinger equation with the modified Pöschl–Teller potential hole $U(x)$ defined in (12.1), treated in §12. Substitution: $\xi = \cosh^2(\alpha x)$, $\psi(x) = u(\xi)$.

$\sigma(\xi)$	$\xi(1-\xi)$
$\widetilde{\sigma}(\xi)$	$\frac{1}{4}\left[\frac{c^2}{\alpha^2}\xi - a(a-1)\right](1-\xi), \quad c^2 = \frac{2m(-E)}{\hbar^2}$
$\widetilde{\tau}(\xi)$	$\frac{1}{2} - \xi$
k	$\frac{1}{4}\left(\frac{c^2}{\alpha^2} + 1 - a^2\right)$
$\pi(\xi)$	$\frac{1-a}{2}(1-\xi)$
$\tau(\xi) = \widetilde{\tau} + 2\pi$	$\frac{3}{2} - a + (a-2)\xi$
$\lambda = k + \pi'$	$\frac{1}{4}\left[\frac{c^2}{\alpha^2} - (a-1)^2\right]$
$\varphi(\xi)$	$\xi^{(1-a)/2}$
$\rho(\xi)$	$\xi^{(1/2)-a}(\xi-1)^{-(1/2)}$
$y_n(\xi)$	$C_n P_n^{(-1/2,\,1/2-a)}(\cosh(2\alpha x))\,,\ n < (a-1)/2$
C_n^2	$\alpha\frac{n!(a-1-2n)\Gamma(a-n-1/2)}{\Gamma(n+1/2)\Gamma(a-n)}$

Moreover, by the normalization condition,

(11.17)

$$\begin{aligned} 1 &= \int_0^{\pi/(2\alpha)} \psi_n^2(x)\,\mathrm{d}x = \frac{C_n^2}{2\alpha}\int_{-1}^1 \left[P_n^{(a-1/2,\,b-1/2)}(\eta)\right]^2 \frac{\varphi^2\,\mathrm{d}\eta}{(1-\eta)^{1/2}(1+\eta)^{1/2}} \\ &= \frac{C_n^2}{(2\alpha)2^{a+b}}\int_{-1}^1 \left[P_n^{(a-1/2,\,b-1/2)}(\eta)\right]^2 (1-\eta)^{a-1/2}(1+\eta)^{b-1/2}\,\mathrm{d}\eta \\ &= \frac{C_n^2}{2\alpha}\frac{\Gamma(a+n+1/2)\Gamma(b+n+1/2)}{n!(a+b+2n)\Gamma(a+b+n)}. \end{aligned}$$

(Here, the value of the squared norm d_n^2 for the Jacobi polynomials has been taken from Table 19.) As a result,

$$C_n^2 = \frac{2\alpha n!(a+b+2n)\Gamma(a+b+n)}{\Gamma(a+n+1/2)\Gamma(b+n+1/2)}. \tag{11.18}$$

12. The modified Pöschl–Teller potential hole

In order to solve the one-dimensional stationary Schrödinger equation (11.1) for the potential

$$U(x) = -\frac{\hbar^2\alpha^2}{2m}\frac{a(a-1)}{\cosh^2(\alpha x)} \qquad (-\infty < x < \infty) \tag{12.1}$$

with $a > 1$, one can use the substitution $\psi(x) = u(\xi)$, where $\xi = \cosh^2(\alpha x)$ [**10**,**30**]. As a result, we arrive at the generalized equation of hypergeometric type

$$\xi(1-\xi)u'' + \left(\frac{1}{2} - \xi\right)u' + \frac{1}{4}\left[\frac{c^2}{\alpha^2} - \frac{a(a-1)}{\xi}\right]u = 0, \quad c^2 = \frac{2m(-E)}{\hbar^2}, \tag{12.2}$$

in which the coefficient functions are

$$\begin{aligned} \sigma(\xi) &= \xi(1-\xi), \qquad \widetilde{\tau}(\xi) = \left(\frac{1}{2} - \xi\right), \\ \widetilde{\sigma}(\xi) &= \frac{1}{4}\left[\frac{c^2}{\alpha^2}\xi - a(a-1)\right](1-\xi). \end{aligned} \tag{12.3}$$

Using the standard substitution $u = \varphi(\xi)y$ with $\varphi(\xi) = \xi^{(1-a)/2}$, one gets a hypergeometric differential equation of the form:

$$\xi(1-\xi)y'' + \left[\frac{3}{2} - a + (a-2)\xi\right]y' + \frac{1}{4}\left[\frac{c^2}{\alpha^2} - (a-1)^2\right]y = 0. \tag{12.4}$$

Here, we concentrate only on the bound states (the continuous spectrum is discussed in [**10**]).

There are a finite number of negative discrete energy levels that are explicitly given by

$$E = E_n = -\frac{\hbar^2\alpha^2}{2m}(1 - a + 2n)^2, \quad n = 0, 1, \ldots < (a-1)/2. \tag{12.5}$$

The corresponding orthonormal wave functions are related to a set of Jacobi polynomials with a negative value of one parameter that are orthogonal on an infinite interval $(1, +\infty)$. They are given by the Rodrigues-type formula (2.8), or in terms of a terminating hypergeometric series as

$$\begin{aligned} y_n(\xi) &= \frac{(1/2)_n}{n!}\,{}_2F_1\left(-n,\ 1-a+n;\ \frac{1}{2};\ 1-\xi\right) \\ &= P_n^{(-1/2,\,1/2-a)}\left(\cosh(2\alpha x)\right), \end{aligned} \tag{12.6}$$

where $(1/2)_n = \Gamma(n+1/2)/\Gamma(1/2)$. Cauchy's beta integral,

$$\int_0^\infty \frac{t^{A-1}}{(1+t)^{A+B}}\,dt = \frac{\Gamma(A)\,\Gamma(B)}{\Gamma(A+B)}, \qquad \Re(A) > 0,\ \Re(B) > 0, \tag{12.7}$$

should be used in order to find the value of the normalization constant (see [**42**, Exercise 1.15] and [**28**, (5.12.3)]). Details of the calculation are presented in Table 10 (see also the corresponding Mathematica file). As a result, for the bound states (12.5), the normalized wave functions are given by

$$\begin{aligned} \psi_n(x) = \sqrt{\alpha}&\left[\frac{n!(a-1-2n)\,\Gamma(a-n-1/2)}{\Gamma(n+1/2)\,\Gamma(a-n)}\right]^{1/2} \\ &\times \cosh^{1-a}(\alpha x)\,P_n^{(-1/2,\,1/2-a)}\left(\cosh(2\alpha x)\right). \end{aligned} \tag{12.8}$$

13. Kratzer's molecular potential

In order to investigate the rotation–vibration spectrum of a diatomic molecule, the potential

$$U(r) = -2D\left(\frac{a}{r} - \frac{1}{2}\frac{a^2}{r^2}\right), \qquad D > 0, \tag{13.1}$$

TABLE 11. Detials for the case of a Schrödinger equation with Kratzer's molecular potential (13.1), treated in §13. Substitution: $x = r/a$, $R(r) = u(x)$; see also (13.2).

$\sigma(x)$	x
$\widetilde{\sigma}(x)$	$2\gamma^2 x - \beta^2 x^2 - \gamma^2 - l(l+1)$
$\widetilde{\tau}(x)$	0
k	$2\gamma^2 \pm \beta(2\nu - 1), \quad \nu = \dfrac{1}{2} + \sqrt{\gamma^2 + \left(l + \dfrac{1}{2}\right)^2}$
$\pi(x)$	$\nu - \beta x$
$\tau(x) = \widetilde{\tau} + 2\pi$	$2(\nu - \beta x)$
$\lambda = k + \pi'$	$2(\gamma^2 - \nu\beta)$
$\varphi(x)$	$x^\nu \exp(-\beta x)$
$\rho(x)$	$x^{2\nu-1} \exp(-2\beta x)$
$y_n(x)$	$C_n \, L_n^{2\nu-1}(2\beta x)$
C_n^2	$\dfrac{(2\beta)^{2\nu+1} n!}{a(2\nu + 2n)\Gamma(2\nu + n)}, \quad \beta = \beta_n = \dfrac{2ma^2}{\hbar^2} \dfrac{D}{\nu + n}$

with a minimum $U(a) = -D$, has been used [**10**]. Once again we are looking for solutions of the Schrödinger equation (5.1) in spherical coordinates as in (9.1), and introduce the dimensionless quantities:

$$x = \frac{r}{a}, \qquad \beta^2 = -\frac{2ma^2}{\hbar^2} E, \qquad \gamma^2 = \frac{2ma^2}{\hbar^2} D, \tag{13.2}$$

together with the standard substitution $R(r) = u(x)$.

For bound states $E < 0$ and $\beta > 0$, and the radial equation takes the form

$$u'' + \left[-\beta^2 + \frac{2\gamma^2}{x} - \frac{\gamma^2 + l(l+1)}{x^2}\right] u = 0. \tag{13.3}$$

Further computational details are presented in Table 11 (see also the corresponding Mathematica file). This case is somewhat similar to the Coulomb and relativistic Coulomb problems.

As a result, the bound states are given by

$$E_{n,l} = -\frac{2ma^2 D^2}{\hbar^2} \frac{1}{(\nu + n)^2}, \tag{13.4}$$

where

$$\nu = \frac{1}{2} + \sqrt{\gamma^2 + \left(l + \frac{1}{2}\right)^2}, \qquad \gamma^2 = \frac{2ma^2}{\hbar^2} D. \tag{13.5}$$

We can obtain exactly the same result with the aid of the Bohr–Sommerfeld quantization rule in the semiclassical approximation (the WKB-method [**3, 22, 27, 40**]).

The dimensionless parameter γ^2 is proportional to the reduced mass m of the nuclei. When $\gamma \gg 1$ and $n \ll \gamma$, $l \ll \gamma$, the energy (13.4) can be expanded as follows:

$$\begin{aligned} E_{n,l} = & -D + \hbar\omega\left(n+\frac{1}{2}\right) + \frac{\hbar^2}{2ma^2}\left(l+\frac{1}{2}\right)^2 \\ & -\frac{2\hbar^2}{ma^2}\left[\frac{3}{4}\left(n+\frac{1}{2}\right)^2 - \frac{1}{\gamma}\left(n+\frac{1}{2}\right)^3\right] \\ & -\frac{3\hbar^2}{2m\gamma a^2}\left(n+\frac{1}{2}\right)\left(l+\frac{1}{2}\right)^2 + \mathrm{O}\left(\gamma^{-4}\right), \qquad \gamma \gg 1 \end{aligned} \tag{13.6}$$

(see the Mathematica notebook). Here

$$\hbar\omega = \frac{2D}{\gamma} = \frac{\hbar}{a}\sqrt{\frac{2D}{m}}. \tag{13.7}$$

The second and third terms in this expression give the energy of the vibrational and rotational motion, respectively. The fourth term takes into account anharmonic oscillations, and finally, the last term gives the correction to the energy due to the interaction between the rotational and vibrational motions of the nuclei [**10, 14**] (we have corrected a couple of typos therein).

The normalized radial wave functions, satisfying

$$\int_0^\infty R^2(r)\, dr = 1, \tag{13.8}$$

are related to the Laguerre polynomials (Table 19):

$$R_n(r) = C_n\, x^\nu \exp(-\beta x) L_n^{2\nu-1}(2\beta x) \qquad \left(x = \frac{r}{a}\right), \tag{13.9}$$

where

$$C_n^2 = \frac{(2\beta)^{2\nu+1} n!}{a(2\nu+2n)\Gamma(2\nu+n)}. \tag{13.10}$$

Once again, we have used the integral (7.13). (See also [**1, 5, 10, 13, 14, 18**] and the Mathematica file for some applications of Kratzer's molecular potential, graphs, and further references.)

14. The Hulthén potential

We now look for solutions of the Schrödinger equation (5.1) in spherical coordinates as in (9.1) with the central field potential

$$U(r) = -V_0 \frac{e^{-r/a}}{1-e^{-r/a}} \qquad (0 \leqslant r < \infty) \tag{14.1}$$

when $l = 0$. (See, for example, (6.2) with $F = R$, and use these data for an explicit form of the corresponding radial equation.)

With the aid of the substitution

$$R(r) = u(\xi), \qquad \xi = e^{-r/a} \tag{14.2}$$

when

$$\alpha^2 = -\frac{2mE}{\hbar^2}a^2 > 0, \qquad \beta^2 = \frac{2mV_0}{\hbar^2}a^2 > 0 \tag{14.3}$$

TABLE 12. The case of a Schrödinger equation with the Hulthén potential (14.1) in the spherically symmetric case $l = 0$, treated in §14. Substitution: $R(r) = u(\xi)$, $\xi = \exp(-r/a)$, $\alpha^2 = -(2mE/\hbar^2)a^2 > 0$, $\beta^2 = (2mV_0/\hbar^2)a^2 > 0$.

$\sigma(\xi)$	$\xi(1-\xi)$
$\widetilde{\sigma}(\xi)$	$(1-\xi)\left[\left(\alpha^2+\beta^2\right)\xi-\alpha^2\right]$
$\widetilde{\tau}(\xi)$	$1-\xi$
k	$\beta^2-\alpha$
$\pi(\xi)$	$\alpha-(\alpha+1)\xi$
$\tau(\xi)=\widetilde{\tau}+2\pi$	$2\alpha+1-(2\alpha+3)\xi$
$\lambda=k+\pi'$	$\beta^2-2\alpha-1=(\gamma-\alpha-1)(\gamma+\alpha+1), \quad \gamma=\sqrt{\alpha^2+\beta^2}$
$\varphi(\xi)$	$\xi^{\alpha}(1-\xi)$
$\rho(\xi)$	$\xi^{2\alpha}(1-\xi)$
E_n	$-V_0\left(\dfrac{\beta^2-n^2}{2\beta n}\right)^2, \qquad n=1,2,3,\ \ldots \quad (n^2<\beta^2)$
$y_n(\xi)$	${}_2F_1(1-n,\ 1+2\alpha+n;\ 2\alpha+1;\ \xi)$
C_n	$\dfrac{(2\alpha)_n}{n!}\sqrt{\dfrac{(\alpha+n)(2\alpha+n)}{(2\alpha)a}}, \qquad \alpha=\alpha_n=\dfrac{\beta^2-n^2}{2n}>0$

with $\alpha > 0$ and $\beta > 0$, one obtains a generalized equation of hypergeometric type,

$$\xi^2\frac{\mathrm{d}^2u}{\mathrm{d}\xi^2}+\xi\frac{\mathrm{d}u}{\mathrm{d}\xi}+\left(-\alpha^2+\beta^2\frac{\xi}{1-\xi}\right)u=0, \tag{14.4}$$

in which the coefficient functions are

$$\sigma(\xi)=\xi(1-\xi), \qquad \widetilde{\tau}(\xi)=1-\xi, \tag{14.5}$$

and

$$\widetilde{\sigma}(\xi)=(1-\xi)\left(\left(\alpha^2+\beta^2\right)\xi-\alpha^2\right). \tag{14.6}$$

The boundary conditions are

$$R(0)=\lim_{r\to 0^+}R(r)=0, \qquad R(\infty)=\lim_{r\to\infty}R(r)=0, \tag{14.7}$$

or

$$u(0)=u(1)=0. \tag{14.8}$$

We have

$$k=\beta^2\mp\alpha, \qquad \pi(\xi)=-\frac{\xi}{2}\pm\left(-\alpha\pm\frac{\xi}{2}+\alpha\xi\right) \tag{14.9}$$

for all four possible sign combinations. The solutions can be found by putting [**10**]

$$u=\varphi(\xi)y(\xi)=\xi^{\alpha}(1-\xi)y(\xi), \tag{14.10}$$

which results in

$$\xi(1-\xi)y''+\left[2\alpha+1-(2\alpha+3)\xi\right]y'-\left(2\alpha+1-\beta^2\right)y=0 \tag{14.11}$$

(details of the calculations are presented in Table 12 and in the complementary Mathematica file).

As one can see, a direct quantization in terms of the classical orthogonal polynomials by Nikiforov and Uvarov's approach is not applicable here, right away, because the second coefficient function

$$\tau(\xi) = 2\alpha + 1 - (2\alpha+3)\xi, \tag{14.12}$$

depends on α and therefore on the energy E. We have to utilize the boundary conditions (14.8) instead, in a somewhat similar way to the consideration of the familiar case of an infinite well.

Equation (14.11) is a special case of the hypergeometric equation [**2, 28**],

$$\xi(1-\xi)w'' + \left[C - (A+B+1)\xi\right]w' - ABw = 0 \tag{14.13}$$

with

$$A = 1+\alpha+\gamma, \quad B = 1+\alpha-\gamma, \quad C = 2\alpha+1; \qquad \gamma = \sqrt{\alpha^2+\beta^2}. \tag{14.14}$$

The required solution, which is bounded at $\xi = 0$, has the form

$$y = {}_2F_1\left(\begin{matrix} A,\ B \\ C \end{matrix};\xi\right) = {}_2F_1\left(\begin{matrix} 1+\alpha+\gamma,\ 1+\alpha-\gamma \\ 2\alpha+1 \end{matrix};\xi\right) \tag{14.15}$$

up to a constant factor, and the first boundary condition is satisfied: $u(0) = 0$, when $\alpha > 0$.

As is known [**28**],

$$\lim_{\xi\to 1^-} (1-\xi)^{A+B-C}\, {}_2F_1\left(\begin{matrix} A,\ B \\ C \end{matrix};\ \xi\right) = \frac{\Gamma(C)\,\Gamma(A+B-C)}{\Gamma(A)\,\Gamma(B)}, \tag{14.16}$$

provided $\Re(C-A-B) < 0$. Thus

$$\begin{aligned} u(1) &= \lim_{\xi\to 1^-} (1-\xi)\ {}_2F_1\left(\begin{matrix} 1+\alpha+\gamma,\ 1+\alpha-\gamma \\ 2\alpha+1 \end{matrix};\xi\right) \\ &= \frac{\Gamma(2\alpha+1)\,\Gamma(1)}{\Gamma(1+\alpha+\gamma)\,\Gamma(1+\alpha-\gamma)} = 0, \end{aligned} \tag{14.17}$$

provided that

$$\alpha - \gamma = -n = -1, -2, -3, \dots, \tag{14.18}$$

$$\alpha = \alpha_n = \frac{\beta^2 - n^2}{2n} > 0. \tag{14.19}$$

As a result, the discrete energy levels are given by

$$E_n = -V_0\left(\frac{\beta^2-n^2}{2\beta n}\right)^2, \qquad n = 1, 2, 3, \dots \quad \left(n^2 < \beta^2\right). \tag{14.20}$$

There exists a minimum size of the potential hole before any energy eigenvalue at all can be obtained, viz., $\beta^2 = 1$. The condition $1 \leqslant n^2 \leqslant \beta^2$ determines the finite number of eigenvalues in a potential hole of a given size [**10**].

The radial wave functions take the form

$$R_n(r) = C_n\ \xi^\alpha (1-\xi)\ {}_2F_1\left(\begin{matrix} 1-n,\ 1+2\alpha+n \\ 2\alpha+1 \end{matrix};\xi\right), \qquad \xi = e^{-r/a}, \tag{14.21}$$

where the hypergeometric series terminates and C_n is a constant to be determined. Thus, the energy levels can be obtained by the condition (2.7) and the corresponding wave functions are derived with the help of the Rodrigues-type formula (2.8), as follows:

(14.22)
$$\left(\xi^{2\alpha+n-1}(1-\xi)^n\right)^{(n-1)} = \frac{\Gamma(2\alpha+n)}{\Gamma(2\alpha+1)}\xi^{2\alpha}(1-\xi)\ {}_2F_1\left(\begin{matrix}1-n,\ 2\alpha+n+1\\ 2\alpha+1\end{matrix};\xi\right).$$

This result follows also, as a special case, from (15.5.9) of [**28**].

Once again, we can use (13.8) for normalization of the radial wave function. Then

$$aC_n^2\int_0^1 \xi^{2\alpha-1}(1-\xi)^2 y_n^2(\xi)\,\mathrm{d}\xi = 1, \tag{14.23}$$

where

$$\xi^{2\alpha}(1-\xi)y_n(\xi) = \frac{\Gamma(2\alpha+1)}{\Gamma(2\alpha+n)}\left[\xi^{2\alpha+n-1}(1-\xi)^n\right]^{(n-1)} \tag{14.24}$$

by (14.22). Moreover,

$$(1-\xi)y_n(\xi) = {}_2F_1\left(\begin{matrix}-n,\ 2\alpha+n\\ 2\alpha+1\end{matrix};\xi\right) \tag{14.25}$$

by a familiar transformation of Euler [**27**]:

$${}_2F_1\left(\begin{matrix}A,\ B\\ C\end{matrix};\xi\right) = (1-\xi)^{C-A-B}\ {}_2F_1\left(\begin{matrix}C-A,\ C-B\\ C\end{matrix};\xi\right) \tag{14.26}$$

with $A=-n$, $B=2\alpha+n$, $C=2\alpha+1$. Therefore,

$$\begin{aligned}\int_0^1 \xi^{2\alpha-1}(1-\xi)^2 y_n^2(\xi)\,\mathrm{d}\xi &= \frac{\Gamma(2\alpha+1)}{\Gamma(2\alpha+n)}\\ &\times\int_0^1\left[\xi^{-1}\ {}_2F_1\left(\begin{matrix}-n,\ 2\alpha+n\\ 2\alpha+1\end{matrix};\xi\right)\right]\left[\xi^{2\alpha+n-1}(1-\xi)^n\right]^{(n-1)}\,\mathrm{d}\xi,\end{aligned} \tag{14.27}$$

and integrating by parts $n-1$ times, one gets

(14.28)

$$\begin{aligned}
&\int_0^1 \xi^{-1}\, {}_2F_1\left(\begin{matrix}-n,\ 2\alpha+n\\ 2\alpha+1\end{matrix};\xi\right)\left[\xi^{2\alpha+n-1}(1-\xi)^n\right]^{(n-1)}\,d\xi\\
&= \left(\xi^{-1}\, {}_2F_1\left(\begin{matrix}-n,\ 2\alpha+n\\ 2\alpha+1\end{matrix};\xi\right)\left[\xi^{2\alpha+n-1}(1-\xi)^n\right]^{(n-2)}\right)\Bigg|_{\xi=0}^{1}\\
&\qquad -\int_0^1 \left[\xi^{-1}\, {}_2F_1\left(\begin{matrix}-n,\ 2\alpha+n\\ 2\alpha+1\end{matrix};\xi\right)\right]'\left[\xi^{2\alpha+n-1}(1-\xi)^n\right]^{(n-2)}\,d\xi\\
&= \cdots =\\
&(-1)^{k-1}\left(\left[\xi^{-1}{}_2F_1\left(\begin{matrix}-n,\ 2\alpha+n\\ 2\alpha+1\end{matrix};\xi\right)\right]^{(k-1)}\left[\xi^{2\alpha+n-1}(1-\xi)^n\right]^{(n-k-1)}\right)\Bigg|_{\xi=0}^{1}\\
&\qquad +(-1)^k\int_0^1 \left[\xi^{-1}\, {}_2F_1\left(\begin{matrix}-n,\ 2\alpha+n\\ 2\alpha+1\end{matrix};\xi\right)\right]^{(k)}\left[\xi^{2\alpha+n-1}(1-\xi)^n\right]^{(n-k-1)}\,d\xi\\
&= \cdots = (-1)^{n-1}\int_0^1 \left[\xi^{-1}\, {}_2F_1\left(\begin{matrix}-n,\ 2\alpha+n\\ 2\alpha+1\end{matrix};\xi\right)\right]^{(n-1)}\xi^{2\alpha+n-1}(1-\xi)^n\,d\xi,
\end{aligned}$$

in view of the boundary conditions (14.7)–(14.8). By the power series expansion

(14.29)

$$\begin{aligned}
\left[\xi^{-1}\, {}_2F_1\left(\begin{matrix}-n,\ 2\alpha+n\\ 2\alpha+1\end{matrix};\xi\right)\right]^{(n-1)} &= \left(\frac{1}{\xi}\right)^{(n-1)} + \frac{(-n)_n(2\alpha+n)_n}{(n!)(2\alpha+1)_n}\left(\xi^{n-1}\right)^{(n-1)}\\
&= (-1)^{n-1}\frac{(n-1)!}{\xi^n} + \frac{(-n)_n(2\alpha+n)_n}{(n!)(2\alpha+1)_n}(n-1)!,
\end{aligned}$$

and our integral evaluation can be completed with the aid of the Euler beta integrals (B.2):

$$\int_0^1 \xi^{2\alpha-1}(1-\xi)^{(n+1)-1}\,d\xi = \frac{\Gamma(2\alpha)(n!)}{\Gamma(2\alpha+n+1)},$$

$$\int_0^1 \xi^{2\alpha+n-1}(1-\xi)^{(n+1)-1}\,d\xi = \frac{\Gamma(2\alpha+n)(n!)}{\Gamma(2\alpha+2n+1)}.$$

The final result is given by

$$C_n = \frac{(2\alpha)_n}{n!}\sqrt{\frac{(\alpha+n)(2\alpha+n)}{(2\alpha)a}}, \qquad \alpha = \frac{\beta^2-n^2}{2n} \qquad (n=1,2,3,\dots), \tag{14.30}$$

as a complementary normalization in (14.21) (Table 12). We were not able to find the value of this constant in the available literature (see, for example, [**10**]).

The Hulthén potential at small values of r behaves like a Coulomb potential $U_C = -V_0 a/r$, whereas for large values of r it decreases exponentially. (See [**10**] for more details and a numerical example.) Section 17 below contains an extension of this potential that is suitable for diatomic molecules.

TABLE 13. Details for the case of a Schrödinger equation with the Morse potential (15.1) in the spherically symmetric case $l = 0$, treated in §15. Substitution: $R(r) = u(\xi)$, $\xi = (2\gamma/\alpha)\exp(-\alpha x)$, $\beta^2 = -2mEr_0^2/\hbar^2 > 0$, $\gamma^2 = 2mDr_0^2/\hbar^2$.

$\sigma(\xi)$	ξ
$\widetilde{\sigma}(\xi)$	$-\dfrac{\beta^2}{\alpha^2} + \dfrac{\gamma}{\alpha}\xi - \dfrac{1}{4}\xi^2$
$\widetilde{\tau}(\xi)$	1
k	$\dfrac{\gamma - \beta}{\alpha}$
$\pi(\xi)$	$\dfrac{\beta}{\alpha} - \dfrac{\xi}{2}$
$\tau(\xi) = \widetilde{\tau} + 2\pi$	$1 + \dfrac{2\beta}{\alpha} - \xi$
$\lambda = k + \pi'$	$\dfrac{\gamma - \beta}{\alpha} - \dfrac{1}{2}$
$\varphi(\xi)$	$\xi^{\beta/\alpha}e^{-\xi/2}$
$\rho(\xi)$	$\xi^{2\beta/\alpha}e^{-\xi}$
E_v	$-D + \dfrac{\hbar^2}{2mr_0^2}\left[2\gamma\alpha\left(v + \dfrac{1}{2}\right) - \alpha^2\left(v + \dfrac{1}{2}\right)^2\right]$
$y_v(\xi)$	$C_v\, L_v^{2\beta/\alpha}(\xi) = C_v\dfrac{\Gamma(2\beta/\alpha + v + 1)}{v!\,\Gamma(2\beta/\alpha + 1)}\ {}_1F_1(-v; 2\beta/\alpha + 1\ ; \xi)$
C_v^2	$\dfrac{(2\beta)v!}{r_0\Gamma(2\beta/\alpha + v + 1)}$

15. The Morse potential

The central field potential

$$U(r) = D\left(e^{-2\alpha x} - 2e^{-\alpha x}\right), \quad x = \frac{r - r_0}{r_0} \quad (0 \leqslant r < \infty) \tag{15.1}$$

is used for the study of vibrations of diatomic molecules [**1, 10, 25**]. The corresponding Schrödinger equation (5.1) can be solved in spherical coordinates as in (9.1) when $l = 0$. Introducing new parameters

$$\beta^2 = -\frac{2mEr_0^2}{\hbar^2} > 0, \qquad \gamma^2 = \frac{2mDr_0^2}{\hbar^2} \tag{15.2}$$

$(\beta, \gamma > 0)$, with the help of the substitution

$$R(r) = u(\xi), \qquad \xi = \frac{2\gamma}{\alpha}e^{-\alpha x}, \quad x = \frac{r - r_0}{r_0}, \tag{15.3}$$

one gets

$$\xi^2\frac{d^2u}{d\xi^2} + \xi\frac{du}{d\xi} + \left(-\frac{\beta^2}{\alpha^2} + \frac{\gamma}{\alpha}\xi - \frac{1}{4}\xi^2\right)u = 0. \tag{15.4}$$

This is a generalized equation of hypergeometric type with coefficient functions

$$\sigma(\xi) = \xi, \qquad \widetilde{\tau}(\xi) = 1, \tag{15.5}$$

and

$$\widetilde{\sigma}(\xi) = -\frac{\beta^2}{\alpha^2} + \frac{\gamma}{\alpha}\xi - \frac{1}{4}\xi^2. \tag{15.6}$$

The substitution

$$u = \varphi(\xi)y(\xi) = \xi^{\beta/\alpha}e^{-\xi/2}y(\xi) \tag{15.7}$$

results in the confluent hypergeometric equation

$$\xi y'' + \left(\frac{2\beta}{\alpha} + 1 - \xi\right)y' + \left(\frac{\gamma-\beta}{\alpha} - \frac{1}{2}\right)y = 0 \tag{15.8}$$

with the parameter values

$$c = 2\frac{\beta}{\alpha} + 1, \qquad a = \frac{1}{2}c - \frac{\gamma}{\alpha} = \frac{1}{2} + \frac{\beta-\gamma}{\alpha} \tag{15.9}$$

(see Table 13 and the corresponding Mathematica file for more details).

The general solution of (15.8) has the form [**27, 28**]:

$$y = A\ {}_1F_1\left(\begin{matrix} a \\ c \end{matrix}; \xi\right) + B\ \xi^{1-c}\ {}_1F_1\left(\begin{matrix} 1+a-c \\ 2-c \end{matrix}; \xi\right). \tag{15.10}$$

Here, the second constant must vanish, i.e., $B = 0$, due to the boundary condition $\lim_{r\to\infty} R(r) = u(0) = 0$, because

$$1 - c + \frac{\beta}{\alpha} = -\frac{\beta}{\alpha} < 0. \tag{15.11}$$

The first constant A is determined by the normalization.

The second boundary condition, namely, $\lim_{r\to 0} R(r) = u(\xi_0) = 0$ with $\xi_0 = (2\gamma/\alpha)e^{\alpha}$, states that

$${}_1F_1\left(\begin{matrix} a \\ c \end{matrix}; \xi_0\right) = 0, \tag{15.12}$$

where both coefficients depend on the energy, due to (15.2) and (15.9). This transcendental equation for the discrete energy levels cannot be solved explicitly, but for all real diatomic molecules, $\xi_0 \gg 1$ [**10**]. This is why one can use the familiar asymptotic

$$\begin{aligned} {}_1F_1\left(\begin{matrix} a \\ c \end{matrix}; \xi\right) &= \frac{\Gamma(c)}{\Gamma(c-a)}(-\xi)^{-a}\left[1 + \mathrm{O}\left(\frac{1}{\xi}\right)\right] \\ &\quad + \frac{\Gamma(c)}{\Gamma(a)}\ e^{\xi}\ \xi^{a-c}\left[1 + \mathrm{O}\left(\frac{1}{\xi}\right)\right], \qquad \xi \to \infty \end{aligned} \tag{15.13}$$

for the confluent hypergeometric function [**27, 28**].

By eliminating the largest asymptotic term with $\Gamma(a) = \infty$, one gets an approximate quantization rule, which states[1]:

$$a = -v; \qquad v = 0, 1, 2, \ldots; \qquad v < \frac{\gamma}{\alpha} - \frac{1}{2} \tag{15.14}$$

[1]The value $v = 0$ can be added because $\exp(-\xi_0/2) \ll 1$; and the upper bound is due to convergence of the normalization integral (15.19).

(more details can be found in [**10**]). The corresponding approximation to the discrete energy levels is given by

$$-\beta^2 = -\gamma^2 + 2\gamma\alpha\left(v + \frac{1}{2}\right) - \alpha^2\left(v + \frac{1}{2}\right)^2. \tag{15.15}$$

This result can also be obtained by the Nikiforov–Uvarov approach, using (2.7). Hence, the approximate energy levels in terms of the vibrational quantum number v are

$$E_v = -D + \frac{\hbar^2}{2mr_0^2}\left[2\gamma\alpha\left(v + \frac{1}{2}\right) - \alpha^2\left(v + \frac{1}{2}\right)^2\right], \tag{15.16}$$

where the last term reflects the anharmonicity correction. This formula can be rewritten as [**10**]
(15.17)

$$E_v = -D + \hbar\omega\left[\left(v + \frac{1}{2}\right) - \frac{\alpha}{2\gamma}\left(v + \frac{1}{2}\right)^2\right], \qquad \hbar\omega = \hbar^2\frac{\alpha\gamma}{mr_0^2} = \hbar\frac{\alpha}{r_0}\sqrt{\frac{2D}{m}}.$$

The first two terms in this formula are in complete agreement with the harmonic oscillator energy levels. The last term reflects the anharmonicity correction, which shows that the anharmonic term never exceeds the harmonic one [**10**].

The corresponding radial wave functions are given in terms of the Laguerre polynomials (Table 19):

$$R(r) = R_v(r) = \sqrt{\frac{(2\beta)\ v!}{r_0\Gamma\left(2\beta/\alpha + v + 1\right)}}\ \xi^{\beta/\alpha}e^{-\xi/2}\ L_v^{2\beta/\alpha}(\xi). \tag{15.18}$$

Here, we have used the normalization

$$\begin{aligned}\int_{r=0}^{\infty} R^2(r)\,\mathrm{d}r &= \frac{r_0}{\alpha}\int_{\xi=0}^{\xi_0} \xi^{2\beta/\alpha-1}e^{-\xi}y^2(\xi)\,\mathrm{d}\xi \\ &\approx C_v^2\frac{r_0}{\alpha}\int_{\xi=0}^{\infty} \xi^{2\beta/\alpha-1}e^{-\xi}\left(L_v^{2\beta/\alpha}(\xi)\right)^2\,\mathrm{d}\xi = 1\end{aligned} \tag{15.19}$$

and the integral

$$I_{-1} = J_{mm,-1}^{\delta\delta} = \int_0^{\infty} e^{-\xi}\xi^{\delta-1}\left(L_m^{\delta}(\xi)\right)^2\,\mathrm{d}\xi = \frac{\Gamma(\delta + m + 1)}{m!\delta}, \tag{15.20}$$

where $\delta = 2\beta/\alpha > 0$ (see [**41, 42**] and Appendix B; further details are left to the reader).

16. The rotation correction of the Morse potential

The standard centrifugal term [**34**]

$$\frac{l(l+1)}{r^2} = \frac{l(l+1)}{r_0^2}\frac{1}{(1+x)^2}, \qquad x = \frac{r - r_0}{r_0} \tag{16.1}$$

can be approximated, in the neighborhood of the minimum of the Morse potential $r = r_0$ (or $x = 0$), as follows [**10**]:

$$\frac{l(l+1)}{r^2} \approx \frac{l(l+1)}{r_0^2}\left(C_0 + C_1e^{-\alpha x} + C_2e^{-2\alpha x}\right), \tag{16.2}$$

TABLE 14. Details for the case of a Schrödinger equation with the rotation correction of the Morse potential (16.2), the coefficients C_0, C_1, and C_2 being given by (16.3), which is treated in §16.

$\sigma(\xi)$	ξ
$\widetilde{\sigma}(\xi)$	$-\frac{\beta_1^2}{\alpha^2}+\frac{\gamma_1^2}{\alpha\gamma_2}\xi-\frac{1}{4}\xi^2$
$\widetilde{\tau}(\xi)$	1
k	$\frac{\gamma_1^2}{\alpha\gamma_2}-\frac{\beta_1}{\alpha}$
$\pi(\xi)$	$\frac{\beta_1}{\alpha}-\frac{\xi}{2}$
$\tau(\xi)=\widetilde{\tau}+2\pi$	$1+\frac{2\beta_1}{\alpha}-\xi$
$\lambda=k+\pi'$	$\frac{\gamma_1^2}{\alpha\gamma_2}-\frac{\beta_1}{\alpha}-\frac{1}{2}$
$\varphi(\xi)$	$\xi^{\beta_1/\alpha}e^{-\xi/2}$
$\rho(\xi)$	$\xi^{2\beta_1/\alpha}e^{-\xi}$

where

$$C_0=1-\frac{3}{\alpha}+\frac{3}{\alpha^2},\quad C_1=\frac{4}{\alpha}-\frac{6}{\alpha^2},\quad C_2=-\frac{1}{\alpha}+\frac{3}{\alpha^2}. \tag{16.3}$$

Indeed,

$$\frac{1}{(1+x)^2}-\left(C_0+C_1e^{-\alpha x}+C_2e^{-2\alpha x}\right)=x^3\left(-\frac{2}{3}\alpha^2+3\alpha-4\right)+\mathrm{O}(x^4),\quad x\to 0. \tag{16.4}$$

This consideration allows one to introduce a rotation correction to the Morse potential without changing the mathematical model much [**11**].

In this approximation, the radial Schrödinger equation (5.1) in spherical coordinates as in (9.1),

$$R''(r)+\left[\frac{2m}{\hbar^2}\left(E-D\left(e^{-2\alpha x}-2e^{-\alpha x}\right)\right)-\frac{l(l+1)}{r^2}\right]R(r)=0, \tag{16.5}$$

with the new variables

$$R(r)=u(\xi),\qquad \xi=\frac{2\gamma_2}{\alpha}e^{-\alpha x},\quad x=\frac{r-r_0}{r_0}, \tag{16.6}$$

and with the modified parameters

$$\beta_1^2=\beta^2+l(l+1)C_0,\qquad \beta^2=-\frac{2mEr_0^2}{\hbar^2}>0, \tag{16.7}$$

$$\gamma_1^2=\gamma^2-\frac{1}{2}l(l+1)C_1,\quad \gamma_2^2=\gamma^2+l(l+1)C_2,\quad \gamma^2=\frac{2mDr_0^2}{\hbar^2}, \tag{16.8}$$

becomes a generalized equation of hypergeometric type:

$$\xi^2 u'' + \xi u' + \left(-\frac{\beta_1^2}{\alpha^2} + \frac{\gamma_1^2}{\alpha\gamma_2}\xi - \frac{1}{4}\xi^2 \right) u = 0. \tag{16.9}$$

Here, the coefficient functions are

$$\sigma(\xi) = \xi, \qquad \widetilde{\tau}(\xi) = 1, \tag{16.10}$$

and

$$\widetilde{\sigma}(\xi) = -\frac{\beta_1^2}{\alpha^2} + \frac{\gamma_1^2}{\alpha\gamma_2}\xi - \frac{1}{4}\xi^2. \tag{16.11}$$

The substitution

$$u = \xi^{\beta_1/\alpha} e^{-\xi/2} y(\xi) \tag{16.12}$$

results, once again, in a confluent hypergeometric equation

$$\xi y'' + \left(\frac{2\beta_1}{\alpha} + 1 - \xi \right) y' + \left(\frac{\gamma_1^2}{\alpha\gamma_2} - \frac{\beta_1}{\alpha} - \frac{1}{2} \right) y = 0 \tag{16.13}$$

with new values of the parameters:

$$c_1 = 2\frac{\beta_1}{\alpha} + 1, \quad a_1 = \frac{1}{2}c_1 - \frac{\gamma_1^2}{\alpha\gamma_2} = \frac{1}{2} + \frac{\beta_1}{\alpha} - \frac{\gamma_1^2}{\alpha\gamma_2} \tag{16.14}$$

(see Table 14 and the corresponding Mathematica file for more details).

An approximate quantization rule states:

$$a_1 = -v; \quad v = 0, 1, 2, \ldots. \tag{16.15}$$

Therefore,

$$-\beta_1^2 = -\left[\frac{\gamma_1^2}{\gamma_2} - \alpha\left(v + \frac{1}{2} \right) \right]^2, \tag{16.16}$$

and, in the energy formula, one has to replace γ by

$$\frac{\gamma_1^2}{\gamma_2} \approx \gamma \left[1 - l(l+1)\frac{C_1 + C_2}{2\gamma^2} \right], \quad \gamma \gg 1. \tag{16.17}$$

As a result, we arrive at the following vibration–rotation energy levels:

$$\begin{aligned} E = E_{vl} = \frac{\hbar^2}{2mr_0^2} \Bigg[& -\gamma^2 + 2\gamma\alpha\left(v + \frac{1}{2} \right) - \alpha^2 \left(v + \frac{1}{2} \right)^2 + l(l+1) \\ & - \frac{3(\alpha-1)}{\alpha\gamma}\left(v + \frac{1}{2} \right) l(l+1) - \frac{9(\alpha-1)^2}{4\alpha^4\gamma^2} l^2(l+1)^2 \Bigg]. \end{aligned} \tag{16.18}$$

This formula can be presented in the form

$$\begin{aligned} E_{vl} = & -D + \hbar\omega \left[\left(v + \frac{1}{2} \right) - \frac{\alpha}{2\gamma}\left(v + \frac{1}{2} \right)^2 \right] + \frac{\hbar^2 l(l+1)}{2mr_0^2} \\ & - \frac{3(\alpha-1)}{2\alpha^2 D}\hbar\omega\left(v + \frac{1}{2} \right)\frac{\hbar^2 l(l+1)}{2mr_0^2} - \frac{9(\alpha-1)^2}{4\alpha^2 D}\left(\frac{\hbar^2 l(l+1)}{2mr_0^2} \right)^2, \end{aligned} \tag{16.19}$$

TABLE 15. Details for the case of a Schrödinger equation with the modified Hulthén potential (17.1) in the spherically symmetric case $l = 0$, treated in §17. Substitution: $R(r) = u(\xi)$, $\xi = \exp(-r/a)$, $\alpha^2 = -(2mE/\hbar^2)\, a^2 > 0$, $\beta^2 = (2mV_0/\hbar^2)\, a^2 > 0$.

$\sigma(\xi)$	$\xi(1-\xi)$
$\widetilde{\sigma}(\xi)$	$-\left(\alpha^2 + b\beta^2\right)\xi^2 + \left(2\alpha^2 + \beta^2\right)\xi - \alpha^2$
$\widetilde{\tau}(\xi)$	$1-\xi$
k	$\beta^2 - 2\alpha\kappa, \qquad \kappa = \sqrt{(1/4) + (b-1)\beta^2}$
$\pi(\xi)$	$\alpha - (\alpha + \kappa + 1/2)\xi$
$\tau(\xi) = \widetilde{\tau} + 2\pi$	$2\alpha + 1 - 2(\alpha + \kappa + 1)\xi$
$\lambda = k + \pi'$	$\beta^2 - (2\alpha+1)(\kappa + 1/2) = (\gamma - \alpha - \kappa - 1/2)(\gamma + \alpha + \kappa + 1/2)$
$\varphi(\xi)$	$\xi^{\alpha}(1-\xi)^{\kappa+1/2}, \qquad \gamma = \sqrt{\alpha^2 + b\beta^2}$
$\rho(\xi)$	$\xi^{2\alpha}(1-\xi)^{2\kappa}$
E_n	$-V_0\left(\dfrac{b\beta^2 - (n+\kappa+1/2)^2}{2\beta(n+\kappa+1/2)}\right)^2, \quad n = 0, 1, 2, \ \ldots \ (n \leqslant n_{\max})$
$y_n(\xi)$	${}_2F_1(-n,\, 2\alpha + 2\kappa + n + 1;\, 2\alpha + 1;\, \xi)$, $\quad \alpha = \alpha_n = \dfrac{b\beta^2 - (n+\kappa+1/2)^2}{2(n+\kappa+1/2)} > 0$
C_n^2	$\dfrac{\Gamma(2\alpha+n+1)\Gamma(2\alpha+2\kappa+n+1)(2\alpha+2\kappa+2n+1)}{a(2\alpha)\Gamma^2(2\alpha)\Gamma(2\kappa+n+1)(2\kappa+2n+1)n!}$

where

$$\hbar\omega = \frac{\hbar^2\alpha\gamma}{mr_0^2} = \hbar\left(\frac{\alpha}{r_0}\sqrt{\frac{2D}{m}}\right). \tag{16.20}$$

The first three terms of this formula are exactly the same as those derived in the previous case; see (15.17). The fourth term can be interpreted as the molecule rotational energy at fixed distance r_0. The next term represents a coupling of the vibrations and rotations, which is negative because at higher vibrational quantum numbers the average nuclear distance increases beyond r_0 in consequence of the anharmonicity [**10**]. The last term can be thought of as a negative second-order correction to the rotation energy.

The corresponding radial wave functions are given in terms of Laguerre polynomials (Table 19):

$$R(r) = R_{vl}(r) = \sqrt{\frac{(2\beta_1)\ v!}{r_0\Gamma(2\beta_1/\alpha + v + 1)}}\ \xi^{\beta_1/\alpha}e^{-\xi/2}\ L_v^{2\beta_1/\alpha}(\xi). \tag{16.21}$$

Once again, we have used a normalization similar to (15.19). Further details on the rotational corrections of Morse formulas are discussed in [**10, 11**].

17. The modified Hulthén potential

Once again, we are looking for solutions of the stationary Schrödinger equation (5.1) in spherical coordinates as in (9.1), now with the central field potential,

$$U(r) = -V_0 \frac{e^{-r/a}\left(1 - be^{-r/a}\right)}{\left(1 - e^{-r/a}\right)^2} \qquad (0 \leqslant r < \infty) \tag{17.1}$$

when $b > 1$ and $l = 0$. (The special case $b = 1$ corresponds to the original Hulthén potential (14.1) above.) By using substitution (14.2) with the same positive parameters (14.3), one obtains a generalized equation of hypergeometric type,

$$\xi^2 \frac{d^2u}{d\xi^2} + \xi \frac{du}{d\xi} + \left[-\alpha^2 + \beta^2 \frac{\xi(1 - b\xi)}{(1 - \xi)^2}\right] u = 0, \tag{17.2}$$

in which the coefficient functions are

$$\sigma(\xi) = \xi(1 - \xi), \qquad \widetilde{\tau}(\xi) = 1 - \xi, \tag{17.3}$$

and

$$\widetilde{\sigma}(\xi) = -\left(\alpha^2 + b\beta^2\right)\xi^2 + \left(2\alpha^2 + \beta^2\right)\xi - \alpha^2. \tag{17.4}$$

For the modified Hulthén potential, let us first analyse the same boundary conditions (14.8) as before. We have

$$k = \beta^2 \mp 2\alpha\kappa, \qquad \pi(\xi) = -\frac{\xi}{2} \pm \left[(\alpha \pm \kappa)\xi - \alpha\right], \tag{17.5}$$

where

$$\kappa = \sqrt{\frac{1}{4} + (b - 1)\beta^2} > \frac{1}{2}, \tag{17.6}$$

for all four possible sign combinations. The solutions can be found by putting:

$$u = \varphi(\xi)y(\xi) = \xi^\alpha (1 - \xi)^{\kappa + 1/2} y(\xi), \tag{17.7}$$

which results in the hypergeometric equation (14.13):

$$\xi(1 - \xi)y'' + \left[2\alpha + 1 - 2(\alpha + \kappa + 1)\xi\right] y' - \left[(2\alpha + 1)\left(\kappa + \frac{1}{2}\right) - \beta^2\right] y = 0 \tag{17.8}$$

with

$$A = \alpha + \kappa + \frac{1}{2} + \gamma, \quad B = \alpha + \kappa + \frac{1}{2} - \gamma, \quad C = 2\alpha + 1, \quad \gamma = \sqrt{\alpha^2 + b\beta^2} \tag{17.9}$$

(details of the calculations are presented in Table 15 and in the complementary Mathematica file).

The required solution, which is bounded at $\xi = 0$, has the form

$$y = {}_2F_1\left(\begin{matrix} 1/2 + \kappa + \alpha + \gamma, \ 1/2 + \kappa + \alpha - \gamma \\ 2\alpha + 1 \end{matrix}; \xi\right) \tag{17.10}$$

up to a constant factor, and the first boundary condition is satisfied: $u(0) = 0$, when $\alpha > 0$.

In view of (14.16), one gets

(17.11)

$$u(1) = \lim_{\xi\to 1^-} (1-\xi)^{-(\kappa-1/2)} \left[(1-\xi)^{2\kappa} {}_2F_1\left(\begin{matrix} 1/2+\kappa+\alpha+\gamma, 1/2+\kappa+\alpha-\gamma \\ 2\alpha+1 \end{matrix} ; \xi \right) \right]$$

$$= \frac{\Gamma(2\alpha+1)\,\Gamma(2\kappa)}{\Gamma(1/2+\kappa+\alpha+\gamma)\,\Gamma(1/2+\kappa+\alpha-\gamma)} \times \lim_{\xi\to 1^-} (1-\xi)^{-(\kappa-1/2)} = \infty.$$

Therefore, one has to look for square-integrable solutions that are not bounded at the origin.

By terminating the hypergeometric series in (17.10), or by using the Nikiforov–Uvarov condition (2.7), we arrive at the following quantization rule:

$$\alpha+\kappa+\frac{1}{2}-\gamma = -n \quad (n=0,1,2,\dots \leqslant n_{\max}), \tag{17.12}$$

or

$$\alpha = \frac{b\beta^2-(n+\kappa+1/2)^2}{2(n+\kappa+1/2)} > 0. \tag{17.13}$$

The corresponding energy levels are given by

$$E_n = -V_0\left(\frac{b\beta^2-(n+\kappa+1/2)^2}{2\beta(n+\kappa+1/2)}\right)^2, \tag{17.14}$$

where

$$b\beta^2 - \left(\kappa+n+\frac{1}{2}\right)^2 = \beta^2-n^2-(2\kappa+1)\left(n+\frac{1}{2}\right) > 0. \tag{17.15}$$

Once again, there exists a minimum size of the potential hole before any energy eigenvalue at all can be obtained, viz., $b\beta^2=(\kappa+1/2)^2$. Equation $(n+\kappa+1/2)^2 \leqslant b\beta^2$ determines the finite number of eigenvalues, $n\leqslant n_{\max}$, in a potential hole of a given size.

The radial wave functions take the form

$$R_n(r) = C_n\, \xi^\alpha (1-\xi)^{\kappa+1/2}\, {}_2F_1\left(\begin{matrix} -n,\ 2\alpha+2\kappa+n+1 \\ 2\alpha+1 \end{matrix} ; \xi \right), \qquad \xi = e^{-r/a}, \tag{17.16}$$

where the hypergeometric series terminates and C_n is a constant to be determined. Once again, this result can also be obtained with the help of the Rodrigues-type formula (2.8):

(17.17)

$$\left[\xi^{2\alpha+n}(1-\xi)^{2\kappa+n}\right]^{(n)} = \frac{\Gamma(2\alpha+n+1)}{\Gamma(2\alpha+1)}\, \xi^{2\alpha}(1-\xi)^{2\kappa} {}_2F_1\left(\begin{matrix} -n,\ 2\alpha+2\kappa+n+1 \\ 2\alpha+1 \end{matrix} ; \xi \right).$$

(It can be thought of as a special case of (15.5.9) in [**28**].)

The normalization condition (13.8) for the radial wave function becomes

$$aC_n^2 \int_0^1 \xi^{2\alpha-1}(1-\xi)^{2\kappa+1} y_n^2(\xi)\, d\xi = 1, \tag{17.18}$$

where

$$\xi^{2\alpha}(1-\xi)^{2\kappa} y_n(\xi) = \frac{\Gamma(2\alpha+1)}{\Gamma(2\alpha+n+1)} \left[\xi^{2\alpha+n}(1-\xi)^{2\kappa+n}\right]^{(n)}. \tag{17.19}$$

Integrating by parts n times, one gets

(17.20)

$$\begin{aligned}
&\int_0^1 (\xi^{-1}-1)\,{}_2F_1\left(\begin{matrix}-n,\ 2\alpha+2\kappa+n+1\\ 2\alpha+1\end{matrix};\xi\right)\left[\xi^{2\alpha+n}(1-\xi)^{2\kappa+n}\right]^{(n)}\,\mathrm{d}\xi\\
&=\left((\xi^{-1}-1)\,{}_2F_1\left(\begin{matrix}-n,\ 2\alpha+2\kappa+n+1\\ 2\alpha+1\end{matrix};\xi\right)\left[\xi^{2\alpha+n}(1-\xi)^{2\kappa+n}\right]^{(n-1)}\right)\Bigg|_{\xi=0}^{1}\\
&\qquad-\int_0^1\left[(\xi^{-1}-1)\,{}_2F_1\left(\begin{matrix}-n,\ 2\alpha+2\kappa+n+1\\ 2\alpha+1\end{matrix};\xi\right)\right]'\left[\xi^{2\alpha+n}(1-\xi)^{2\kappa+n}\right]^{(n-1)}\,\mathrm{d}\xi\\
&=\cdots=\\
&(-1)^{k-1}\left(\left[(\xi^{-1}-1)\,{}_2F_1\left(\begin{matrix}-n,\ 2\alpha+2\kappa+n+1\\ 2\alpha+1\end{matrix};\xi\right)\right]^{(k-1)}\left[\xi^{2\alpha+n}(1-\xi)^{2\kappa+n}\right]^{(n-k)}\right)\Bigg|_{\xi=0}^{1}\\
&+(-1)^k\int_0^1\left[(\xi^{-1}-1)\,{}_2F_1\left(\begin{matrix}-n,\ 2\alpha+2\kappa+n+1\\ 2\alpha+1\end{matrix};\xi\right)\right]^{(k)}\left[\xi^{2\alpha+n}(1-\xi)^{2\kappa+n}\right]^{(n-k)}\,\mathrm{d}\xi\\
&=\cdots=(-1)^n\int_0^1\left[(\xi^{-1}-1)\,{}_2F_1\left(\begin{matrix}-n,\ 2\alpha+2\kappa+n+1\\ 2\alpha+1\end{matrix};\xi\right)\right]^{(n)}\xi^{2\alpha+n}(1-\xi)^{2\kappa+n}\,\mathrm{d}\xi.
\end{aligned}$$

By the power series expansion,

$$\begin{aligned}
&\left[(\xi^{-1}-1)\,{}_2F_1\left(\begin{matrix}-n,\ 2\alpha+2\kappa+n+1\\ 2\alpha+1\end{matrix};\xi\right)\right]^{(n)}\\
&\qquad=\left(\frac{1}{\xi}\right)^{(n)}-\frac{(-n)_n(2\alpha+2\kappa+n+1)_n}{(n!)(2\alpha+1)_n}(\xi^n)^{(n)}\\
&\qquad=(-1)^n\frac{(n!)}{\xi^{n+1}}-\frac{(-n)_n(2\alpha+2\kappa+n+1)_n}{(n!)(2\alpha+1)_n}(n!),
\end{aligned}$$

and the integral evaluation can be completed with the aid of the Euler beta integrals (B.2):

$$\begin{aligned}
\int_0^1 \xi^{2\alpha-1}(1-\xi)^{(2\kappa+n+1)-1}\,\mathrm{d}\xi&=\frac{\Gamma(2\alpha)\Gamma(2\kappa+n+1)}{\Gamma(2\alpha+2\kappa+n+1)},\\
\int_0^1 \xi^{2\alpha+n}(1-\xi)^{2\kappa+n}\,\mathrm{d}\xi&=\frac{\Gamma(2\alpha+n+1)\Gamma(2\kappa+n+1)}{\Gamma(2\alpha+2\kappa+2n+2)}.
\end{aligned}$$

The final result is given by

$$C_n^2=\frac{\Gamma(2\alpha+n+1)\Gamma(2\alpha+2\kappa+n+1)(2\alpha+2\kappa+2n+1)}{a(2\alpha)\Gamma^2(2\alpha)\Gamma(2\kappa+n+1)(2\kappa+2n+1)n!}. \tag{17.21}$$

Our modification of the Hulthén potential can be used for studying the vibrations of diatomic molecules, when m is the reduced mass of the two atoms. Therefore, it is informative to compare the classical Morse potential and the modified Hulthén one. Suppose that at the common point of the potential minimum

TABLE 16. Comparison of the parameters for the Morse potential and the modified Hulthén potential: $E(\text{eV}) = E(\text{cm}^{-1}) \times 1.2398 \times 10^{-4}$ according to [**10**].

Molecule	$\dfrac{\hbar^2}{2mr_0^2}$ (cm^{-1})	D (cm^{-1})	α	b	V_0 (cm^{-1})
H_2	60.8296	38292	1.440	1.5904	67394
HCl	10.5930	37244	2.380	4.51744	524010
I_2	0.0374	12550	4.954	68.848	1984 90

$r = r_{\min}$ the following conditions hold:

$$\begin{aligned} U_{\text{Morse}}(r_{\min}) &= U_{\text{ModHulthén}}(r_{\min}), \\ \frac{\mathrm{d}}{\mathrm{d}r}U_{\text{Morse}}(r_{\min}) &= \frac{\mathrm{d}}{\mathrm{d}r}U_{\text{ModHulthén}}(r_{\min}) = 0, \\ \frac{\mathrm{d}^2}{\mathrm{d}r^2}U_{\text{Morse}}(r_{\min}) &= \frac{\mathrm{d}^2}{\mathrm{d}r^2}U_{\text{ModHulthén}}(r_{\min}). \end{aligned} \tag{17.22}$$

Then $r_{\min} = r_0$, $\exp(r_0/a) = 2b - 1$, and

$$D = \frac{V_0}{4(b-1)}, \qquad 2D\frac{\alpha^2}{r_0^2} = \frac{(2b-1)^2 V_0}{8a^2(b-1)^3} \tag{17.23}$$

(see the Mathematica file for more details). The Morse potential (15.1) can be rewritten in the equivalent form

(17.24)

$$\begin{aligned} U_{\text{Morse}}(r) &= \frac{V_0}{4(b-1)}(2b-1)^{\frac{2b-1}{2(b-1)}} \\ &\times \left[(2b-1)^{\dfrac{2b-1}{2(b-1)}} \exp\left(-\frac{(2b-1)r}{(b-1)a}\right) - 2\exp\left(-\frac{(2b-1)r}{2(b-1)a}\right)\right], \end{aligned}$$

where the parameters of the modified Hulthén potential (17.1) have been utilized.

A graphical example, comparing both potentials, when $V_0 = a = 1$, $b = 2$, and $r_{\min} = \ln 3 \simeq 1.09861$, is presented in Figure 2 and is further discussed in the Mathematica notebook for the reader's convenience, as well as potentials for molecules H_2, HCl, and I_2, with a completion of the data from [**10**] in Table 16 (see also Figure 3).

In order to find the parameters of the modified Hulthén potential in terms of the Morse ones, we have solved numerically the following equation:

$$\alpha = \varphi(b) := \frac{(2b-1)\ln(2b-1)}{2(b-1)}, \qquad b \geqslant 1 \tag{17.25}$$

for the monotone function $\varphi(b)$. Then $V_0 = 4(b-1)D$ and $a = r_0/\ln(2b-1)$. (More details are provided in the Mathematica file.)

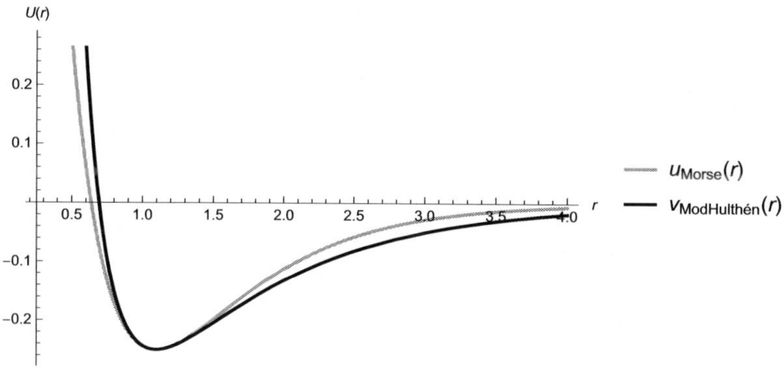

FIGURE 2. An example, comparing the Morse potential and the modified Hulthén potential, when $V_0 = a = 1$, $b = 2$, and $r_{\min} = \ln 3 \simeq 1.09861$.

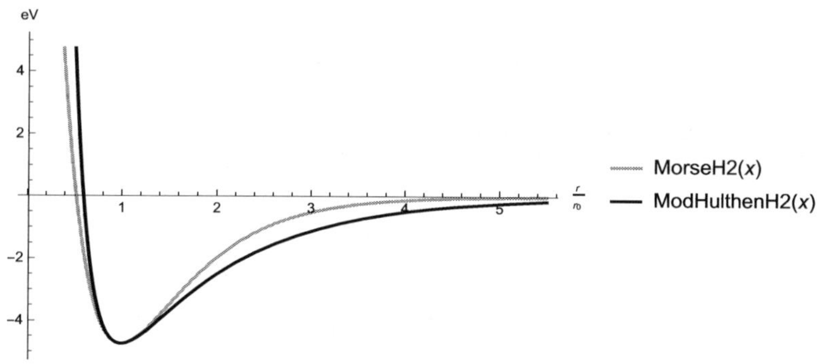

FIGURE 3. The Morse potential [**10**] and the modified Hulthén potential, for the molecule H_2.

REMARK. The so-called generalized Morse potential is usually introduced as

$$U(r) = D\left(1 - \frac{e^{\alpha r_0} - 1}{e^{\alpha r} - 1}\right)^2 \tag{17.26}$$

(D, α, r_0 are parameters), with the following asymptotic:

$$U(\infty) = \lim_{r\to\infty} U(r) = D > 0 \tag{17.27}$$

(see [**9, 31**], and the references therein). The difference

$$U(r) - U(\infty) = -2D\left(e^{\alpha r_0} - 1\right) \frac{1 - e^{-\alpha r}\left(1 + e^{\alpha r_0}\right)/2}{\left(1 - e^{-\alpha r}\right)^2} e^{-\alpha r} \tag{17.28}$$

has the form of our modified Hulthén potential (17.1) with the parameters given by

$$V_0 = 2D\left(e^{\alpha r_0} - 1\right), \quad b = \frac{1}{2}\left(1 + e^{\alpha r_0}\right), \quad a = \frac{1}{\alpha}. \tag{17.29}$$

Therefore, this case has been studied here as well (see also Section 19 for an independent computer algebra approach).

TABLE 17. Details for the case of a Schrödinger equation with the rotation correction of the modified Hulthén potential (18.5), its coefficients C_0, C_1, and C_2 being given in (18.4), which is treated in §18.

$\sigma(\xi)$	$\xi(1-\xi)$
$\widetilde{\sigma}(\xi)$	$-\left(\alpha_1^2+b_1\beta_1^2\right)\xi^2+\left(2\alpha_1^2+\beta_1^2\right)\xi-\alpha_1^2$
$\widetilde{\tau}(\xi)$	$1-\xi$
k	$\beta_1^2-2\alpha_1\kappa_1, \quad \kappa_1=\sqrt{(1/4)+(b_1-1)\beta_1^2}$
$\pi(\xi)$	$\alpha_1-(\alpha_1+\kappa_1+1/2)\xi$
$\tau(\xi)=\widetilde{\tau}+2\pi$	$(2\alpha_1+1)-2(\alpha_1+\kappa_1+1)\xi$
$\lambda=k+\pi'$	$\beta_1^2-(2\alpha_1+1)(\kappa_1+1/2)$
$\varphi(\xi)$	$\xi^{\alpha_1}(1-\xi)^{\kappa_1+1/2}$
$\rho(\xi)$	$\xi^{2\alpha_1}(1-\xi)^{2\kappa_1}$

18. The rotation correction of the modified Hulthén potential

In view of

$$\frac{\mathrm{d}U(r)}{\mathrm{d}r}=V_0\, e^{r/a}\frac{1-2b+e^{r/a}}{a\left(e^{r/a}-1\right)^3}=0, \tag{18.1}$$

one sees that the minimum of the modified Hulthén potential (17.1) occurs when

$$e^{r_{\min}/a}=2b-1>1. \tag{18.2}$$

By letting $r=a(1+x)x_0$ in the neighborhood of this minimum $r\approx r_{\min}$ (or $x\approx 0$), with the aid of Mathematica, we derive the expansion

$$\begin{aligned}&\frac{1}{(1+x)^2}-C_0-e^{-(1+x)x_0}\frac{C_1+C_2e^{-(1+x)x_0}}{\left(1-e^{-(1+x)x_0}\right)^2}\\&=\left(-4+\frac{3bx_0}{b-1}-\frac{1-2b+4b^2}{6(b-1)^2}x_0^2\right)x^3+\mathrm{O}(x^4),\qquad x\to 0,\end{aligned} \tag{18.3}$$

where

$$\begin{aligned}C_0&=1+4(b-1)\frac{3(b-1)-(3b-1)x_0}{(2b-1)^2x_0^2},\\C_1&=8(b-1)^2\frac{6(1-b)+(4b-1)x_0}{(2b-1)^2x_0^2},\\C_2&=8(b-1)^2\frac{6b(b-1)+(1-2b(b+1))x_0}{(2b-1)^2x_0^2},\end{aligned} \tag{18.4}$$

and $e^{x_0}=2b-1>1$ (see the complementary Mathematica file).

Therefore,

$$\frac{l(l+1)}{r^2}\approx\frac{l(l+1)}{r_{\min}^2}\left[C_0+e^{-r/a}\frac{C_1+C_2e^{-r/a}}{\left(1-e^{-r/a}\right)^2}\right]\qquad(r_{\min}=ax_0), \tag{18.5}$$

and the equation (17.2) for the modified Hulthén potential holds, once again, but with the following modified values of the parameters:

$$(18.6)\qquad \begin{aligned} \alpha^2 \to \alpha_1^2 &= \alpha^2 + \frac{l(l+1)}{x_0^2}C_0, & \beta^2 \to \beta_1^2 &= \beta^2 - \frac{l(l+1)}{x_0^2}C_1, \\ b \to b_1 &= \frac{\beta_2^2}{\beta_1^2}, & \beta_2^2 &= b\beta^2 + \frac{l(l+1)}{x_0^2}C_2, \end{aligned}$$

namely,

$$(18.7)\qquad \xi^2\frac{\mathrm{d}^2u}{\mathrm{d}\xi^2} + \xi\frac{\mathrm{d}u}{\mathrm{d}\xi} + \left[-\alpha_1^2 + \beta_1^2\,\frac{\xi\,(1-b_1\xi)}{(1-\xi)^2}\right]u = 0.$$

As a result, we have arrived at a generalized equation of hypergeometric type (see Table 17 for a summary of details), and our analysis in the previous section is valid, say, up to a proper change of parameters.

For example, the vibrational and rotational energy levels are given by

$$(18.8)\qquad \begin{aligned} E_{v,l} &= -V_0\left(\frac{\beta_2^2-(\kappa_1+v+1/2)^2}{2\beta\,(\kappa_1+v+1/2)}\right)^2 \\ &\quad + \frac{\hbar^2 l(l+1)}{2mr_{\min}^2}\left[1+4(b-1)\frac{3(b-1)-(3b-1)x_0}{(2b-1)^2x_0^2}\right] \end{aligned}$$

($v = 0, 1, 2, \ldots \leqslant v_{\max}$). Here,

$$(18.9)\qquad \beta_2^2 - \left(\kappa_1 + v + \frac{1}{2}\right)^2 = \beta_1^2 - v^2 - (2\kappa_1+1)\left(v+\frac{1}{2}\right) > 0.$$

The corresponding normalized wave functions are the same as in (17.16) and (17.21) but with the new values of parameters given in (18.6) (we leave further details to the reader).

Our study of the rotation correction of the modified Hulthén potential will be continued elsewhere (see also the Mathematica file).

19. The generalized Morse potential

The potential of the form

$$(19.1)\qquad U(r) = D\left(1-\frac{\gamma}{e^{ar}-1}\right)^2, \quad \gamma = e^{ar_0}-1 \qquad (0 \leqslant r < \infty)$$

is reduced by (17.28) to the modified Hulthén potential (17.1), up to the proper change of parameters (17.29). This is why we can use solutions from Section 17.

On second thought, the standard change of variables [**9**]:

$$(19.2)\qquad R(r) = \sqrt{a}u(\eta), \qquad \eta = (e^{ar}-1)^{-1}$$

results in a generalized equation of hypergeometric type with the coefficient functions

$$(19.3)\qquad \sigma(\eta) = \eta(1+\eta), \qquad \widetilde{\sigma}(\eta) = \epsilon - \kappa(1-\gamma\eta)^2, \qquad \widetilde{\tau}(\eta) = 2\eta+1,$$

where

$$(19.4)\qquad \kappa = \frac{2mD}{a^2\hbar^2}, \quad \epsilon = \frac{2mE}{a^2\hbar^2}; \qquad \alpha^2 = \kappa - \epsilon > 0, \quad \beta^2 = \kappa(\gamma+1)^2 - \epsilon > 0$$

TABLE 18. Details for the case of a Schrödinger equation with the generalized Morse potential (19.1), treated in §19. Parameters are given in (19.4).

$\sigma(\eta)$	$\eta(1+\eta)$
$\widetilde{\sigma}(\eta)$	$\epsilon-\kappa\left(1-\gamma\eta\right)^2, \qquad \kappa\gamma^2=\delta(\delta-1)$
$\widetilde{\tau}(\eta)$	$2\eta+1$
k	$2\left(\alpha^2-\alpha\beta+\kappa\gamma\right)=(\alpha-\beta)^2-\delta(\delta-1)$
$\pi(\eta)$	$\alpha+(\alpha-\beta)\eta, \quad \alpha=\sqrt{\kappa-\epsilon}\,,\ \beta=\sqrt{\kappa(\gamma+1)^2-\epsilon}$
$\tau(\eta)=\widetilde{\tau}+2\pi$	$2\alpha+1+2(\alpha-\beta+1)\eta$
$\lambda=k+\pi'$	$\alpha-\beta+2\alpha^2-2\alpha\beta+2\kappa\gamma=(\alpha-\beta+\delta)(\alpha-\beta-\delta+1)$
$\varphi(\eta)$	$\eta^{\alpha}(1+\eta)^{-\beta}$
$\rho(\eta)$	$\eta^{2\alpha}(1+\eta)^{-2\beta}$

(see Table 18 and the Mathematica file). The substitution

$$u=\varphi(\eta)y(\eta)=\frac{\eta^{\alpha}}{(1+\eta)^{\beta}}y(\eta) \tag{19.5}$$

results in the hypergeometric equation of Gauss (14.13) in the variable $-\eta$ with the following values of parameters:

$$A=\alpha-\beta+\delta,\quad B=\alpha-\beta-\delta+1,\quad C=2\alpha+1;\quad \delta=\frac{1}{2}+\sqrt{\frac{1}{4}+\kappa\gamma^2}. \tag{19.6}$$

Bound states correspond to polynomial solutions, for which $A=\alpha-\beta+\delta=-n$ with n some nonnegative integer. Thus the parameters α and β are n-dependent, once again, and satisfy

$$\beta_n-\alpha_n=n+\delta,\qquad \beta_n^2-\alpha_n^2=\kappa\gamma(\gamma+2). \tag{19.7}$$

One finds that

$$\alpha_n=\frac{1}{2}\left(\frac{\kappa\gamma(\gamma+2)}{n+\delta}-n-\delta\right),\quad \beta_n=\frac{1}{2}\left(\frac{\kappa\gamma(\gamma+2)}{n+\delta}+n+\delta\right), \tag{19.8}$$

and the energy levels have the form [**9**]

$$E_n=D-\frac{a^2\hbar^2}{8m}\left(n+\delta-\frac{\kappa\gamma(\gamma+2)}{n+\delta}\right)^2. \tag{19.9}$$

The corresponding normalized wave functions are given by

$$\begin{aligned} R_n(r)&=\sqrt{a}C_ne^{-\beta_nar}\left(e^{ar}-1\right)^{\beta_n-\alpha_n}\ {}_2F_1\left(\begin{matrix}-n,\ 1-n-2\delta\\ 2\alpha_n+1\end{matrix};\frac{1}{1-e^{ar}}\right)\\ &=\sqrt{a}C_ne^{-\alpha_nar}\left(1-e^{-ar}\right)^{\delta}\ {}_2F_1\left(\begin{matrix}-n,\ 2\alpha_n+2\delta+n\\ 2\alpha_n+1\end{matrix};e^{-ar}\right)\end{aligned} \tag{19.10}$$

[we have used (15.8.1) of [**28**]) in order to obtain the standard form (17.16)], with the normalization coefficients found in [**9**], as follows:

$$C_n^2 = \frac{a(\alpha_n + n + \delta)\Gamma(2\alpha_n + n + 1)\Gamma(2\alpha_n + n + 2\delta)}{n!(n+\delta)\Gamma(2\alpha_n)\Gamma(2\alpha_n + 1)\Gamma(n + 2\delta)}, \tag{19.11}$$

due to the normalization condition

$$\int_0^\infty R_n^2(r)\, \mathrm{d}r = 1. \tag{19.12}$$

There is only a finite set of energy levels:

$$n = 0, 1, 2, \ldots, n_{\max} \leqslant \sqrt{\kappa\gamma(\gamma+2)} - \delta. \tag{19.13}$$

One can use our results from Section 17 in order to verify all these formulas, originally presented in [**9**] (see also [**1**, **31**], and the references therein). We leave further details to the reader.

In a similar fashion, one can consider the Wood–Saxon potential and the motion of an electron in a magnetic field (see [**1**], [**9**], [**10**], [**15**] [**23**], and [**27**]).

Acknowledgments

We are grateful to Dr. Steven Baer, Dr. Kamal Barley, Dr. Sergey Kryuchkov, Dr. Eugene Stepanov, Dr. José Vega-Guzmán, and Dr. Alexei Zhedanov for valuable discussions and help.

A. Data for the classical orthogonal polynomials

The basic information about classical orthogonal polynomials, namely the Jacobi $P_n^{(\alpha,\beta)}(x)$, Laguerre $L_n^\alpha(x)$, and Hermite $H_n(x)$ polynomials, is presented for the reader's convenience in Table 19. It contains the coefficients of the differential equation (2.3), the intervals of orthogonality (a, b), the weight functions $\rho(x)$, and constants B_n in the Rodrigues-type formula (2.8), the leading terms

$$y_n(x) = a_n x^n + b_n x^{n-1} + \ldots \tag{A.1}$$

for these polynomials, their squared norms

$$d_n^2 = \int_a^b y_n^2(x)\rho(x)\, \mathrm{d}x, \tag{A.2}$$

and the coefficients of the three-term recurrence relation:

$$x\, y_n(x) = \alpha_n\, y_{n+1}(x) + \beta_n\, y_n(x) + \gamma_n\, y_{n-1}(x), \tag{A.3}$$

where

$$\alpha_n = \frac{a_n}{a_{n+1}}, \quad \beta_n = \frac{b_n}{a_n} - \frac{b_{n+1}}{a_{n+1}}, \quad \gamma_n = \alpha_{n-1}\, \frac{d_n^2}{d_{n-1}^2}. \tag{A.4}$$

(More details can be found in [**2**], [**26**], [**27**], [**28**], and [**42**].)

TABLE 19. Data for the Jacobi $P_n^{(\alpha,\beta)}(x)$, Laguerre $L_n^\alpha(x)$, and Hermite $H_n(x)$ polynomials.

$y_n(x)$	$P_n^{(\alpha,\beta)}(x)\,(\alpha>-1,\beta>-1)$	$L_n^\alpha(x)\,(\alpha>-1)$	$H_n(x)$
(a,b)	$(-1,1)$	$(0,\infty)$	$(-\infty,\infty)$
$\rho(x)$	$(1-x)^\alpha(1+x)^\beta$	$x^\alpha\,e^{-x}$	e^{-x^2}
$\sigma(x)$	$1-x^2$	x	1
$\tau(x)$	$\beta-\alpha-(\alpha+\beta+2)\,x$	$1+\alpha-x$	$-2x$
λ_n	$n(\alpha+\beta+n+1)$	n	$2n$
B_n	$\dfrac{(-1)^n}{2^n n!}$	$\dfrac{1}{n!}$	$(-1)^n$
a_n	$\dfrac{\Gamma(\alpha+\beta+2n+1)}{2^n n!\Gamma(\alpha+\beta+n+1)}$	$\dfrac{(-1)^n}{n!}$	2^n
b_n	$\dfrac{(\alpha-\beta)\Gamma(\alpha+\beta+2n)}{2^n(n-1)!\Gamma(\alpha+\beta+n+1)}$	$(-1)^{n-1}\dfrac{\alpha+n}{(n-1)!}$	0
d^2	$\dfrac{2^{\alpha+\beta+1}\Gamma(\alpha+n+1)\Gamma(\beta+n+1)}{n!(\alpha+\beta+2n+1)\Gamma(\alpha+\beta+n+1)}$	$\dfrac{\Gamma(\alpha+n+1)}{n!}$	$2^n n!\sqrt{\pi}$
α_n	$\dfrac{2(n+1)(\alpha+\beta+n+1)}{(\alpha+\beta+2n+1)(\alpha+\beta+2n+2)}$	$-(n+1)$	$\dfrac{1}{2}$
β_n	$\dfrac{\beta^2-\alpha^2}{(\alpha+\beta+2n)(\alpha+\beta+2n+2)}$	$\alpha+2n+1$	0
γ_n	$\dfrac{2(\alpha+n)(\beta+n)}{(\alpha+\beta+2n)(\alpha+\beta+2n+1)}$	$-(\alpha+n)$	n

B. An integral evaluation

The useful integral

$$\begin{aligned} J_{nms}^{\alpha\beta} &= \int_0^\infty e^{-x}x^{\alpha+s}\,L_n^\alpha(x)L_m^\beta(x)\,dx \\ &= (-1)^{n-m}\frac{\Gamma(\alpha+s+1)\,\Gamma(\beta+m+1)\,\Gamma(s+1)}{m!\,(n-m)!\,\Gamma(\beta+1)\,\Gamma(s-n+m+1)} \\ &\quad\times {}_3F_2\left(\begin{matrix}-m,\ s+1,\ \beta-\alpha-s\\ \beta+1,\quad n-m+1\end{matrix};1\right), \qquad n\geqslant m, \end{aligned} \tag{B.1}$$

where the parameter s may take some integer value and ${}_3F_2(1)$ is the generalized hypergeometric series [**2**,**28**], has been evaluated in [**41**] and [**42**] (see also [**13**,**35**]). Special cases have been used above for the normalization of wave functions; see (7.13) and (15.20).

We use also throughout the manuscript the familiar Euler beta and gamma integrals:

$$B(\alpha,\ \beta)=\int_0^1 t^{\alpha-1}(1-t)^{\beta-1}\,\mathrm{d}t=\frac{\Gamma(\alpha)\Gamma(\beta)}{\Gamma(\alpha+\beta)}, \tag{B.2}$$

when $\Re(\alpha)>0$ and $\Re(\alpha)>0$,

$$\Gamma(\alpha)=\int_0^\infty t^{\alpha-1}e^{-t}\,\mathrm{d}t, \qquad \Re(\alpha)>0. \tag{B.3}$$

(See [**2**], [**27**], and [**28**] for more details.)

C. The Mathematica file

We have discussed basic potentials of nonrelativistic and relativistic quantum mechanics that can be integrated in the Nikiforov–Uvarov paradigm, with the aid of the Mathematica computer algebra system.

(The corresponding Mathematica notebook is available from the authors, by request. It is also posted on the Wolfram community `https://community.wolfram.com/groups/-/m/t/2897057`, and is featured in the editorial columns `https://community.wolfram.com/content?curTag=staff+picks` .)

In Section 2 the general formulas are derived. In the notebook, they are stored in global variables that are used in each of Sections 3–19. For this purpose, one allows Mathematica to evaluate all initialization cells. After that, one can run each case independently of the others. The results, which include most integrable cases that are available in the literature, are presented in the Tables 1–18.

The Mathematica package `HolonomicFunctions` can be downloaded from the webpage of the third-named author: `http://www.koutschan.de/`.

References

[1] A. B. Al-Othman and A. S. Sandouqa, *Comparison study of bound states for diatomic molecules using Kratzer, Morse, and modified Morse potentials,* Physica Scripta, **97**(3), 035401, 2022. `https://doi.org/10.1088/1402-4896/ac514c`.

[2] G. E. Andrews, R. Askey, and R. Roy, *Special functions*, Encyclopedia of Mathematics and its Applications, vol. 71, Cambridge University Press, Cambridge, 1999, DOI 10.1017/CBO9781107325937. MR1688958

[3] K. Barley, J. Vega-Guzmán, A. Ruffing, and S. K. Suslov, *Discovery of the relativistic Schrödinger equation,* Physics–Uspekhi, **69**(1), 90–103, 2022 [in English]; **192**(1), 100–114, 2022 [in Russian]; DOI 10.3367/UFNr.2021.06.039000.

[4] H. A. Bethe and E. Salpeter, *Quantum Mechanics of One- and Two-Electron Atoms*, Dover Publications, Mineola, New York, 2008.

[5] C. Berkdemir, A. Berkdemir, and J. Han, *Bound state solutions of the Schrödinger equation for modified Kratzer's potential*, Chemical Physics Letters **417**(4–6), 326–329, 2006. `https://www.sciencedirect.com/science/article/abs/pii/S0009261405015812`.

[6] D. I. Blokhintsev, *Quantum mechanics*, D. Reidel Publishing Co., Dordrecht, 1964. MR169559

[7] A. S. Davydov, *Quantum Mechanics*, Pergamon Press, Oxford and New York, 1965.

[8] C. G. Darwin, *The wave equations of the electron*, Proceedings of the Royal Society A: Mathematical, Physical and Engineering Sciences **118**(780), 654–680, 1928. DOI 10.1098/rspa.1928.0076

[9] A. Del Sol Mesa, C. Quesne, and Yu. F. Smirnov, *Generalized Morse potential: symmetry and satellite potentials*, J. Phys. A **31** (1998), no. 1, 321–335, DOI 10.1088/0305-4470/31/1/028. MR1620265

[10] S. Flügge, *Practical quantum mechanics*, Reprint of the 1994 English edition, Classics in Mathematics, Springer-Verlag, Berlin, 1999. Translated from the 1947 German original. MR1746199

[11] S. Flügge, P. Walger, and A. Weiguny, *A generalization of the Morse potential*, Journal of Molecular Spectroscopy **23**(3), 243–257, 1967. DOI 10.1016/S0022-2852(67)80013-4

[12] V. A. Fock, *Fundamentals of Quantum Mechanics*, Moscow: Mir Publishers, 1978. `https://mirtitles.org/2013/01/01/fock-fundamentals-of-quantum-mechanics/`.

[13] E. Fues, *Das Eigenschwingungsspektrum zweiatomiger Moleküle in der Undulationsmechanik*, Annalen der Physik **385**(12), 367–396, 1926. DOI 10.1002/andp.19263851204 [in German].

[14] I. I. Gol'dman and V. D. Krivchenkov, *Problems in Quantum Mechanics*, Dover Publications, Inc., New York, 1993.

[15] B. Gönül and K. Köksal, *Solutions for a generalized Woods-Saxon potential*, Phys. Scr. **76** (2007), no. 5, 565–570, DOI 10.1088/0031-8949/76/5/026. MR2393460

[16] W. Gordon, *Die Energieniveaus des Wasserstoffatoms nach der Diracschen Quantentheorie des Elektrons*, Zeitschrift für Physik **48**(1), 11–14, 1928. DOI 10.1007/BF01351570 [in German]

[17] W. Greiner, *Relativistic quantum mechanics*, 2nd ed., Springer-Verlag, Berlin, 1997. Wave equations; Translated from the second German (1987) edition; With a foreword by D. A. Bromley, DOI 10.1007/978-3-662-03425-5. MR1474235

[18] M. Karplus and R. N. Porter, *Atoms & Molecules: An Introduction for Students of Physical Chemistry*, The Benjamin/Cummings Company, Menlo Park, California, 1970.

[19] C. Koutschan, E. Suazo, and S. K. Suslov, *Fundamental laser modes in paraxial optics: from computer algebra and simulations to experimental observation*. Applied Physics B **121**(3), 315–336, 2015. DOI 10.1007/s00340-015-6231-9

[20] C. Koutschan and D. Zeilberger, *The 1958 Pekeris-Accad-WEIZAC ground-breaking collaboration that computed ground states of two-electron atoms (and its 2010 redux)*, Math. Intelligencer **33** (2011), no. 2, 52–57, DOI 10.1007/s00283-010-9192-1. MR2813264

[21] S. I. Kryuchkov, S. K. Suslov, and J. M. Vega-Guzmán, *The minimum-uncertainty squeezed states for atoms and photons in a cavity*. Journal of Physics B: Atomic, Molecular and Optical Physics **46**(10), 104007 (2013). DOI 10.1088/0953-4075/46/10/104007

[22] R. E. Langer, *On the connection formulas and the solutions of the wave equation*, *Physical Review* **51** no. 8 (1937), 669–676. `https://journals.aps.org/pr/abstract/10.1103/PhysRev.51.669`

[23] L. D. Landau and E. M. Lifshitz, *Quantum Mechanics: Non-Relativistic Theory*, 3rd ed., Butterworth–Heinemann, Oxford, 1998.

[24] R. M. López, S. K. Suslov, and J. M. Vega-Guzmán, *On a hidden symmetry of quantum harmonic oscillators*, J. Difference Equ. Appl. **19** (2013), no. 4, 543–554, DOI 10.1080/10236198.2012.658384. MR3040814

[25] P. M. Morse, *Diatomic molecules according to the wave mechanics. II*. Physical Review **34**(1), 57–64, 1929. DOI 10.1103/PhysRev.34.57

[26] A. F. Nikiforov, S. K. Suslov, and V. B. Uvarov, *Classical orthogonal polynomials of a discrete variable*, Springer Series in Computational Physics, Springer-Verlag, Berlin, 1991, DOI 10.1007/978-3-642-74748-9. MR1149380

[27] A. F. Nikiforov and V. B. Uvarov, *Special functions of mathematical physics: A unified introduction with applications*, Birkhäuser Verlag, Basel, 1988, DOI 10.1007/978-1-4757-1595-8. MR922041

[28] *NIST Handbook of Mathematical Functions*, (F. W. J. Olver and D. W. Lozier, eds.), Cambridge University Press, New York, 2010. `https://dlmf.nist.gov/`.

[29] C. L. Pekeris, *Ground state of two-electron atoms*, Physical Review (2) **112** (1958), 1649–1958. MR101783

[30] G. Pöschl and E. Teller, *Bemerkungen zur Quantenmechanik des anharmonischen Oszillators*. Zeitschrift für Physik **83** March issue, 143–151 (1933), DOI 10.1007/BF01331132.

[31] Z. Rong, H. G. Kjaergaard, and M. L. Sage, *Comparison of the Morse and Deng–Fan potentials for X–H bonds in small molecules*, Molecular Physics **101**(14), 2285–2294, 2003. DOI 10.1080/0026897031000137706

[32] N. Rosen and P. M. Morse, *On the vibrations of polyatomic molecules*, Physical Review **42** (2), 210–217 (1932), DOI 10.1103/PhysRev.42.210.

[33] E. Schrödinger, *Quantisation as a problem of proper values (Part I)*, in *Collected Papers on Wave Mechanics*, New York, Providence, Rhode Island: AMS Chelsea Publishing, 2010, (Original: Annalen der Physik (4), vol. 79(6), pp. 489–527, 1926 [in German]), pp. 1–12. DOI 10.1002/andp.19263840404

[34] E. Schrödinger, *Quantisation as a problem of proper values (Part II)*, in *Collected Papers on Wave Mechanics*, New York, Providence, Rhode Island: AMS Chelsea Publishing, 2010, (Original: Annalen der Physik (4), vol. 79(6), pp. 489–527, 1926 [in German]), pp. 13–40. DOI 10.1002/andp.19263840602

[35] E. Schrödinger, *Quantisation as a problem of proper values (Part III)*, in *Collected Papers on Wave Mechanics*, New York, Providence, Rhode Island: AMS Chelsea Publishing, 2010, (Original: Annalen der Physik (4), vol. 386 # 18, pp. 109–139, 1926 [in German]), pp. 62–101. DOI 10.1002/andp.19263851302.

[36] E. Schrödinger, *The continuous transition from micro-to macro-mechanics*, in *Collected Papers on Wave Mechanics*, **28**, New York, Providence, Rhode Island: AMS Chelsea Publishing, 2010, (Original: Die Naturwissenschaften, vol. 28, pp. 664–666, 1926 [in German]), pp. 41–44. DOI 10.1007/BF01507634

[37] E. Schrödinger, *Quantisation as a problem of proper values (Part IV)*, in *Collected Papers on Wave Mechanics*, New York, Providence, Rhode Island: AMS Chelsea Publishing, 2010, (Original: Annalen der Physik (4), vol. 81, pp. 109–139, 1926 [in German]), pp. 102–123. DOI 10.1002/andp.19263861802.

[38] E. Schrödinger, *An undulatory theory of the mechanics of atoms and molecules*, Physical Review **28**(6), 1049–1070, 1926. DOI 10.1103/PhysRev.28.1049.

[39] L. I. Schiff, *Quantum Mechanics*, 3rd edn. International series in pure and applied physics. McGraw-Hill, Inc., New York, 1968.

[40] A. Sommerfeld, *Atombau und Spektrallinien*, 2 ed., **1**, Friedrich Vieweg & Sohn, Braunschweig, German, 1951. [in German]

[41] S. K. Suslov and B. Trey, *The Hahn polynomials in the nonrelativistic and relativistic Coulomb problems*, J. Math. Phys. **49** (2008), no. 1, 012104, 51, DOI 10.1063/1.2830804. MR2385251

[42] S. K. Suslov, J. M. Vega-Guzmán, and K. Barley, *An introduction to special functions with some applications to quantum mechanics*, Orthogonal polynomials, Tutor. Sch. Workshops Math. Sci., Birkhäuser/Springer, Cham, 2020, pp. 517–628, DOI 10.1007/978-3-030-36744-2_21. MR4174720

[43] D. A. Varshalovich, A. N. Moskalev, and V. K. Khersonskiĭ, *Quantum theory of angular momentum: Irreducible tensors, spherical harmonics, vector coupling coefficients, 3nj symbols*, World Scientific Publishing Co., Inc., Teaneck, NJ, 1988, DOI 10.1142/0270. MR1022665

[44] G. N. Watson, *A treatise on the theory of Bessel functions*, Cambridge Mathematical Library, Cambridge University Press, Cambridge, 1995. Reprint of the second (1944) edition. MR1349110

[45] E. T. Whittaker and G. N. Watson, *A course of modern analysis: An introduction to the general theory of infinite processes and of analytic functions; with an account of the principal transcendental functions*, Cambridge Mathematical Library, Cambridge University Press, Cambridge, 1996, DOI 10.1017/CBO9780511608759. MR1424469

Dept. of Mathematics and Statistics, Northern Arizona University, P. O. Box 5717, Flagstaff, Arizona 86011
Email address: linabellis@gmail.com

Dept. of Chemistry, College of Science and Engineering, University of Minnesota, Smith Hall, 207 Pleasant St. SE, Minneapolis, Minnesota 55455-0431
Email address: elli0687@umn.edu

Johann Radon Institute for Computational and Applied Mathematics, Austrian Academy of Sciences, Altenberger Strasse 69, 4040 Linz, Austria
Email address: christoph.koutschan@ricam.oeaw.ac.at

School of Mathematical and Statistical Sciences, Arizona State University, P. O. Box 871804, Tempe, Arizona 85287-1804
Email address: sergei@asu.edu

Contemporary Mathematics
Volume **819**, 2025
https://doi.org/10.1090/conm/819/16389

On the Mellin symbol of singular integral operators associated with the biharmonic equation in infinite sectors

Jeongsu Kyeong, Irina Mitrea, and Katharine Ott

Abstract. This paper is concerned with the study of singular integral operators associated with the Neumann problem for the bi-Laplacian in infinite sectors in the plane. The main goal is to explicitly describe the Mellin symbol of such singular integral operators for arbitrary apertures $\theta \in (0, 2\pi)$ and Poisson ratios $\eta \in [-1, 1)$. The analysis carried out here uses Mellin transform techniques and properties of hypergeometric functions of gamma, beta, Ferrers and Gauss type. The explicit Mellin symbol formulas open the door to obtaining information on the range of integrability exponents $p \in (1, \infty)$ for which the Neumann problem for Δ^2 is well-posed in this geometric setting.

1. Introduction

The focus of this paper is on the Mellin analysis of singular integral operators associated with the treatment of the Neumann problem for the bi-Laplacian in infinite sectors in $\mathbb{R}^2$ via the layer potential method. To set the stage, fix $\theta \in (0, 2\pi)$ and let $\Omega \subseteq \mathbb{R}^2$ be the region above the graph of the function $\phi \colon \mathbb{R} \to \mathbb{R}$ defined by

$$\phi(x) := |x| \cot(\theta/2), \quad x \in \mathbb{R}. \tag{1.1}$$

For each $X \in \partial\Omega \setminus \{0\}$ there exist outward unit normal and unit tangential vectors denoted by $\nu(X)$ and $\tau(X)$ respectively. Then the Neumann problem for the bi-Laplacian in Ω with L^p boundary datum, with integrability index $p \in (1, \infty)$, can be formulated as:

$$(NBH_p) \quad \begin{cases} \triangle^2 u = 0 \quad \text{in } \Omega, \\ \mathcal{N}(\nabla^2 u) \in L^p(\partial\Omega), \\ \eta \triangle u\Big|_{\partial\Omega}^{n.t.} + (1-\eta)\dfrac{\partial^2 u}{\partial \nu^2} = f \in L^p(\partial\Omega), \\ \dfrac{\partial \triangle u}{\partial \nu} + (1-\eta)\dfrac{\partial}{\partial \tau}\Big[\dfrac{\partial^2 u}{\partial \tau \partial \nu}\Big] = \Lambda \in \big(\dot{L}_1^{p'}(\partial\Omega)\big)^*, \end{cases} \tag{1.2}$$

2020 *Mathematics Subject Classification.* Primary 31A10, 35C15, 35J08, 35J47, 42B37; Secondary 28A75, 31A20, 35J15, 35J57, 33C05, 33D90, 33B15.

Key words and phrases. Bilaplacian, Neumann problem, nontangential maximal function, nontangential trace, jump formulas, Hardy kernels, beta function, gamma function, Ferrers function, hypergeometric function, special functions.

Supported in part by a Temple University Summer Research Award and by the Simons Foundation Grant MP-TSM-00002820.

where $1 < p' < \infty$ is such that $1/p + 1/p' = 1$. Here, with $\mathcal{H}^1$ standing for the 1-dimensional Hausdorff measure in $\mathbb{R}^2$, set $\sigma := \mathcal{H}^1 \lfloor \partial\Omega$ for the surface measure on $\partial\Omega$. Then the space $L^p(\partial\Omega)$ is that of Lebesgue p-integrable functions on $\partial\Omega$ with respect to σ and $\big(\dot{L}_1^{p'}(\partial\Omega)\big)^*$ is the dual space of all locally integrable functions on $\partial\Omega$ with tangential derivative in $L^{p'}(\partial\Omega)$. In addition, $\mathcal{N}$ denotes the non-tangential maximal function defined in (2.2) and by $\cdot\Big|_{\partial\Omega}^{\rm n.t.}$ we have denoted the nontangential restriction to the boundary introduced in (2.3). Finally, in the formulation of the boundary conditions in (1.2) the convention for repeated normal and/or tangential derivatives as introduced in (2.5) has been used, while $\eta \in [-1,1)$ is a given real constant called the Poisson ratio.

The formulation of the Neumann problem (1.2) follows from the Kirchhoff–Love theory of thin plates and the problem has been previously studied in the literature in a number of important cases. J. Giroire and J. C. Nédélec, see [**5**], considered the geometric setting of bounded smooth planar domains and a variational setting formulation, and their well-posdeness approach relied on establishing coerciveness of the corresponding boundary bilinear form. C. Nazaret, see [**8**], extended their results (in the variational setting) to bounded polygonal domains. The case when $\Omega \subseteq \mathbb{R}^n$, $n \geqslant 2$, is a bounded Lipschitz domain and the integrability index p is near 2 has been treated by G. Verchota in [**13**], employing layer potential theory and Rellich-type identities. Shortly thereafter, Z. Shen proved in [**11**] L^p versions of Verchota's results when Ω is a bounded Lipschitz domain in $\mathbb{R}^n$, $n \geqslant 4$, and $\frac{2(n-1)}{n+1} - \varepsilon < p < 2$, for some $\varepsilon > 0$ depending on Ω. A different formulation of the Neumann problem for the bi-Laplacian, with boundary datum from duals of Whitney–Lebesgue spaces, in bounded Lipschitz domains with connected boundary can be found in [**6**, Theorem 6.55, p. 388]. There the authors establish well-posedness whenever $p \in (2-\varepsilon, \infty)$ if $n \in \{2,3\}$ and $p \in (2-\varepsilon, \frac{2(n-1)}{n-3} + \varepsilon)$ if $n \geqslant 4$, where again $\varepsilon > 0$ depends on Ω.

To put the formulation (1.2) of the Neumann problem for Δ^2 into perspective, in connection with the study of the problem (1.2), for each $r \in \mathbb{R}$, consider the bilinear form $\mathcal{B}_r$ associated to Δ^2 in $\mathbb{R}^2$ given by

$$\begin{aligned} \mathcal{B}_r(u,v) &:= c_r \sum_{i,j=1}^{2} \int_\Omega \big[(\partial_i\partial_j + r\,\delta_{ij}\Delta)u\big](X)\big[(\partial_i\partial_j + r\,\delta_{ij}\Delta)v\big](X)\,dX, \\ c_r &:= \frac{1}{1+2r+2r^2}, \end{aligned} \tag{1.3}$$

where $u, v \in W^{2,2}(\Omega)$ (the L^2 based Sobolev space with smoothness index 2) and δ_{ij} stands for the standard Kronecker symbol. It can be checked that, indeed,

$$\mathcal{B}_r(u,v) = \int_\Omega (\Delta^2 u)(X) v(X)\,dX \quad \text{for all } u, v \in \mathscr{C}_0^\infty(\Omega) \text{ and all } r \in \mathbb{R}. \tag{1.4}$$

If u is a biharmonic function in Ω such that $\mathcal{N}(\nabla^2 u), \mathcal{N}(\nabla^3 u) \in L^2(\partial\Omega)$, and $v \in \mathscr{C}_0^\infty(\mathbb{R}^n)$, integrating by parts one obtains

$$\int_{\partial\Omega} \mathcal{B}_r(u,v)\,d\sigma = \int_{\partial\Omega} \Big(M_r(u)\frac{\partial v}{\partial\nu} - N_r(u)v\Big)\,d\sigma, \tag{1.5}$$

where

$$N_r(u) := \frac{\partial(\Delta u)}{\partial\nu} + c_r\frac{\partial}{\partial\tau}\Big[\frac{\partial^2 u}{\partial\tau\partial\nu}\Big], M_r(u) \quad := c_r(2r+2r^2)\,\Delta u\Big|_{\partial\Omega}^{\text{n.t.}} + c_r\frac{\partial^2 u}{\partial\nu^2}. \tag{1.6}$$

Note that, setting $\eta := \dfrac{2(r+r^2)}{1+2r+2r^2} \in [-1,1)$, the boundary conditions in (1.2) can be rephrased as $M_r(u) = f \in L^p(\partial\Omega)$, and $N_r(u) = \Lambda \in \Big(\dot{L}_1^{p'}(\partial\Omega)\Big)^*$, respectively.

In Section 2.1 we employ the layer potential method for the treatment of (1.2). Concretely, we show how seeking a solution of (1.2) as a single layer potential leads to the consideration of the invertibility properties of a boundary-to-boundary matrix-valued integral operator $T\colon \big(L^p(\partial\Omega)\big)^2 \longrightarrow \big(L^p(\partial\Omega)\big)^2$ defined below. Specifically, for each $F, G \in L^p(\partial\Omega)$, set

$$T(F,G) := \begin{pmatrix} -\eta K_1 + (\eta-1)K_2 & -\frac{1}{2}I - \eta K_3 + (\eta-1)K_4 \\ -K_5 + (\eta-1)K_6 - \frac{1}{2}I & K_7 + (\eta-1)K_8 \end{pmatrix}\begin{pmatrix} G \\ F \end{pmatrix}, \tag{1.7}$$

where I stands for the identity operator, and, for $f \in L^p(\partial\Omega)$,

$$K_i f(P) := \int_{\partial\Omega} k_i(P,Q)f(Q)\,d\sigma(Q) \quad \text{for each } i \in \{1,\dots,8\} \tag{1.8}$$

with the kernels k_i, for $i \in \{1,\dots,8\}$, defined on $\partial\Omega \times \partial\Omega \setminus \operatorname{diag}(\partial\Omega)$ as follows:

$$k_1(P,Q) = k_7(P,Q) := \frac{\partial}{\partial\tau(Q)}[E(P-Q)], \tag{1.9}$$

$$k_2(P,Q) := \frac{\partial^3}{\partial\nu^2(P)\partial\tau(Q)}[B(P-Q)], \tag{1.10}$$

$$k_3(P,Q) = k_5(P,Q) := \frac{\partial}{\partial\nu(Q)}[E(P-Q)], \tag{1.11}$$

$$k_4(P,Q) := \frac{\partial^3}{\partial\nu^2(P)\partial\nu(Q)}[B(P-Q)], \tag{1.12}$$

$$k_6(P,Q) := \frac{\partial^3}{\partial\tau(P)\partial\nu(P)\partial\tau(Q)}[B(P-Q)], \tag{1.13}$$

$$k_8(P,Q) := \frac{\partial^3}{\partial\tau(P)\partial\nu(P)\partial\nu(Q)}[B(P-Q)]. \tag{1.14}$$

Above, E is the classical radial fundamental solution for the Laplacian and B is the classical radial fundamental solution for the bi-Laplacian in the plane. Specifically

$$E\colon \mathbb{R}^2\setminus\{0\} \longrightarrow \mathbb{R}, \quad E(X) := \frac{1}{2\pi}\log|X| \qquad \forall X \in \mathbb{R}^2\setminus\{0\}, \tag{1.15}$$

and

$$B\colon \mathbb{R}^2\setminus\{0\} \longrightarrow \mathbb{R}, \quad B(X) := -\frac{1}{8\pi}|X|^2(1-\log|X|) \qquad \forall X \in \mathbb{R}^2\setminus\{0\}. \tag{1.16}$$

In particular, in the sense of distributions in $\mathbb{R}^2$ there holds that $\Delta E = \Delta^2 B = \delta$, the Dirac delta distribution with mass at the origin, and that $\Delta B = E$.

A preliminary step in the analysis of the operator T from (1.7) is "transporting" this operator from one acting on functions defined on $\partial\Omega$ to an operator acting on functions defined on $\mathbb{R}_+ := [0,\infty)$. This is done by exploiting the fact that Ω is the region above the graph of the function $\phi\colon \mathbb{R} \to \mathbb{R}$ from (1.1), and as such $\partial\Omega = \partial\Omega_1 \cup \partial\Omega_2$, where

$$\partial\Omega_1 := \big\{s\vec{v}_1(\theta) : s \in \mathbb{R}_+\big\} \text{ and } \partial\Omega_2 := \{s\vec{v}_2(\theta) : s \in \mathbb{R}_+\} \tag{1.17}$$

with

$$\vec{v}_1(\theta) := \big(-\sin\tfrac{\theta}{2}, \cos\tfrac{\theta}{2}\big) \text{ and } \vec{v}_2(\theta) := \big(\sin\tfrac{\theta}{2}, \cos\tfrac{\theta}{2}\big). \tag{1.18}$$

Note that the vectors

$$\begin{aligned} &\vec{v}_3(\theta) := \big(-\cos\tfrac{\theta}{2}, -\sin\tfrac{\theta}{2}\big) \text{ and } \vec{v}_4(\theta) := \big(\cos\tfrac{\theta}{2}, -\sin\tfrac{\theta}{2}\big) \text{ satisfy} \\ &\vec{v}_1(\theta) \perp \vec{v}_3(\theta) \text{ and } \vec{v}_2(\theta) \perp \vec{v}_4(\theta), \quad \text{and point outwardly to } \Omega. \end{aligned} \tag{1.19}$$

The two pieces of the boundary $\partial\Omega_j$, $j = 1, 2$, are then identified with $\mathbb{R}_+$ via the mappings

$$\iota_j\colon \partial\Omega_j \longrightarrow \mathbb{R}_+, \quad \iota_j(P) := |P| \quad \text{for each } P \in \partial\Omega_j. \tag{1.20}$$

As a result, for each $p \in (1,\infty)$ the space $L^p(\partial\Omega)$ is in turn identified with the product space $L^p(\mathbb{R}_+) \times L^p(\mathbb{R}_+)$ via the mapping $\mathscr{I} : L^p(\partial\Omega) \longrightarrow L^p(\mathbb{R}_+) \times L^p(\mathbb{R}_+)$ given by

$$\mathscr{I}(f) := \begin{pmatrix} f\Big|_{\partial\Omega_1} \circ \iota_1^{-1} \\ f\Big|_{\partial\Omega_2} \circ \iota_2^{-1} \end{pmatrix} \qquad \forall f \in L^p(\partial\Omega). \tag{1.21}$$

This and explicit trigonometrical calculations lead us to consider the operator $\mathcal{T}$

$$\begin{aligned} &\mathcal{T}\colon \Big(L^p(\mathbb{R}_+) \times L^p(\mathbb{R}_+)\Big)^2 \longrightarrow \Big(L^p(\mathbb{R}_+) \times L^p(\mathbb{R}_+)\Big)^2, \\ &\mathcal{T}\begin{pmatrix} f_1 \\ f_2 \\ g_1 \\ g_2 \end{pmatrix}(s) = \int_0^\infty R(s,t) \cdot \begin{pmatrix} f_1(t) \\ f_2(t) \\ g_1(t) \\ g_2(t) \end{pmatrix} dt, \end{aligned} \tag{1.22}$$

where the kernel $R\colon \mathbb{R}_+ \times \mathbb{R}_+ \setminus \operatorname{diag}(\mathbb{R}_+) \longrightarrow M_{4\times 4}(\mathbb{R})$ is given by

$$R := \begin{pmatrix} -\eta\kappa_1 + (\eta-1)\kappa_2 & -\frac{\delta}{2}I_{2\times 2} - \eta\kappa_3 + (\eta-1)\kappa_4 \\ -\kappa_5 + (\eta-1)\kappa_6 - \frac{\delta}{2}I_{2\times 2} & \kappa_7 + (\eta-1)\kappa_8 \end{pmatrix}, \tag{1.23}$$

with the matrix-valued functions κ_i, $i \in \{1,\dots,8\}$, explicitly described in (3.16) and Lemma 3.3 in Section 3, where δ is the Dirac delta distribution (with mass at s). Since the operator $\mathcal{T}$ is merely a re-interpretation of the operator T, there holds

$$\begin{aligned} &T\colon \big(L^p(\partial\Omega)\big)^2 \longrightarrow \big(L^p(\partial\Omega)\big)^2 \text{ is invertible if and only if} \\ &\mathcal{T}\colon \Big(L^p(\mathbb{R}_+) \times L^p(\mathbb{R}_+)\Big)^2 \longrightarrow \Big(L^p(\mathbb{R}_+) \times L^p(\mathbb{R}_+)\Big)^2 \text{ is invertible.} \end{aligned} \tag{1.24}$$

A further analysis of $\mathcal{T}$ reveals that this operator belongs to the algebra of integral operators generated by Hardy kernel operators and the Hilbert transform, making the analysis of its invertibility properties amenable to Mellin transform techniques. This is discussed in Subsection 2.2, where we record the invertibility criterion in Theorem 2.7. Keeping (1.24) in mind, Theorem 2.7 brings into focus the necessity of computing the determinant of the Mellin symbol of the kernel R, i.e., $\det\big(\mathcal{M}R(\cdot,1)\big)(z)$ for $z \in \Gamma_{0,1}$ (an open vertical strip in $\mathbb{C}$ defined as in (2.34)). This symbol is explicitly derived in Section 4 and the calculations leading to the formula for $\det\big(\mathcal{M}R(\cdot,1)\big)(z)$ obtained in Theorem 4.14 involve hypergeometric functions of gamma, beta, and Ferrers types, as well as Gauss' hypergeometric functions.

Theorem 4.14 is the main result of this paper and its power lies in the fact that it establishes explicit formulas for $\det\big(\mathcal{M}R(\cdot,1)\big)(z)$ for *all* aperture values $\theta \in (0, 2\pi)$. In addition, we show that $\det\big(\mathcal{M}R(\cdot,1)\big)(z)$ is the product of two simpler expressions $(\Phi_\theta(z,\eta) + \Psi_\theta(z,\eta))$ and $(\Phi_\theta(z,\eta) - \Psi_\theta(z,\eta))$, where $\Phi_\theta(z,\eta)$ and $\Psi_\theta(z,\eta)$ are as in (4.81) and (4.82). Generally speaking, finding the zeroes of $\det\big(\mathcal{M}R(\cdot,1)\big)(z)$ is a very difficult task, thus obtaining a factorization of $\det\big(\mathcal{M}R(\cdot,1)\big)(z)$ helps with this analysis. Indeed, utilizing the results obtained in this paper, in **[2]** the authors have been successful in solving $\det\big(\mathcal{M}R(\cdot,1)\big)(z) = 0$ in $\Gamma_{0,1}$ in a number of particular cases such as $\theta = \pi$, $\theta = \pi/2$, and $\theta = \pi/4$, ultimately obtaining solvability results for the problem (1.2) of the following type.

THEOREM 1.1. *Let Ω be the interior of an infinite upright sector of aperture $\frac{\pi}{2}$ in the plane. Then for $\eta \in [-1,1)$ and for $p \in (1,\infty)$, the following implication holds:*

$$p \in (1,\infty) \setminus \{\alpha,\beta\} \Longrightarrow (NBH_p) \ \text{ has a solution,} \tag{1.25}$$

where

$$\alpha^{-1} \ \text{ is the unique real root of } \ z - 1 + \cos\Big(\frac{3\pi z}{2}\Big) = 0 \ \text{ in } \ \Gamma_{0,1}, \tag{1.26}$$

$$\beta^{-1} \ \text{ is the unique real root of } \ z - 1 - \cos\Big(\frac{3\pi z}{2}\Big) = 0 \ \text{ in } \ \Gamma_{0,1}. \tag{1.27}$$

In particular, $\alpha, \beta \in \mathbb{R}$ and $\alpha \in (9,12)$ while $\beta \in (2,3)$.

Validated numerics considerations show that the critical index $\alpha \in (10.92, 10.94)$, while the critical index $\beta \in (2.194, 2.196)$.

2. Preliminaries

In this section we introduce notation and record a number of useful results. Throughout, $\theta \in (0, 2\pi)$ is fixed and $\Omega \subseteq \mathbb{R}^2$ is the upper graph of the function ϕ from (1.1). If $\operatorname{dist}(\cdot, \partial\Omega)$ is the distance function to $\partial\Omega$ and the parameter $a > 1$ is fixed, for each $X \in \partial\Omega$ the non-tangential approach region associated with point X is defined as

$$\Upsilon_a(X) := \big\{Y \in \Omega : |X - Y| < a \cdot \operatorname{dist}(Y, \partial\Omega)\big\}\,. \tag{2.1}$$

For a fixed $a > 1$ and each function $w\colon \Omega \to \mathbb{R}$, the non-tangential maximal function of w, denoted by $\mathcal{N}w$, is set to be

$$\mathcal{N}w(X) := \sup_{Y \in \Upsilon_a(X)} |w(Y)| \quad \text{for each } X \in \partial\Omega\,, \tag{2.2}$$

and the non-tangential limit of w at $X \in \partial\Omega$, denoted by $w\Big|_{\partial\Omega}^{n.t.}(X)$, is defined as

$$w\Big|_{\partial\Omega}^{n.t.}(X) := \lim_{\substack{Y \to X \\ Y \in \Upsilon_a(X)}} w(Y)\,, \tag{2.3}$$

whenever the limit in the right-hand side of (2.3) exists.

Regarding directional derivatives, if w is a differentiable function in Ω and if $\langle\cdot,\cdot\rangle$ is the canonical inner product in $\mathbb{R}^n$, we have set

$$\frac{\partial w}{\partial \nu} := \Big\langle (\nabla w)\Big|_{\partial\Omega}^{n.t.}, \nu \Big\rangle \tag{2.4}$$

with ν being the outward unit normal vector to Ω. In addition, in (1.2) and (1.6) the following convention for repeated normal and/or tangential derivatives has been used. If u is a sufficiently smooth function in Ω, given a collection of vectors $V := \{v_1,\dots,v_m\} \subset \mathbb{R}^n$ on $\partial\Omega$ such that $v_j = (v_j^1, v_j^2,\dots,v_j^n)$ for each $j \in \{1,\dots,m\}$, and $f(V)$ is an expression depending solely on $v_1,\dots,v_m$, then for all $j \in \{1,\dots,m\}$ we set

$$\frac{\partial}{\partial v_j}\Big[f(V)u\Big] := \sum_{k=1}^{n} v_j^k f(V)(\partial_k u)\Big|_{\partial\Omega}^{n.t.}. \tag{2.5}$$

Thus, the iterated normal and tangential derivatives appearing in the formulation of the Neumann problem (1.2) take the form

$$\frac{\partial^2 u}{\partial \nu^2} = \sum_{i,j=1}^{2} \nu_i\nu_j(\partial_i\partial_j u)\Big|_{\partial\Omega}^{n.t.}, \tag{2.6}$$

$$\frac{\partial \Delta u}{\partial \nu} = \nu_1(\partial_1^3 u)\Big|_{\partial\Omega}^{n.t.} + \nu_1(\partial_1\partial_2^2 u)\Big|_{\partial\Omega}^{n.t.} + \nu_2(\partial_1^2\partial_2 u)\Big|_{\partial\Omega}^{n.t.} + \nu_2(\partial_2^3 u)\Big|_{\partial\Omega}^{n.t.}, \tag{2.7}$$

$$\frac{\partial^2 u}{\partial \tau^2} = \nu_2^2(\partial_1^2 u)\Big|_{\partial\Omega}^{n.t.} - 2\nu_1\nu_2(\partial_1\partial_2 u)\Big|_{\partial\Omega}^{n.t.} + \nu_1^2(\partial_2^2 u)\Big|_{\partial\Omega}^{n.t.}, \tag{2.8}$$

$$\frac{\partial^2 u}{\partial \tau\partial \nu} = -\nu_1\nu_2(\partial_1^2 u)\Big|_{\partial\Omega}^{n.t.} + (\nu_1^2-\nu_2^2)(\partial_1\partial_2 u)\Big|_{\partial\Omega}^{n.t.} + \nu_1\nu_2(\partial_2^2 u)\Big|_{\partial\Omega}^{n.t.}, \tag{2.9}$$

$$\begin{aligned}\frac{\partial^3 u}{\partial \tau^2\partial \nu} &= \nu_1\nu_2^2(\partial_1^3 u)\Big|_{\partial\Omega}^{n.t.} + (\nu_2^3-\nu_1^2\nu_2)(\partial_1^2\partial_2 u)\Big|_{\partial\Omega}^{n.t.} \\ &\quad + (\nu_1^3-\nu_2^2\nu_1)(\partial_2^2\partial_1 u)\Big|_{\partial\Omega}^{n.t.} + \nu_1^2\nu_2(\partial_2^3 u)\Big|_{\partial\Omega}^{n.t.}.\end{aligned} \tag{2.10}$$

2.1. The single layer potential for the bi-Laplacian. Here we shall introduce the (modified) single layer potential operator associated with the bi-Laplacian and record some of its basic properties such as jump formulas and non-tangential maximal function estimates. To get started, recall B, the fundamental solution of the bi-Laplacian in $\mathbb{R}^2$ from (1.16). Fix $X_o \in \Omega$ and for each $Y \in \partial\Omega$ consider the following affine (in the variable $X \in \mathbb{R}^2 \setminus \partial\Omega$) correction of B:

$$\mathcal{B}(X,Y) := B(X-Y) - \big[B(X_o-Y) + \langle(\nabla B)(X_o-Y), X_o-X\rangle\big]. \tag{2.11}$$

A double application of the Fundamental Theorem of Calculus yields

$$\mathcal{B}(X,Y) = \sum_{j,k=1}^{2} (X-X_o)_j(X-X_o)_k \int_0^1\!\!\int_0^1 t(\partial_j\partial_k B)(X_o+st(X-X_o)-Y)\,ds\,dt. \tag{2.12}$$

For each $p \in (1,\infty)$, consider the modified single layer potential operator $\mathcal{S}$ acting on a pair $(\Lambda, F) \in \dot{L}^p_{-1}(\partial\Omega) \times L^p(\partial\Omega)$ given at each point $X \in \mathbb{R}^2 \setminus \partial\Omega$ by

$$\mathcal{S}(\Lambda, F)(X) := \langle \mathcal{B}(X,\cdot), \Lambda\rangle - \int_{\partial\Omega} \partial_{\nu(Q)}[\mathcal{B}(X,Q)]F(Q)d\sigma(Q), \tag{2.13}$$

where the two terms are understood in the following manner. Given a functional $\Lambda \in \dot{L}^p_{-1}(\partial\Omega) \equiv \big(\dot{L}^{p'}_1(\partial\Omega)\big)^*$ and $G \in L^p(\partial\Omega)$ such that $\partial_\tau G = \Lambda$, the pairing $\langle \mathcal{B}(X,\cdot), \Lambda\rangle$ is interpreted as
(2.14)

$$\begin{aligned}\langle \mathcal{B}(X,\cdot), \Lambda\rangle := -\int_{\partial\Omega} \sum_{j,k=1}^{2} &(X - X_o)_j (X - X_o)_k G(Y)\\ &\times \left(\int_0^1\!\!\int_0^1 t\partial_{\tau(Y)}\left[(\partial_j\partial_k B)(X_o + st(X - X_o) - Y)\right] ds\, dt\right) d\sigma(Y).\end{aligned}$$

Note that in (2.14) the kernel $\nabla^3 B \in L^{p'}(\partial\Omega)$, with $1/p + 1/p' = 1$, in light of (3.3)–(3.4). Also, using (2.12) and (3.3)–(3.4), the integrand in the second term in the right hand side of (2.13) is absolutely integrable. This makes the formula of $\mathcal{S}$ in (2.13) meaningful. In addition, it is straightforward to check that there holds

$$\Delta^2(\mathcal{S}(\Lambda, F)) = 0 \quad \text{on } \mathbb{R}^2 \setminus \partial\Omega, \tag{2.15}$$

and, using [**13**, Lemma 8.4, p. 241], there holds $\mathcal{N}(\nabla^2(\mathcal{S}(\Lambda, F))) \in L^p(\partial\Omega)$.

Next, for $r \in \mathbb{R}$ set $\eta := \dfrac{2(r + r^2)}{1 + 2r + 2r^2}$ and recall the operators M_r, N_r from (1.6). Then the following jump-formulas hold for σ-a.e. $P \in \partial\Omega$ (see [**13**, Section 9]):

$$\begin{aligned}M_r(\mathcal{S}(\Lambda, F))(P) = &-\frac{1}{2}F(P) - \eta \int_{\partial\Omega} \frac{\partial}{\partial\nu(Q)}[E(P - Q)]F(Q)\, d\sigma(Q)\\ &+ (\eta - 1)\int_{\partial\Omega} \frac{\partial^3}{\partial\nu^2(P)\partial\nu(Q)}[B(P - Q)]F(Q)\, d\sigma(Q)\\ &+ \eta\,\langle E(P - \cdot), \Lambda\rangle + (1 - \eta)\left\langle \frac{\partial^2}{\partial\nu^2(P)}[B(P - \cdot)], \Lambda\right\rangle,\end{aligned} \tag{2.16}$$

and
(2.17)

$$\begin{aligned}N_r(\mathcal{S}(\Lambda, F))(P) = &-\frac{1}{2}\Lambda(P) - \int_{\partial\Omega} \frac{\partial^2}{\partial\nu(P)\partial\nu(Q)}[E(P - Q)]F(Q)\, d\sigma(Q)\\ &- (1 - \eta)\int_{\partial\Omega} \frac{\partial^4}{\partial\tau^2(P)\partial\nu(P)\partial\nu(Q)}[B(P - Q)]F(Q)\, d\sigma(Q)\\ &+ \langle \partial_{\nu(P)}[E(P - \cdot)], \Lambda\rangle + (1 - \eta)\left\langle \frac{\partial^3}{\partial\tau^2(P)\partial\nu(P)}[B(P - \cdot)], \Lambda\right\rangle.\end{aligned}$$

Next, for each $\Lambda \in \dot{L}^p_{-1}(\partial\Omega)$ let

$$G \in L^p(\partial\Omega) \quad \text{be such that } \Lambda = \partial_\tau G. \tag{2.18}$$

Using this, integrating by parts in (2.16) and in (2.17), and the properties of the duality pairing between $\dot{L}^p_{-1}(\partial\Omega)$ and $\dot{L}^{p'}_1(\partial\Omega)$, we obtain that for σ-a.e. $P \in \partial\Omega$ there hold

$$\begin{aligned}
M_r(\mathcal{S}(\Lambda, F))(P) = &-\frac{1}{2}F(P) - \eta \int_{\partial\Omega} \frac{\partial}{\partial\nu(Q)}[E(P-Q)]F(Q)\,d\sigma(Q) \\
&+ (\eta - 1)\int_{\partial\Omega} \frac{\partial^3}{\partial\nu^2(P)\partial\nu(Q)}[B(P-Q)]F(Q)\,d\sigma(Q) \\
&- \eta \int_{\partial\Omega} \frac{\partial}{\partial\tau(Q)}[E(P-Q)]G(Q)\,d\sigma(Q) \\
&+ (\eta - 1)\int_{\partial\Omega} \frac{\partial^3}{\partial\tau(Q)\partial\nu^2(P)}[B(P-Q)]G(Q)\,d\sigma(Q),
\end{aligned} \tag{2.19}$$

and
(2.20)

$$\begin{aligned}
N_r(\mathcal{S}(\Lambda, F))(P) = &-\frac{1}{2}\partial_\tau G(P) - \int_{\partial\Omega} \frac{\partial^2}{\partial\nu(P)\partial\nu(Q)}[E(P-Q)]F(Q)\,d\sigma(Q) \\
&+ (\eta - 1)\int_{\partial\Omega} \frac{\partial^4}{\partial\tau^2(P)\partial\nu(P)\partial\nu(Q)}[B(P-Q)]F(Q)\,d\sigma(Q) \\
&- \int_{\partial\Omega} \frac{\partial^2}{\partial\tau(Q)\partial\nu(P)}[E(P-Q)]G(Q)\,d\sigma(Q) \\
&+ (\eta - 1)\int_{\partial\Omega} \frac{\partial^4}{\partial\tau(Q)\partial\tau^2(P)\partial\nu(P)}[B(P-Q)]G(Q)\,d\sigma(Q).
\end{aligned}$$

At this point let us observe that a direct calculation based on (2.5) shows that for σ-a.e. points $P, Q \in \partial\Omega$ such that $P \neq Q$, there holds

$$\frac{\partial^2}{\partial\nu(P)\partial\tau(Q)}[E(P-Q)] = \frac{\partial^2}{\partial\tau(P)\partial\nu(Q)}[E(P-Q)], \tag{2.21}$$

$$\frac{\partial^2}{\partial\nu(P)\partial\nu(Q)}[E(P-Q)] = -\frac{\partial^2}{\partial\tau(P)\partial\tau(Q)}[E(P-Q)]. \tag{2.22}$$

Thus
(2.23)

$$\int_{\partial\Omega} \frac{\partial^2}{\partial\nu(P)\partial\tau(Q)}[E(P-Q)]G(Q)\,d\sigma(Q) = \partial_{\tau(P)} \int_{\partial\Omega} \partial_{\nu(Q)}[E(P-Q)]G(Q)\,d\sigma(Q)$$

and
(2.24)

$$\int_{\partial\Omega} \frac{\partial^2}{\partial\nu(P)\partial\nu(Q)}[E(P-Q)]F(Q)\,d\sigma(Q) = -\partial_{\tau(P)} \int_{\partial\Omega} \partial_{\tau(Q)}[E(P-Q)]F(Q)\,d\sigma(Q).$$

Recall next the kernels (1.9)–(1.14). In this notation, formula (2.19) becomes

$$\begin{aligned} M_r(\mathcal{S}(\Lambda, F))(P) = &-\frac{1}{2}F(P) + (\eta-1)\int\limits_{\partial\Omega}\big(k_2(P,\cdot)G + k_4(P,\cdot)F\big)\,d\sigma \\ &-\eta\int\limits_{\partial\Omega}\big(k_1(P,\cdot)G + k_3(P,\cdot)F\big)\,d\sigma, \end{aligned} \tag{2.25}$$

while (2.20) can be written as

$$\begin{aligned} N_r(\mathcal{S}(\Lambda, F))(P) = &-\frac{1}{2}\partial_{\tau(P)}G(P) + (\eta-1)\partial_{\tau(P)}\Big(\int\limits_{\partial\Omega}\big(k_6(P,\cdot)G + k_8(P,\cdot)F\big)\,d\sigma\Big) \\ &-\partial_{\tau(P)}\Big(\int\limits_{\partial\Omega}\big(k_5(P,\cdot)G + k_7(P,\cdot)F\big)\,d\sigma\Big). \end{aligned} \tag{2.26}$$

Consequently, with I standing for the identity operator, there holds

$$\begin{aligned} &\begin{pmatrix} M_r(\mathcal{S}(\Lambda, F)) \\ N_r(\mathcal{S}(\Lambda, F)) \end{pmatrix} = \\ &\begin{pmatrix} -\eta K_1 + (\eta-1)K_2 & -\frac{1}{2}I - \eta K_3 + (\eta-1)K_4 \\ \partial_\tau(-K_5 + (\eta-1)K_6 - \frac{1}{2}I) & \partial_\tau(K_7 + (\eta-1)K_8) \end{pmatrix}\begin{pmatrix} G \\ F \end{pmatrix}, \end{aligned} \tag{2.27}$$

where for a generic function $f \in L^p(\partial\Omega)$, and k_i, $i \in \{1,\dots,8\}$, as in (1.9)–(1.14), we have set

$$K_i f(P) := \int\limits_{\partial\Omega} k_i(P,Q)f(Q)\,d\sigma(Q) \quad \text{for each } i \in \{1,\dots,8\}. \tag{2.28}$$

Recall the operator $T\colon \big(L^p(\partial\Omega)\big)^2 \longrightarrow \big(L^p(\partial\Omega)\big)^2$ from (1.7). Given a functional $\Lambda \in \big(\dot{L}^{p'}_1(\partial\Omega)\big)^* \equiv \dot{L}^p_{-1}(\partial\Omega)$, consider

$$G \in L^p(\partial\Omega) \ \text{ such that } \ \partial_\tau G = \Lambda. \tag{2.29}$$

Then, the jump relations (2.27) can be re-written as

$$\begin{pmatrix} M_r(\mathcal{S}(\Lambda, F)) \\ N_r(\mathcal{S}(\Lambda, F)) \end{pmatrix} = T(F, G). \tag{2.30}$$

In light of this discussion, given $f \in L^p(\partial\Omega)$ and $\Lambda \in \big(\dot{L}^{p'}_1(\partial\Omega)\big)^*$, seeking a solution of the Neumann problem (NBH_p) in (1.2) as a single layer potential operator, $\mathcal{S}(\tilde{\Lambda}, H)$, with $H \in L^p(\partial\Omega)$ and $\tilde{\Lambda} \in \big(\dot{L}^{p'}_1(\partial\Omega)\big)^*$, is then reduced to finding a solution of the boundary integral equation

$$T(H, \tilde{G}) = (f, G), \tag{2.31}$$

where G is as in (2.29), and $\tilde{G}$ is associated with $\tilde{\Lambda}$ also as in (2.29). Indeed, if $(H, \tilde{G}) \in L^p(\partial\Omega) \times L^p(\partial\Omega)$ is a solution of (2.31), then $u := \mathcal{S}(\partial_\tau\tilde{G}, H)$ solves the Neumann problem (NBH_p) in (1.2).

2.2. Hardy kernel operators and the Mellin transform. The goal here is to recall the class of Hardy kernels and Hardy kernel operators on $L^p(\mathbb{R}_+)$, for $p \in [1,\infty)$, and to introduce the Mellin transform as a tool to study spectral properties of Hardy kernel operators.

DEFINITION 2.1. Let $k\colon \mathbb{R}_+ \times \mathbb{R}_+ \to \mathbb{R}$ be a Lebesgue measurable function. Then k is a Hardy kernel for $L^p(\mathbb{R}_+)$ for some $1 \leqslant p < \infty$ provided that:

(1) k is a homogeneous function of degree -1, i.e., for any $\lambda > 0$ and any $x, y \in \mathbb{R}_+$, there holds $k(\lambda x, \lambda y) = \lambda^{-1} k(x,y)$;

(2) $\displaystyle\int_0^\infty |k(1,y)| y^{-1/p} dy < \infty.$

Furthermore, if $\ell, m \in \mathbb{N}$, a matrix-valued function $k\colon \mathbb{R}_+ \times \mathbb{R}_+ \to \mathbb{R}^{\ell\times m}$ whose entries are measurable is called a Hardy kernel for $(L^p(\mathbb{R}_+))^m$ provided that each entry k_{ij}, for $i \in \{1,\ldots,\ell\}$ and for $j \in \{1,\ldots,m\}$, is a Hardy kernel for $L^p(\mathbb{R}_+)$.

With Definition 2.1 in hand, let k be a Hardy kernel for $L^p(\mathbb{R}_+)$, $1 \leqslant p < \infty$ and for any $f \in L^p(\mathbb{R}_+)$ define the action of the Hardy kernel operator $\mathfrak{T}$ with kernel k on the function f by setting

$$\mathfrak{T}f(x) := \int_0^\infty k(x,y) f(y) dy, \quad x \in \mathbb{R}_+. \tag{2.32}$$

The setup of the vector-valued case follows a similar blueprint. Concretely, fix two integers $\ell, m \in \mathbb{N}$ and let $k = (k_{ij})_{1\leqslant i\leqslant \ell, 1\leqslant j\leqslant m}$ be a Hardy kernel for $(L^p(\mathbb{R}_+))^m$. For any vector-valued function $\vec{f} \in (L^p(\mathbb{R}_+))^m$, define the action of the operator $\mathfrak{T}$, called a Hardy kernel operator with kernel k, on $\vec{f}$ by setting

$$\mathfrak{T}\vec{f}(s) := \int_0^\infty k(s,t) \cdot \vec{f}(t)\, dt, \quad s \in \mathbb{R}_+, \tag{2.33}$$

where $\cdot$ denotes matrix multiplication.

Next we recall the Mellin transform. Given $b, c \in \mathbb{R}$ satisfying $b < c$, we shall denote by $\Gamma_{b,c}$ the infinite strip

$$\Gamma_{b,c} := \{z \in \mathbb{C} : \mathfrak{Re}\, z \in (b,c)\}, \tag{2.34}$$

where $\mathfrak{Re}$ denotes real part.

DEFINITION 2.2. Let $a, b, c \in \mathbb{R}$ such that $b < c$ and set

$$\mathcal{A}_a := \Big\{f\colon \mathbb{R}_+ \to \mathbb{C} \text{ measurable function} : \int_0^\infty |f(x)| x^{a-1} dx < \infty\Big\}, \tag{2.35}$$

and

$$\mathcal{A}_{(b,c)} := \bigcap_{a\in(b,c)} \mathcal{A}_a. \tag{2.36}$$

DEFINITION 2.3. Assume that $a \in \mathbb{R}$ and that $f \in \mathcal{A}_a$. Set

$$\mathcal{M}f(z) := \int_0^\infty x^{z-1} f(x) dx \quad \text{for each } z \in a + i\mathbb{R}. \tag{2.37}$$

$\mathcal{M}f(z)$ is called the Mellin transform of the function f at z. The fundamental strip of $\mathcal{M}f$ is the largest strip $\Gamma_{b,c}$ with $b, c \in \mathbb{R}$, $b < c$, such that $f \in \mathcal{A}_{(b,c)}$.

Consider next the Hilbert transform of the positive semi-axis. Concretely, for a fixed $y \in \mathbb{R}_+$, set

$$h\colon \mathbb{R}_+ \setminus \{y\} \to \mathbb{R}, \quad h(x) := \frac{1}{x-y}. \tag{2.38}$$

Note that $h \notin \mathcal{A}_c$ for any $c \in \mathbb{R}$ because for a fixed $c \in \mathbb{R}$ and for $0 < \varepsilon < y$ there holds

$$\lim_{t\to y^-} \int_\varepsilon^t \frac{x^{c-1}}{x-y} dx = -\infty. \tag{2.39}$$

However $\lim_{\varepsilon\to 0^+} \int_{\mathbb{R}_+\setminus B(y,\varepsilon)} h(x)x^{z-1}dx$ exists whenever $z \in \Gamma_{0,1}$. In particular, since the function h does not fall under the scope of Definition 2.3, it is meaningful to set

$$\mathcal{M}h(z) := \lim_{\varepsilon\to 0^+} \int_{\mathbb{R}_+\setminus B(y,\varepsilon)} h(x)x^{z-1}dx \quad \text{for each } z \in \Gamma_{0,1}. \tag{2.40}$$

In relation to this, the following result can be found in [**9**].

PROPOSITION 2.4. Fix $y \in \mathbb{R}_+$ and let h be as in (2.38). Then,

$$\mathcal{M}h(z) = -\pi y^{z-1}\cot(\pi z) \quad \text{for each } z \in \Gamma_{0,1}. \tag{2.41}$$

The following result, see [**3**] and [**4**], allows one to determine explicitly the spectrum of bounded linear operators which belong to the algebra generated by Hardy kernel operators and the Hilbert transform on $\mathbb{R}_+$.

THEOREM 2.5. *Let $k = (k_{ij})_{1\leqslant i,j\leqslant m}$ be a Hardy kernel for $(L^p(\mathbb{R}_+))^m$, for some $p \in (1,\infty)$ and $m \in \mathbb{N}$, and let A, B be $m\times m$ matrices with real entries. Consider the operator $\mathcal{R}$ defined as*

$$\mathcal{R}f(s) := Af + \int_0^\infty \mathfrak{R}(s,t)f(t)\,dt, \quad s \in \mathbb{R}_+ \tag{2.42}$$

with

$$\mathfrak{R}(s,t) := k(s,t) + \frac{1}{s-t}B \qquad \forall s,t \in \mathbb{R}_+ \text{ such that } s \neq t. \tag{2.43}$$

Then the operator

$$\mathcal{R}\colon (L^p(\mathbb{R}_+))^m \to (L^p(\mathbb{R}_+))^m \tag{2.44}$$

is well defined, linear and bounded, with spectrum

$$\sigma(\mathcal{R}; (L^p(\mathbb{R}_+))^m) = \overline{\mathcal{O}}, \tag{2.45}$$

where $\overline{\mathcal{O}}$ stands for the closure in $\mathbb{C}$ of the set $\mathcal{O}$, which is defined as

$$\mathcal{O} := \Big\{\omega \in \mathbb{C} : \det\big(\omega I - (A + \mathcal{M}\mathfrak{R}(\cdot,1))(\tfrac{1}{p} + i\xi)\big) = 0 \text{ for some } \xi \in \mathbb{R}\Big\}. \tag{2.46}$$

A useful consequence of Theorem 2.5 is singled out below (see, e.g., [**7**] for a proof).

COROLLARY 2.6. Retain the setting of Theorem 2.5 and make the additional assumption that $\det(A - \pi iB) \neq 0$. Then the operator $\mathcal{R}$ is invertible on $(L^p(\mathbb{R}_+))^m$ if and only if

$$\det\big(A + \mathcal{M}\mathfrak{R}(\cdot,1)(\tfrac{1}{p} + i\xi)\big) \neq 0 \quad \text{for each } \xi \in \mathbb{R}. \tag{2.47}$$

Corollary 2.6 ultimately allows us to establish the following invertibility result for the operator $\mathcal{T}$ from (1.22).

THEOREM 2.7. *Let Ω be the interior of an infinite sector of aperture $\theta \in (0, 2\pi)$ and fix $\eta \in [-1, 1)$ and $p \in (1, \infty)$. Consider the operator $\mathcal{T}$ from (1.22) with matrix-valued kernel R as in (1.23). Then*

$$\begin{aligned}&\mathcal{T} \text{ is invertible on } \big(L^p(\mathbb{R}_+) \times L^p(\mathbb{R}_+)\big)^2 \text{ if and only if}\\ &\text{for each } \xi \in \mathbb{R} \text{ there holds } \det\big(\mathcal{M}R(\cdot, 1)\big)\big(\tfrac{1}{p} + i\xi\big) \neq 0.\end{aligned} \tag{2.48}$$

PROOF. Using Lemma 3.3 the operator $\mathcal{T}$ is of the form discussed in Theorem 2.5 for

$$B := \tfrac{\eta+1}{4\pi} \begin{pmatrix} -1 & 0 & 0 & 0 \\ 0 & 1 & 0 & 0 \\ 0 & 0 & -1 & 0 \\ 0 & 0 & 0 & 1 \end{pmatrix} \quad \text{and} \quad A := -\tfrac{1}{2} \begin{pmatrix} O_{2\times 2} & I_{2\times 2} I_{2\times 2} & O_{2\times 2} \end{pmatrix}, \tag{2.49}$$

where $O_{2\times 2}$ is the 2×2 zero matrix and $I_{2\times 2}$ is the 2×2 identity matrix (and A is understood as a 4×4 matrix). With (2.49) in hand, a simple calculation shows $\det(A - \pi i B) = \Big(-\frac{(\eta+1)^2}{16} + \frac{1}{4}\Big)^2 \neq 0$ for all $\eta \in [-1, 1)$. Thus Corollary 2.6 applies and gives (2.48). □

3. Explicit formulas for the operator $\mathcal{T}$

A preliminary step in the analysis of the operator T from (1.7) is determining explicit expressions for the kernel functions k_i for $i \in \{1, \ldots, 8\}$ from (1.9)–(1.14), which then via the identifications (1.20) lead to a more explicit formula for the operator $\mathcal{T}$ when Ω is the infinite sector above the graph of the function $\phi \colon \mathbb{R} \to \mathbb{R}$ in (1.1) of arbitrary aperture $\theta \in (0, 2\pi)$.

Using formulas (1.15)–(1.16), it is straightforward to compute at each point $X = (X_1, X_2) \in \mathbb{R}^2 \setminus \{0\}$ and for each $j \in \{1, 2\}$ that

$$\partial_j E(X) = \frac{X_j}{2\pi |X|^2}, \qquad \partial_j B(X) = -\frac{X_j}{8\pi}\left(1 - \log(|X|^2)\right), \tag{3.1}$$

$$\partial_1 \partial_2 B(X) = \frac{X_1 X_2}{4\pi |X|^2}, \qquad \partial_j^2 B(X) = -\frac{1}{8\pi}\Big(1 - \log(|X|^2) - \frac{2X_j^2}{|X|^2}\Big), \tag{3.2}$$

and that

$$\partial_1^2 \partial_2 B(X) = \frac{X_2^3 - X_1^2 X_2}{4\pi |X|^4}, \qquad \partial_1 \partial_2^2 B(X) = \frac{X_1^3 - X_1 X_2^2}{4\pi |X|^4}, \tag{3.3}$$

$$\partial_1^3 B(X) = \frac{X_1^3 + 3X_1 X_2^2}{4\pi |X|^4}, \qquad \partial_2^3 B(X) = \frac{X_2^3 + 3X_1^2 X_2}{4\pi |X|^4}. \tag{3.4}$$

Then (2.5), (1.9)–(1.14) and (3.1)–(3.4) allow us to obtain the following formulas.

LEMMA 3.1. *Assume $\Omega \subset \mathbb{R}^2$ is a Lipschitz domain with outward unit normal vector $\nu = (\nu_1, \nu_2)$. For each $P = (P_1, P_2) \in \partial\Omega$ and $Q = (Q_1, Q_2) \in \partial\Omega$, $P \neq Q$,* *the functions k_i, $i \in \{1, \dots, 8\}$, introduced in* (1.9)–(1.14) *satisfy*

$$k_1(P,Q) = k_7(P,Q) = \frac{\nu_2(Q)(P_1 - Q_1) - \nu_1(Q)(P_2 - Q_2)}{2\pi|P-Q|^2}, \tag{3.5}$$

$$\begin{aligned} k_2(P,Q) = {} & (P_1 - Q_1)^3 \cdot \frac{\nu_2(Q) - 2\nu_1(Q)\nu_1(P)\nu_2(P)}{4\pi|P-Q|^4} \\ & + (P_2 - Q_2)^3 \cdot \frac{-\nu_1(Q) + 2\nu_2(Q)\nu_1(P)\nu_2(P)}{4\pi|P-Q|^4} \\ & + (P_1 - Q_1)(P_2 - Q_2)^2 \cdot \frac{\nu_2(Q)(3\nu_1^2(P) - \nu_2^2(P))}{4\pi|P-Q|^4} \\ & + (P_1 - Q_1)^2(P_2 - Q_2) \cdot \frac{\nu_1(Q)(-3\nu_2^2(P) + \nu_1^2(P))}{4\pi|P-Q|^4} \\ & - \frac{(P_1 - Q_1)(P_2 - Q_2)\nu_1(P)\nu_2(P)}{|P-Q|^2} \cdot k_1(P,Q), \end{aligned} \tag{3.6}$$

$$k_3(P,Q) = k_5(P,Q) = -\frac{\langle P - Q, \nu(Q)\rangle}{2\pi|P-Q|^2}, \tag{3.7}$$

$$\begin{aligned} k_4(P,Q) = {} & -(P_1 - Q_1)^3 \cdot \frac{\nu_1(Q) + 2\nu_1(P)\nu_2(P)\nu_2(Q)}{4\pi|P-Q|^4} \\ & - (P_2 - Q_2)^3 \cdot \frac{\nu_2(Q) + 2\nu_1(P)\nu_2(P)\nu_1(Q)}{4\pi|P-Q|^4} \\ & - (P_1 - Q_1)(P_2 - Q_2)^2 \cdot \frac{\nu_1(Q)(3\nu_1^2(P) - \nu_2^2(P))}{4\pi|P-Q|^4} \\ & - (P_1 - Q_1)^2(P_2 - Q_2) \cdot \frac{\nu_2(Q)(3\nu_2^2(P) - \nu_1^2(P))}{4\pi|P-Q|^4} \\ & - \frac{(P_1 - Q_1)(P_2 - Q_2)\nu_1(P)\nu_2(P)}{|P-Q|^2} \cdot k_3(P,Q), \end{aligned} \tag{3.8}$$

$$\begin{aligned} k_6(P,Q) = {} & -(P_1 - Q_1)^3 \frac{(\nu_1^2(P) - \nu_2^2(P))\nu_1(Q)}{4\pi|P-Q|^4} \\ & + \frac{2(P_1 - Q_1)(P_2 - Q_2)\nu_1(P)\nu_2(P)}{|P-Q|^2} \cdot k_3(P,Q) \\ & - \frac{(P_1 - Q_1)(P_2 - Q_2)(\nu_1^2(P) - \nu_2^2(P))}{2|P-Q|^2} \cdot k_1(P,Q) \\ & + (P_2 - Q_2)^3 \frac{(\nu_1^2(P) - \nu_2^2(P))\nu_2(Q)}{4\pi|P-Q|^4}, \end{aligned} \tag{3.9}$$

and

$$
\begin{aligned}
k_8(P,Q) = &-(P_1-Q_1)^3\frac{(\nu_1^2(P)-\nu_2^2(P))\nu_2(Q)}{4\pi|P-Q|^4} \\
&-(P_2-Q_2)^3\frac{(\nu_1^2(P)-\nu_2^2(P))\nu_1(Q)}{4\pi|P-Q|^4} \\
&-\frac{2(P_1-Q_1)(P_2-Q_2)\nu_1(P)\nu_2(P)}{|P-Q|^2}\cdot k_1(P,Q) \\
&-\frac{(P_1-Q_1)(P_2-Q_2)(\nu_1^2(P)-\nu_2^2(P))}{2|P-Q|^2}\cdot k_3(P,Q).
\end{aligned}
\tag{3.10}
$$

The following remark will be useful in the sequel.

REMARK 3.2. Let Ω be the set above the graph of the function ϕ from (1.1). With $(\partial\Omega_1)$ and $(\partial\Omega)_2$ being the left and right rays of $\partial\Omega$, respectively, with corresponding direction vectors $v_1(\theta)$ and $v_2(\theta)$ as in (1.17)–(1.18), pick two points $P,Q\in\partial\Omega=(\partial\Omega)_1\cup(\partial\Omega)_2$ such that $P\neq Q$ and let $s:=|P|$ and $t:=|Q|$. Recall the outward unit normal vectors $\vec{v}_3(\theta)$ and $\vec{v}_4(\theta)$ as in (1.19). Then clearly:

$$\text{if } P,Q\in(\partial\Omega)_1:\ P=s\vec{v}_1(\theta),\ Q=t\vec{v}_1(\theta)\ \text{ and }\ \nu(P)=\nu(Q)=\vec{v}_3(\theta), \tag{3.11}$$

$$\text{if } P,Q\in(\partial\Omega)_2:\ P=s\vec{v}_2(\theta),\ Q=t\vec{v}_2(\theta)\ \text{ and }\ \nu(P)=\nu(Q)=\vec{v}_4(\theta), \tag{3.12}$$

$$\text{if } P\in(\partial\Omega)_1,\ Q\in(\partial\Omega)_2:\ \begin{cases} P=s\vec{v}_1(\theta),\\ Q=t\vec{v}_2(\theta),\end{cases} \text{ and } \begin{cases}\nu(P)=\vec{v}_3(\theta),\\ \nu(Q)=\vec{v}_4(\theta),\end{cases} \tag{3.13}$$

$$\text{if } P\in(\partial\Omega)_2,\ Q\in(\partial\Omega)_1:\ \begin{cases} P=s\vec{v}_2(\theta),\\ Q=t\vec{v}_1(\theta),\end{cases} \text{ and } \begin{cases}\nu(P)=\vec{v}_4(\theta),\\ \nu(Q)=\vec{v}_3(\theta).\end{cases} \tag{3.14}$$

Moving on, for each $j\in\{1,\dots,8\}$ and each $l,m\in\{1,2\}$, we shall introduce

$$
\begin{aligned}
&k_j^{lm}:\mathbb{R}_+\times\mathbb{R}_+\setminus\operatorname{diag}(\mathbb{R}_+)\longrightarrow\mathbb{R}\ \text{ given by }\ k_j^{lm}(s,t):=k_j(P,Q)\\
&\text{for }\ P\in(\partial\Omega)_l\ \text{ and }\ Q\in(\partial\Omega)_m\ \text{ such that }\ |P|=s\ \text{ and }\ |Q|=t,
\end{aligned}
\tag{3.15}
$$

where the functions k_j, $j\in\{1,\dots,8\}$, are as in (1.9)–(1.14). With this notation in hand, consider the matrix-valued functions κ_j acting on $\mathbb{R}_+\times\mathbb{R}_+\setminus\operatorname{diag}(\mathbb{R}_+)$ with values in $M_{2\times2}(\mathbb{R})$ given by

$$\kappa_j:=\begin{pmatrix} k_j^{11} & k_j^{12}\\ k_j^{21} & k_j^{22}\end{pmatrix},\qquad j\in\{1,\dots,8\}. \tag{3.16}$$

A direct consequence (whose elementary proof we omit) of Lemma 3.1 and (3.11)–(3.14) in Remark 3.2 is the following.

LEMMA 3.3. *Fix $\theta\in(0,2\pi)$ and for each $j\in\{1,\dots,8\}$ and $l,m\in\{1,2\}$ let k_j^{lm} be as in (3.15) and κ_j as in (3.16). Then, corresponding to $j=1$ and $j=2$,* *for each* $s,t\in\mathbb{R}_+$ *with* $s\neq t$ *there hold*

$$\begin{cases} k_1^{11}(s,t) = -k_1^{22}(s,t) = \dfrac{1}{2\pi(s-t)}, \\ -k_1^{12}(s,t) = k_1^{21}(s,t) = \dfrac{s\cos\theta - t}{2\pi(s^2-2st\cos\theta+t^2)}, \end{cases} \tag{3.17}$$

$$\begin{cases} k_2^{11}(s,t) = -k_2^{22}(s,t) = \dfrac{1}{4\pi(s-t)}, \\ -k_2^{12}(s,t) = k_2^{21}(s,t) = \dfrac{f(s,t)}{4\pi(s^2-2st\cos\theta+t^2)^2}, \end{cases} \tag{3.18}$$

where

$$\begin{aligned} f(s,t) :=& -(s+t)^3(1+2\cos^2(\theta/2))\sin^4(\theta/2) + (s-t)^3(1+2\sin^2(\theta/2))\cos^4(\theta/2) \\ &- 2s(s+t)(s-t)\sin^2(\theta/2)\cos^2(\theta/2)\cos\theta. \end{aligned} \tag{3.19}$$

Corresponding to $j=3$ *and* $j=4$, *for each* $s,t\in\mathbb{R}_+$ *with* $s\neq t$ *there hold*

$$\begin{cases} k_3^{11}(s,t) = k_3^{22}(s,t) = 0, \\ k_3^{12}(s,t) = k_3^{21}(s,t) = \dfrac{s\sin\theta}{2\pi(s^2-2st\cos\theta+t^2)}, \end{cases} \tag{3.20}$$

$$\begin{cases} k_4^{11}(s,t) = k_4^{22}(s,t) = 0, \\ k_4^{12}(s,t) = k_4^{21}(s,t) = \dfrac{g(s,t)}{4\pi(s^2-2st\cos\theta+t^2)^2}, \end{cases} \tag{3.21}$$

where

$$\begin{aligned} g(s,t) :=&(s+t)^3\sin^3(\theta/2)\cos(\theta/2)\cos\theta - (s-t)^3\sin(\theta/2)\cos^3(\theta/2)\cos\theta \\ &+ (s+t)^2(s-t)\sin^2(\theta/2)\cos(\theta/2)[\sin(\theta/2)\cos^2(\theta/2)+3\sin^3(\theta/2)] \\ &+ (s+t)(s-t)^2\sin(\theta/2)\cos^2(\theta/2)[\sin^2(\theta/2)\cos(\theta/2)+3\cos^3(\theta/2)]. \end{aligned} \tag{3.22}$$

Corresponding to $j=6$, *for each* $s,t\in\mathbb{R}_+$ *with* $s\neq t$ *there holds*

$$\begin{cases} k_6^{11}(s,t) = k_6^{22}(s,t) = 0, \\ k_6^{12}(s,t) = k_6^{21}(s,t) = \dfrac{h(s,t)}{4\pi(s^2-2st\cos\theta+t^2)^2}, \end{cases} \tag{3.23}$$

where

$$\begin{aligned} h(s,t) :=&(s+t)^3\sin^3(\theta/2)\cos(\theta/2)\cos\theta - (s-t)^3\sin(\theta/2)\cos^3(\theta/2)\cos\theta \\ &- (s+t)^2(s-t)\sin^2(\theta/2)\cos(\theta/2)\big[3\sin(\theta/2)\cos^2(\theta/2)+\sin^3(\theta/2)\big] \\ &- (s+t)(s-t)^2\sin(\theta/2)\cos^2(\theta/2)\big[3\sin^2(\theta/2)\cos(\theta/2)+\cos^3(\theta/2)\big]. \end{aligned} \tag{3.24}$$

Corresponding to $j = 8$, for each $s, t \in \mathbb{R}_+$ with $s \neq t$ there holds

$$\begin{cases} k_8^{11}(s,t) = -k_8^{22}(s,t) = \dfrac{1}{4\pi(s-t)}, \\ -k_8^{12}(s,t) = k_8^{21}(s,t) = \dfrac{\ell(s,t)}{4\pi(s^2 - 2st\cos\theta + t^2)^2}, \end{cases} \tag{3.25}$$

where

$$\begin{aligned} \ell(s,t) :=& (s+t)^3 \sin^4(\theta/2)\cos\theta + (s-t)^3\cos^4(\theta/2)\cos\theta \\ & - \tfrac{1}{2}s(s+t)(s-t)\sin^2\theta\cos\theta \\ & - (s+t)^2(s-t)\sin^2\theta\sin^2(\theta/2) + (s+t)(s-t)^2\sin^2\theta\cos^2(\theta/2). \end{aligned} \tag{3.26}$$

Corresponding to $j = 5$ and $j = 7$, for each $s, t \in \mathbb{R}_+$ with $s \neq t$ there hold

$$\kappa_5(s,t) = \kappa_3(s,t) \quad \textit{and} \quad \kappa_7(s,t) = \kappa_1(s,t). \tag{3.27}$$

Moreover, for each $j \in \{1, \dots, 8\}$ and each $l, m \in \{1, 2\}$, the function k_j^{lm} is either a Hardy kernel or a scalar multiple of the kernel of the Hilbert transform.

4. Mellin symbol computations

The main goal of this section is to compute the Mellin symbol of the matrix-valued functions κ_i, $i \in \{1, \dots, 8\}$, introduced in (3.15)–(3.16), and explicitly calculated in Lemma 3.3.

PROPOSITION 4.1. Fix a parameter $\theta \in (0, 2\pi)$ and consider the functions $F_j \colon \mathbb{R}_+ \to \mathbb{R}$, $j \in \{1, \dots, 4\}$, given at $x \in \mathbb{R}_+$ by

$$\begin{aligned} F_1(x) &:= \frac{1}{x^2 + 2x\cos(\pi - \theta) + 1}, \\ F_2(x) := xF_1(x), \quad F_3(x) &:= F_1'(x) \text{ and } F_4(x) := F_2'(x). \end{aligned} \tag{4.1}$$

Then,

$$F_1 \in \mathcal{A}_{(0,2)}, \ F_2 \in \mathcal{A}_{(-1,1)}, \ F_3 \in \mathcal{A}_{(1,3)} \text{ and } F_4 \in \mathcal{A}_{(0,2)}, \tag{4.2}$$

and, if $\theta \in (0, 2\pi) \setminus \{\pi\}$,

$$\mathcal{M}F_1(z) = \pi \csc\theta \csc(\pi z)\sin(\theta + z(\pi - \theta)) \quad \text{for each } z \in \Gamma_{0,2}, \tag{4.3}$$

$$\mathcal{M}F_2(z) = \pi \csc\theta \csc(\pi z)\sin(z(\pi - \theta)) \quad \text{for each } z \in \Gamma_{-1,1}, \tag{4.4}$$

$$\mathcal{M}F_3(z) = -\frac{\pi(z-1)\csc\theta}{\sin(\pi z)}\sin(2\theta + z(\pi - \theta)) \quad \text{for each } z \in \Gamma_{1,3}, \tag{4.5}$$

$$\mathcal{M}F_4(z) = -\frac{\pi(z-1)\csc\theta}{\sin(\pi z)}\sin(z(\pi - \theta) + \theta) \quad \text{for each } z \in \Gamma_{0,2}. \tag{4.6}$$

When $\theta = \pi$, the following hold:

$$\mathcal{M}F_1(z) = -\frac{\pi(z-1)}{\sin(\pi z)} \quad \text{for each } z \in \Gamma_{0,2}, \tag{4.7}$$

$$\mathcal{M}F_2(z) = \frac{\pi z}{\sin(\pi z)} \quad \text{for each } z \in \Gamma_{-1,1}, \tag{4.8}$$

$$\mathcal{M}F_3(z) = -\frac{\pi(z-2)}{\sin(\pi z)} \quad \text{for each } z \in \Gamma_{1,3}, \tag{4.9}$$

$$\mathcal{M}F_4(z) = -\frac{\pi(z-1)}{\sin(\pi z)} \quad \text{for each } z \in \Gamma_{0,2}. \tag{4.10}$$

PROOF. When $\theta \neq \pi$, that $F_1 \in \mathcal{A}_{(0,2)}$ and that (4.3) holds are consequences of [**9**, formula (2.54), p. 23] applied for $\pi - \theta \in (-\pi, \pi) \setminus \{0\}$ in place of θ. The case (4.7) when $\theta = \pi$, follows from [**9**, formula (2.5), p. 13]. Regarding F_2, since $F_2(x) = xF_1(x)$ and $F_1 \in \mathcal{A}_{(0,2)}$, basic properties of the Mellin transform yield that $F_2 \in \mathcal{A}_{(-1,1)}$ and that

$$\mathcal{M}F_2(z) = \mathcal{M}F_1(z+1) \quad \text{for each } z \in \Gamma_{-1,1}. \tag{4.11}$$

This together with (4.3), (4.7) and the fact that $\sin(\pi(z+1)) = -\sin(\pi z)$ gives (4.4) and (4.8).

Turning our attention to F_3 we notice first that

$$\lim_{x\to 0} F_1(x)x^{z-1} = 0 \quad \text{and} \quad \lim_{x\to\infty} F_1(x)x^{z-1} = 0 \quad \text{for each } z \in \Gamma_{0,2}. \tag{4.12}$$

Thus, [**9**, formula (1.9), p. 11] guarantees that $F_3 \in \mathcal{A}_{(1,3)}$ and

$$\mathcal{M}F_3(z) = -(z-1)\mathcal{M}F_1(z-1) \quad \text{for each } z \in \Gamma_{1,3}. \tag{4.13}$$

Using this, (4.3), and (4.7), we obtain (4.5) and (4.9). The function F_2 satisfies similar properties to those of F_1 in (4.12) formulated this time in the strip $\Gamma_{-1,1}$. Appealing again to [**9**, formula (1.9), p. 11] it follows that $F_4 \in \mathcal{A}_{(0,2)}$ and

$$\mathcal{M}F_4(z) = -(z-1)\mathcal{M}F_2(z-1) \quad \text{for each } z \in \Gamma_{0,2}. \tag{4.14}$$

This together with (4.4) and (4.8) yield (4.6) and (4.10). □

Going further, for $z \in \Gamma_{0,1}$ and $\theta \in (0, 2\pi)$ set

$$a(z) := \frac{\cos(\pi z)}{2\sin(\pi z)}, \quad b_\theta(z) := \frac{\cos\big((\pi-\theta)z\big)}{2\sin(\pi z)}, \quad d_\theta(z) := \frac{\sin\big((\pi-\theta)z\big)}{2\sin(\pi z)}. \tag{4.15}$$

With this notation in hand, we provide next explicit formulas for the Mellin transform of the kernels κ_1, κ_3, κ_5, and κ_7.

LEMMA 4.2. *Let $\theta \in (0, 2\pi)$ be fixed and recall the matrix-valued functions κ_i for $i \in \{1, \ldots, 8\}$ from (3.16). For each $z \in \Gamma_{0,1}$ there holds*

$$\mathcal{M}(\kappa_1(\cdot, 1))(z) = \mathcal{M}(\kappa_7(\cdot, 1))(z) = \begin{pmatrix} -a(z) & b_\theta(z) \\ -b_\theta(z) & a(z) \end{pmatrix}, \tag{4.16}$$

and

$$\mathcal{M}(\kappa_3(\cdot, 1))(z) = \mathcal{M}(\kappa_5(\cdot, 1))(z) = d_\theta(z) \cdot \begin{pmatrix} 0 & 1 \\ 1 & 0 \end{pmatrix}. \tag{4.17}$$

PROOF. With an eye towards proving (4.16), recall the functions F_1, F_2 from (4.1) and let $h\colon \mathbb{R}_+ \longrightarrow \mathbb{R}_+$ be given by $h(x) := 1/x$. Notice that using the first set of identities in (3.17), we may write

$$k_1^{11}(s,1) = -k_1^{22}(s,1) = \frac{1}{2\pi}\, h(s-1) \quad \text{for each } s \in \mathbb{R}_+. \tag{4.18}$$

Using the second set of identities in (3.17), there holds

$$-k_1^{12}(s,1) = k_1^{21}(s,1) = \tfrac{1}{2\pi}((\cos\theta)F_2(s) - F_1(s)) \quad \text{for each } s \in \mathbb{R}_+. \tag{4.19}$$

Appealing now to (4.18)–(4.19), (4.3) and (4.7), and (4.4) and (4.8), together with Proposition 2.4 and the first equality in (3.27), we obtain that (4.16) is valid for each $z \in \Gamma_{0,1}$.

Turning our attention to the statement made in (4.17), using (3.20) we have

$$\left.\begin{aligned} k_3^{11}(s,1) = k_3^{22}(s,1) = 0,\\ k_3^{12}(s,1) = k_3^{21}(s,1) = \frac{\sin\theta}{2\pi}\, F_2(s), \end{aligned}\right\} \quad \text{for each } s \in \mathbb{R}_+. \tag{4.20}$$

Next, using (4.20) and formulas (4.4) and (4.8) from Proposition 4.1 along with the second equality in (3.27), we obtain that (4.17) holds for each $z \in \Gamma_{0,1}$. This completes the proof of the Lemma. □

The next result will be useful when computing the Mellin transform of the remaining kernels.

LEMMA 4.3. *For $\theta \in (-\pi, \pi)$, consider the function $g\colon \mathbb{R}_+ \to \mathbb{R}_+$, given by*

$$g(x) := \frac{1}{(x^2 + 2x\cos\theta + 1)^2}. \tag{4.21}$$

Then $g \in \mathcal{A}_{(0,4)}$. In addition, whenever $z \in \Gamma_{0,4} \setminus \{1,2,3\}$, and $\theta \in (-\pi,\pi)\setminus\{0\}$, there hold

$$\mathcal{M}g(z) = \frac{\pi}{4(\sin^3\theta)\sin(\pi z)}\,[(z-3)\sin((z-1)\theta) - (z-1)\sin((z-3)\theta)] \tag{4.22}$$

and, when $\theta = 0$,

$$\mathcal{M}g(z) = -\pi\,\frac{(z-1)(z-2)(z-3)}{6\sin(\pi z)}. \tag{4.23}$$

PROOF. Using [**9**, formula (2.58), p. 24], the function g satisfies $g \in \mathcal{A}_{(0,4)}$. Next, let us observe that if $\theta = 0$, then $g(x) = (x+1)^{-4}$ and on grounds of [**9**, formula (2.5), p. 13], there holds

$$\mathcal{M}g(z) = -\pi\,\frac{(z-1)(z-2)(z-3)}{6\sin(\pi z)}. \tag{4.24}$$

When $\theta \in (-\pi,\pi)\setminus\{0\}$ we invoke again [**9**, formula (2.58), p. 24] to write that for each $z \in \Gamma_{0,4}$

$$\mathcal{M}g(z) = 2^{3/2}(\sin\theta)^{-3/2}\Gamma(\tfrac{5}{2})\cdot B(z, 4-z)\cdot \mathsf{P}^{-3/2}_{z-5/2}(\cos\theta), \tag{4.25}$$

where Γ is the gamma function, B is the beta function, and P^{μ}_{ν} is the Ferrers function (evaluated at $\cos\theta \in (-1,1)$). Since $B(z,4-z) = \frac{\Gamma(z)\Gamma(4-z)}{\Gamma(4)}$, using Euler's reflection formula (here we use that $z \notin \mathbb{Z}$) we obtain

$$B(z, 4-z) = \pi\,\frac{(3-z)(2-z)(1-z)}{6\sin(\pi z)}. \tag{4.26}$$

Appealing to [**9**, p. 262] and elementary trigonometry, we may write

$$\begin{aligned}\mathsf{P}^{-3/2}_{z-5/2}(\cos\theta) &= \frac{1}{\Gamma(\frac{5}{2})}\left[\frac{1+\cos\theta}{1-\cos\theta}\right]^{-3/4} {}_2F_1(-z+\tfrac{5}{2}, z-\tfrac{3}{2}; \tfrac{5}{2}; \tfrac{1-\cos\theta}{2}) \\ &= \frac{1}{\Gamma(\frac{5}{2})}\left[\frac{\sin(\theta/2)}{\cos(\theta/2)}\right]^{3/2} {}_2F_1(-z+\tfrac{5}{2}, z-\tfrac{3}{2}; \tfrac{5}{2}; \sin^2(\tfrac{\theta}{2})),\end{aligned} \tag{4.27}$$

where ${}_2F_1$ is Gauss' hypergeometric function. At this point we find it useful to recall the two following formulas from [**1**, formula (15.2.6), p. 557] (see also [**10**, (15.5.E6)]) and [**12**, formula (6c), p. 2] (see also [**10**, (15.4.E16)]):

$${}_2F_1\left(\tfrac{1}{2}+a, \tfrac{1}{2}-a; \tfrac{3}{2}; \sin^2\alpha\right) = \frac{\sin(2a\alpha)}{2a\sin\alpha} \quad \text{for any } \alpha \in (-\tfrac{\pi}{2}, \tfrac{\pi}{2}), \tag{4.28}$$

and, when c is not a non-positive integer and $|x| < 1$,

$$\frac{d}{dx}[(1-x)^{a+b-c}{}_2F_1(a,b;c;x)] = \tfrac{(c-a)(c-b)}{c}(1-x)^{a+b-c-1}{}_2F_1(a,b;c+1;x). \tag{4.29}$$

Using (4.28), with $a = 2-z$ and $\alpha = \theta/2$, gives

$${}_2F_1(-z+\tfrac{5}{2}, z-\tfrac{3}{2}; \tfrac{3}{2}; \sin^2(\tfrac{\theta}{2})) = \frac{\sin((2-z)\theta)}{2(2-z)\sin\frac{\theta}{2}}. \tag{4.30}$$

Letting $F(x) := {}_2F_1\left(-z+\frac{5}{2}, z-\frac{3}{2}; \frac{3}{2}; x\right)$ and $x := \sin^2(\theta/2)$, using the chain rule we may write $\frac{dF}{dx}(x) = \frac{dF}{d\theta}(x)\frac{d\theta}{dx}(x)$. Differentiating with respect to θ in (4.30) and using that $\frac{d\theta}{dx}(x) = 2/\sin\theta$, this gives

$$\frac{dF}{dx}(x) = \frac{(2-z)\cos((2-z)\theta)\sin\frac{\theta}{2} - \frac{1}{2}\sin((2-z)\theta)\cos\frac{\theta}{2}}{(2-z)(\sin\theta)\sin^2\frac{\theta}{2}}. \tag{4.31}$$

Applying (4.29) for the parameters $a = -z+\frac{5}{2}$, $b = z-\frac{3}{2}$ and $c = \frac{3}{2}$ gives that, on the one hand

$$\frac{d}{dx}[(1-x)^{-1/2}F(x)] = \frac{(z-1)(3-z)}{(3/2)}(1-x)^{-3/2}{}_2F_1(-z+\tfrac{5}{2}, z-\tfrac{3}{2}; \tfrac{5}{2}; x). \tag{4.32}$$

Using the chain rule, the definition of F, (4.31), and elementary trigonometry, we conclude on the other hand that

$$\begin{aligned}\frac{d}{dx}[(1-x)^{-1/2}F(x)]\Big|_{x=\sin^2(\frac{\theta}{2})} &= \frac{2}{(2-z)\sin^3\theta} \\ &\quad\times [(2-z)\cos((2-z)\theta)\sin\theta - \sin((2-z)\theta)\cos\theta].\end{aligned} \tag{4.33}$$

Combining (4.32) (evaluated at $x = \sin^2(\frac{\theta}{2})$) with (4.33) implies that

$$\begin{aligned}{}_2F_1(-z+\tfrac{5}{2}, z-\tfrac{3}{2}; \tfrac{5}{2}; \sin^2(\tfrac{\theta}{2})) &= \frac{3}{8(1-z)(2-z)(3-z)\sin^3(\theta/2)} \\ &\quad\times [\sin((2-z)\theta)\cos\theta - (2-z)\cos((2-z)\theta)\sin\theta].\end{aligned} \tag{4.34}$$

Now (4.34) and (4.27) imply

$$\begin{aligned}\mathsf{P}^{-3/2}_{z-5/2}(\cos\theta) &= \frac{1}{\Gamma(5/2)}\frac{3}{(1-z)(2-z)(3-z)\sin^3\theta} \\ &\quad\times [\sin((2-z)\theta)\cos\theta - (2-z)\cos((2-z)\theta)\sin\theta].\end{aligned} \tag{4.35}$$

Combining (4.25), (4.26) and (4.35) we can finally conclude that when the parameter $\theta \in (-\pi, \pi) \setminus \{0\}$, the formula for $\mathcal{M}g(z)$ from (4.22) holds, completing the proof of the Lemma 4.3. □

REMARK 4.4. Note that (4.22)–(4.23) imply that $\mathcal{M}g$ is continuous with respect to $\theta \in (-\pi, \pi)$.

LEMMA 4.5. *For each parameter $\theta \in (0, 2\pi)$ and each number $m \in \{0, \ldots, 3\}$, consider the function $g_m : \mathbb{R}_+ \to \mathbb{R}_+$ given by*

$$g_m(x) := \frac{x^m}{(x^2 - 2x\cos\theta + 1)^2}. \tag{4.36}$$

Then for each $m \in \{0, \ldots, 3\}$ there holds that $g_m \in \mathcal{A}_{(-m,4-m)}$. In addition, if $\theta \in (0, 2\pi) \setminus \{\pi\}$, and $z \in \Gamma_{-m,4-m} \setminus \mathbb{Z}$ for each of the formulas for $\mathcal{M}g_m(z)$ below, there hold:

$$\mathcal{M}g_0(z) = \frac{\pi\left[(z-3)\sin((z-1)(\pi-\theta)) - (z-1)\sin((z-3)(\pi-\theta))\right]}{4(\sin^3\theta)\sin(\pi z)}, \tag{4.37}$$

$$\mathcal{M}g_1(z) = -\frac{\pi\left[(z-2)\sin(z(\pi-\theta)) - z\sin((z-2)(\pi-\theta))\right]}{4(\sin^3\theta)\sin(\pi z)}, \tag{4.38}$$

$$\mathcal{M}g_2(z) = \frac{\pi\left[(z-1)\sin((z+1)(\pi-\theta)) - (z+1)\sin((z-1)(\pi-\theta))\right]}{4(\sin^3\theta)\sin(\pi z)}, \tag{4.39}$$

$$\mathcal{M}g_3(z) = -\frac{\pi\left[z\sin((z+2)(\pi-\theta)) - (z+2)\sin(z(\pi-\theta))\right]}{4(\sin^3\theta)\sin(\pi z)}. \tag{4.40}$$

Finally, when $\theta = \pi$ and $z \in \Gamma_{-m,4-m} \setminus \mathbb{Z}$ for each of the formulas for $\mathcal{M}g_m(z)$ below, there hold:

$$\mathcal{M}g_0(z) = -\frac{\pi(z-1)(z-2)(z-3)}{6\sin(\pi z)}, \tag{4.41}$$

$$\mathcal{M}g_1(z) = \frac{\pi z(z-1)(z-2)}{6\sin(\pi z)}, \tag{4.42}$$

$$\mathcal{M}g_2(z) = -\frac{\pi z(z-1)(z+1)}{6\sin(\pi z)}, \tag{4.43}$$

$$\mathcal{M}g_3(z) = \frac{\pi z(z+1)(z+2)}{6\sin(\pi z)}. \tag{4.44}$$

PROOF. These statements immediately follow from Lemma 4.3 and elementary properties of the Mellin transform. □

LEMMA 4.6. *Fix $\theta \in (0, 2\pi)$ and recall the kernel κ_2 in (3.18). For each $z \in \Gamma_{0,1}$ there holds*

$$\mathcal{M}\left(\kappa_2(\cdot, 1)\right)(z) = \begin{pmatrix} -\frac{1}{2}a(z) & A_\theta(z) \\ -A_\theta(z) & \frac{1}{2}a(z) \end{pmatrix}, \tag{4.45}$$

where, when $\theta \neq \pi$,
(4.46)

$$\begin{aligned}A_\theta(z) := \frac{1}{16(\sin^3\theta)\sin(\pi z)} & \{(\cos\theta)\,[z\sin((z+2)(\pi-\theta)) - (z+2)\sin(z(\pi-\theta))] \\ & + 3\,[(z-1)\sin((z+1)(\pi-\theta)) - (z+1)\sin((z-1)(\pi-\theta))] \\ & + (3+2\sin^2\theta)(\cos\theta)\,[(z-2)\sin(z(\pi-\theta)) - z\sin((z-2)(\pi-\theta))] \\ & + [(z-3)\sin((z-1)(\pi-\theta)) - (z-1)\sin((z-3)(\pi-\theta))]\},\end{aligned}$$

and when $\theta = \pi$,

$$A_\pi(z) := \frac{1}{4\sin(\pi z)}. \tag{4.47}$$

PROOF. Using Proposition 2.4 along with the definition of a from (4.15), we may write for each $z \in \Gamma_{0,1}$

$$\mathcal{M}(k_2^{11}(\cdot,1))(z) = -\mathcal{M}(k_2^{22}(\cdot,1))(z) = -\frac{\cos(\pi z)}{4\sin(\pi z)} = -\frac{1}{2}a(z), \tag{4.48}$$

taking care of the diagonal entries in (4.45) for each $\theta \in (0, 2\pi)$. From (3.18), (3.19), straightforward algebra, and the definition of the functions g_m, $m \in \{0,\dots,3\}$, from (4.36), note that for each $s \in \mathbb{R}_+$ we have

$$\begin{aligned}& -k_2^{12}(s,1) = k_2^{21}(s,1) \\ & = \frac{1}{4\pi}\left[(\cos\theta)g_3(s) - 3g_2(s) + (3+2\sin^2\theta)(\cos\theta)g_1(s) - g_0(s)\right].\end{aligned} \tag{4.49}$$

Consequently
(4.50)

$$\begin{aligned}& -\mathcal{M}(k_2^{12}(\cdot,1))(z) = \mathcal{M}(k_2^{21}(\cdot,1))(z) \\ & = \frac{1}{4\pi}\left[(\cos\theta)\mathcal{M}g_3(z) - 3\mathcal{M}g_2(z) + (3+2\sin^2\theta)(\cos\theta)\mathcal{M}g_1(z) - \mathcal{M}g_0(z)\right].\end{aligned}$$

In concert, (4.50), formulas (4.37)–(4.40), and straightforward algebra allow us to obtain that, for $\theta \in (0, 2\pi) \setminus \{\pi\}$, there holds

$$-\mathcal{M}(k_2^{12}(\cdot,1))(z) = \mathcal{M}(k_2^{21}(\cdot,1))(z) = -A_\theta(z) \qquad \forall z \in \Gamma_{0,1}, \tag{4.51}$$

where $A_\theta(z)$ is as in (4.46), completing the proof of (4.45) when $\theta \in (0, 2\pi) \setminus \{\pi\}$. Finally, the formulas for the off diagonal entries in (4.45) when $\theta = \pi$ follow from (4.50) and formulas (4.41)–(4.44). □

LEMMA 4.7. *Fix* $\theta \in (0, 2\pi)$ *and recall the kernel* κ_4 *from* (3.21). *Then, for each* $z \in \Gamma_{0,1}$ *there holds*

$$\mathcal{M}(\kappa_4(\cdot,1))(z) = B_\theta(z) \cdot \begin{pmatrix} 0 & 1 \\ 1 & 0 \end{pmatrix}, \tag{4.52}$$

where, when $\theta \neq \pi$,
(4.53)

$$\begin{aligned}B_\theta(z) := \frac{-1}{16(\sin^2\theta)\sin(\pi z)} & \{[z\sin((z+2)(\pi-\theta)) - (z+2)\sin(z(\pi-\theta))] \\ & - (2+\cos(2\theta))\,[(z-2)\sin(z(\pi-\theta)) - z\sin((z-2)(\pi-\theta))] \\ & - 2(\cos\theta)\,[(z-3)\sin((z-1)(\pi-\theta)) - (z-1)\sin((z-3)(\pi-\theta))]\},\end{aligned}$$

and when $\theta = \pi$,

$$B_\pi(z) := 0. \tag{4.54}$$

PROOF. In light of (3.21), there holds that $k_4^{11}(s,1) = k_4^{22}(s,1) = 0$ for each $s \in \mathbb{R}_+$, taking care immediately of the diagonal entries in (4.52) for each $\theta \in (0, 2\pi)$. Going further, (3.21), and elementary algebra and trigonometry give that

$$k_4^{12}(s,1) = k_4^{21}(s,1) = \frac{\sin\theta}{4\pi}\left[g_3(s) - (2+\cos(2\theta))g_1(s) + 2(\cos\theta)g_0(s)\right], \tag{4.55}$$

where the functions g_m, $m \in \{0,\dots,3\}$, are as in (4.36). Consequently, for each $z \in \Gamma_{0,1}$,

$$\begin{aligned} \mathcal{M}(k_4^{12}(\cdot,1))(z) &= \mathcal{M}(k_4^{21}(\cdot,1))(z) \\ &= \frac{\sin\theta}{4\pi}\left[\mathcal{M}g_3(z) - (2+\cos(2\theta))\mathcal{M}g_1(z) + 2(\cos\theta)\mathcal{M}g_0(z)\right], \end{aligned} \tag{4.56}$$

and applying formulas (4.37)–(4.40), after straightforward algebra we obtain

$$\mathcal{M}(k_4^{12}(\cdot,1))(z) = \mathcal{M}(k_4^{21}(\cdot,1))(z) = B_\theta(z) \qquad \forall z \in \Gamma_{0,1}, \tag{4.57}$$

where $B_\theta(z)$ is as in (4.53). This proves (4.52) whenever $\theta \in (0,2\pi)\setminus\{\pi\}$. To deal with the off diagonal entries in (4.52) when $\theta = \pi$, appeal again to (4.56) and notice that its right-hand side vanishes in this case. □

LEMMA 4.8. *Fix* $\theta \in (0, 2\pi)$ *and recall the kernel* κ_6 *from* (3.23). *For each* $z \in \Gamma_{0,1}$ *there holds*

$$\mathcal{M}(\kappa_6(\cdot,1))(z) = C_\theta(z) \cdot \begin{pmatrix} 0 & 1 \\ 1 & 0 \end{pmatrix}, \tag{4.58}$$

where, when $\theta \neq \pi$,

$$\begin{aligned} C_\theta(z) := &\frac{1}{16(\sin^2\theta)\sin(\pi z)}\left\{\left[z\sin((z+2)(\pi-\theta)) - (z+2)\sin(z(\pi-\theta))\right]\right. \\ &+ 2(\cos\theta)\left[(z-1)\sin((z+1)(\pi-\theta)) - (z+1)\sin((z-1)(\pi-\theta))\right] \\ &\left.+ \cos(2\theta)\left[(z-2)\sin(z(\pi-\theta)) - z\sin((z-2)(\pi-\theta))\right]\right\}, \end{aligned} \tag{4.59}$$

and when $\theta = \pi$,

$$C_\pi(z) := 0. \tag{4.60}$$

PROOF. Formula (3.23) gives that $k_6^{11}(s,1) = k_6^{22}(s,1) = 0$ for each $s \in \mathbb{R}_+$, justifying the diagonal entries in the right hand side of (4.58). In addition,

$$k_6^{12}(s,1) = k_6^{21}(s,1) = \frac{\sin\theta}{4\pi}\left[-g_3(s) + 2(\cos\theta)g_2(s) - \cos(2\theta)g_1(s)\right], \tag{4.61}$$

where the functions g_m, $m \in \{0,\dots,3\}$, are as in (4.36). Consequently, for each $z \in \Gamma_{0,1}$,

$$\begin{aligned} \mathcal{M}(k_6^{12}(\cdot,1))(z) &= \mathcal{M}(k_6^{21}(\cdot,1))(z) \\ &= \frac{\sin\theta}{4\pi}\Big[-\mathcal{M}g_3(z) + 2(\cos\theta)\mathcal{M}g_2(z) - \cos(2\theta)\mathcal{M}g_1(z)\Big]. \end{aligned} \tag{4.62}$$

Now, (4.62) combined with (4.38)–(4.40) and straightforward algebra and trigonometry yield

$$\mathcal{M}(\kappa_6^{12}(\cdot,1))(z) = \mathcal{M}(\kappa_6^{21}(\cdot,1))(z) = C_\theta(z) \qquad \forall z \in \Gamma_{0,1}, \tag{4.63}$$

where $C_\theta(z)$ is as in (4.59). If $\theta = \pi$, the right-hand side of (4.62) clearly vanishes, finishing the proof. □

LEMMA 4.9. *Fix $\theta \in (0, 2\pi)$ and recall the kernel κ_8 from (3.25). For each $z \in \Gamma_{0,1}$ there holds*

$$\mathcal{M}(\kappa_8(\cdot,1))(z) = \begin{pmatrix} -\frac{1}{2}a(z) & D_\theta(z) \\ -D_\theta(z) & \frac{1}{2}a(z) \end{pmatrix}, \tag{4.64}$$

where, when $\theta \neq \pi$,

(4.65)
$$\begin{aligned} D_\theta(z) :=& \frac{1}{16(\sin^3\theta)\sin(\pi z)} \big\{ (\cos\theta) \left[z \sin((z+2)(\pi-\theta)) - (z+2)\sin(z(\pi-\theta)) \right] \\ &+ (2+\cos(2\theta)) \left[(z-1)\sin((z+1)(\pi-\theta)) - (z+1)\sin((z-1)(\pi-\theta)) \right] \\ &+ (\cos\theta)(2+\cos(2\theta)) \left[(z-2)\sin(z(\pi-\theta)) - z\sin((z-2)(\pi-\theta)) \right] \\ &+ \cos(2\theta) \left[(z-3)\sin((z-1)(\pi-\theta)) - (z-1)\sin((z-3)(\pi-\theta)) \right] \big\}, \end{aligned}$$

and when $\theta = \pi$,

$$D_\pi(z) := \frac{1}{4\sin(\pi z)}. \tag{4.66}$$

PROOF. Combining Proposition 2.4, the definition of a from (4.15), and (3.25), we obtain that for $z \in \Gamma_{0,1}$

$$\mathcal{M}(k_8^{11}(\cdot,1))(z) = -\mathcal{M}(k_8^{22}(\cdot,1))(z) = -\frac{\cos(\pi z)}{4\sin(\pi z)} = -\frac{1}{2}a(z), \tag{4.67}$$

taking care of the diagonal entries in (4.64). For the other entries in $\kappa_8(\cdot,1)$, appeal again to (3.25) and elementary trigonometry to write for each $s \in \mathbb{R}_+$,

$$\begin{aligned} &- k_8^{12}(s,1) = k_8^{21}(s,1) \\ &= \frac{1}{4\pi} \Big[(\cos\theta) g_3(s) - (2+\cos(2\theta)) g_2(s) + (\cos\theta)(2+\cos(2\theta)) g_1(s) \\ &\qquad - \cos(2\theta) g_0(s) \Big], \end{aligned} \tag{4.68}$$

where the functions g_0, g_1, g_2 and g_3 are as in (4.36). Consequently

$$\begin{aligned} &- \mathcal{M}(k_8^{12}(\cdot,1))(z) = \mathcal{M}(k_8^{21}(\cdot,1))(z) \\ &= \frac{1}{4\pi} \big[(\cos\theta)\mathcal{M}g_3(z) - (2+\cos(2\theta))\mathcal{M}g_2(z) \\ &\qquad + (\cos\theta)(2+\cos(2\theta))\mathcal{M}g_1(z) - \cos(2\theta)\mathcal{M}g_0(z) \big]. \end{aligned} \tag{4.69}$$

Combining (4.69) with (4.37)–(4.40), and using straightforward algebra and trigonometry, we ultimately arrive at

$$-\mathcal{M}(k_8^{12}(\cdot,1))(z) = \mathcal{M}(k_8^{21}(\cdot,1))(z) = -D_\theta(z) \qquad \forall z \in \Gamma_{0,1},$$

where $D_\theta(z)$ as in (4.65). This completes the proof of (4.64) when $\theta \in (0, 2\pi) \setminus \{\pi\}$. Finally, when $\theta = \pi$, invoke again (4.69) and formulas (4.41)–(4.44). □

REMARK 4.10. Straightforward calculations show that, with A_θ as in (4.46)–(4.47), B_θ as in (4.53)–(4.54), C_θ as in (4.59)–(4.60), and D_θ as in (4.65)–(4.66), there holds

$$\lim_{\theta\to\pi} A_\theta = A_\pi,\ \lim_{\theta\to\pi} B_\theta = B_\pi,\ \lim_{\theta\to\pi} C_\theta = C_\pi,\ \text{ and } \lim_{\theta\to\pi} D_\theta = D_\pi. \tag{4.70}$$

Having computed explicitly $\mathcal{M}(\kappa_j(\cdot,1))(z)$ for each $j \in \{1,\dots,8\}$ and each $z \in \Gamma_{0,1}$, we are ready to compute the Mellin symbol of the kernel matrix R from (1.23).

LEMMA 4.11. *Fix $\theta \in (0,2\pi)$ and $\eta \in \mathbb{R}$, and consider the matrix R as in (1.23). Then*

$$\mathcal{M}(R(\cdot,1))(z) = \begin{pmatrix} M_{11}(z) & M_{12}(z) \\ M_{21}(z) & M_{22}(z) \end{pmatrix} \qquad \forall z \in \Gamma_{0,1}, \tag{4.71}$$

where, with a, b_θ, d_θ as in (4.15), A_θ as in (4.46)–(4.47), B_θ as in (4.53)–(4.54), C_θ as in (4.59)–(4.60), and D_θ as in (4.65)–(4.66),

$$M_{11}(z) := \begin{pmatrix} \frac{1+\eta}{2}\, a(z) & -\eta \cdot b_\theta(z) + (\eta-1)\cdot A_\theta(z) \\ \eta \cdot b_\theta(z) - (\eta-1)\cdot A_\theta(z) & -\frac{1+\eta}{2}\, a(z) \end{pmatrix}, \tag{4.72}$$

$$M_{12}(z) := \begin{pmatrix} -\frac{1}{2} & -\eta \cdot d_\theta(z) + (\eta-1)\cdot B_\theta(z) \\ -\eta \cdot d_\theta(z) + (\eta-1)\cdot B_\theta(z) & -\frac{1}{2} \end{pmatrix}, \tag{4.73}$$

$$M_{21}(z) := \begin{pmatrix} -\frac{1}{2} & -d_\theta(z) + (\eta-1)\cdot C_\theta(z) \\ -d_\theta(z) + (\eta-1)\cdot C_\theta(z) & -\frac{1}{2} \end{pmatrix}, \tag{4.74}$$

and

$$M_{22}(z) := \begin{pmatrix} -\frac{1+\eta}{2}\, a(z) & b_\theta(z) + (\eta-1)\cdot D_\theta(z) \\ -b_\theta(z) - (\eta-1)\cdot D_\theta(z) & \frac{1+\eta}{2}\, a(z) \end{pmatrix}. \tag{4.75}$$

PROOF. This follows directly from (1.23), Lemma 4.6, Lemmal 4.7, Lemma 4.8, and Lemma 4.9. □

The following straightforward linear algebra result will be useful for our further analysis.

LEMMA 4.12. *Consider 2×2 matrices $A, D \in \operatorname{span}\left\{\begin{pmatrix} 1 & 0 \\ 0 & -1\end{pmatrix}, \begin{pmatrix} 0 & 1 \\ -1 & 0\end{pmatrix}\right\}$ and $B, C \in \operatorname{span}\left\{\begin{pmatrix} 1 & 0 \\ 0 & 1\end{pmatrix}, \begin{pmatrix} 0 & 1 \\ 1 & 0\end{pmatrix}\right\}$. Then*

$$\det\begin{pmatrix} A & B \\ C & D \end{pmatrix} = \det(A\cdot D + B\cdot \widetilde{C}), \tag{4.76}$$

where

$$\widetilde{C} := \begin{pmatrix} -1 & 0 \\ 0 & 1\end{pmatrix} \cdot C \cdot \begin{pmatrix} 1 & 0 \\ 0 & -1\end{pmatrix}. \tag{4.77}$$

Our next results are focused on properties of the determinant of the Mellin symbol of the matrix-valued kernel R from (1.23).

LEMMA 4.13. *Let $\theta \in (0, 2\pi)$ and let R be as in (1.23). Then, with $M_{ij}(z)$, $i, j \in \{1, 2\}$, as in (4.72)–(4.75), the following holds:*

$$\det\left(\mathcal{M}(R(\cdot,1))(z)\right) = \det\big(M_{11}(z)\cdot M_{22}(z) + M_{12}(z)\cdot \widetilde{M}_{21}(z)\big) \qquad \forall z \in \Gamma_{0,1}, \tag{4.78}$$

where

$$\widetilde{M}_{21}(z) := \begin{pmatrix} \frac{1}{2} & -d_\theta(z) + (\eta-1)\cdot C_\theta(z) \\ -d_\theta(z) + (\eta-1)\cdot C_\theta(z) & \frac{1}{2} \end{pmatrix} \tag{4.79}$$

PROOF. This is a direct application of Lemma 4.11 and Lemma 4.12. □

THEOREM 4.14. *Fix $\theta \in (0, 2\pi)$ and $\eta \in \mathbb{R}$ and let R be as in (1.23). Then for each $z \in \Gamma_{0,1}$ there holds*

$$\det(\mathcal{M}(R(\cdot,1)))(z) = \big(\Phi_\theta(z,\eta) + \Psi_\theta(z,\eta)\big)\big(\Phi_\theta(z,\eta) - \Psi_\theta(z,\eta)\big), \tag{4.80}$$

where, with $a(z), b_\theta(z), d_\theta(z)$ as in (4.15), and A_θ as in (4.46)–(4.47), B_θ as in (4.53)–(4.54), C_θ as in (4.59)–(4.60), and D_θ as in (4.65)–(4.66),

$$\begin{aligned} \Phi_\theta(z,\eta) := &-\frac{1}{4}\left[(1+\eta)^2 a^2(z) + 1\right] \\ &+ (\eta\cdot b_\theta(z) - (\eta-1)\cdot A_\theta(z))\cdot(b_\theta(z) + (\eta-1)\cdot D_\theta(z)) \\ &+ (\eta\cdot d_\theta(z) - (\eta-1)\cdot B_\theta(z))\cdot(d_\theta(z) - (\eta-1)\cdot C_\theta(z)), \end{aligned} \tag{4.81}$$

and

$$\Psi_\theta(z,\eta) := \frac{1-\eta^2}{2} a(z)\left[b_\theta(z) - A_\theta(z) - D_\theta(z)\right] + \frac{1-\eta}{2}\left[d_\theta(z) - B_\theta(z) + C_\theta(z)\right]. \tag{4.82}$$

PROOF. With $z \in \Gamma_{0,1}$ and the matrices $M_{11}(z)$ and $M_{22}(z)$ as in (4.72) and (4.75), first note that performing the matrix multiplication $M_{11}(z)\cdot M_{22}(z)$ and elementary algebra give

$$M_{11}(z)\cdot M_{22}(z) = \begin{pmatrix} U(z) & V(z) \\ V(z) & U(z) \end{pmatrix}, \tag{4.83}$$

where

$$U(z) := -\frac{(1+\eta)^2}{4} a^2(z) + (\eta\cdot b_\theta(z) - (\eta-1)\cdot A_\theta(z))\cdot(b_\theta(z) + (\eta-1)\cdot D_\theta(z)), \tag{4.84}$$

and

$$V(z) := \frac{1-\eta^2}{2} a(z)\cdot\left[b_\theta(z) - (A_\theta(z) + D_\theta(z))\right]. \tag{4.85}$$

Second, multiplying the matrices $M_{12}(z)$ and $\widetilde{M}_{21}(z)$ from (4.73) and (4.79) yields

$$M_{12}(z)\cdot \widetilde{M}_{21}(z) = \begin{pmatrix} \widetilde{U}(z) & \widetilde{V}(z) \\ \widetilde{V}(z) & \widetilde{U}(z) \end{pmatrix}, \tag{4.86}$$

where

$$\widetilde{U}(z) := -\frac{1}{4} + (\eta\cdot d_\theta(z) - (\eta-1)\cdot B_\theta(z))\cdot(d_\theta(z) - (\eta-1)\cdot C_\theta(z)) \tag{4.87}$$

and

$$\widetilde{V}(z) := \frac{1}{2}\left[(1-\eta) d_\theta(z) + (\eta-1)\cdot(B_\theta(z) - C_\theta(z))\right]. \tag{4.88}$$

Using now formula (4.78) in Lemma 4.11 along with identities (4.83) and (4.86), for each $z \in \Gamma_{0,1}$ we may write

$$\det\left(\mathcal{M}(R(\cdot,1))(z) = \det\begin{pmatrix} U(z)+\widetilde{U}(z) & V(z)+\widetilde{V}(z) \\ V(z)+\widetilde{V}(z) & U(z)+\widetilde{U}(z)\end{pmatrix}. \tag{4.89}$$

Finally (4.80) follows as soon as we notice that

$$\Phi_\theta(z,\eta) = U(z)+\widetilde{U}(z) \quad\text{and}\quad \Psi_\theta(z,\eta) = V(z)+\widetilde{V}(z), \tag{4.90}$$

finishing the proof of Theorem 4.14. □

This ultimately yields the following implication of the solvability range of the integrability exponent p for the Neumann problem in (1.2) when the domain is an infinite sector of an angle $\theta \in (0,2\pi)$.

THEOREM 4.15. *Let Ω be the interior of an infinite upright sector of aperture $\theta \in (0,2\pi)$ in the plane, fix $\eta \in [-1,1)$, and recall the operator R as in (1.23). Then if $p \in (1,\infty)$, the following implication holds:*

(4.91)

$$\det\big(\mathcal{M}R(\cdot,1)\big)(z) \neq 0 \qquad \forall z \in \Gamma_{0,1},\ \Re\mathfrak{e}\, z = \tfrac{1}{p} \implies (NBH_p) \textit{ has a solution.}$$

PROOF. Fix $p \in (1,\infty)$ and assume that

$$\det\big(\mathcal{M}R(\cdot,1)\big)(z) \neq 0 \qquad \forall z \in \Gamma_{0,1},\ \Re\mathfrak{e}\, z = \frac{1}{p}, \tag{4.92}$$

and consider the boundary value problem (NBH_p) from (1.2) with given datum $(f_0,f_1) \in L^p(\partial\Omega) \times \big(\dot{L}_1^{p'}(\partial\Omega)\big)^*$ where $1/p+1/p'=1$. Since $f_1 \in \dot{L}^p_{-1}(\partial\Omega)$ there exists $g \in L^p(\partial\Omega)$ such that $f_1 = \partial_\tau g$ and the conditions on p from the hypothesis ensure that Theorem 2.7 applies. Combining this with (1.24) allows us to define $(h_0,h_1) \in L^p(\partial\Omega) \times L^p(\partial\Omega)$ via

$$\begin{pmatrix} h_0 \\ h_1 \end{pmatrix} := T^{-1}\begin{pmatrix} f_0 \\ g \end{pmatrix}. \tag{4.93}$$

With this in hand, set $u := \mathcal{S}(\partial_\tau h_0, h_1)$ in Ω, where $\mathcal{S}$ is the bi-harmonic single layer potential operator introduced in (2.13). Then, as discussed in Section 2.1,

$$\Delta^2 u = 0 \text{ in } \Omega \quad\text{and}\quad \mathcal{N}(\nabla^2 u) \in L^p(\partial\Omega). \tag{4.94}$$

Pick next $r \in \mathbb{R}$ such that $\eta = \dfrac{2(r+r^2)}{1+2r+2r^2} \in [-1,1)$(note that for $\eta \in [-1,1)$ this is always possible), and recall the boundary operators M_r and N_r from (1.6). Using the jump formulas (2.27) we have

$$\begin{pmatrix} M_r(u) \\ N_r(u) \end{pmatrix} = \begin{pmatrix} -\eta K_1 - 2K_2 & -\frac{1}{2}I - \eta K_3 - 2K_4 \\ \partial_\tau(-K_5 - 2K_6 - \frac{1}{2}I) & \partial_\tau(K_7 - 2K_8) \end{pmatrix}\begin{pmatrix} h_0 \\ h_1 \end{pmatrix}. \tag{4.95}$$

Let

$$\begin{aligned} &\mathfrak{t}\colon L^p(\partial\Omega) \times L^p(\partial\Omega) \longrightarrow L^p(\partial\Omega) \times \dot{L}^p_{-1}(\partial\Omega) \\ &\text{be given by } \mathfrak{t}(F,G) := (F,\partial_\tau G) \qquad \forall (F,G) \in L^p(\partial\Omega) \times L^p(\partial\Omega). \end{aligned} \tag{4.96}$$

Then using the definition of T from (1.7) together with (4.95), we obtain that

$$\begin{pmatrix} M_r(u) \\ N_r(u) \end{pmatrix} = \mathfrak{t} \circ T \begin{pmatrix} h_0 \\ h_1 \end{pmatrix}. \tag{4.97}$$

This and (4.93) ultimately yield

$$M_r(u) = f_0 \text{ and } N_r(u) = f_1, \qquad \sigma\text{-a.e. on } \partial\Omega. \tag{4.98}$$

Finally, (4.94), (4.98), and the definition of M_r and N_r show that u is a solution for (NBH_p) with datum $(f_0, f_1) \in L^p(\partial\Omega) \times \big(\dot{L}_1^{p'}(\partial\Omega)\big)^*$, completing the proof of Theorem 4.15. □

References

[1] M. Abramowitz and I. A. Stegun, *Handbook of mathematical functions with formulas, graphs, and mathematical tables*, National Bureau of Standards Applied Mathematics Series, No. 55, U. S. Government Printing Office, Washington, DC, 1964. MR167642

[2] H. Awala, J. Kyeong, I. Mitrea, K. Ott and W. Tucker, On the solvability of the Neumann problem for the bi-Laplacian in infinite sectors, preprint, 2023.

[3] D. W. Boyd, *Spectra of convolution operators*, Acta Sci. Math. (Szeged) **35** (1973), 31–37. MR326493

[4] E. B. Fabes, M. Jodeit Jr., and J. E. Lewis, *On the spectra of a Hardy kernel*, J. Functional Analysis **21** (1976), no. 2, 187–194, DOI 10.1016/0022-1236(76)90076-8. MR394311

[5] J. Giroire and J.-C. Nédélec, *A new system of boundary integral equations for plates with free edges*, Math. Methods Appl. Sci. **18** (1995), no. 10, 755–772, DOI 10.1002/mma.1670181002. MR1343389

[6] I. Mitrea and M. Mitrea, *Multi-layer potentials and boundary problems for higher-order elliptic systems in Lipschitz domains*, Lecture Notes in Mathematics, vol. 2063, Springer, Heidelberg, 2013, DOI 10.1007/978-3-642-32666-0. MR3013645

[7] I. Mitrea, K. Ott, and W. Tucker, *Invertibility properties of singular integral operators associated with the Lamé and Stokes systems on infinite sectors in two dimensions*, Integral Equations Operator Theory **89** (2017), no. 2, 151–207, DOI 10.1007/s00020-017-2396-4. MR3717826

[8] C. Nazaret, *A system of boundary integral equations for polygonal plates with free edges*, Math. Methods Appl. Sci. **21** (1998), no. 2, 165–185, DOI 10.1002/(SICI)1099-1476(19980125)21:2¡165::AID-MMA943¿3.0.CO;2-C. MR1600018

[9] F. Oberhettinger, *Tables of Mellin transforms*, Springer-Verlag, New York-Heidelberg, 1974. MR352890

[10] NIST Digital Library of Mathematical Functions. `https://dlmf.nist.gov/` (release 1.1.11 of 2023-09-15). F. W. J. Olver, A. B. Olde Daalhuis, D. W. Lozier, B. I. Schneider, R. F. Boisvert, C. W. Clark, B. R. Miller, B. V. Saunders, H. S. Cohl, and M. A. McClain, eds.

[11] Z. Shen, *On estimates of biharmonic functions on Lipschitz and convex domains*, J. Geom. Anal. **16** (2006), no. 4, 721–734, DOI 10.1007/BF02922138. MR2271951

[12] C. Snow, *Hypergeometric and Legendre functions with applications to integral equations of potential theory*, National Bureau of Standards Applied Mathematics Series, No. 19, U. S. Government Printing Office, Washington, DC, 1952. MR48145

[13] G. C. Verchota, *The biharmonic Neumann problem in Lipschitz domains*, Acta Math. **194** (2005), no. 2, 217–279, DOI 10.1007/BF02393222. MR2231342

Dept. of Mathematics, Syracuse University, Syracuse, NY 13244
Email address: `jkyeong@syr.edu`

Dept. of Mathematics, Temple University, Philadelphia, PA 19122
Email address: `imitrea@temple.edu`

Dept. of Mathematics, Bates College, Lewiston, Maine 04240
Email address: `kott@bates.edu`

Contemporary Mathematics
Volume **819**, 2025
https://doi.org/10.1090/conm/819/16390

Gravitational potentials of ellipsoidal masses

Norman Lebovitz

ABSTRACT. This paper revisits the problem of calculating the gravitational potential of ellipsoids. We begin with general considerations of potential theory but, starting in the second section entitled "Ellipsoids", we emphasize the procedure of Ferrers [N. M. Ferrers, On the potentials of ellipsoids, ellipsoidal shells, elliptic laminae, and elliptic rings, of variable densities, Quart. J. Pure Appl. Math., 14:1–22, 1877] for the explicit calculation of potentials when the density function is a polynomial in Cartesian coordinates.

Note from the editors: The following beautiful manuscript found on Norman Lebovitz's web page has not previously been published. Norman Lebovitz passed away on December 28, 2022 at the age of 87. The manuscript is published posthumously with the permission of the University of Chicago Mathematics Department (which maintained the web page) and his family (who were able to obtain a LaTeX version from his laptop). Although not mentioned in this manuscript, the theory of the Newtonian potential of three-dimensional triaxial ellipsoidal mass distributions is governed by important special functions called incomplete elliptic integrals. These are generalizations of hypergeometric functions and therefore fit nicely in this AMS volume.

Signed: The editors of this Contemporary Mathematics Volume,
Howard S. Cohl, Roberto S. Costas-Santos, Robert S. Maier.

CONTENTS

Key words and phrases. Newtonian potential; ellipsoidal masses; multinomials; elliptic integrals.

1. Introduction

The gravitational potential $\Psi(\mathbf{x})$ at a point $\mathbf{x}$, due to a distribution of matter of density ρ distributed throughout a domain D of three-dimensional Euclidean space $\mathbb{R}^3$, is defined by the expression

$$\Psi(\mathbf{x}) = \int_D \frac{\rho(\mathbf{y})}{|\mathbf{x}-\mathbf{y}|}\, d\mathbf{y}, \tag{1.1}$$

where $d\mathbf{y}$ is the element of volume and $|\mathbf{x}-\mathbf{y}| = \left(\Sigma (x_i - y_i)^2\right)^{\frac{1}{2}}$ is the Euclidean distance between the points $\mathbf{x}$ and $\mathbf{y}$. The universal constant of gravitation G, which should multiply the expression on the right, has been suppressed. In the present chapter we present certain results regarding the Newtonian potential Ψ and its dependence on the density function ρ and the domain D.

From the standpoint of physics the more directly relevant quantity is the force per unit mass $\mathbf{f} = \nabla\Psi$. In Cartesian coordinates we find the components of $\mathbf{f}$ to be

$$f_k(\mathbf{y}) = -\int_D \frac{\rho(\mathbf{y})(x_k - y_k)}{|\mathbf{x}-\mathbf{y}|^3}\, d\mathbf{y}, \qquad k = 1,2,3. \tag{1.2}$$

In deriving equation (1.2) we have taken it for granted that we may differentiate the expression for Ψ above under the integral sign. That this is permissible is one of the regularity results of classical potential theory. We present in Section 1.2 a survey of those regularity results that we shall need subsequently, not only for the volume integrals of equations (1.1) and (1.2) but also for the surface integrals

$$V(\mathbf{x}) = \int_S \frac{\sigma(\mathbf{y})}{|\mathbf{x}-\mathbf{y}|}\, dS_y \tag{1.3}$$

and

$$W(\mathbf{x}) = \int_S \mu(\mathbf{y}) \left(\frac{\partial}{\partial n}\frac{1}{|\mathbf{x}-\mathbf{y}|}\right) dS_y. \tag{1.4}$$

Here S is a surface and dS_y is the element of surface area, the subscript y indicating with respect to which variable the integration is extended. The derivative $\partial/\partial n$ is shorthand for $\hat{\mathbf{n}}(\mathbf{y})\cdot\partial/\partial\mathbf{y}$. These are called *single-layer* and *double-layer* potentials, respectively.

We begin with special results (going back to Newton) regarding the spherically symmetric case.

1.1. Spherical symmetry. Consider first the force $\mathbf{f}$ of attraction of a spherical surface S of radius R, of constant surface mass-density σ, on a unit mass situated at a point $\mathbf{x}$ not on the S. Choosing the coordinate system so that the positive x_3 axis is the ray from the center of the sphere through the given point $\mathbf{x}$, i.e., so that its components are $(0,0,r)$, we easily find that $f_1 = f_2 = 0$ by virtue of symmetry about the x_3 axis. For f_3 we have, by equation (1.3) above,

$$\begin{aligned} f_3(\mathbf{x}) &= -\int_S \frac{\sigma(x_3 - y_3)}{|\mathbf{x}-\mathbf{y}|^3}\, dS_y \\ &= -R^2\sigma \int_S \frac{x_3 - R\cos\theta}{(R^2 + r^2 - 2rR\cos\theta)^{\frac{3}{2}}} \sin\theta\, d\theta\, d\phi, \end{aligned}$$

where we have chosen to integrate in polar coordinates in which the direction of the vector $\mathbf{x}$ is the polar axis (so $\mathbf{x} \cdot \mathbf{y} = rR\cos\theta$). Introducing $\mu = \cos\theta$ we find for the last integral

$$-2\pi\sigma R^2 \int_{-1}^{1} \frac{r - R\mu}{(R^2 + r^2 - 2Rr\mu)^{\frac{3}{2}}}\, d\mu.$$

This integral is elementary and the result is

$$f_3(\mathbf{x}) = -2\pi\sigma \frac{R^2}{r^2}\left(1 + \frac{r - R}{|r - R|}\right). \tag{1.5}$$

Consider the case when the point $\mathbf{x}$ is located in the interior of the spherical domain, so $r < R$. Equation (1.5) then shows that $f_3 = 0$ and therefore that $\mathbf{f}$ vanishes in the interior. If instead of a spherical surface we consider a spherical shell of inner and outer radii R_1 and $R_2 > R_1 > 0$, say, filled with matter having a radially symmetric density function $\rho(r)$, and if $\mathbf{x}$ lies interior to the inner shell, i.e., $r = |\mathbf{x}| < R_1$, it is clear that the force exerted on a particle at $\mathbf{x}$ must vanish.

On the other hand, suppose that $\mathbf{x}$ lies outside the spherical surface of radius R. Then equation (1.5) implies that $f_3 = -4\pi\sigma R^2 r^{-2}$ or, since $4\pi R^2\sigma$ is the mass of the surface, the force is directed toward the center of the sphere and has magnitude M/r^2. The same conclusion now easily follows for a spherical shell with a radially symmetric distribution of matter (ρ independent of angle). The following is Newton's theorem.

THEOREM 1.1 (Newton). *Let $\mathbf{x}$ be an interior point of spherical surface of constant surface density, or of a spherical shell of radially symmetric volume density. Then the gravitational force vanishes at $\mathbf{x}$, i.e., the gravitational potential is independent of $\mathbf{x}$ in the interior. If $\mathbf{x}$ is an exterior point of a spherical surface of constant surface density, or of a spherical shell or ball of radially symmetric volume density, the force is directed toward the center of the sphere and has magnitude M/r^2 where $r = |\mathbf{x}|$*[1] *and M is the mass of the surface, shell or ball.*

From this we obtain the force due to a radially symmetric distribution of matter acting on a unit mass at an arbitrary point $\mathbf{x}$. The force is directed toward the center of the sphere and has magnitude $M(r)/r^2$, where $M(r)$ is the mass enclosed in a sphere of radius $r = |\mathbf{x}|$:

$$M(r) = 4\pi \int_0^r \rho(s)s^2\, ds.$$

If r exceeds the radius R of the sphere, i.e., if the density vanishes for $r > R$, then $M(r) = M(R)$ becomes constant and equal to the total mass of the figure. With the definition (1.1) we find for the potential in the interior the expression

$$\Psi(r) = \frac{M(r)}{r} + 4\pi \int_r^R s\rho(s)\, ds, \tag{1.6}$$

whereas, for $r > R$, the potential is M/r, where M is the total mass.

We can now give examples of radially symmetric, self-gravitating, fluid masses. It is natural to expect that, if the velocity field vanishes, the configuration *must* be radially symmetric[2] and it is easy to construct examples of this kind. We

[1] As measured from the origin, located at the center of the spherical surface.

[2] This is true but we do not pursue it here.

assume the fluid is inviscid and consider the Euler equations for vanishing **v** and radial symmetry. The equation of mass conservation is trivially satisfied, and the equation of momentum conservation becomes

$$\frac{1}{\rho}\frac{dp}{dr} = -\frac{M(r)}{r^2},$$

where $M(r)$ is as given above.[3] We encounter in this special context the necessity of providing a further relation between p and ρ in order to proceed. We shall make some simple choices to avoid the complexity attending the energy equation.

(1) ($\rho =$ constant) In this case we find $M(r) = 4\pi r^3\rho/3$ and, requiring that p vanish at the boundary $r = R$, we obtain

$$p(r) = \frac{2\pi}{3}\rho^2\left(R^2 - r^2\right),$$

and this completely specifies the solution in this case.

(2) (barotropes) Instead of specifying ρ explicitly, we specify that $p = f(\rho)$ for some function f. Substituting for p in the equation above, multiplying through by r^2, and differentiating, we obtain

$$\frac{d}{dr}\left(\frac{f'(\rho)}{\rho}\frac{d\rho}{dr}\right) = -4\pi r^2\rho.$$

This is a second-order, (in general) nonlinear, ordinary differential equation for ρ. It must be solved with the condition that p vanish at the boundary. This can be viewed in the following way: seek a solution for which ρ and $p = f(\rho)$ are positive on an interval $[0, R)$ but the latter vanishes at $r = R$. For an arbitrary function f there may be no such solution, but we expect that there will be if f is chosen appropriately. A traditional choice has been $p = K\rho^{1+1/n}$ where K and n are constants, or, expressing p and ρ in terms of a third, dimensionless variable θ,

$$\rho = \rho_c\theta^n \text{ and } p = p_c\theta^{n+1}. \tag{1.7}$$

The variable θ is proportional to the temperature if one further assumes that the ideal-gas law is satisfied. By assumption, $\theta(0) = 1$, so the constants ρ_c and p_c represent the central values of density and pressure, respectively. Introducing the spatial scaling

$$r = \alpha\xi, \quad \text{where } \alpha = \sqrt{\frac{(n+1)K}{4\pi\rho_c^{1-1/n}}},$$

we find in place of the differential equation above

$$\frac{d}{d\xi}\left(\xi^2\frac{d\theta}{d\xi}\right) = -\xi^2\theta^n. \tag{1.8}$$

This equation does indeed have solutions satisfying $\theta(0) = 1$ and vanishing for some positive value ξ_n, provided $0 \leqslant n < 5$. They are called *polytropes*, and have been studied extensively (cf. Chandrasekhar (1939) [**1**]). In a few cases equation (1.8) can be integrated explicitly:

(a) ($n = 0$) The solution is $\theta = 1 - \frac{1}{6}\xi^2$, and is a rescaled version of the solution given above for $\rho =$ constant.

(b) ($n = 1$) The equation is linear and its solution is $\theta = (\sin\xi)/\xi$.

[3]The variable p represents a pressure function.

(c) ($n = 5$) The solution is $\theta = \left(1 + \frac{1}{3}\xi^2\right)^{-\frac{1}{2}}$. It does *not* vanish for any finite value of ξ, but falls off as $\xi \to \infty$ sufficiently rapidly when its mass is finite.

Other values of the *polytropic index* n require numerical integration or other approximation techniques. The value $n = 3$, called the *standard model*, is often used to simulate the run of density and pressure in a common class of stars.

An initially radially symmetric configuration remains so if the velocity field is in the radial direction only and depends only on radius. We can easily formulate the Euler equations in this case. They are

$$\frac{Dv}{Dt} = -\frac{1}{\rho}\frac{\partial p}{\partial r} - \frac{M(r,t)}{r^2}$$

and

$$\frac{D\rho}{Dt} + \frac{\rho}{r^2}\frac{\partial}{\partial r}\left(r^2 v\right) = 0,$$

where $v = v(r,t)$ is the radial component of velocity, and $D/Dt = \partial/\partial t + v\partial/\partial r$. Since $M(r,t)$ is an increasing function of r, the equation

$$m = \int_0^r s^2 \rho(s,t)\, ds$$

can be inverted giving $r = r_*(m,t)$ and expressing the other variables v, p, ρ as functions of m and t. This is useful since one can specify the mass of a star (so m has fixed limits) but not its radius (which is not specified in advance in general, and is a function of time in the present example).

The equation

$$4\pi r_*^2 \frac{\partial r_*}{\partial m}\rho_*(m,t) = 1,$$

which follows on differentiating the preceding expression with respect to m, allows us to express $\rho(r,t) = \rho_*(m,t)$ in terms of r_* and its derivative. Differentiation of the same expression for m above with respect to t and use of the chain rule shows that $D/Dt = \partial/\partial t|_m$. This is not surprising since m is clearly a Lagrangian variable in this case. The Euler equations are now reduced to the single second-order equation

$$\frac{\partial^2 r_*}{\partial t^2} = -4\pi r_*^2 \frac{\partial p_*}{\partial m} - \frac{m}{r_*^2}. \tag{1.9}$$

If we further specify a barotropic relationship $p = f(\rho)$, this becomes a second-order, nonlinear, partial differential equation for the single function $r_*(m,t)$. Its solution is sought on the space-time region $0 \leqslant m \leqslant M$, $t > 0$, where M is the total mass. Initial data for r_* and $\partial r_*/\partial t$ must be supplied. Boundary data are that $r_*(0,t) = 0$, whereas $p_*(M,t) = 0$.

While there are many numerical treatments of more realistic (and more complicated) versions of this nonlinear initial-boundary value problem, a mathematical analysis does not seem to have been attempted.

1.2. General properties. In this section we summarize the principal regularity properties of Newtonian potentials, including not only those given by volume integrals but also those given by the surface integrals introduced above in equations (1.3) and (1.4). We follow Kellogg [**4**].

We consider *regular* regions of three-dimensional Euclidean space $\mathbb{R}^3$ and *regular* surfaces embedded in $\mathbb{R}^3$. A regular surface is one that can be constructed of finitely many patches, each of which has the description $z = f(x,y)$ for an appropriate orientation of the coordinate axes; it is assumed that the function f is continuously differentiable in its domain, and that descriptions agree on overlapping patches (the surface is piecewise C^1). A regular region is a bounded region of space whose boundary is a regular surface. If $\mathbf{F}$ is a C^1 vector field in a regular region R, the divergence theorem holds:

$$\int_R \operatorname{div} \mathbf{F}\, d\mathbf{x} = \int_{\partial R} \mathbf{F} \cdot \hat{\mathbf{n}}\, dS,$$

where $\hat{\mathbf{n}}$ represents the outward normal to ∂R.

Two easily verifiable facts dominate the analysis. One is that

$$\Delta_{\mathbf{x}} \frac{1}{|\mathbf{x} - \mathbf{y}|} = 0, \tag{1.10}$$

at points where $\mathbf{x} \neq \mathbf{y}$. Here

$$\Delta_{\mathbf{x}} = \frac{\partial^2}{\partial x_1^2} + \frac{\partial^2}{\partial x_2^2} + \frac{\partial^2}{\partial x_3^2}$$

is the Laplace operator (expressed in Cartesian coordinates x_1, x_2 and x_3). The second is that the integral

$$\int_D \frac{1}{|\mathbf{x} - \mathbf{y}|^\beta}\, d\mathbf{y}$$

is well defined provided $\beta < 3$ even if $\mathbf{x}$ is an interior point of D.

It is straightforward to prove the following theorem.

THEOREM 1.2. *Suppose ρ is piecewise continuous in the closure of the region D. Then the functions $\Psi(\mathbf{x})$ and $\mathbf{f}(\mathbf{x})$ given by equations* (1.1) *and* (1.2), *respectively, are continuous throughout $\mathbb{R}^3$, and $\mathbf{f} = \nabla\Psi$.*

1.2.1. *Hölder condition and Poisson's equation.* A further assumption regarding the density is needed to establish *Poisson's equation* (equation (1.12) below): ρ satisfies a *Hölder condition* in D:

$$|\rho(\mathbf{y}) - \rho(\mathbf{x})| \leqslant k\, |\mathbf{y} - \mathbf{x}|^\alpha, \tag{1.11}$$

where k and α are positive constants, and the inequality is supposed to hold for all $\mathbf{y}$ in a sufficiently small neighborhood of the point $\mathbf{x}$.

THEOREM 1.3. *At points where ρ satisfies a Hölder condition, the potential Ψ satisfies Poisson's equation*

$$\Delta\Psi = -4\pi\rho. \tag{1.12}$$

The theorem is intended for interior points of the domain D, but holds also, in the form of *Laplace's equation*

$$\Delta\Psi = 0 \tag{1.13}$$

for points outside D by setting $\rho = 0$ there; Laplace's equation follows for exterior points directly from equation (1.10).

Theorem 1.3 is easily verified for a constant density ρ and a spherical domain D with the aid of equation (1.6) and the expression for the Laplace operator in spherical coordinates. Its proof in general may be outlined as follows. Suppose we wish to verify equation (1.12) at particular interior point $\mathbf{x_0}$. Write

$$\Psi(\mathbf{x}) = \int_{B_{\mathbf{x_0}}} \frac{\rho(\mathbf{y})}{|\mathbf{x}-\mathbf{y}|}\, d\mathbf{y} + \int_{D\backslash B_{\mathbf{x_0}}} \frac{\rho(\mathbf{y})}{|\mathbf{x}-\mathbf{y}|}\, d\mathbf{y} = \Psi_1 + \Psi_2,$$

say, where $B_{\mathbf{x_0}}$ is a ball of small radius about the point $\mathbf{x_0}$. For $\mathbf{x}$ confined to $B_{\mathbf{x_0}}$, $\Delta\Psi_2 = 0$. Further write

$$\Psi_1(\mathbf{x}) = \int_{B_{\mathbf{x_0}}} \frac{\rho(\mathbf{x_0})}{|\mathbf{x}-\mathbf{y}|}\, d\mathbf{y} + \int_{B_{\mathbf{x_0}}} \frac{\kappa(\mathbf{y})}{|\mathbf{x}-\mathbf{y}|}\, d\mathbf{y},$$

where $\kappa(\mathbf{y}) = \rho(\mathbf{y}) - \rho(\mathbf{x_0})$. The Laplacian of the first term gives the required value $-4\pi\rho(\mathbf{x_0})$ and it remains to show that the Laplacian of the second term above vanishes at $\mathbf{x_0}$. Formally differentiating this term twice with respect to x_1 gives

$$\int_{B_{\mathbf{x_0}}} \kappa(\mathbf{y}) \left(\frac{3(x_{01} - y_1)^2}{r_0^5} - \frac{1}{r_0^3} \right) d\mathbf{y},$$

where $r_0 = |\mathbf{x_0} - \mathbf{y}|$. The factor of κ becomes infinite like r_0^3, but κ vanishes like r_0^α by virtue of the Hölder condition, so the integral converges. It therefore vanishes in the limit as the diameter of $B_{\mathbf{x_0}}$ tends to zero. Making these formal remarks rigorous leads to Theorem 1.3.

Turning now to the surface integrals V and W of equations (1.3) and (1.4) above, we record the following result.

THEOREM 1.4. *At a point of the surface where the normal is defined and continuous and σ and μ are continuous,*

$$\left[\frac{\partial V}{\partial n}\right] = -4\pi\sigma \tag{1.14}$$

and

$$[W] = 4\pi\mu, \tag{1.15}$$

where the brackets $[\phi]$ represent the difference of the values of ϕ at upper and lower sides of the surface S (upper in the sense of the direction of the outward normal).

Some examples will help clarify the meaning of these results.

(1) (full spherical surface) At a spherical surface with a constant surface density σ, equation (1.5) gives the derivative $\partial V/\partial r = \partial V/\partial n$. This derivative is -4π at the upper edge of the surface and 0 at the lower edge, in agreement with equation (1.14)

(2) (disk of radius a) Imagine a disk of radius a lying in the xy plane with its center at the origin. Calculating V for points along the z axis, we easily obtain

$$V(z) = 2\pi\left(\sqrt{a^2 + z^2} - |z|\right).$$

From this one easily calculates $\partial V/\partial z = \partial V/\partial n = \mp 2\pi$, where the minus sign refers to the "top" of the disk and the plus sign to the "bottom." This example plays a role in the exact analysis analogous to that played by the radially symmetric distribution of matter in the analysis leading to Poisson's equation.

(3) (flat surface) For the double-layer potential W we observe that if the surface is flat (say it lies in the x_1x_2 plane)

$$\begin{aligned} W &= \int_S \mu(\mathbf{y}) \left(\frac{\partial}{\partial y_3} \frac{1}{|\mathbf{x}-\mathbf{y}|} \right) dS \\ &= -\int_S \mu(\mathbf{y}) \left(\frac{\partial}{\partial x_3} \frac{1}{|\mathbf{x}-\mathbf{y}|} \right) dS = -\frac{\partial}{\partial n} V, \end{aligned}$$

where V is the single-layer potential with surface density μ; therefore equation (1.15) holds (at least in this special case) by virtue of equation (1.14) for the single-layer potential.

It will be useful to characterize a given volume distribution by its uniqueness properties. Consider the potential Ψ given by equation (1.1) above, with ρ Hölder continuous in the bounded, regular domain D. It satisfies the following conditions:

(1) $\Delta \Psi = -4\pi\rho$ in D,
(2) $\Delta \Psi = 0$ in $\mathbb{R}^3 \backslash \overline{D}$ and $\Psi \to 0$ as $|\mathbf{x}| \to \infty$,
(3) Ψ is C^1 in all of $\mathbb{R}^3$.

THEOREM 1.5. *The conditions* 1, 2 *and* 3 *above uniquely determine the function* Ψ *in* $\mathbb{R}^3$.

PROOF. Suppose there is a second function Φ defined on $\mathbb{R}^3$ and satisfying these same conditions there, and put $\chi = \Psi - \Phi$. Then

$$0 = \int_D \chi \Delta \chi \, dV = \int_{\partial D} \chi \frac{\partial \chi}{\partial n} \, dS - \int_D |\nabla \chi|^2 \, dV$$

and

$$0 = \int_{\mathbb{R}^3 \backslash \overline{D}} \chi \Delta \chi \, dV = -\int_{\partial D} \chi \frac{\partial \chi}{\partial n} \, dS - \int_{\mathbb{R}^3 \backslash \overline{D}} |\nabla \chi|^2 \, dV,$$

where we have noted that the outward normal to the outer region is $-\hat{\mathbf{n}}$. By virtue of the continuity of χ and $\partial\chi/\partial n$ we infer that

$$\int_{\mathbb{R}^3} |\nabla \chi|^2 \, dV = 0,$$

so χ is constant in all of $\mathbb{R}^3$. Since it tends to zero at large distances, it is zero. □

We now introduce certain formulas for integration on specialized domains.

1.2.2. *An integration formula.* We provide a formula, in equation (1.20) below, applicable to the case when the domain in question is stratified by a particular family of surfaces.

Consider a volume integral over a domain D

$$\int_D f(\mathbf{x}) \, d\mathbf{x} \tag{1.16}$$

and suppose we introduce the smooth, invertible change of variables $\mathbf{x} = \mathbf{X}(\boldsymbol{\xi})$ with new coordinates $\boldsymbol{\xi} = (\xi_1, \xi_2, \xi_3)$ in a domain $\mathcal{D}$ for $\mathbf{x} \in D$. Then

$$\int_D f(\mathbf{x}) \, d\mathbf{x} = \int_{\mathcal{D}} \phi(\boldsymbol{\xi}) \det(\mathbf{X}_{\boldsymbol{\xi}}) \, d\boldsymbol{\xi}, \tag{1.17}$$

where we have written $\phi(\boldsymbol{\xi}) = f(\mathbf{X}(\boldsymbol{\xi}))$ and have assumed that the Jacobian determinant $\det(\mathbf{X}_{\boldsymbol{\xi}})$ is everywhere positive.[4]

Consider now a set on which $\xi_3 =$ constant, say, and suppose that this describes a surface embedded in this region. The element of area on this surface is given by the expression

$$d\mathbf{S} = \hat{\mathbf{n}}\, dS = \frac{\partial \mathbf{X}}{\partial \xi_1} \times \frac{\partial \mathbf{X}}{\partial \xi_2}\, d\xi_1 d\xi_2$$

where $\hat{\mathbf{n}}$ is the unit normal vector to the surface. If the region D_{ab} is swept out by such surfaces with ξ_3 running from a to b, rewrite the integral (1.17) in the iterated form

$$\int_{D_{ab}} f(\mathbf{x})\, d\mathbf{x} = \int_a^b d\xi_3 \int_{S_{\xi_3}} \phi(\boldsymbol{\xi}) \frac{\partial \mathbf{X}}{\partial \xi_1} \times \frac{\partial \mathbf{X}}{\partial \xi_2} \cdot \frac{\partial \mathbf{X}}{\partial \xi_3}\, d\xi_1 d\xi_2, \tag{1.18}$$

where we have expanded the Jacobian determinant by the third row. By the preceding expression for the area element, this becomes

$$\int_{D_{ab}} f(\mathbf{x})\, d\mathbf{x} = \int_a^b d\xi_3 \int_{S_{\xi_3}} \phi(\boldsymbol{\xi}) \frac{\partial \mathbf{X}}{\partial \xi_3} \cdot \hat{\mathbf{n}}\, dS. \tag{1.19}$$

We now view this for a domain in $\mathbb{R}^3$ through each point of which passes one and only one surface of a family of closed surfaces $S_u\colon F(\mathbf{x}, u) = 0$. If $\mathbf{x} = \mathbf{X}(u)$ is a family of solutions of the equation $F = 0$, then

$$\nabla F \cdot \mathbf{X}_u + F_u = 0.$$

Since ∇F gives the direction of the normal,[5] this can be written

$$|\nabla F| \hat{\mathbf{n}} \cdot \mathbf{X}_u = -F_u.$$

If ΔS represents an element of area on a surface of constant u, we can infer that an element of volume contained between two surfaces on which the parameters are u and $u + \Delta u$ has the form

$$\Delta V = \Delta S \hat{\mathbf{n}} \cdot \Delta \mathbf{x} = -\Delta S \frac{F_u}{|\nabla F|} \Delta u.$$

We now assume that u and ξ_3 are constant on the same set of surfaces $\xi_3 = \psi(u)$ for some appropriate function ψ and therefore that

$$\hat{\mathbf{n}} \cdot \frac{\partial \mathbf{X}}{\partial \xi_3} = -\frac{F_u}{|\nabla F|} \frac{du}{d\xi_3}.$$

Equation (1.19) takes the form

$$\int_{D_{ab}} f(\mathbf{x})\, d\mathbf{x} = -\int_a^b \left\{ \int_{S_u} f(\mathbf{x}) F_u |\nabla F|^{-1}\, dS \right\} du. \tag{1.20}$$

The advantage of this formulation is that it isn't necessary to specify a full change of variables $\mathbf{x} = \mathbf{X}(\boldsymbol{\xi})$.

---

[4] It is reasonable to assume it is everywhere of one sign, since this is the usual assumption guaranteeing invertibility.

[5] The outward normal if $F_u < 0$, which we shall provisionally assume.

2. Ellipsoids

Among bounded, three-dimensional domains containing matter, the next in simplicity to the sphere, from the standpoint of calculating its gravitational potential, is the ellipsoid. We develop in the following sections the theory of the gravitational attraction of matter in domains bounded by ellipsoids.

2.1. Two geometrical lemmas. Denote by $D = D(a_1, a_2, a_3)$ the domain bounded by the ellipsoid with semiaxes a_1, a_2 and a_3. In coordinates aligned with the principal axes,

$$\frac{x_1^2}{a_1^2} + \frac{x_2^2}{a_2^2} + \frac{x_3^2}{a_3^2} < 1. \tag{2.1}$$

The ellipsoid is the surface $E = E(a_1, a_2, a_3) = \partial D$ obtained by replacing the inequality with equality in this equation. It will be convenient to have a notation for the left-hand side:

$$b(\mathbf{x}) = b(\mathbf{x}, a_1, a_2, a_3) = \frac{x_1^2}{a_1^2} + \frac{x_2^2}{a_2^2} + \frac{x_3^2}{a_3^2}. \tag{2.2}$$

An ellipsoid with semiaxes ua_1 ua_2 and ua_3 $(u > 0)$ is said to be *similar* to $E(a_1, a_2, a_3)$. Two ellipsoids that are concentric and whose principal axes are parallel will be said to be *similarly placed.* In the same coordinates in which E is given by $b(\mathbf{x}, a_1, a_2, a_3) = 1$, a similar and similarly placed ellipsoid has the equation $b(\mathbf{x}, a_1, a_2, a_3) = u^2$. An ellipsoid is *confocal* with the ellipsoid $E(a_1, a_2, a_3)$ if, in the coordinates aligned with the semiaxes of the latter, its equation is

$$\frac{x_1^2}{a_1^2 + \lambda} + \frac{x_2^2}{a_2^2 + \lambda} + \frac{x_3^2}{a_3^2 + \lambda} = 1, \tag{2.3}$$

i.e., if its semiaxes are $a_i' = \sqrt{a_i^2 + \lambda}$ for $i = 1, 2, 3$. For this confocal figure to be an ellipsoid, λ must exceed the numbers $-a_1^2$ $- a_2^2$ and $-a_3^2$.

Two kinds of ellipsoidal shells will be encountered, *homoeoids*[6] and *focaloids.* A homoeoid is the region between two similar and similarly placed ellipsoids. A focaloid is the region between two confocal ellipsoids. The basic gravitational theorem regarding homoeoids (Newton's Theorem 2.3, below) depends on the following geometric property of ellipsoids.

LEMMA 2.1 (Lemma on Homoeoids). *Define a homoeoid H via its inner and outer boundaries*[7]

$$b(\mathbf{x}) = u^2 < 1 \quad \textit{and} \quad b(\mathbf{x}) = 1 \tag{2.4}$$

and let p be a point in the domain bounded by the inner ellipsoid, i.e., the coordinates of p satisfy

$$\frac{x_1^2}{a_1^2} + \frac{x_2^2}{a_2^2} + \frac{x_3^2}{a_3^2} < u^2.$$

Let a straight line ℓ pass through p, intersecting the inner boundary in points ξ, η and the outer boundary in points ξ', η', where ξ and ξ' lie on one ray through p, η and η' on the other. Then the lengths of the segments $\overline{\xi\xi'}$ and $\overline{\eta\eta'}$ are the same.

[6]The name homoeoid was coined by Lord Kelvin (William Thomson) and Peter Tait.

[7]Recall the definition (2.2).

PROOF. The line can be expressed in the form

$$x_i = \alpha_i + \beta_i t, \quad i = 1, 2, 3$$

for a parameter t, intersecting the inner ellipsoid at points ξ and η with parameter values t_ξ and t_η and the outer ellipsoid at points ξ' and η' with parameter values $t_{\xi'}$ and $t_{\eta'}$. We may assume that the parameter values are increasing in the order $t_{\eta'}, t_\eta, t_\xi, t_{\xi'}$. From this we infer that the distance from ξ to ξ' is $\|\beta\|(t_{\xi'} - t_\xi)$ and that from η' to η is $\|\beta\|(t_\eta - t_{\eta'})$. The result is therefore proved if we can show that

$$t_{\xi'} - t_\xi = t_\eta - t_{\eta'}. \tag{2.5}$$

For this we use the equations (2.4) of the ellipsoids. For example the point ξ' satisfies the equation

$$\sum_{i=1}^{3} a_i^{-2}(\alpha_i + \beta_i t_{\xi'})^2 = A + 2Bt_{\xi'} + Ct_{\xi'}^2 = 1,$$

where the coefficients A, B, C are given by simple formulas left to the reader; $t_{\eta'}$ satisfies the same equation. The points t_ξ and t_η satisfy a similar equation, the only difference being that the right-hand side has u^2 in place of 1. Solving these quadratics for the t values and recalling their order from $t_{\eta'}$ to $t_{\xi'}$ we find that[8]

$$t_{\xi'} - t_\xi = C^{-1}\left(\sqrt{C + B^2 - AC} - \sqrt{u^2C + B^2 - AC}\right)$$

and that $t_\eta - t_{\eta'}$ is given by the same expression, verifying equation (2.5) and proving the theorem. □

The second geometrical lemma requires a further definition. Let $E(a)$ and $E(a')$ be concentric ellipsoids having their principal axes parallel. Suppose p lies on $E(a)$; let its coordinates in a system aligned along the principal axes be (x_1, x_2, x_3). The point p' with cooordinates given by the equations

$$x_i' = \frac{a_i'}{a_i} x_i, \quad i = 1, 2, 3 \tag{2.6}$$

lies on $E(a')$. It is called the *corresponding point* on $E(a')$ to the point p on $E(a)$. $E(a)$ and $E(a')$ enclose a focaloid if $a_i'^2 = a_i^2 + \lambda$, $i = 1, 2, 3$.

LEMMA 2.2 (Ivory's Lemma on Focaloids). *Let p, p' be one pair of corresponding points on the confocal ellipsoids $E(a)$ and $E(a')$, and q, q' be a second. Then $\overline{pq'} = \overline{p'q}$.*

PROOF. Let the position vectors of p, p' be $\mathbf{x}$, $\mathbf{x}'$ and those of q, q' be $\mathbf{y}$, $\mathbf{y}'$. For the difference $\overline{pq'}^2 - \overline{p'q}^2$ we find

$$\begin{aligned}\sum_{i=1}^{3}\left((x_i - y_i')^2 - (x_i' - y_i)^2\right) &= \sum_{i=1}^{3}\left(\frac{a_i'^2}{a_i^2} - 1\right)(y_i^2 - x_i^2)\\ &= \lambda\sum_{i=1}^{3}\frac{y_i^2}{a_i^2} - \lambda\sum_{i=1}^{3}\frac{x_i^2}{a_i^2},\end{aligned}$$

where we have used the equation $a_i'^2 = a_i^2 + \lambda$ expressing the confocal property. Since each of the sums in the final expression is one, the conclusion follows. □

[8]Note from the editors: This expression which originally read $t_{\xi'} - t_\xi = C^{-1}\left(\sqrt{1 + B^2 - A} - \sqrt{u^2 + B^2 - A}\right)$ has been corrected.

2.2. Consequences of the lemma on homoeoids. From Lemma 2.1 on homoeoids one now obtains the following theorem.[9]

THEOREM 2.3 (Newton's Theorem). *The attraction in the domain inside the inner boundary of a homoeoid of uniform density* $\rho = 1$ *vanishes.*

PROOF. Let p be an arbitrary point in the interior and consider a typical component of the attraction at p, say

$$f_3(p) = -\int_H \frac{x_3(p) - y_3}{|\mathbf{x}(p) - \mathbf{y}|^3}\, d\mathbf{y}.$$

If we shift the origin of coordinates to the point p via the transformation $\mathbf{y} = \mathbf{x}(p) + \eta$ and then transform to spherical coordinates

$$\eta_1 = r \sin\theta \cos\phi, \quad \eta_2 = r \sin\theta \sin\phi, \quad \eta_3 = r \cos\theta$$

with the η_3 axis as the polar axis, this becomes

$$\begin{aligned} f_3(p) &= \int_0^{2\pi} \int_0^{\pi} \int_{R_u(\theta,\phi)}^{R_1(\theta,\phi)} \cos\theta \sin\theta \, dr\, d\theta\, d\phi \\ &= \int_0^{2\pi} \int_0^{\pi} F(\theta,\phi)\, d\theta\, d\phi, \end{aligned}$$

where

$$F(\theta,\phi) = (R_1(\theta,\phi) - R_u(\theta,\phi)) \cos\theta \sin\theta$$

and the function R_u[10] can easily be read off from the expression for an ellipsoid:

$$R_u(\theta,\phi) = u \left\{ \sin^2\theta \left(\frac{\cos^2\phi}{a_1^2} + \frac{\sin^2\phi}{a_2^2} \right) + \frac{\cos^2\theta}{a_3^2} \right\}^{-1/2}.$$

The transformation $\theta \to \pi - \theta$, $\phi \to \phi + \pi$ takes $\mathbf{y}$ to $-\mathbf{y}$ and therefore leaves the term $R_1 - R_u$ unchanged, by the Lemma 2.1. The integral becomes

$$-\int_{-\pi}^{\pi} \int_0^{\pi} F(\theta',\phi')\, d\theta'\, d\phi'.$$

The function F is 2π-periodic in ϕ so the ϕ integration above, on $(-\pi,\pi)$, is equal to that above, on $(0,2\pi)$, but has the opposite sign, showing that it is equal to its negative and hence vanishes. We chose the polar axis to agree with the ξ_3 but could have chosen it to agree with ξ_1 or ξ_2, so the force in any direction vanishes. □

It follows of course that the potential is constant in the interior of the homoeoid.

Let $H(u_1, u_2)$ be a homoeoid of uniform density $\rho = 1$, $u_1^2 < b(\mathbf{x}) < u_2^2$ where $b(x)$ is defined above in equation (2.2). Refer to equation (1.20) above,

---

[9]Chandrasekhar referred to this theorem as Newton's theorem [**2**, Chapter 3, Theorem 1]. The book by Chandrasekhar (1969), *Ellipsoidal Figures of Equilibrium*, will henceforth be referred to as EFE [**2**].

[10]The following expression for R_u is valid only when $\mathbf{x} = (0,0,0)$, but $\mathbf{x}$ can be any point inside the ellipsoid in this proof. The correct formula for R_u would be much more complicated. But we do not need any formula for R_u in this proof.

with $F(x,u) = b(x) - u^2$. Then one easily finds that

$$\int_{H(u_1,u_2)} f(\mathbf{x})\, d\mathbf{x} = \int_{u_1}^{u_2} v \left\{ \int_{b(\mathbf{x})=v^2} f(\mathbf{x})\sigma(\mathbf{x})\, dS \right\} dv \tag{2.7}$$

where

$$\sigma(\mathbf{x}) = \left(\frac{x_1^2}{a_1^4} + \frac{x_2^2}{a_2^4} + \frac{x_3^2}{a_3^4} \right)^{-1/2}. \tag{2.8}$$

These formulas facilitate the formal proofs of the following deductions from Newton's theorem.[11]

COROLLARY 2.3.1. *Let D_1 be the region interior to the ellipsoid $E_1\colon b(\mathbf{x}) = u_1^2$. Then*

(1) *The single-layer potential $V(\mathbf{x}, u_1)$[12] due to matter of surface density $\sigma(\mathbf{x})$ on E_1 is constant in D_1.*
(2) *The potential $\Psi(\mathbf{x})$, due to matter of density $f(b(\mathbf{x}))$ filling the homoeoid bounded by E_1 and $E_2\colon b(\mathbf{x}) = u_2^2$, $u_2^2 > u_1^2$, is constant in D_1, for any choice of the function f.*

PROOF. First consider the homoeoid $H(u_1, u)$ for $u > u_1 > 0$ and density $\rho = 1$. Let Φ be the potential:

$$\begin{aligned} \Phi(\mathbf{x}, u_1, u) &= \int_{H(u_1,u)} \frac{1}{|\mathbf{x}-\mathbf{y}|}\, d\mathbf{y} \\ &= \int_{u_1}^{u} v \left\{ \int_{b(\mathbf{y})=v^2} \frac{\sigma(\mathbf{y})}{|\mathbf{x}-\mathbf{y}|}\, dS_y \right\} dv. \end{aligned}$$

To arrive at this formula we have used equation (2.7) above using $\mathbf{y}$ as integration variable in place of $\mathbf{x}$ and reintroducing $\mathbf{x}$ as a parameter in the function $f(\mathbf{y}) = 1/|\mathbf{x}-\mathbf{y}|$. By Newton's theorem, this is independent of $\mathbf{x}$ in D_1. Calling it $\Phi(u_1, u)$, we have

$$\frac{\partial \Phi(u_1,u)}{\partial u} = u \int_{b(\mathbf{y})=u^2} \frac{\sigma(\mathbf{y})}{|\mathbf{x}-\mathbf{y}|}\, dS_y = uV(u) \tag{2.9}$$

which is likewise independent of $\mathbf{x}$ for $u \in [u_1, u_2]$. Since the right-hand side is, except for a factor of u, the potential of the single layer with density σ, this verifies statement 1.

Next consider the potential Ψ described in the second statement. We have

$$\begin{aligned} \Psi(\mathbf{x}) &= \int_{u_1}^{u_2} v \left\{ \int_{b(\mathbf{y})=v^2} \frac{f(b(\mathbf{y}))\,\sigma(\mathbf{y})}{|\mathbf{x}-\mathbf{y}|}\, dS_y \right\} dv \\ &= \int_{u_1}^{u_2} v f(v^2) \left\{ \int_{b(\mathbf{y})=v^2} \frac{\sigma(\mathbf{y})}{|\mathbf{x}-\mathbf{y}|}\, dS_y \right\} dv \\ &= \int_{u_1}^{u_2} v f(v^2) V(v)\, dv, \end{aligned} \tag{2.10}$$

where in passing from the second to the third step we have used equation (2.9). The last expression shows that Ψ is independent of $\mathbf{x}$ in D_1, verifying statement 2. □

[11] Corollary to EFE [**2**, Theorem 1, Chapter 3].

[12] Recall the definition in equation (1.3) above.

Since the potential $\Phi(u_1, u)$ appearing in the proof of the preceding theorem is constant in the interior, it can be evaluated by choosing $\mathbf{x} = 0$. It will prove useful to have the explicit expression for this potential as a function of the semiaxes a_1, a_2, a_3. Using spherical coordinates we have

$$\Phi\left(u_1, u\right) = \int_{H(u_1,u)} \frac{d\mathbf{y}}{|\mathbf{y}|} = 8 \int_0^{\pi/2} \int_0^{\pi/2} \int_{R(u_1,\theta,\phi)}^{R(u,\theta,\phi)} r \sin\theta \, dr \, d\theta \, d\phi,$$

where we have observed that, in view of the eight-fold symmetry of the integral, it suffices to integrate in the first octant and multiply by eight. Since in these coordinates the expression for the ellipsoid $b(\mathbf{x}) = u^2$ takes the form

$$r^2 \left(\sin^2\theta \left(\frac{\cos^2\phi}{a_1^2} + \frac{\sin^2\phi}{a_2^2} \right) + \frac{\cos^2\theta}{a_3^2} \right) = u^2,$$

we find, on performing the r-integration in the last integral above EFE [**2**, Theorem 2, Chapter 3],

$$\Phi\left(u_1, u\right) = 4\left(u^2 - u_1^2\right) \int_0^{\pi/2} \int_0^{\pi/2} \frac{\sin\theta \, d\theta \, d\phi}{\sin^2\theta \left(\frac{\cos^2\phi}{a_1^2} + \frac{\sin^2\phi}{a_2^2} \right) + \cos^2\theta \frac{1}{a_3^2}}. \tag{2.11}$$

With the definitions

$$a = \frac{\cos^2\theta}{a_3^2} + \frac{\sin^2\theta}{a_1^2} \quad \text{and} \quad b = \frac{\cos^2\theta}{a_3^2} + \frac{\sin^2\theta}{a_2^2} \tag{2.12}$$

and the substitution $t - \tan\phi$, the ϕ integration takes the form

$$\int_0^\infty \frac{dt}{a + bt^2} = \frac{\pi}{2\sqrt{ab}} \equiv j_0(a, b), \tag{2.13}$$

where we have used the integration formula

$$\int^t \frac{dt}{a + bt^2} = \frac{1}{\sqrt{ab}} \tan^{-1}\left(\sqrt{\tfrac{b}{a}} t \right). \tag{2.14}$$

This brings the integral (2.11) to the form

$$2\pi(u^2 - u_1^2) \int_0^{\pi/2} \frac{\sin\theta}{\sqrt{\left(\frac{\sin^2\theta}{a_1^2} + \frac{\cos^2\theta}{a_3^2} \right) \left(\frac{\sin^2\theta}{a_2^2} + \frac{\cos^2\theta}{a_3^2} \right)}} d\theta.$$

By means of the further substitution $s = a_3^2 \tan^2\theta$, this can be put into the standard form

$$\pi(u^2 - u_1^2) a_1 a_2 a_3 \int_0^\infty \frac{ds}{\Delta(s)},$$

where

$$\Delta(s) = \sqrt{\left(a_1^2 + s\right)\left(a_2^2 + s\right)\left(a_3^2 + s\right)}. \tag{2.15}$$

Define EFE [**2**, Chapter 3, equation (15)]

$$I\left(a_1, a_2, a_3\right) = a_1 a_2 a_3 \int_0^\infty \frac{ds}{\Delta(s)}. \tag{2.16}$$

Then the expression for the constant interior potential $\Phi(u_1, u)$ due to matter of density $\rho = 1$ filling the homoeoid $u_1^2 < b(\mathbf{x}) < u^2$ is[13]

$$\Phi\left(u_1, u\right) = \pi\left(u^2 - u_1^2\right) I\left(a_1, a_2, a_3\right). \tag{2.17}$$

The expression for the constant interior value of the single layer potential of the ellipsoid $E(a)$ with surface density $\sigma(\mathbf{x})$ is obtained from the equation above by differentiating with respect to u and then setting $u = 1$ (equation (2.9) above). It is $2\pi I(a_1, a_2, a_3)$. The corresponding expression for the constant interior potential due to a homoeoid $H(u_1, u_2)$ of density $\rho = f(b^2)$ is obtained from equation (2.10) and equation (2.17). It is

$$\begin{aligned}
\Psi\left(0\right) &= \int_{u_1}^{u_2} f\left(v^2\right) 2\pi v I(a_1, a_2, a_3) dv \\
&= 2\pi I(a_1, a_2, a_3) \int_{u_1}^{u_2} v f(v^2)\, dv \\
&= \pi I(a_1, a_2, a_3) \int_{u_1^2}^{u_2^2} f(s)\, ds.
\end{aligned} \tag{2.18}$$

It will be useful to rephrase these results, and to augment Corollary 2.3.1 above, as follows.

COROLLARY 2.3.2. *Let D_1 be the region interior to the ellipsoid $E_1\colon b(\mathbf{x}) = u_1^2$. Then*

(1) *The constant value of the single-layer potential in D_1, due to matter of surface density $\sigma(\mathbf{x})$ on E_1, is*

$$2\pi I\left(a_1, a_2, a_3\right). \tag{2.19}$$

(2) *The constant value of the potential $\Psi(\mathbf{x})$, due to matter of density $f(b(\mathbf{x}))$ filling the homoeoid bounded by E_1 and E_2, is*

$$\pi I\left(a_1, a_2, a_3\right) \int_{u_1^2}^{u_2^2} f\left(s\right)\, ds, \tag{2.20}$$

where I is given by equation (2.16) *above.*

2.3. Consequences of the lemma on focaloids. We turn now to a theorem, due to Ivory[14], relating the single-layer potentials of two confocal ellipsoids, from which we can infer the potentials exterior to ellipsoids from our knowledge of those interior to ellipsoids. Recall the definition of *corresponding points* (equation (2.6) above).

THEOREM 2.4 (Ivory). *Suppose $E = E(a)$ and $E' = E(a')$ are confocal ellipsoids, E' the larger: ${a_i'}^2 = a_i^2 + \lambda$, $i = 1, 2, 3$, with $\lambda > 0$. Let their respective surface densities be*

$$\sigma\left(\mathbf{x}\right) = \left(\sum_{i=1}^{3} \frac{x_i^2}{a_i^4}\right)^{-1/2}, \qquad \sigma'\left(\mathbf{x}\right) = \left(\sum_{i=1}^{3} \frac{x_i^2}{{a_i'}^4}\right)^{-1/2}. \tag{2.21}$$

[13]Recall that we are suppressing factors of the density ρ and the gravitational constant G.

[14]James Ivory (February 17, 1765–September 21, 1842) was a British mathematician.

Denote the potentials due to these simple layers by $V_E(\mathbf{x})$ *and* $V_{E'}(\mathbf{x})$*, respectively. Then if* $\mathbf{x} \in E$ *and* $\mathbf{x}' \in E'$ *are corresponding points,*

$$\frac{V_{E'}(\mathbf{x})}{a_1' a_2' a_3'} = \frac{V_E(\mathbf{x}')}{a_1 a_2 a_3}. \tag{2.22}$$

PROOF. The expressions for the potentials are

$$V_E(\mathbf{x}') = \int_E \frac{\sigma(\mathbf{y})}{|\mathbf{x}' - \mathbf{y}|}\, dS_y \text{ and } V_{E'}(\mathbf{x}) = \int_{E'} \frac{\sigma'(\mathbf{y}')}{|\mathbf{x} - \mathbf{y}'|}\, dS_{y'}.$$

The point $\mathbf{x}$ is an arbitrary point of E but is held fixed in the following considerations; the corresponding point $\mathbf{x}'$ on E' is therefore likewise held fixed.

We can transform the integral over E' to an integral over E via the corresponding-points transformation. We shall express this surface integral in Cartesian coordinates. To do this we divide the surface into patches on each of which one of the Cartesian coordinates has a fixed sign while the other two serve as integration parameters. For example, let E_0' be a patch of the surface E' on which $y_3' > 0$. Using y_1', y_2' as integration variables, we have

$$\int_{E_0'} \frac{\sigma(\mathbf{y}')}{|\mathbf{x} - \mathbf{y}'|}\, dS_{y'} = \int_{d_0'} \frac{\sigma(\mathbf{y}')}{|\mathbf{x} - \mathbf{y}'|} \sec\gamma'\, dy_1' dy_2',$$

where d_0' is the projection of E_0' on the $y_1'y_2'$-plane, and

$$\sec\gamma' = \sqrt{1 + \frac{\partial y_3'}{\partial y_1'}^2 + \frac{\partial y_3'}{\partial y_2'}^2} = \frac{a_3'^2}{y_3'} \left(\sum_{i=1}^{3} \frac{y_i'^2}{a_i'^4}\right)^{1/2} = \frac{a_3'^2}{y_3'} \sigma'(\mathbf{y}')^{-1}.$$

The last expressions above depend on the explicit expression

$$y_3' = a_3' \left(1 - \frac{y_1'^2}{a_1'^2} - \frac{y_2'^2}{a_2'^2}\right)^{1/2} \tag{2.23}$$

that holds under present circumstances and on the particular definition of the surface density σ'. Therefore

$$\int_{E_0'} \frac{\sigma(\mathbf{y}')}{|\mathbf{x} - \mathbf{y}'|}\, dS_{y'} = \int_{d_0'} \frac{a_3'^2/y_3'}{|\mathbf{x} - \mathbf{y}'|}\, dy_1' dy_2'.$$

The transformation $y_1 = a_1 y_1'/a_1'$, $y_2 = a_2 y_2'/a_2'$ carries the planar domain d_0' to a planar domain d_0, transforming the integral above to

$$\frac{a_1' a_2'}{a_1 a_2} \int_{d_0} \frac{a_3'^2/y_3'}{|\mathbf{x} - \mathbf{y}'|}\, dy_1 dy_2,$$

wherein equation (2.23) implies that

$$\frac{y_3'}{a_3'} = \left(1 - \frac{y_1^2}{a_1^2} - \frac{y_2^2}{a_2^2}\right)^{1/2} = \frac{y_3}{a_3}, \tag{2.24}$$

i.e., it maps E_0', in a one-to-one fashion, to a patch E_0 on E on which $y_3 > 0$. By Ivory's Lemma on Focaloids 2.2, $|\mathbf{x} - \mathbf{y}'| = |\mathbf{x}' - \mathbf{y}|$ under this transformation. This and the preceding formula take the integral above to

$$\frac{a_1' a_2' a_3'}{a_1 a_2 a_3} \int_{d_0} \frac{a_3^2/y_3}{|\mathbf{x}' - \mathbf{y}|}\, dy_1 dy_2.$$

But the last integral is precisely the expression for the contribution from the patch E_0 to the integral for $V_E(\mathbf{x}')$. Adding the contributions from a union of patches making up E', we arrive at the stated conclusion. □

The preceding theorem enables us to find the potential $V_E(\mathbf{x}')$ at an arbitrary point $\mathbf{x}'$ outside a given ellipsoidal surface $E(a)$ on which the surface density σ is given by the first formula of equation (2.21) above. The basic observation is that, by the corollary to Newton's theorem, the potential $V_{E'}(\mathbf{x})$ is constant in the interior and therefore equal to the central value expressed in the formula (2.19), wherein a_i' must be substituted for a_i, $i = 1, 2, 3$. We can now prove the following theorem.

THEOREM 2.5. *The potential V_E at a point $\mathbf{x}'$ external to the ellipsoidal surface $E(a)$, on which the surface density σ is that given by equation* (2.21), *is*

$$V_E(\mathbf{x}') = 2\pi a_1 a_2 a_3 \int_{\lambda(\mathbf{x}')}^{\infty} \frac{ds}{\Delta(s)}, \tag{2.25}$$

where Δ is given by equation (2.15), *and λ is the positive root of the equation*

$$\sum_{i=1}^{3} \frac{x_i'^2}{a_i^2 + \lambda} = 1. \tag{2.26}$$

PROOF. Equation (2.26) is easily reduced to a polynomial equation of degree three having real roots. Two of these are negative and the third is positive if we require $\mathbf{x}'$ to lie outside the ellipsoid $E(a)$, as we do. By the preceding theorem, the required potential

$$V_E(\mathbf{x}') = \frac{a_1 a_2 a_3}{a_1' a_2' a_3'} V_{E'}(\mathbf{x}).$$

The potential $V_{E'}$ is, by Corollary 2.3.1 to Newton's theorem, constant for values $\mathbf{x}$ in its interior and, by Corollary 2.3.2, that constant value is

$$V_{E'}(\mathbf{x}) = 2\pi a_1' a_2' a_3' \int_0^{\infty} \frac{ds}{\left((a_1'^2 + s)(a_2'^2 + s)(a_3'^2 + s)\right)^{1/2}}.$$

Since $a_i'^2 = a_i^2 + \lambda(\mathbf{x}')$, this gives to $V_E(\mathbf{x}')$ the value (2.25) after a change of integration variable. □

For the purpose of keeping track of all the variables in this problem, it will be useful to repeat the conclusion of the preceding theorem in more explicit form. We continue to assume that the point $\mathbf{x}'$ lies outside the ellipsoid $\sum y_i^2/a_i^2 = 1$. Then equation (2.25) may be written

$$\begin{aligned} &\int_{\sum y_i^2/a_i^2 = 1} \frac{\left(\sum y_i^2/a_i^4\right)^{-1/2}}{|\mathbf{x}' - \mathbf{y}|} dS_y \\ &\quad = 2\pi a_1 a_2 a_3 \int_{\lambda(\mathbf{x}')}^{\infty} \frac{ds}{\sqrt{(a_1^2 + s)(a_2^2 + s)(a_3^2 + s)}}, \end{aligned} \tag{2.27}$$

where $\lambda(\mathbf{x}')$ is the positive root of the equation

$$\sum \frac{x_i'^2}{a_i^2 + \lambda} = 1.$$

2.4. Exterior potential. With the aid of the preceding theorem we can express the potential Ψ of a solid ellipsoid D of uniform density $\rho = 1$ at an exterior point $\mathbf{x}'$, i.e., we can find an explicit formula for

$$\Psi(\mathbf{x}') = \int_{D_1} \frac{d\mathbf{x}}{|\mathbf{x}-\mathbf{x}'|} \quad \text{where } D_1 = \{\mathbf{x} : \sum_{i=1}^{3} \frac{x_i^2}{a_i^2} < 1\}. \tag{2.28}$$

For this we again exploit the integration formula (1.20) (cf. equations (2.7) and (2.8) above) to write

$$\Psi(\mathbf{x}') = \int_0^1 v \left\{ \int_{E_v} \frac{\sigma(\mathbf{x})}{|\mathbf{x}-\mathbf{x}'|} dS_x \right\} dv, \qquad E_v = \{\mathbf{x} : \sum_{i=1}^{3} \frac{x_i^2}{a_1^2} = v^2\}. \tag{2.29}$$

The inner integral above, under the transformation of variables $x_i = vy_i$, $i = 1, 2, 3$,[15] takes the form

$$\int_{E_1} \frac{\sigma(\mathbf{y})}{|\mathbf{y}-v^{-1}\mathbf{x}'|} dS_y = V_E(v^{-1}\mathbf{x}') = 2\pi a_1 a_2 a_3 \int_{\lambda(v^{-1}\mathbf{x}')}^{\infty} \frac{ds}{\Delta(s)}, \tag{2.30}$$

where we have applied the formula (2.25) of Theorem 2.5. Equation (2.29) now takes the form

$$\Psi(\mathbf{x}') = 2\pi a_1 a_2 a_3 \int_0^1 \left\{ \int_{\lambda(v^{-1}\mathbf{x}')}^{\infty} \frac{ds}{\Delta(s)} \right\} v\, dv. \tag{2.31}$$

In the equation above the variables v and λ are related through the equation

$$\sum_{i=1}^{3} \frac{x_i'^2}{a_i^2 + \lambda} = v^2$$

in which x_i' and a_i are held fixed for $i = 1, 2, 3$. The relationship between λ and v is easily seen to be monotone, with λ ranging from a positive minimum of $\lambda(\mathbf{x}')$ when $v = 1$ to $+\infty$ when $v = 0$, and the preceding expression implies that

$$2v\frac{dv}{d\lambda} = \frac{d}{d\lambda} \sum_{i=1}^{3} \frac{x_i'^2}{a_i^2 + \lambda};$$

substitution of this in the expression above to exchange integration with respect to λ for integration with respect to v, followed by an integration by parts, provides the following.

THEOREM 2.6. *The gravitational potential of an ellipsoid of uniform density ρ at an external point $\mathbf{x}'$ is*

$$\Psi(\mathbf{x}') = \pi G \rho a_1 a_2 a_3 \left\{ \int_{\lambda(\mathbf{x}')}^{\infty} \frac{ds}{\Delta(s)} - \sum_{i=1}^{3} x_i^2 \int_{\lambda(\mathbf{x}')}^{\infty} \frac{ds}{(a_i^2 + s)\Delta(s)} \right\} \tag{2.32}$$

where $\lambda(\mathbf{x}')$ is the positive root of equation (2.26).

In this theorem factors G and ρ, sometimes suppressed above, have been restored for physical and dimensional accuracy.

[15] On a two-dimensional surface there are only two surface parameters, not three. However, if we take these to be (say) y_1 and y_2 and set $x_1 = vy_1, x_2 = vy_2$ then, for the special case of an ellipsoid E_v, we find that $x_3 = \pm vy_3$ where y lies on E_1.

2.5. Interior potential. The potential in the interior of an ellipsoid of uniform density can be obtained from results found above. It has the comparatively simple structure of a polynomial of degree two.

THEOREM 2.7. *The gravitational potential at a point* $\mathbf{x}$ *in the interior of the basic ellipsoid* $b(\mathbf{x}) < 1$ *where* b *is given by equation* (2.2), *is*

$$\Psi(\mathbf{x}) = \pi G\rho a_1 a_2 a_3 \left\{ \int_0^\infty \frac{ds}{\Delta(s)} - \sum_{i=1}^3 x_i^2 \int_0^\infty \frac{ds}{(a_i^2+s)\Delta(s)} \right\}. \tag{2.33}$$

Proof: The integration formulas (2.7) and (2.8) imply that

$$\Psi(\mathbf{x}) = \int_0^1 v \left\{ \int_{b(\mathbf{y})=v^2} \frac{\sigma(\mathbf{y})}{|\mathbf{x}-\mathbf{y}|} dS_y \right\} dv, \tag{2.34}$$

where the function σ is given by equation (2.21) above. Define the number $w \in [0,1]$ by the formula $w^2 = b(\mathbf{x})$. The surface integral in equation (2.34) has a different outcome according as $v < w$ or $v > w$ so we break up the v integral above accordingly into two integrals, $\Psi_{v<w}$ and $\Psi_{v>w}$. For the first of these, on the interval $[0, w)$, the point $\mathbf{x}$ is an external point to the uniform ellipsoid with axes (wa_1, wa_2, wa_3) and we find, from Theorem (2.32) above,

$$\Psi_{v<w} = \pi G\rho a_1 a_2 a_3 \left\{ w^2 \int_0^\infty \frac{ds}{\Delta(s)} - \sum_{i=1}^3 x_i^2 \int_0^\infty \frac{ds}{(a_i^2+s)\Delta(s)} \right\}.$$

The other contribution is that of the homoeoid enclosed between ellipsoidal surfaces $b(\mathbf{x}) = w^2$ and $b(\mathbf{x}) = 1$, evaluated at the point $\mathbf{x}$ of the inner surface. Since this is the constant value of the interior potential of the homoeoid, its value is given by constant $V(1, w)$ of equation (2.17) above:

$$\Psi_{v>w} = \pi G\rho a_1 a_2 a_3 (1-w^2) \int_0^\infty \frac{ds}{\Delta(s)}. \tag{2.35}$$

Adding these expressions leads to equation (2.17) and completes the proof.

With the definitions of EFE [**2**, Chapter 3, equations (15) and (18)],

$$I = a_1 a_2 a_3 \int_0^\infty \frac{ds}{\Delta(s)}, \quad A_i = a_1 a_2 a_3 \int_0^\infty \frac{ds}{(a_i^2+s)\Delta(s)}, \quad i = 1, 2, 3, \tag{2.36}$$

we can rewrite equation (2.33) more compactly as

$$\Psi(\mathbf{x}) = \pi G\rho \left(I - \sum_{i=1}^3 A_i x_i^2 \right). \tag{2.37}$$

This derivation of the interior potential rests on that for the exterior potential and is different from that in EFE [**2**]. The latter derivation is provided below, with comments, in Appendix D.

3. Ellipsoidal stratifications of density

We begin with distributions of mass that are constant on similar and similarly placed ellipsoids, that is, density functions of the form

$$\rho(\mathbf{x}) = f(b(\mathbf{x})) \quad \text{where} \quad b(\mathbf{x}) = \sum_{i=1}^3 \frac{x_i^2}{a_i^2} \tag{3.1}$$

on the domain

$$D\colon b(x) < 1. \tag{3.2}$$

3.1. Exterior points. Consider first the case when the point $\mathbf{x}'$ lies outside the ellipsoidal figure, i.e., $b(\mathbf{x}') > 1$. Then in the formula

$$\Psi(\mathbf{x}') = \int_0^1 vf(v^2)\left\{\int_{b(\mathbf{y})=v^2} \frac{\sigma(\mathbf{y})}{|\mathbf{x}'-\mathbf{y}|}dS_y\right\}dv \tag{3.3}$$

we use the expression for the inner, surface integral provided by equation (2.25) above, or, more explicitly, by equation (2.27). This leads to the formula

$$\Psi(\mathbf{x}') = 2\pi a_1a_2a_3\int_0^1 vf(v^2)\left\{\int_{\lambda(v^{-1}\mathbf{x}')}^\infty \frac{ds}{\Delta(s)}\right\}dv, \tag{3.4}$$

where $\lambda(v^{-1}\mathbf{x}')$ is the unique positive root of the equation

$$\sum_{i=1}^3 \frac{(v^{-1}x_i')^2}{a_i^2+\lambda} = 1 \quad \text{or} \quad \sum_{i=1}^3 \frac{x_i'^2}{a_i^2+\lambda} = v^2. \tag{3.5}$$

Define

$$G(u) = \int_u^1 f(t)\,dt \qquad \text{so that } \frac{d}{dv}G(v^2) = -2f(v^2)v. \tag{3.6}$$

Then

$$\Psi(\mathbf{x}') = -\pi a_1a_2a_3\int_0^1 \left\{\left(\frac{d}{dv}G(v^2)\right)\left(\int_{\lambda(v^{-1}\mathbf{x}')}^\infty \frac{ds}{\Delta(s)}\right)\right\}dv.$$

Integrating by parts and observing that $\lambda\to\infty$ as $v\to 0$ we find

$$\Psi(\mathbf{x}') = \pi a_1a_2a_3\int_0^1 G(v^2)\left(\frac{d}{dv}\int_{\lambda(v^{-1}\mathbf{x}')}^\infty \frac{ds}{\Delta(s)}\right)dv. \tag{3.7}$$

Changing the integration variable from v to λ via equation (3.5), we obtain the following.

THEOREM 3.1. *Let the density in the ellipsoidal domain D of equation (3.2) be given by equation (3.1). Then the potential at an exterior point $\mathbf{x}'$ is given by the expression*

$$\Psi(\mathbf{x}') = \pi a_1a_2a_3\int_{\lambda(\mathbf{x}')}^\infty G\left(\sum_{i=1}^3 \frac{x_i'^2}{a_i^2+s}\right)\frac{ds}{\Delta(s)}, \tag{3.8}$$

where the function G is defined by equation (3.6) above and the lower limit $\lambda(\mathbf{x}')$ is the (unique) positive solution of the equation

$$\sum_{k=1}^3 \frac{x_k'^2}{a_k^2+\lambda} = 1. \tag{3.9}$$

In this theorem any factor of the dimension of density is included in the definition of the functions f and G.

Verifications

1. Ψ is harmonic:

Since Ψ represents the exterior potential, it must be a harmonic function. We verify this here. For this purpose let's skip the prime on $\mathbf{x}$ and ignore the constant in front of the expression above as well: the function

$$\psi(\mathbf{x}) = \int_{\lambda(\mathbf{x})}^{\infty} G\left(\sum_{i=1}^{3} \frac{x_i^2}{a_i^2+s}\right) \frac{ds}{\Delta(s)} \tag{3.10}$$

should be harmonic. In the expressions below, the summation convention is suspended: if we wish to indicate a sum we do so explicitly. For $i = 1, 2, 3$,

$$\begin{aligned}\frac{\partial \psi}{\partial x_i} &= -\frac{\partial \lambda}{\partial x_i} G\left(\sum_k \frac{x_k^2}{a_k^2+\lambda(\mathbf{x})}\right) \frac{1}{\Delta(\lambda(\mathbf{x}))} \\ &\quad + \int_{\lambda(\mathbf{x})}^{\infty} G'\left(\sum_k \frac{x_k^2}{a_k^2+s}\right) \frac{2x_i}{a_i^2+s} \frac{ds}{\Delta(s)} \\ &= -2x_i \int_{\lambda(\mathbf{x})}^{\infty} f\left(\sum_k \frac{x_k^2}{a_k^2+s}\right) \frac{ds}{(a_i^2+s)\Delta(s)}\end{aligned} \tag{3.11}$$

because, in the first term above, the argument of G is 1 and $G(1) = 0$. Then for $i = 1, 2, 3$,

$$\begin{aligned}\frac{\partial^2 \psi}{\partial x_i^2} &= -2 \int_{\lambda}^{\infty} f\left(\sum_k \frac{x_k^2}{a_k^2+s}\right) \frac{ds}{(a_i^2+s)\Delta(s)} \\ &\quad - 4x_i^2 \int_{\lambda}^{\infty} f'\left(\sum_k \frac{x_k^2}{a_k^2+s}\right) \frac{ds}{(a_i^2+s)^2\Delta(s)} \\ &\quad + 2x_i \frac{\partial \lambda}{\partial x_i} f\left(\sum_k \frac{x_k^2}{a_k^2+\lambda}\right) \frac{1}{(a_i^2+\lambda)\Delta(\lambda)},\end{aligned} \tag{3.12}$$

where we have written $\lambda = \lambda(\mathbf{x})$. In the last term of the equation above we make two observations. One is that the argument of the function f is 1, by virtue of equation (3.9). The other is that we can use that same equation to express $\partial \lambda / \partial x_i$:

$$\frac{\partial \lambda}{\partial x_i} = \frac{2x_i/(a_i^2+\lambda)}{\sum_k x_k^2/(a_k^2+\lambda)^2}.$$

Equation (3.12) then takes the form

$$\begin{aligned}\frac{\partial^2 \psi}{\partial x_i^2} &= -2 \int_{\lambda}^{\infty} f\left(\sum_k \frac{x_k^2}{a_k^2+s}\right) \frac{ds}{(a_i^2+s)\Delta(s)} \\ &\quad - 4x_i^2 \int_{\lambda}^{\infty} f'\left(\sum_k \frac{x_k^2}{a_k^2+s}\right) \frac{ds}{(a_i^2+s)^2\Delta(s)} \\ &\quad + \frac{4x_i^2/(a_i^2+\lambda)}{\sum_k x_k^2/(a_k^2+\lambda)^2} \frac{f(1)}{(a_i^2+\lambda)\Delta(\lambda)},\end{aligned} \tag{3.13}$$

where we continue to write λ for $\lambda(\mathbf{x})$. Note the identity

$$\frac{d}{ds} f\left(\sum_k \frac{x_k^2}{a_k^2+s}\right) = -f'\left(\sum_k \frac{x_k^2}{a_k^2+s}\right) \sum_i \frac{x_i^2}{(a_i^2+s)^2}.$$

We use this while summing over i to get

$$\begin{aligned}\nabla^2 \psi &= -2 \int_{\lambda(\mathbf{x})}^{\infty} f\left(\sum_k \frac{x_k^2}{a_k^2+s}\right) \left(\sum_i \frac{1}{a_i^2+s}\right) \frac{ds}{\Delta(s)} \\ &\quad + 4 \int_{\lambda(\mathbf{x})}^{\infty} \frac{d}{ds} f\left(\sum_k \frac{x_k^2}{a_k^2+s}\right) \frac{ds}{\Delta(s)} + 4\frac{f(1)}{\Delta(\lambda)}.\end{aligned} \tag{3.14}$$

Integration by parts in the middle term now leads to

$$\begin{aligned}\nabla^2\psi = &-2\int_{\lambda(\mathbf{x})} f\left(\sum_k \frac{x_k^2}{a_k^2+s}\right)\left(\sum_i \frac{1}{a_i^2+s}\right)\frac{ds}{\Delta(s)}\\ &-4\int_{\lambda(\mathbf{x})} f\left(\sum_k \frac{x_k^2}{a_k^2+s}\right)\left(\frac{d}{ds}\Delta(s)^{-1}\right)ds,\end{aligned} \tag{3.15}$$

where we have noted the cancellation of the final term of the previous equation against the endpoint contribution from the integration by parts. Now recalling the definition (2.15) of $\Delta(s)$, we find that the integrands of the remaining two integrals are equal in magnitude and opposite in sign. This cancellation completes the verification that $\nabla^2\psi = 0$.

2. The spherical limit:
We know that if $a_1 = a_2 = a_3 = a$, the gravitational potential in the exterior $(r > a)$ is $M(r)/r$ where $M(r)$ is the mass enclosed in a ball of radius r. In this limit $\lambda(\mathbf{x})$ is defined by the equation

$$\sum_{i=1}^{3} \frac{x_i^2}{a^2+\lambda} = 1 \quad \text{or} \quad \lambda = r^2 - a^2.$$

The function ψ of equation (3.10) above takes the form

$$\psi(\mathbf{x}) = \int_{r^2-a^2}^{\infty} G\left(\frac{r^2}{a^2+s}\right)\left(a^2+s\right)^{-3/2} ds. \tag{3.16}$$

Observing that

$$\left(a^2+s\right)^{-3/2} = -2\frac{d}{ds}\left(a^2+s\right)^{-1/2} \tag{3.17}$$

and integrating by parts leads to

$$2\int_{r^2-a^2}^{\infty} (a^2+s)^{-1/2}\frac{d}{ds}G\left(\frac{r^2}{a^2+s}\right) ds,$$

since the integrated part vanishes at both endpoints. Recalling (equation (3.1)) that $G' = -f$ where $f(r^2/a^2) = \rho(r)$ is the density, we find that

$$\psi(\mathbf{x}) = 2r^2\int_{r^2-a^2}^{\infty} (a^2+s)^{-5/2} f\left(\frac{r^2}{a^2+s}\right) ds.$$

Make the change of variable

$$\frac{r^2}{a^2+s} = \frac{u^2}{a^2}$$

from s to u in the integration. Then restore the suppressed factor πa^3. After straightforward manipulations we find indeed that

$$\Psi(\mathbf{x}') = M/r \quad \text{where } r = |\mathbf{x}'|,$$

as expected.

3.2. Interior points. Now suppose $\mathbf{x}$ is an interior point of the basic ellipsoidal domain, say

$$b(\mathbf{x}) = w^2 \quad \text{where } 0 \leqslant w < 1.$$

With the density again given by equation (3.1), we write the potential function as the sum $\Psi_{v<w} + \Psi_{v>w}$ where

$$\Psi_{v<w} = \int_0^w vf(v^2)\left\{\int_{b(\mathbf{y})=v^2} \frac{\sigma(\mathbf{y})}{|\mathbf{x}-\mathbf{y}|}dS_y\right\} dv \tag{3.18}$$

and

$$\Psi_{v>w} = \int_w^1 vf(v^2)\left\{\int\int_{b(\mathbf{y})=v^2} \frac{\sigma(\mathbf{y})}{|\mathbf{x}-\mathbf{y}|}dS_y\right\} dv. \tag{3.19}$$

The point $\mathbf{x}$ is a limiting exterior point for the ellipsoid of semiaxes wa_1, wa_2, wa_3 and a limiting interior point for the homoeoid $w^2 < b(\mathbf{y}) < 1$.

The term $\Psi_{v<w}$ can be obtained with the aid of equation (2.25) which allows us to infer that

$$\int_{b(\mathbf{y})=v^2} \frac{\sigma(\mathbf{y})}{|\mathbf{x}-\mathbf{y}|}dS_y = \int_{b(\mathbf{y})=1} \frac{\sigma(\mathbf{y})}{|v^{-1}\mathbf{x}-\mathbf{y}|}dS_y = 2\pi a_1 a_2 a_3 \int_{\lambda(v^{-1}\mathbf{x})}^\infty \frac{ds}{\Delta(s)},$$

where the lower limit $\lambda(v^{-1}\mathbf{x})$ is the algebraically largest solution of the equation

$$\sum_{i=1}^3 \frac{x_i^2}{a_i^2+\lambda} = v^2. \tag{3.20}$$

Equation (3.18) now takes the form

$$\begin{aligned}\Psi_{v<w}(\mathbf{x}) &= 2\pi a_1a_2a_3\int_0^w vf(v^2)\left(\int_{\lambda(v^{-1}\mathbf{x})}^\infty \frac{ds}{\Delta(s)}\right) dv\\ &= -\pi a_1a_2a_3 \int_0^w \left(\frac{d}{dv}G(v^2)\right)\left(\int_{\lambda(v^{-1}\mathbf{x})}^\infty \frac{ds}{\Delta(s)}\right) dv\end{aligned}$$

where equation (3.6) above provides the relation between f and G. Integrating by parts leads to

$$\begin{aligned}\Psi_{v<w}(\mathbf{x}) = -\pi a_1a_2a_3 &\left\{\left[G(v^2)\int_{\lambda(v^{-1}\mathbf{x})}^\infty \frac{ds}{\Delta(s)}\right]_0^w\right.\\ &\left. - \int_0^w G(v^2)\left(\frac{d}{dv}\int_{\lambda(v^{-1}\mathbf{x})}^\infty \frac{ds}{\Delta(s)}\right) dv\right\}.\end{aligned} \tag{3.21}$$

The integrated term above vanishes when evaluated at the lower limit since $\lambda \to \infty$ as $v \to 0$, according to equation (3.20), and reduces to

$$G(w^2)\int_0^\infty \frac{ds}{\Delta(s)}$$

at the upper limit. The remaining integral over v can be reduced to an integral over λ with the change of variable (3.20), with the result

$$\Psi_{v<w}(\mathbf{x}) = \pi a_1a_2a_3\int_0^\infty \left[G\left(\sum_k \frac{x_k^2}{a_k^2+s}\right) - G(w^2)\right]\frac{ds}{\Delta(s)}. \tag{3.22}$$

For the equation (3.19) we use part 2 of Corollary 2.3.2 to infer

$$\Psi_{v>w} = \pi a_1a_2a_3\left(\int_0^\infty \frac{ds}{\Delta(s)}\right) G(w^2).$$

Adding these two we find the following.

THEOREM 3.2. *Let the density in the ellipsoidal domain D of equation* (3.2) *be given by equation* (3.1). *Then the potential at an interior point* $\mathbf{x}$ *is given by the expression*

$$\Psi(\mathbf{x}) = \pi a_1 a_2 a_3 \int_0^\infty G\left(\sum_{i=1}^3 \frac{x_i^2}{a_i^2 + s}\right) \frac{ds}{\Delta(s)}, \tag{3.23}$$

where the function G is defined by equation (3.6) *above.*

We have as usual omitted the factor of the gravitational constant.

Verifications: We'll make frequent use of the definitions (3.1) and (3.6) above.

Poisson's equation: In equation (3.23) we provisionally introduce ψ by the equation $\Psi = \pi a_1 a_2 a_3 \psi$ to save some writing, and we suspend the summation convention. Then

$$\frac{\partial \psi}{\partial x_i} = -2x_i \int_0^\infty f\left(\sum_k \frac{x_k^2}{a_k^2 + s}\right) \frac{ds}{(a_i^2 + s)\Delta(s)} \tag{3.24}$$

and

$$\begin{aligned} \frac{\partial^2 \psi}{\partial x_i^2} &= -2 \int_0^\infty f\left(\sum_k \frac{x_k^2}{a_k^2 + s}\right) \frac{ds}{(a_i^2 + s)\Delta(s)} \\ &\quad - 4x_i^2 \int_0^\infty f'\left(\sum_k \frac{x_k^2}{a_k^2 + s}\right) \frac{ds}{(a_i^2 + s)^2\Delta(s)}. \end{aligned} \tag{3.25}$$

Since

$$\frac{d}{ds} f\left(\sum_k \frac{x_k^2}{a_k^2 + s}\right) = -f'\left(\sum_k \frac{x_k^2}{a_k^2 + s}\right) \sum_j \frac{x_j^2}{(a_j^2 + s)^2}$$

we can use this with equation (3.25) to obtain

$$\begin{aligned} \nabla^2 \psi &= -2 \int_0^\infty f\left(\sum_k \frac{x_k^2}{a_k^2 + s}\right) \left(\sum_i \frac{1}{a_i^2 + s}\right) \frac{ds}{\Delta(s)} \\ &\quad + 4 \int_0^\infty \left[\frac{d}{ds} f\left(\sum_k \frac{x_k^2}{a_k^2 + s}\right)\right] \frac{ds}{\Delta(s)}. \end{aligned} \tag{3.26}$$

If we integrate the second term by parts, the integrated part is

$$-4f\left(\sum_k \frac{x_k^2}{a_k^2}\right) \frac{1}{a_1 a_2 a_3}$$

whereas the remaining integral exactly cancels the first term of equation (3.26) and we are left with only the expression above. Recalling that

$$\rho(\mathbf{x}) = f\left(\sum_k \frac{x_k^2}{a_k^2}\right)$$

and restoring the factor $\pi a_1 a_2 a_3$ suppressed above, we find

$$\nabla^2 \Psi = -4\pi\rho, \tag{3.27}$$

which is indeed Poisson's equation.

The spherical limit:

The formula (1.6) expresses the potential for the spherically symmetric case. We check to see that we can recover this from equation (3.23) if $a_1 =$

$a_2 = a_3 = a$. With these values for the semiaxes the latter equation takes the form

$$\begin{aligned}\Psi &= \pi a^3 \int_0^\infty G\left(\frac{r^2}{a^2+s}\right)\left(-2\frac{d}{ds}(a^2+s)^{-1/2}\right) ds \\ &= -2\pi a^3 \left\{ G\left(\frac{r^2}{a^2+s}\right)(a^2+s)^{-1/2}\right]_0^\infty \\ &\quad - r^2 \int_0^\infty (a^2+s)^{-5/2} f\left(\frac{r^2}{a^2+s}\right) ds\right\}.\end{aligned} \tag{3.28}$$

The first term above, resulting from an integration by parts, vanishes at the upper limit, and contributes

$$2\pi a^2 G(r^2/a^2)$$

to the sum on the right. This may be written

$$2\pi a^2 \int_{r^2/a^2}^1 f(s)\, ds = 4\pi \int_r^a t\rho(t)\, dt, \tag{3.29}$$

where the last expression is found by making the change of variable $s = t^2/a^2$ and recalling that $f(t^2/a^2) = \rho(t)$. In the final term of equation (3.28) we change the integration variable from s to t via the transformation $a^2 + s = a^2r^2/t^2$. This term is then found, after a brief calculation, to contribute

$$\frac{4\pi}{r}\int_0^r t^2\rho(t)\, dt = \frac{M(r)}{r}.$$

Adding the two contributions and comparing with equation (1.6), we indeed find agreement.

We note in passing that the condition that the potential and its gradient are everywhere continuous is satisfied, in agreement with Theorem 1.5 above. It's clear for points $\mathbf{x}$ that do not lie on the boundary of the ellipsoid. For points that do lie on the boundary it is clear from the formulas (3.8) and (3.23) that the function is continuous, and it is clear from the formulas (3.11) and (3.24) that the gradients are continuous.

4. Polynomial distributions

We consider here density functions that are represented as polynomials in the Cartesian coordinates. Since the potential is a linear functional of the density and polynomials form a complete set, any density function restricted to an ellipsoidal domain can be approximated arbitrarily well in this manner.

The general procedure depends on the following fact.

LEMMA 4.1. *If ρ is a C^1 density function that vanishes at the boundary of a domain D and Ψ the corresponding potential corresponding to the density $\partial\rho/\partial x_j$ is the potential $\partial\Psi/\partial x_j$.*

PROOF. Explicitly,

$$\begin{aligned}\frac{\partial\Psi(\mathbf{x})}{\partial x_j} &= \int_D \frac{\partial}{\partial x_j}\frac{\rho(\mathbf{y})}{|\mathbf{x}-\mathbf{y}|}\, dV_y = -\int_D \rho(\mathbf{y})\frac{\partial}{\partial y_j}\frac{1}{|\mathbf{x}-\mathbf{y}|}\, dV_y \\ &= -\int_D \left[\frac{\partial}{\partial y_j}\frac{\rho(\mathbf{y})}{|\mathbf{x}-\mathbf{y}|} - \frac{\partial\rho/\partial y_j}{|\mathbf{x}-\mathbf{y}|}\right] dV_y \\ &= \int_D \frac{\partial\rho/\partial y_j}{|\mathbf{x}-\mathbf{y}|}\, dV_y - \int_{\partial D}\frac{\rho(\mathbf{y})}{|\mathbf{x}-\mathbf{y}|} n_j\, dS,\end{aligned}$$

where we have employed the divergence theorem. Noting that the surface term vanishes we have

$$\frac{\partial \Psi(\mathbf{x})}{\partial x_j} = \int_D \frac{\partial \rho(\mathbf{y})/\partial y_j}{|\mathbf{x}-\mathbf{y}|} dV_y. \tag{4.1}$$ □

Note: The argument $\mathbf{x}$ may be any point of $\mathbb{R}^3$ in this lemma, but in the current section we shall restrict it to lie in D.

Consider an example:

EXAMPLE 4.2. Suppose we wish to evaluate

$$\Phi(\mathbf{x}) = \int_D \frac{y_1}{|\mathbf{x}-\mathbf{y}|}\, dV_y, \tag{4.2}$$

where the vector $\mathbf{x}$ is an interior point of the ellipsoid

$$D\colon \sum_k \frac{y_k^2}{a_k^2} \leqslant 1.$$

We begin by considering the function $f(t) = 1 - t$ (as in equation (3.1) above) so that the density function is

$$\rho(\mathbf{x}) = 1 - \sum_k \frac{x_k^2}{a_k^2}$$

which vanishes at the boundary. Denote by Ψ the gravitational potential associated with this density. Its values in the interior are provided by equation (3.23) above wherein the function (see equation (3.6)) $G(t) = \frac{1}{2}(1-t)^2$:

$$\begin{aligned} \Psi(\mathbf{x}) &= \int_D \frac{1 - \sum_k y_k^2/a_k^2}{|\mathbf{x}-\mathbf{y}|}\, dV_y \\ &= \frac{1}{2}\pi a_1 a_2 a_3 \int_0^\infty \left(1 - \sum_{k=1}^3 \frac{x_k^2}{a_k^2+s}\right)^2 \frac{ds}{\Delta(s)}. \end{aligned} \tag{4.3}$$

Since $\Phi = -\frac{1}{2}a_1^2 \partial\Psi/\partial x_1$ the required potential is

$$\Phi(\mathbf{x}) = \pi a_1^3 a_2 a_3\, x_1 \int_0^\infty \left(1 - \sum_k \frac{x_k^2}{a_k^2+s}\right) \frac{ds}{(a_1^2+s)\Delta(s)}.$$

Formulas in which the ‘density’ y_1 is replaced by y_2 or y_3 are obvious.

We can put this formula in a somewhat more manageable form by introducing coefficients along the lines of those introduced above (equation (2.36)):

$$A_{ij} = a_1 a_2 a_3 \int_0^\infty \frac{ds}{(a_i^2+s)(a_j^2+s)\Delta(s)}. \tag{4.4}$$

Then

$$\Phi(\mathbf{x}) = \pi a_1^2 x_1 \left(A_1 - \sum_{k=1}^3 A_{1k} x_k^2\right). \tag{4.5}$$

A necessary condition for this to be the correct formula is for the equation $\nabla^2\Phi = -4\pi x_1$ to hold. Since

$$\nabla^2\Phi = -2\pi a_1^2\,(3A_{11} + A_{12} + A_{13})\,x_1,$$

it must be true that the term in parentheses adds up to $2/a_1^2$. This can be checked (see Appendix C, but beware altered notation).

In considering the formula (4.1) above, it is clear that we can iterate it: if both ρ and $\partial\rho/\partial x_i$ vanish on the boundary, then

$$\frac{\partial^2\Psi(\mathbf{x})}{\partial x_i \partial x_j} = \int_D \frac{\partial^2\rho(\mathbf{y})/(\partial y_i \partial y_j)}{|\mathbf{x}-\mathbf{y}|}\, dV_y.$$

Let's carry this out in some detail in a second example.

EXAMPLE 4.3. Suppose we wish to evaluate the six potentials

$$\int_D \frac{y_i y_j}{|\mathbf{x}-\mathbf{y}|} dV_y. \tag{4.6}$$

If we define

$$P(\mathbf{x}) = \int_D \frac{(1-\sum_k y_k^2/a_k^2)^2}{|\mathbf{x}-\mathbf{y}|}\, dV_y \tag{4.7}$$

we can apply Lemma 4.1 successively and expect to attain our objective. Let's see how that goes. The first and second partial derivatives of P are

$$\frac{\partial P(\mathbf{x})}{\partial x_i} = -\frac{4}{a_i^2}\int_D \frac{y_i(1-\sum_k y_k^2/a_k^2)}{|\mathbf{x}-\mathbf{y}|}\, dV_y$$

and

$$\begin{aligned} \frac{\partial^2 P(\mathbf{x})}{\partial x_i \partial x_j} &= \frac{8}{a_i^2 a_j^2}\int_D \frac{y_i y_j}{|\mathbf{x}-\mathbf{y}|}\, dV_y - \frac{4}{a_i^2}\delta_{ij}\int_D \frac{1-\sum_k y_k^2/a_k^2}{|\mathbf{x}-\mathbf{y}|}\, dV_y \\ &= \frac{8}{a_i^2 a_j^2}\int_D \frac{y_i y_j}{|\mathbf{x}-\mathbf{y}|}\, dV_y - \frac{4}{a_i^2}\delta_{ij}\Psi(\mathbf{x}), \end{aligned} \tag{4.8}$$

where Ψ is expressed in equation (4.3) of the preceding example.

An alternative, explicit expression for the potential P is provide by Theorem 3.2. Since it is defined for the density function $f(t) = (1-t)^2$, its potential function $P(\mathbf{x})$ is determined by the function $G(t) = \frac{1}{3}(1-t)^3$:

$$P(\mathbf{x}) = \frac{1}{3}\pi a_1 a_2 a_3 \int_0^\infty \left(1 - \sum_k \frac{x_k^2}{a_k^2+s}\right)^3 \frac{ds}{\Delta(s)}$$

and therefore (no sum on repeated indices!)

$$\frac{\partial P(\mathbf{x})}{\partial x_i} = -2\pi a_1 a_2 a_3 x_i \int_0^\infty \left(1 - \sum_k \frac{x_k^2}{a_k^2+s}\right)^2 \frac{ds}{(a_i^2+s)\Delta(s)}$$

and

$$\begin{aligned} \frac{1}{2\pi a_1 a_2 a_3}\frac{\partial^2 P(\mathbf{x})}{\partial x_i \partial x_j} &= 4 x_i x_j \int_0^\infty \left(1 - \sum_k \frac{x_k^2}{a_k^2+s}\right) \frac{ds}{(a_i^2+s)(a_j^2+s)\Delta(s)} \\ &\quad - \delta_{ij}\int_0^\infty \left(1 - \sum_k \frac{x_k^2}{a_k^2+s}\right)^2 \frac{ds}{(a_i^2+s)\Delta(s)}. \end{aligned} \tag{4.9}$$

Equating the two expressions for $\partial^2 P(\mathbf{x})/(\partial x_i \partial x_j)$, from equations (4.8) and (4.9) then provides the information needed to determine the six integrals represented by equation (4.6). For $i \neq j$

$$\int_D \frac{y_i y_j}{|\mathbf{x}-\mathbf{y}|}\, dV_y = C a_i^2 a_j^2 x_i x_j \int_0^\infty \left(1 - \sum_k \frac{x_k^2}{a_k^2+s}\right) \frac{ds}{(a_i^2+s)(a_j^2+s)\Delta(s)} \tag{4.10}$$

and for $i = j$,

$$\begin{aligned} \int_D \frac{y_i^2}{|\mathbf{x}-\mathbf{y}|}\, dV_y = C a_i^4 \Bigg\{ x_i^2 \int_0^\infty \left(1 - \sum_k \frac{x_k^2}{a_k^2+s}\right) \frac{ds}{(a_i^2+s)^2\Delta(s)} \\ - \frac{1}{4}\int_0^\infty \left(1 - \sum_k \frac{x_k^2}{a_k^2+s}\right)^2 \frac{ds}{(a_i^2+s)\Delta(s)} \Bigg\} \\ + \frac{C a_i^2}{4}\int_0^\infty \left(1 - \sum_k \frac{x_k^2}{a_k^2+s}\right)^2 \frac{ds}{\Delta(s)}, \end{aligned} \tag{4.11}$$

where we have set $C = \pi a_1 a_2 a_3$.

In terms of the symbols $A_i, A_{ij}, \ldots$ we may rewrite these as

$$\int_D \frac{y_i y_j}{|\mathbf{x}-\mathbf{y}|}\, dV_y = \pi a_i^2 a_j^2\, x_i x_j \left(A_{ij} - \sum_k A_{ijk} x_k^2\right) \quad \text{for } i \neq j \tag{4.12}$$

and

$$\begin{aligned} \int_D \frac{y_i^2}{|\mathbf{x}-\mathbf{y}|}\, dV_y = \frac{\pi a_i^4}{4}\Bigg\{ 4x_i^2 \left[A_{ii} - \sum_k A_{iik} x_k^2\right] \\ + a_i^{-2}\left[B_i - 2\sum_k B_{ik} x_k^2 + \sum_{k,j} B_{ijk} x_k^2 x_j^2\right]\Bigg\}, \end{aligned} \tag{4.13}$$

where we have introduced new symbols:

$$B_{ijk\ldots} = A_{jk\ldots} - a_i^2 A_{ijk\ldots}.$$

Checking that these satisfy the corresponding Poisson equations is an exercise facilitated by the identities in Appendix C. We find, if i, j, k is an even permutation of $1, 2, 3$, that the Laplacian applied to the right-hand side of equation (4.12) yields

$$-2\pi a_i^2 a_j^2 x_i x_j \left\{3A_{iij} + 3A_{ijj} + A_{ijk}\right\} = -4\pi x_i x_j$$

on relating the symbols A_{ijk} to the symbols $A_{[k_1 k_2 k_3]}$ and appealing to (C.3). Verifying that applying the Laplace operator to the right-hand side of equation (4.13) produces the result $-4\pi x_i^2$ is rather arduous (I have checked it for $i = 1$).

We need to approach more systematically the (essentially algebraic) problem of expressing a monomial in terms of powers of $x_1^2/a_1^2 + x_2^2/a_2^2 + x_3^2/a_3^2 - 1$ and their derivatives .

4.1. Ferrers' expression of monomials. Let $\kappa = (\kappa_1, \kappa_2, \kappa_3)$ and denote the monomial $x_1^{\kappa_1} x_2^{\kappa_2} x_3^{\kappa_3}$ by $\mathbf{x}^\kappa$.

DEFINITION 4.4. Multi-indices such as $\mu = (\mu_1, \mu_2, \mu_3)$ are denoted by Greek letters and have non-negative integer components: $\mu_i \geqslant 0$, $i = 1, 2, 3$.

We have written out the potentials due to such monomial densities above for the cases when $|\kappa| = \kappa_1 + \kappa_2 + \kappa_3$ is one or two. We denote the ellipsoid with semiaxes $a = (a_1, a_2, a_3)$ by E_a and seek, for arbitrary κ, a formula for the integral

$$V_\kappa(\mathbf{x}) = \int_{E_a} \frac{\mathbf{y}^\kappa}{|\mathbf{x} - \mathbf{y}|} \, dV_y \tag{4.14}$$

via a method originally due to Ferrers [**3**] and reconsidered in my paper, referred to below as L79 [**5**]. I want to recapitulate and extend the results of that paper in a manner suitable to programming the results for computation. We consider here points $\mathbf{x}$ lying inside the ellipsoid E_a.

The first objective is to represent the monomial $\mathbf{x}^\kappa$ as a sum of terms of the form $D^\nu \rho_n(\mathbf{x})$ since the potential of the latter can be expressed in polynomials, as in the examples above, if the density ρ_n is chosen appropriately. We'll choose it in the polynomial form

$$\rho_n(x) = \left(\sum_k \frac{x_k^2}{a_k^2} - 1 \right)^n. \tag{4.15}$$

4.1.1. *Polynomials in one variable.* Consider the one-dimensional problem of expressing a power of the single variable ξ in terms of derivatives, up to order N, of $(\xi^2 - h^2)^n$ for some $h \neq 0$ and another integer n. If $N > 2n$ this results in the zero function so we need consider only the cases when $N \leqslant 2n$. We want to represent ξ^n as the sum of derivatives of terms of the form $D^k(\xi^2 - h^2)^j$ where h is independent of ξ and we have written $d/d\xi = D$. Experimentation shows that

$$\xi = \frac{1}{2} D(\xi^2 - h^2), \qquad \xi^2 = \frac{1}{8} D^2(\xi^2 - h^2)^2 - \frac{1}{2}(\xi^2 - h^2). \tag{4.16}$$

If ξ is raised to an odd power only odd-order derivatives will appear in its representation and if it is raised to an even power only even-order derivatives will appear and we expect a representation like that of Ferrers [**5**, equation (15)]:

$$\xi^n = \sum_{k=0}^{[n/2]} A_{n,k} D^{n-2k} \left(\xi^2 - h^2\right)^{n-k}. \tag{4.17}$$

In discussing this representation the following, which we state without proof, will be helpful.

PROPOSITION 4.5. *For non-negative integers n, N and $D = d/d\xi$, the Nth-order derivative evaluated at $\xi = h$ is*

$$D^N \left(\xi^2 - h^2\right)^n |_{\xi = h} = \frac{n!\, N!}{(N-n)!(2n-N)!} (2h)^{2n-N}$$

if $n \leqslant N \leqslant 2n$. If $N < n$ or $N > 2n$ the right-hand side is replaced by zero.

A sufficient condition for the representation (4.17) to hold is that, for each n, the *Ferrers polynomials*

$$F_k(\xi) = D^{n-2k} \left(\xi^2 - h^2\right)^{n-k}, \quad k = 0, 1, \dots, [n/2]$$

be linearly independent on the interval $[-h,h]$ since then, for given n they will form a basis for polynomials having the parity of n. These polynomials are all of degree n; they are even if n is even and odd if n is odd. Let's consider the even case and put $n=2m$. There are $m+1$ such polynomials in that case and we can write the relation of linear dependence as $\sum_k c_k F_k = 0$ or

$$c_0 D^{2m}\left(\xi^2-h^2\right)^{2m} + c_1 D^{2m-2}\left(\xi^2-h^2\right)^{2m-1} + \cdots + c_m\left(x^2-h^2\right)^m = 0.$$

According to Proposition 4.5 all the functions appearing in the preceding expression vanish at $\xi = h$ with the exception of the first, showing that $c_0 = 0$. Expressing this above and differentiating the remaining expression once more, we find

$$c_1 D^{2m-1}\left(\xi^2-h^2\right)^{2m-1} + \cdots + c_m D\left(x^2-h^2\right)^m = 0.$$

A further application of the Proposition shows that $c_1 = 0$. Proceeding in this way we see that the coefficients all vanish and the Ferrers polynomials are linearly independent. The proof is the same if n is odd. This proves the following lemma.

LEMMA 4.6. *Any even (odd) polynomial of degree n can be represented as a linear sum of Ferrers polynomials of the same parity.*

We will need the coefficients $A_{n,k}$ of equation (4.17) explicitly. To find these coefficients we shall proceed by evaluating $D^l\xi^n$ at $\xi = h$ for $l = 0,1,\ldots$. If we do this for all values of l from 0 to n, we will have an overdetermined system of $n+1$ equations in $[n/2]+1$ unknowns. This system must be compatible since we know from the preceding lemma that the coefficients exist. We'll find that the first $[n/2]+1$ of these equations determine these $[n/2]+1$ coefficients. Equation (4.17) implies that

$$D^l\xi^n|_h = \frac{n!\,h^{n-l}}{(n-l)!} = \sum_{k=0}^{[n/2]+1} A_{n,k}\left[D^{n+l-2k}\left(\xi^2-h^2\right)^{n-k}\right]_h. \tag{4.18}$$

Using Proposition 4.6 we see that the coefficient of $A_{n,k}$ vanishes if $k>l$ so the matrix of this system of equations is triangular and therefore nonsingular if the diagonal entries are nonzero. These are

$$D^{n-l}\left(\xi^2-h^2\right)^{n-l}|_{\xi=h} = (n-l)!(2h)^{n-l}$$

which are all nonzero, so this system is indeed nonsingular and determines the coefficients $A_{n,k}$ for $k=1,2,\ldots,[n/2]+1$. With the aid of the Proposition we find

$$\sum_{k=0}^{l} \frac{(n-k)!(n+l-2k)!}{(l-k)!(n-l)!}(2h)^{n-l}A_{n,k} = \frac{n!}{(n-l)!}h^{n-l}$$

or, cancelling factors of $(n-l)!$ and h^{n-l},

$$\sum_{k=0}^{l} \frac{(n-k)!(n+l-2k)!}{(l-k)!}A_{n,k} = \frac{n!}{2^{n-l}}, \quad l = 0,1,\ldots,[n/2]. \tag{4.19}$$

There is a tidy expression for the coefficient $A_{n,k}$: equation (4.22) below. We can guess it from the calculation of the first few of these for some fixed, sufficiently large value of n and low values of k:

For some given value of n we can reduce the amount of writing by defining $B_{n,k} = 2^n A_{n,k}/n!$ so that the defining equations (4.19) take the form

(4.20) $$\sum_{k=0}^{l} \frac{(n-k)!(n+l-2k)!}{(l-k)!} B_{n,k} = 2^l, \quad l = 0, 1, \ldots, [n/2].$$

For low values of l we calculate (somewhat tediously):

$$B_{n,0} = \frac{1}{(n!)^2}.$$

If $l = 1$ the sum has two terms and we find

$$(n+1)!\, n! B_{n,0} + [(n-1)!]^2 B_{n,1} = 2 \quad \text{or} \quad B_{n,1} = -\frac{1}{(n-1)!\,(n-2)!}.$$

Similarly we discover that

$$B_{n,2} = \frac{1}{2} \frac{1}{(n-2)!\,(n-4)!} \quad \text{and} \quad B_{n,3} = -\frac{1}{3!} \frac{1}{(n-3)!\,(n-6)!}.$$

These lead to the conjecture that

(4.21) $$B_{n,k} = (-1)^k \frac{1}{k!\,(n-k)!\,(n-2k)!}, \quad k = 0, 1, \ldots, [n/2].$$

This conjecture is correct and is proved in [**5**]. I had hoped to find a more intuitive proof for it than that presented there but have not been able to do so. The final formula for $A_{n,k}$ is

(4.22) $$A_{n,k} = \frac{n!}{2^n} \frac{(-1)^k}{k!\,(n-k)!\,(n-2k)!}, \quad k = 0, 1, \ldots, [n/2].$$

This together with equation (4.17) allows us to express any power ξ^n in terms of powers of polynomials in ξ and their derivatives. We need to extend this to functions of three variables.

4.1.2. *Multinomials.* If there are three independent variables $\xi = (\xi_1, \xi_2, \xi_3)$ we can go part of the way toward generalizing the examples of § 4.1.1 by putting $h^2 = 1 - \xi_2^2 - \xi_3^2$ in equation (4.17), thereby expressing ξ_1^n as a sum of derivatives of powers of the combination $\xi_1^2 + \xi_2^2 + \xi_3^2 - 1$:

(4.23) $$\xi_1^n = \sum_{k=0}^{[n/2]} A_{n,k} \frac{\partial^{n-2k} \sigma_{n-k}(\xi)}{\partial \xi_1^{n-2k}},$$

where

(4.24) $$\sigma_l(\xi) = \left(\sum_{i=1}^{3} \xi_i^2 - 1 \right)^l.$$

We'll want to change variables via a linear transformation $\xi = T\eta$; we can choose the matrix T to be orthogonal and thereby leave the function σ_l unchanged. This can be done in such a way that

(4.25) $$\xi_1 = \|t\|^{-1}(t_1\eta_1 + t_2\eta_2 + t_3\eta_3) \quad \text{where} \quad \|t\| = \sqrt{t_1^2 + t_2^2 + t_3^2}$$

for arbitrary real numbers t_1, t_2, t_3.

Digression on orthonormality:

Specifically, given a real vector $t = (t_1, t_2, t_3)$ we can find two further vectors u, v so that the three are mutually orthogonal. We can use the Gram–Schmidt procedure to produce

$$u = (-t_2, t_1, 0) \quad \text{and} \quad v = (t_1 t_3, t_2 t_3, -t_1^2 - t_2^2)$$

which fills the bill (indeed are linearly independent unless $t_1 = t_2 = 0$). Normalizing these provides the rows of an orthonormal matrix whose first row is

$$\|t\|^{-1}(t_1, t_2, t_3).$$

This can *always* be chosen as the first row of an orthonormal matrix since, in the exceptional case when $t_1 = t_2 = 0$, we may choose

$$u = (0, -t_3, 0) \quad \text{and} \quad v = (t_3, 0, 0).$$

With this choice the left-hand side of equation (4.23) takes the form

$$\xi_1^n = \|t\|^{-n}(t_1\eta_1 + t_2\eta_2 + t_3\eta_3)^n = (t_1^2 + t_2^2 + t_3^2)^{-n/2} \sum_{|\nu|=n} \frac{n!}{\nu!} t^\nu \eta^\nu, \tag{4.26}$$

where ν is the multi-index (ν_1, ν_2, ν_3), $\eta^\nu = \eta_1^{\nu_1}\eta_2^{\nu_2}\eta_3^{\nu_3}$ and we define $|\nu| = \nu_1 + \nu_2 + \nu_3$. Here we have invoked the multinomial theorem, a generalization of the binomial theorem (Appendix A below).

Return now to equation (4.23). Under the change of variables (4.25) we find that the left-hand side of this equation takes the form given by equation (4.26) above and now need to convert the right-hand side in the same way. We have arranged that $\sigma_k(\xi) = \sigma_k(\eta)$ by choosing the transformation to be orthogonal. The expression for ξ as a function of η may be written as $\xi = T\eta$ where T, which depends on the parameters t_1, t_2, t_3, is orthogonal, and therefore $\eta = T^t\xi$. From this we find

$$\frac{\partial}{\partial \xi_1} = \sum_{j=1}^{3} \frac{\partial \eta_j}{\partial \xi_1}\frac{\partial}{\partial \eta_j} = \sum_{j=1}^{3} T_{j1}^t \frac{\partial}{\partial \eta_j} = \|t\|^{-1}\sum_{j=1}^{3} t_j \frac{\partial}{\partial \eta_j}.$$

Any higher derivative may be expressed as

$$\frac{\partial^m}{\partial \xi_1^m} = \|t\|^{-m}\left(t_1\frac{\partial}{\partial \eta_1} + t_2\frac{\partial}{\partial \eta_2} + t_3\frac{\partial}{\partial \eta_3}\right)^m,$$

where the superscript m implies m-fold application of the differential expression inside the parentheses. Since the terms $t_i\partial/\partial\eta_i$ satisfy the same algebraic rules of combination as does the m-fold product of terms $t_i\eta_i$ in equation (4.26), we find

$$\begin{aligned} \frac{\partial^m}{\partial \xi_1^m} &= \|t\|^{-m} \sum_{|\mu|=m} \frac{m!}{\mu!} t^\mu \frac{\partial^m}{\partial \eta^\mu} \\ &\equiv \|t\|^{-m} \sum_{|\mu|=m} \frac{m!}{\mu_1!\mu_2!\mu_3!} t_1^{\mu_1}t_2^{\mu_2}t_3^{\mu_3} \frac{\partial^m}{\partial \eta_1^{\mu_1}\partial \eta_2^{\mu_2}\partial \eta_3^{\mu_3}}. \end{aligned} \tag{4.27}$$

Within the expression on the right-hand side of equation (4.23) this operator is applied to a function of the form

$$\left(\sum_{i=1}^{3} \eta_i^2 - 1\right)^n,$$

where the index m of the differential operator is not greater than the power n of the operand. Equation (4.23) now takes the form

$$
\begin{aligned}
&\sum_{\nu_1+\nu_2+\nu_3=n} \frac{n!}{\nu_1!\nu_3!\nu_3!} t_1^{\nu_1} t_2^{\nu_2} t_3^{\nu_3} \eta_1^{\nu_1} \eta_2^{\nu_2} \eta_3^{\nu_3} \equiv \sum_{|\nu|=n} \frac{n!}{\nu!} t^\nu \eta^\nu \\
&= \sum_{k=0}^{[n/2]} A_{n,k} \|t\|^{2k} \sum_{|\mu|=n-2k} \frac{(n-2k)!}{\mu!} t^\mu \frac{\partial^{n-2k}}{\partial \eta^\mu} \left(\sum_j \eta_j^2 - 1 \right)^{n-k}
\end{aligned} \tag{4.28}
$$

after extracting a common factor of $\|t\|^{-n}$ from each side. Since this holds for arbitrary real values of t_1, t_2, t_3 the factors of $t_1^{\nu_1} t_2^{\nu_2} t_3^{\nu_3}$ on either side of the equation must agree, and this will express an arbitrary monomial $\eta_1^{\nu_1} \eta_2^{\nu_2} \eta_3^{\nu_3}$ in terms of derivatives of powers of $\sum_k \eta_k^2 - 1$. For example:

If $n = 1$ equation (4.28) reduces to

$$
\eta_1 t_1 + \eta_2 t_2 + \eta_3 t_3 = A_{1,0} \left(t_1 \frac{\partial}{\partial \eta_1} + t_2 \frac{\partial}{\partial \eta_2} + t_3 \frac{\partial}{\partial \eta_3} \right) \left(\sum_k \eta_k^2 - 1 \right)
$$

or, for example,

$$
\eta_1 = A_{1,0} \frac{\partial}{\partial \eta_1} \left(\sum_k \eta_k^2 - 1 \right).
$$

Since $A_{1,0} = \frac{1}{2}$, this is indeed so, and we don't need complicated machinery to figure out that η_1 can be written this way. The machinery makes more sense for the next example, $n = 2$:

$$
\begin{aligned}
&t_1^2 \eta_1^2 + t_2^2 \eta_2^2 + t_3^2 \eta_3^2 + 2t_1 t_2 \eta_1 \eta_2 + 2t_2 t_3 \eta_2 \eta_3 + 2t_3 t_1 \eta_3 \eta_1 \\
&= A_{2,0} \left\{ t_1^2 \frac{\partial^2}{\partial \eta_1^2} \left(\sum_k \eta_k^2 - 1 \right)^2 + 2t_1 t_2 \frac{\partial^2}{\partial \eta_1 \partial \eta_2} \left(\sum_k \eta_k^2 - 1 \right)^2 + \text{c.p.} \right\} \\
&\quad + A_{2,1} \|t\|^2 \left(\sum_k \eta_k^2 - 1 \right)
\end{aligned}
$$

where we have written "c.p." in the curly braces to indicate missing terms obtained via cyclic permutations of those explicitly indicated. For this to be an identity we need to verify, for example, that

$$
\eta_1^2 = A_{2,0} \frac{\partial^2}{\partial \eta_1^2} \left(\sum_k \eta_k^2 - 1 \right)^2 + A_{2,1} \left(\sum_k \eta_k^2 - 1 \right)^2
$$

and

$$
\eta_1 \eta_2 = A_{2,0} \frac{\partial^2}{\partial \eta_1 \partial \eta_2} \left(\sum_k \eta_k^2 - 1 \right)^2.
$$

Carrying out these operations and using the values $A_{2,0} = \frac{1}{8}$ and $A_{2,1} = -\frac{1}{2}$ derived from equation (4.22) above, we find that these are indeed identities. The remaining four verifications follow by cyclic permutations.

More generally, in order to fully express an arbitrary power η^μ in terms of derivatives of powers of $(\sum_k \eta_k^2 - 1)$, we need to re-express the right-hand side of equation (4.28) in the form

$$
\sum_{|\nu|=n} D_\nu t^\nu,
$$

where D_ν is a function not only of the multi-index ν but also of the real-vector variable η and of the order n. This will imply that

$$\eta^\nu = \frac{\nu!}{n!} D_\nu, \tag{4.29}$$

accomplishing the goal of rendering the left-hand side above as a linear combination of powers of $(\sum_k \eta_k^2 - 1)$ and their derivatives; the final goal of replacing η_k by η_k/a_k will then be immediate.

The right-hand side of equation (4.28) may be written

$$\sum_{k=0}^{[n/2]} A_{n,k}(n-2k)!\, \|t\|^{2k} \sum_{|\mu|=n-2k} b_{n,\mu,k} t^\mu, \tag{4.30}$$

where

$$b_{n,\mu,k} = \frac{1}{\mu!} \frac{\partial^{n-2k}}{\partial \eta^\mu} \left(\sum_j \eta_j^2 - 1 \right)^{n-k} \quad \text{if} \quad |\mu| = n - 2k. \tag{4.31}$$

This can be brought in line with the notation of [**5**] if we introduce the notations

$$D^\mu = \frac{\partial^{n-2k}}{\partial \eta_1^{\mu_1} \partial \eta_2^{\mu_2} \partial \eta_3^{\mu_3}} \quad \text{if } |\mu| = n - 2k; \; f_l(\eta) = \left(\sum_k \eta_k^2 - 1 \right)^l. \tag{4.32}$$

Equation (4.31) then takes the more manageable form

$$b_{n,\mu,k} = \frac{1}{\mu!} D^\mu f_{n-k} \quad \text{if} \quad |\mu| = n - 2k. \tag{4.33}$$

The product

$$\|t\|^{2k} \sum_{|\mu|=n-2k} b_{n,\mu,k} t^\mu \tag{4.34}$$

appearing above in equation (4.30) is a sum of terms t^ν of total degree $|\nu| = n$. For a particular multi-index ν the coefficient of t^ν can be found as follows. Note first that

$$\|t\|^{2k} = \sum_{|\kappa|=k} \frac{k!}{\kappa!} t^{2\kappa}.$$

Then the coefficient of t^ν is easily seen to be

$$\sum_{\mu,\kappa \in I_{\nu,k}} \frac{k!}{\kappa!} b_{n,\mu,k},$$

where the multi-indices κ, μ range over the index set $I_{\nu,k}$:

$$I_{\nu,k} = \{\kappa, \mu : |\kappa| = k, \mu + 2\kappa = \nu\}. \tag{4.35}$$

It is understood (see Definition 4.4) that the multi-indices μ, κ (like all multi-indices in this development) have only non-negative components.

The term of equation (4.30) that multiplies t^ν may be written

$$\sum_{k=0}^{[n/2]} A_{n,k}(n-2k)!k! \sum_{\mu,\kappa \in I_{\nu,k}} \frac{1}{\kappa!} b_{n,\mu,k}.$$

It then follows from equation (4.30) above that

$$\eta^\nu = \frac{\nu!}{n!} \sum_{k=0}^{[n/2]} A_{n,k}(n-2k)!\, k! \sum_{\mu,\kappa \in I_{\nu,k}} \frac{1}{\kappa!} b_{n,\mu,k}.$$

This simplifies on using the expression (4.22) for $A_{n,k}$:

$$\eta^\nu = \frac{\nu!}{2^n} \sum_{k=0}^{[n/2]} \frac{(-1)^k}{(n-k)!} \sum_{\mu,\kappa \in I_{\nu,k}} \frac{1}{\kappa!} b_{n,\mu,k}, \quad \text{where } n = |\nu| \tag{4.36}$$

or, more explicitly with the aid of equation (4.31),

$$\eta^\nu = \frac{\nu!}{2^n} \sum_{k=0}^{[n/2]} \frac{(-1)^k}{(n-k)!} \sum_{\mu,\kappa \in I_{\nu,k}} \frac{1}{\mu!\kappa!} \frac{\partial^{n-2k}}{\partial \eta^\mu} \left(\sum_j \eta_j^2 - 1 \right)^{n-k}. \tag{4.37}$$

This is converted to a form appropriate to ellipsoids by making the transformation $\eta_i = y_i/a_i$ for $i = 1, 2, 3$. The preceding equation then becomes

$$y^\nu = a^\nu \frac{\nu!}{2^n} \sum_{k=0}^{[n/2]} \frac{(-1)^k}{(n-k)!} \sum_{\mu,\kappa \in I_{\nu,k}} \frac{a^\mu}{\mu!\kappa!} \frac{\partial^{n-2k}}{\partial y^\mu} \left(\sum_j \frac{y_j^2}{a_j^2} - 1 \right)^{n-k}, \tag{4.38}$$

where we recall that $n = |\nu|$. This is Ferrers' expression of an arbitrary monomial in terms of derivatives of the ellipsoidal distributions $\{\rho_m(\mathbf{x})\}$ (cf. equation (4.15) above).

We can say a little more about the index set $I_{\nu,k}$ (4.35). The order of the multi-indices is important when we write $\mu, \kappa \in I_{\nu,k}$: it wouldn't do to write $|\mu| = k, \kappa + 2\mu = \nu$ when we mean $|\kappa| = k, \mu + 2\kappa = \nu$. We will therefore have to be careful and explicit as to the ordering of κ and μ in designating a sum like that above: if we write $\lambda, \zeta \in I_{\nu,k}$, this will mean that the second multi-index ζ satisfies $|\zeta| = k$ and the first $\lambda = \nu - 2\zeta$.

The maximum number of terms in the sum over $I_{\nu,k}$ is easily found to be $(k+1)(k+2)/2$ and, if we write $K = [|\nu|/2]$, the maximum number of terms in the full sum (5.1) is $(K+1)(K+2)(K+3)/6$. This maximum need not be achieved. For $k = 0$ there is always exactly one term, $\kappa = (0,0,0), \mu = \nu$, and the maximum is achieved. For larger values of k the number is variable, depending on ν. For $k = 1$ for example and $\nu = (1,1,0)$, $I_{\nu,1}$ is empty whereas if $\nu = (2,0,0)$ it has one pair: $\kappa = (1,0,0), \mu = (0,0,0)$.

Within the index set $I_{\nu,k}$ for fixed k, there is no prescribed order of terms. We'll choose the following ordering if we need to choose:

$$\begin{gathered}(0,0,k);\ (0,1,k-1),(1,0,k-1);\ (0,2,k-2),(1,1,k-2),\\ (2,0,k-2);\ \ldots;\ (0,k,0),(1,k-1,0),\ldots,(k,0,0).\end{gathered}$$

This begins with the maximal value of κ_3 and for each κ_3 treats κ_1, κ_2 lexicographically.

5. Interior potentials of monomials

Ferrers' method for finding an expression $V_\nu(\mathbf{x})$ for the potential of an arbitrary monomial distribution (equation (4.14) above) in the interior of the ellipsoid follows from the preceding development.

5.1. Potential of an arbitrary monomial. The general formula is

$$\begin{aligned} V_\nu(\mathbf{x}) &\equiv \int_{E_a} \frac{\mathbf{y}^\nu}{|\mathbf{x}-\mathbf{y}|}\,d\mathbf{y} \\ &= a^\nu \frac{\nu!}{2^n} \sum_{k=0}^{[n/2]} \frac{(-1)^k}{(n-k)!} \sum_{\mu,\kappa\in I_{\nu,k}} \frac{a^\mu}{\mu!\kappa!} \frac{\partial^{n-2k}}{\partial \mathbf{x}^\mu} V[\rho_{n-k}](\mathbf{x}), \end{aligned} \tag{5.1}$$

where $\rho_j(\mathbf{x})$ is given by the expression (4.15), we have introduced the notation

$$V[\rho](\mathbf{x}) = \int_{E_a} \frac{\rho(\mathbf{y})}{|\mathbf{x}-\mathbf{y}|}\,dy, \tag{5.2}$$

and we have reversed the order of integration and differentiation by repeated applications of Lemma 4.1 above. In the notation of equation (5.2), $V_\nu(\mathbf{x}) = V[\rho](\mathbf{x})$ if $\rho(\mathbf{x}) = \mathbf{x}^\nu$.

Equation (5.1) can be put into a more complete and explicit form by employing equations (3.1) and (3.6) above to express $V[\rho_l](\mathbf{x})$:

$$V[\rho_l](\mathbf{x}) = -\frac{\pi a_1 a_2 a_3}{l+1} \int_0^\infty \left(\sum_{j=1}^3 \frac{x_j^2}{a_j^2+s} - 1 \right)^{l+1} \frac{ds}{\Delta(s)}. \tag{5.3}$$

This is a polynomial of degree $2l+2$ whose coefficients are given by definite integrals. In the formula (5.1) this implies that each term in the sum over k is a polynomial of degree $n+2$ (and therefore that V_ν is likewise a polynomial of degree $n+2$). To write these polynomials explicitly we'll need to express

$$\frac{\partial^{|\mu|}}{\partial \mathbf{x}^\mu} V[\rho_{n-k}](\mathbf{x}), \quad |\mu| = n-2k$$

as such a polynomial. We start by expanding the right-hand side of equation (5.3):

$$\begin{aligned} V[\rho_l](\mathbf{x}) &= \pi a_1 a_2 a_3 (-1)^l l! \sum_{t=0}^{l+1} \frac{(-1)^t}{t!(l+1-t)!} \\ &\quad \times \sum_{|\tau|=t} \frac{t!}{\tau!} \int_0^\infty \left(\frac{x_1^2}{a_1^2+s}\right)^{\tau_1} \left(\frac{x_2^2}{a_2^2+s}\right)^{\tau_2} \left(\frac{x_3^2}{a_3^2+s}\right)^{\tau_3} \frac{ds}{\Delta(s)} \\ &= \pi(-1)^l l! \sum_{t=0}^{l+1} \frac{(-1)^t}{(l+1-t)!} \sum_{|\tau|=t} \frac{A(\tau)}{\tau!} \mathbf{x}^{2\tau}, \end{aligned} \tag{5.4}$$

where we have borrowed the notation $A(\tau) = A_0(a,\tau)$ from equation (C.1) below.

Verification in spherical case:
I have checked this in the spherical case $a_1 = a_2 = a_3 = a$ and find

$$V[\rho_l](\mathbf{x}) = \pi(-1)^l l! \sum_{q=0}^{l+1} \frac{(-1)^q a^{-2q+2}}{(l+1-q)!q!(q+1/2)} r^{2q}.$$

From this we calculate $\nabla^2 V$ and, after a brief calculation, find that

$$\nabla^2 V = -4\pi\left(\frac{r^2}{a^2}-1\right)^l,$$

verifying that the expression in question is consistent in the limiting, spherical case.

We now find from equation (5.4)

$$\begin{aligned}
\frac{\partial^{|\mu|}}{\partial \mathbf{x}^\mu}V[\rho_l](\mathbf{x}) &\equiv \frac{\partial^{|\mu|}}{\partial \mathbf{x}^\mu}\int_{E_a}\frac{\rho_l(\mathbf{y})}{|\mathbf{x}-\mathbf{y}|}\,d\mathbf{y} \\
&= \pi(-1)^l l!\sum_{r=0}^{l+1}\frac{(-1)^r}{(l+1-r)!}\sum_{|\tau|=r,\ 2\tau_i\geqslant\mu_i}\frac{(2\tau)!}{\tau!(2\tau-\mu)!}A(\tau)\mathbf{x}^{2\tau-\mu} \quad (5.5)\\
&= \pi(-1)^l l!\sum_{r=[m/2]}^{l+1}\frac{(-1)^r}{(l+1-r)!}\sum_{|\tau|=r,\ 2\tau_i\geqslant\mu_i}\frac{(2\tau)!}{\tau!(2\tau-\mu)!}A(\tau)\mathbf{x}^{2\tau-\mu},
\end{aligned}$$

where $[m := |\mu|]$ and the last term has been rewritten to emphasize that only terms of degree $r \geqslant [m/2]$ survive the differentiations. We note the value of this expression at the origin.

PROPOSITION 5.1.

$$\frac{\partial^{|\mu|}V[\rho_l](\mathbf{x})}{\partial \mathbf{x}^\mu}|_{x=0} = 0 \quad \textit{unless } \mu = 2\tau \textit{ with } |\tau| \leqslant l+1$$

and, if $\mu = 2\tau$ *with* $|\tau| \leqslant l+1$,

$$\frac{\partial^{|2\tau|}V[\rho_l](\mathbf{x})}{\partial \mathbf{x}^{2\tau}}|_{x=0} = \pi\, l!\,\frac{(-1)^{l+|\tau|}}{(l+1-|\tau|)!}\frac{(2\tau)!}{\tau!}A(\tau).$$

Note from the editors: *In the review of this manuscript, a reviewer indicated that for Proposition* 5.1, *the author might mean that the term in* (5.5) *involving* $\mathbf{x}^{2\tau-\mu}$ *is zero at* $\mathbf{x} = 0$ *unless* $\mu = 2\tau$, *which is trivially true.*

We know (cf. L79 [**5**, equation (26)][16]) that

$$V_\nu(\mathbf{x}) = \sum_{|\lambda|\leqslant|\nu|+2} V_{\nu\lambda}\mathbf{x}^\lambda, \quad (5.6)$$

and now seek the coefficients $\{V_{\nu\lambda}\}$. It is clear that

$$V_{\nu\lambda} = \frac{1}{\lambda!}\frac{\partial^{|\lambda|}V_\nu}{\partial \mathbf{x}^\lambda}|_{\mathbf{x}=0}. \quad (5.7)$$

We can express the right-hand side of equation (5.7) in terms of the semiaxes of the ellipsoid. From equation (5.1) we obtain

$$\frac{\partial^{|\lambda|}V_\nu}{\partial \mathbf{x}^\lambda}(\mathbf{x}) = \frac{a^\nu \nu!}{2^{|\nu|}}\sum_{k=0}^{[|\nu|/2]}\frac{(-1)^k}{(|\nu|-k)!}\sum_{\kappa,\mu\in I_{\nu,k}}\frac{a^\mu}{\kappa!\mu!}\frac{\partial^{|\lambda+\mu|}}{\partial \mathbf{x}^{\lambda+\mu}}V[\rho_{|\nu|-k}](\mathbf{x}) \quad (5.8)$$

and evaluate this at $\mathbf{x} = 0$. Consider any nonzero term of the inner sum. For such a term,

$$\mu + 2\kappa = \nu. \quad (5.9)$$

[16]There is a misprint in this equation in L79 [**5**]: μ should be written in place of ν.

Proposition 5.1 implies that the factor

$$\frac{\partial^{|\lambda+\mu|}}{\partial^{\lambda+\mu}} V[\rho_{|\nu|-k}]|_{\mathbf{x}=0} = 0$$

unless

$$\lambda + \mu = 2\omega \tag{5.10}$$

for a multi-index ω. If equations (5.9) and (5.10) both hold, then

$$\omega = \frac{\lambda+\nu}{2} - \kappa, \tag{5.11}$$

providing for each term of the inner sum a unique nonzero value for each choice of the multi-index κ. Mindful of Proposition 5.1, we now find that

$$\frac{\partial^{|\lambda|} V_\nu}{\partial \mathbf{x}^\lambda}|_{\mathbf{x}=0} = \pi \frac{a^\nu \nu!}{2^{|\nu|}} \sum_{k=0}^{[|\nu|/2]} \sum_{\kappa,\mu,\omega \in J_{\lambda,\nu,k}} \frac{a^\mu}{\kappa!\mu!} \frac{(-1)^{|\nu|-|\omega|}}{(|\nu|-|\omega|-k+1)!} \frac{(2\omega)!}{\omega!} A(\omega) \tag{5.12}$$

with

$$J_{\lambda,\nu,k} = \left\{\kappa, \mu, \omega : |\kappa| = k, \mu = \nu - 2\kappa, \omega = \frac{\lambda+\nu}{2} - \kappa\right\}, \tag{5.13}$$

where, as discussed at the end of the last section, the order of the multi-indices is important.

We know that V_ν contains terms up to degree $|\nu|+2$ but not higher. We can see how this is manifested in this formula by examining the denominator $(|\nu|+1-|\omega|-k)!$ that occurs in the inner sum. For $|\omega|$ we have the expression $\frac{1}{2}|\lambda| + \frac{1}{2}|\nu| - k$ giving to the denominator the value $(\frac{1}{2}(|\nu|+2-|\lambda|)!$, which is infinite if $|\lambda| > |\nu|+2$.

Recall that a multi-index is a triple of *non-negative* integers. The inner sum is extended over all multi-indices κ such that $\sum_i \kappa_i = k$ and for which $\nu - 2\kappa$ and $(\lambda+\nu)/2 - \gamma$ are multi-indices. In particular this requires that λ and ν belong to the same symmetry class as defined in Section 5.2 below: if they should belong to *different* symmetry classes the set J would be empty and the right-hand side above vanish.

The typical coefficient $V_{\nu\lambda}$ in the power-series expansion (5.6) of V_ν is now found by dividing the expression (5.12) by $\lambda!$:

$$V_{\nu\lambda} = (-1)^{|\nu|} \pi \frac{a^\nu \nu!}{2^{|\nu|}\lambda!} \sum_{k=0}^{[|\nu|/2]} \sum_{\kappa,\mu,\omega \in J_{\lambda,\nu,k}} \frac{a^\mu}{\kappa!\mu!\omega!} \frac{(-1)^{|\omega|}(2\omega)!}{(|\nu|-|\omega|-k+1)!} A(\omega). \tag{5.14}$$

For any choice of the multi-index ν most of the coefficients $\{V_{\nu\lambda}\}$ vanish for reasons of symmetry. We address this next.

5.2. Symmetry classes. The potentials $\{V_\nu(\mathbf{x})\}$ fall into one or another of the eight symmetry classes of the ellipsoid which we may illustrate, following the notation of Whittaker and Watson [**6**] in their chapter on ellipsoidal harmonics, by the pattern

$$\left\{ 1 \quad \begin{matrix} x_1 \\ x_2 \\ x_3 \end{matrix} \quad \begin{matrix} x_2x_3 \\ x_3x_1 \\ x_1x_2 \end{matrix} \quad x_1x_2x_3 \right\},$$

which presents the simplest example in each symmetry class. Since any monomial $\mathbf{x}^\nu$ lies in one of these classes, so does $V_\nu(\mathbf{x})$, as can easily be inferred from the invariance of the kernel $1/|\mathbf{x}-\mathbf{y}|$ under orthogonal transformations. As a result, for any choice of ν most of the coefficients $\{V_{\nu\kappa}\}$ of equation (5.6) vanish, and it will be convenient to consider separately the more compact expressions obtained when we specify a symmetry class.

The elements of the preceding array can be written $\mathbf{x}^\sigma$ where $\sigma=\sigma(\nu)$ is one of the multi-indices

$$\left\{\begin{matrix} & (1,0,0) & (0,1,1) & \\ (0,0,0) & (0,1,0) & (1,0,1) & (1,1,1) \\ & (0,0,1) & (1,1,0) & \end{matrix}\right\} \tag{5.15}$$

Then if we write the indices ν,λ of equation (5.6) above as

$$\nu=\sigma+2\rho, \qquad \lambda=\sigma+2\zeta, \tag{5.16}$$

we have

$$V_\nu=\mathbf{x}^\sigma\sum_{k=0}^{|\rho|+1}\sum_{|\zeta|=k}U_{\rho\zeta}\mathbf{x}^{2\zeta}, \quad \text{where} \quad \sigma=\sigma(\nu) \quad \text{and} \quad \rho=\rho(\nu). \tag{5.17}$$

In this formula, for any choice of symmetry class – and therefore of $\sigma(\nu)$ –

$$V_{\nu\lambda}=V_{(\sigma+2\rho)(\sigma+2\zeta)}=U_{\rho\zeta}$$

and therefore

$$\begin{aligned}\frac{1}{(\sigma+2\zeta)!}\frac{\partial^{|\sigma+2\zeta|}V_{\sigma+2\rho}}{\partial x^{\sigma+2\zeta}}|_{x=0}&=U_{\rho\zeta}\\ &=\frac{\pi}{(\sigma+2\zeta)!}\frac{a^{\sigma+2\rho}(\sigma+2\rho)!}{2^{|\sigma+2\rho|}}\\ &\quad\times\sum_{k=0}^{|\rho|}\sum_{\kappa,\mu,\omega\in J}\frac{a^\mu}{\kappa!\mu!}\frac{(-1)^{|\sigma|+|\omega|}}{(|\sigma|+2|\rho|+1-|\omega|-k)!}\frac{(2\omega)!}{\omega!}A(\omega),\end{aligned}$$

where the indexing set $J=J_{\lambda,\nu,k}$ takes the form[17]

$$J_{\lambda,\nu,k}=\{\kappa,\mu,\omega:|\kappa|=k,\ \mu=\sigma+2(\rho-\kappa),\ \omega=\sigma+\zeta+\rho-\kappa\}. \tag{5.18}$$

Using this above now gives the expression

$$\begin{aligned}U_{\rho\zeta}&=\pi(-1)^{|\sigma|}\frac{a^{\sigma+2\rho}(\sigma+2\rho)!}{2^{|\sigma+2\rho|}(\sigma+2\zeta)!(1+|\rho|-|\zeta|)!}\\ &\quad\times\sum_{k=0}^{|\rho|}\sum_{\kappa,\mu,\omega\in J_{\lambda,\nu,k}}\frac{a^\mu(-1)^{|\omega|}}{\kappa!\mu!\omega!}(2\omega)!A(\omega).\end{aligned} \tag{5.19}$$

The basic equations specifying the potential $V_\nu(x)$ for any choice of multi-index ν and any interior point x are equations (5.17), (5.18) and (5.19).

It might be thought that the upper limit $|\rho|$ on the outer sum above should be $|\rho|+1$ in the symmetry classes for which $|\sigma|=2$ or 3 since, by equation (5.8) above, the upper limit for k is $|\rho|+1$ in these cases. However, if we consider the set $J_{\lambda,\nu,k}$ as given by equation (5.18) we see that $\mu_i=\sigma_i+2(\rho_i-\kappa_i)$ for $i=1,2,3$, and

[17] The index variables are now σ,ρ,ζ in place of $\nu=\sigma+2\rho$, $\lambda=\sigma+2\zeta$ but we retain the former notation to try keep the proliferation of notation in check.

therefore if κ_i exceeds ρ_i the corresponding value of μ_i is negative and is therefore disallowed. Therefore k can not exceed $|\rho|$.

5.3. Examples. Some examples are now given to check these formulas in different symmetry classes, identified by the value of the multi-index σ of that symmetry class.

- $\sigma = (0,0,0)$: The expression for the coefficient $U_{\rho\zeta}$ becomes

$$U_{\rho\zeta} = \frac{\pi a^{2\rho}(2\rho)!}{4^{|\rho|}(2\zeta)!(1+|\rho|-|\zeta|)!}\sum_{k=0}^{|\rho|}\sum_{J}\frac{(-1)^{|\omega|}a^{\mu}}{\kappa!\mu!\omega!}(2\omega)!A(\omega) \tag{5.20}$$

where

$$J = J_{\zeta,\rho,k} = \{\kappa,\mu,\omega : |\kappa| = k,\ \mu = 2(\rho-\kappa),\ \omega = \zeta+\rho-\kappa\}.$$

◇ density $= 1$
I have checked this for $V_{(0,0,0)}$ with density 1 and the gravitational constant $G = 1$ and obtain

$$\begin{aligned} V_{(0,0,0)}(x) = \pi\,&\big(A(0,0,0) - A(1,0,0)x_1^2 - A(0,1,0)x_2^2 \\ &- A(0,0,1)x_3^2\big), \end{aligned}$$

which is the known, correct expression in the current notation.

◇ density $= x_1^2$
For this choice of density function,

$$\begin{aligned} V_{(2,0,0)}(x) &= \sum_{k=0}^{2}\sum_{|\kappa|=k} U_{(100)\kappa}\mathbf{x}^{2\kappa} \\ &= U_{(100)(000)} + U_{(100)(100)}x_1^2 \\ &\quad + U_{(100)(010)}x_2^2 + U_{(100)(000)}x_3^2 \\ &\quad + U_{(100)(200)}x_1^4 + U_{(100)(020)}x_2^4 \\ &\quad + U_{(100)(002)}x_3^4 + U_{(100)(110)}x_1^2x_2^2 \\ &\quad + U_{(100)(011)}x_2^2x_3^2 + U_{(100)(101)}x_1^2x_3^2, \end{aligned}$$

where we find from equation (5.20) above that the coefficients are as follows:

$$\begin{aligned} U_{(100)(000)} &= \frac{\pi}{4}a_1^2\left(A(0,0,0) - a_1^2A(1,0,0)\right), \\ U_{(100)(100)} &= \frac{\pi}{2}a_1^2\left(3a_1^2A(2,0,0) - A(1,0,0)\right), \\ U_{(100)(010)} &= \frac{\pi}{2}a_1^2\left(a_1^2A(1,1,0) - A(0,1,0)\right), \\ U_{(100)(001)} &= \frac{\pi}{2}a_1^2\left(a_1^2A(1,0,1) - A(0,0,1)\right), \\ U_{(100)(110)} &= \frac{\pi}{2}a_1^2\left(A(1,1,0) - 3a_1^2A(2,1,0)\right), \\ U_{(100)(011)} &= \frac{\pi}{2}a_1^2\left(A(0,1,1) - a_1^2A(1,1,1)\right), \\ U_{(100)(101)} &= \frac{\pi}{2}a_1^2\left(A(1,0,1) - 3a_1^2A(2,0,1)\right), \\ U_{(100)(200)} &= \frac{\pi}{4}a_1^2\left(A(2,0,0) - 5a_1^2A(3,0,0)\right), \\ U_{(100)(020)} &= \frac{\pi}{4}a_1^2\left(A(0,2,0) - a_1^2A(1,2,0)\right), \\ U_{(100)(002)} &= \frac{\pi}{4}a_1^2\left(A(0,0,2) - a_1^2A(1,0,2)\right). \end{aligned}$$

The verification that $\nabla^2 V_{(2,0,0)} = -4\pi$ is now accomplished with the aid of the identities among the coefficients given in equation (C.3) below.

- $\sigma = (1, 0, 0)$: We use equations (5.19) and (5.18) with the current value of σ. We record an example.

  Let the density function be x_1. Then we use equation (5.19) with $\rho = (0, 0, 0)$. The values of $|\zeta|$ can be zero and one. We find

  $$V_{(1,0,0)} = \pi a_1^2 x_1 \left(A(1,0,0) - x_1^2 A(2,0,0) - x_2^2 A(1,1,0)\right. \\ \left. - x_3^2 A(1,0,1)\right)$$

  in agreement with equation (4.5) above.
- $\sigma = (0, 1, 1)$: The gravitational potential due to density $x_2 x_3$ works out, by the formulas above, to be

$$V_{(0,1,1)}(x) = \pi a_2^2 a_3^2 A(0,1,1) x_2 x_3 \\ - \pi a_2^2 a_3^2 \left[A(1,1,1) x_1^2 + A(0,2,1) x_2^2 + A(0,1,2) x_3^2\right] x_2 x_3.$$

  This is seen to satisfy Poisson's equation $\Delta V_{(0,1,1)} = -4\pi x_2 x_3$ with the aid of the identity (C.3) below.
- $\sigma = (1, 1, 1)$: The gravitational potential due to density $x_1 x_2 x_3$ is seen by the formulas above to take the form

$$V_{(1,1,1)}(x) = \pi a_1^2 a_2^2 a_3^2 x_1 x_2 x_3 \\ \times \left(A(1,1,1) - A(2,1,1) x_1^2 - A(1,2,1) x_2^2 - A(1,1,2) x_3^2\right).$$

  Appealing again to the identities (C.3) we find that $\Delta V_{(1,1,1)} = -4\pi x_1 x_2 x_3$, providing a check on this formula.

5.4. Surface integrals. We describe here the potential of the surface distribution

$$H_\mu(\mathbf{x}) = \int_{\partial E_a} \frac{\mathbf{y}^\mu}{|\mathbf{x} - \mathbf{y}|} p dS \tag{5.21}$$

in terms of the coefficients $\{V_{\mu\lambda}\}$ (equation (5.6) above) that serve to describe the potentials of volume distributions. For this we may appeal to L79 [**5**, Theorem 5] to conclude the following:

$$H_\mu(\mathbf{x}) = (\mu + 2) V_\mu - \sum_{l=1}^{3} x_l \frac{\partial V_\mu}{\partial x_l}. \tag{5.22}$$

We supply a proof slightly different from those given in L79 [**5**]. Since on the surface of the ellipsoid we may replace pdS by $y_k dS_k$ we find by the divergence theorem that

$$H_\mu(\mathbf{x}) = \int_{E_a} \frac{\partial}{\partial y_k} \left(\frac{y_k \mathbf{y}^\mu}{|\mathbf{x} - \mathbf{y}|}\right) dy \\ = \int_{E_a} \left[(|\mu| + 3) \frac{\mathbf{y}^\mu}{|\mathbf{x} - \mathbf{y}|} + \mathbf{y}^\mu y_k \frac{\partial}{\partial y_k} \frac{1}{|\mathbf{x} - \mathbf{y}|}\right] dy.$$

The following identity is easily checked:

$$y_k \frac{\partial}{\partial y_k} \frac{1}{|\mathbf{x} - \mathbf{y}|} = -\frac{1}{|\mathbf{x} - \mathbf{y}|} - x_k \frac{\partial}{\partial x_k} \frac{1}{|\mathbf{x} - \mathbf{y}|}. \tag{5.23}$$

Using this in the previous expression gives L79 [**5**, equation (36)]

$$H_\mu(\mathbf{x}) = (|\mu| + 2) V_\mu(\mathbf{x}) - x_k \frac{\partial V_\mu}{\partial x_k}. \tag{5.24}$$

Since

$$x_k \frac{\partial \mathbf{x}^\lambda}{\partial x_k} = |\lambda| \mathbf{x}^\lambda,$$

we find in the representation

$$H_\mu(\mathbf{x}) = \sum_{|\lambda| \leqslant |\mu|} H_{\mu\lambda} \mathbf{x}^\lambda \tag{5.25}$$

that the coefficients in the polynomial expansion of H_μ are

$$H_{\mu\lambda} = (|\mu| + 2 - |\lambda|) V_{\mu\lambda}. \tag{5.26}$$

This shows that they are polynomials of degree $|\mu|$ and provides expressions for their coefficients.

Appendix A. A first multinomial theorem

For positive integers m, r[18] and real or complex numbers $x_1, x_2, \ldots, x_m$,

$$\left(\sum_{i=1}^m x_i \right)^r = \sum_{|\rho|=r} \frac{r!}{\rho!} \mathbf{x}^\rho \tag{A.1}$$

where ρ is the multi-index $\rho = (\rho_1, \ldots, \rho_m)$, $|\rho| = \rho_1 + \cdots + \rho_m$, $\mathbf{x}^\rho$ denotes the product $x_1^{\rho_1} \cdots x_m^{\rho_m}$ and $\rho!$ denotes the product $\rho_1! \cdots \rho_m!$. The summation is extended over all multinomials for which $|\rho| = r$.

1. Proof by induction on r.

We'll fix m and do it by induction on r, the power to which the sum is raised. It's an identity if $r = 1$. Suppose that it's true for r and consider

$$\left(\sum_{i=1}^m x_i \right)^{r+1} = \left(\sum_{i=1}^m x_i \right) \left(\sum_{i=1}^m x_i \right)^r = \left(\sum_{i=1}^m x_i \right) \sum_{|\rho|=r} \frac{r!}{\rho!} \mathbf{x}^\rho$$

where we used the induction hypothesis in the last term. The typical term in the last sum may be written out as

$$\frac{r!}{\rho!} \mathbf{x}^\rho = \frac{r!}{\rho_1! \rho_2! \cdots \rho_m!} x_1^{\rho_1} x_2^{\rho_2} \cdots x_m^{\rho_m}$$

so multiplying it by $\sum_i x_i$ results in the sum of m distinct monomials

$$\frac{r!}{\rho_1! \rho_2! \cdots \rho_m!} \left(x_1^{\rho_1+1} x_2^{\rho_2} \cdots x_m^{\rho_m} + \cdots + x_1^{\rho_1} x_2^{\rho_2} \cdots x_m^{\rho_m+1} \right).$$

Concentrating on the first term in parentheses we note that this monomial appears in as many as $(m-1)$ further terms: for example, the product of x_2 with the monomial $x_1^{\rho_1+1} x_2^{\rho_2-1} \cdots x_m^{\rho_n}$ leads to this term but with the coefficient $r!/((\rho_1+1)!(\rho_2-1)! \cdots (\rho_m)!)$. For the coefficient of $x_1^{\rho_1+1} x_2^{\rho_2} \cdots x_m^{\rho_m}$ we therefore obtain

$$\begin{aligned} &\frac{r!}{\rho_1! \rho_2! \cdots \rho_m!} + \frac{r!}{(\rho_1+1)!(\rho_2-1)! \cdots \rho_m!} \\ &+ \cdots + \frac{r!}{(\rho_1+1)! \rho_2! \cdots (\rho_m-1)!} \end{aligned} \tag{A.2}$$

[18] m for sum and r for power.

This sum is easily seen to be

$$\frac{r!}{(\rho_1+1)!\rho_2!\cdots\rho_m!}((\rho_1+1)+\rho_2+\cdots+\rho_m)=\frac{(r+1)!}{(\rho_1+1)!\rho_2!\cdots\rho_m!},$$

where we have recalled that $\sum_{i=1}^m \rho_i = r$. This is indeed the coefficient of the monomial $x_1^{\rho_1+1}x_2^{\rho_2}\cdots x_m^{\rho_m}$ and *almost* completes the proof.

Almost because we seem to have assumed that there are exactly m terms to be added up in the equations above, but there may be fewer, leading to an apparent gap in the reasoning. For example, consider the first term in the sum on the right-hand side of equation (A.1) above: x_1^n. When multiplied by $\sum_i x_i$ it yields a contribution to the coefficient of x_1^{n+1}, and no other term contributes to this. This can be handled at the level of equation (A.2) by noting that, for the current example, in all but one of the terms in that sum there is a negative (and therefore undefined) factorial, and that the correct result is restored by ignoring such terms. All such cases for which fewer than m terms contribute to the coefficient are correctly handled by formally including them all but then omitting those with a negative factorial.

This procedure can be achieved by adding to the familiar convention that $0! = 1$ the further convention that $(-1)! = \infty$.[19]

2. Proof by induction on m.

Now fix the power r and observe that if $m = 1$ then equation (A.1) is an identity; we further suppose it is true for $m = 2$ since it then reduces to the binomial theorem. In general, suppose it is true for some positive integer m and write

$$\sum_{i=1}^{m+1} x_i = \sum_{i=1}^{m} x_i + x_{m+1} = \sum_{i=1}^{m} \hat{x}_i + x_{m+1},$$

where we have written $x_i = \hat{x}_i$ for $i = 1, \ldots, m$: the reason for this is that we will want to distinguish the multinomial $\mathbf{x}^\rho$ of $m+1$ factors from the multinomial $\hat{\mathbf{x}}^\mu$ of n factors. By the binomial theorem

$$\left(\sum_{i=1}^{m} \hat{x}_i + x_{m+1}\right)^r = \sum_{\lambda_1+\lambda_2=r} \frac{r!}{\lambda_1!\lambda_2!}\left[\sum_{i=1}^{m} \hat{x}_i\right]^{\lambda_1} x_{m+1}^{\lambda_2}.$$

By the induction hypothesis this becomes

$$\sum_{\lambda_1+\lambda_2=r} \frac{r!}{\lambda_1!\lambda_2!}\left(\sum_{|\mu|=\lambda_1} \frac{\lambda_1!}{\mu!}\hat{\mathbf{x}}^\mu\right) x_{m+1}^{\lambda_2} = \sum_{\lambda_1+\lambda_2=r,|\mu|=\lambda_1} \frac{r!}{\mu!\lambda_2!}\hat{\mathbf{x}}^\mu x_{r+1}^{\lambda_2}.$$

The last term is exactly the sum

$$\sum_{|\rho|=r} \frac{r!}{\rho!}\mathbf{x}^\rho,$$

completing the proof.

---

[19]These conventions are consistent with the relation between the Gamma function and factorials.

Appendix B. A second multinomial theorem

We'll need the formula for the product of two multinomial sums. We write them as

$$f(\mathbf{x}) = \sum_{k=0}^{K} \sum_{|\kappa|=k} a_\kappa \mathbf{x}^\kappa = \sum_{k=0}^{K} A_k(\mathbf{x}), \tag{B.1}$$

where we have abbreviated the sum of terms of degree k by $A_k(\mathbf{x})$; and

$$g(\mathbf{x}) = \sum_{l=0}^{L} \sum_{|\lambda|=l} b_\lambda \mathbf{x}^\lambda = \sum_{l=0}^{L} B_l(\mathbf{x}), \tag{B.2}$$

where again we have abbreviated the terms of degree l. The product of these two multinomials is

$$f(\mathbf{x})g(\mathbf{x}) = \sum_{k=0}^{K} \sum_{l=0}^{L} A_k(\mathbf{x})B_l(\mathbf{x}) = \sum_{m=0}^{K+L} \sum_{k+l=m} A_k(\mathbf{x})B_l(\mathbf{x}),$$

where in the last term we have simply reordered the sum by collecting first all terms of total degree m and then summing over m. Next we observe that

$$A_k(\mathbf{x})B_l(\mathbf{x}) = \sum_{|\kappa|=k} a_\kappa \mathbf{x}^\kappa \sum_{|\lambda|=l} b_\lambda \mathbf{x}^\lambda = \sum_{|\mu|=l+k} \left(\sum_{\kappa+\lambda=\mu} a_\kappa b_\lambda \right) \mathbf{x}^\mu.$$

This leads to the equation

$$\left(\sum_{k=0}^{K} \sum_{|\kappa|=k} a_\kappa \mathbf{x}^\kappa \right) \left(\sum_{l=0}^{L} \sum_{|\lambda|=l} b_\lambda \mathbf{x}^\lambda \right) = \sum_{m=0}^{K+L} \sum_{|\mu|=m} c_\mu \mathbf{x}^\mu, \tag{B.3}$$

where

$$c_\mu = \sum_{\kappa+\lambda=\mu} a_\kappa b_\lambda. \tag{B.4}$$

In some applications we need to multiply multinomials like $f(\mathbf{x})$ and $g(\mathbf{x})$ above when these functions result from the truncation at degree K of an infinite series. In these cases the definition (B.3) contains contributions from terms of the expansion higher than K, i.e., requires coefficients beyond those included in the original functions f and g. In those cases we'll rewrite the expression above for $f(\mathbf{x})g(\mathbf{x})$ as

$$f(\mathbf{x})g(\mathbf{x}) \simeq \sum_{m=1}^{K} \sum_{|\mu|=m} c_\mu \mathbf{x}^\mu, \qquad c_\mu = \sum_{\kappa+\lambda=\mu} a_\kappa b_\lambda. \tag{B.5}$$

This partial summation now requires only coefficients appearing in the originally truncated sums. Further products, like $f(\mathbf{x})g(\mathbf{x})h(\mathbf{x})$, are defined by repeated applications of equation (B.5).

Appendix C. Gravitational coefficients

The gravitational coefficients that were originally labeled $A_i, A_{ij}, \dots$ and $B_{ij}, \dots$[20] will require a more systematic notation as in L79 [**5**]. We refer to §5 there and repeat some of that section here.

$$A_j(a,\kappa) = a_1 a_2 a_3 \int_0^\infty \frac{s^j\, ds}{(a_1^2+s)^{\kappa_1}(a_2^2+s)^{\kappa_2}(a_3^2+s)^{\kappa_3}\Delta(s)}, \tag{C.1}$$

which we will denote by $A_j(\kappa)$ if it is clear to which value of $a = (a_1, a_2, a_3)$ it refers. The integral is convergent if $|\kappa| \geqslant j$. The value $j = 0$ plays a special role and we shall write

$$A_0(a,\kappa) = A(a,\kappa). \tag{C.2}$$

If the intended values of the semiaxes are understood we'll suppress the dependence on a. By these definitions what was originally called A_1 is now $A(1,0,0)$, etc. By integrating the expression

$$\frac{d}{ds}\left[s^j(a_1^2+s)^{-\kappa_1-1/2}(a_2^2+1)^{-\kappa_2-1/2}(a_3^2+1)^{-\kappa_3-1/2}\right]$$

from 0 to ∞ we generate the family of identities L79 [**5**, equation (48)]

$$\begin{aligned}(2\kappa_1+1)A_j(\kappa_1+1,\kappa_2,\kappa_3) + (2\kappa_2+1)A_j(\kappa_1,\kappa_2+1,\kappa_3)\\ + (2\kappa_3+1)A_j(\kappa_1,\kappa_2,\kappa_3+1) = \begin{cases} 2a^{-2\kappa} & \text{if } j=0\\ 2jA_{j-1}(\kappa) & \text{if } j\geqslant 1,\end{cases}\end{aligned} \tag{C.3}$$

for integers $\kappa_1, \kappa_2, \kappa_3$ greater than or equal to zero.

Similarly, to see that L79 [**5**, equation (45)] holds, consider the case $\mu = (1,0,0)$ and note that

$$\begin{aligned}&A_j(\kappa_1,\kappa_2,\kappa_3) - a_1^2 A_j(\kappa_1+1,\kappa_2,\kappa_3)\\ &= \int_0^\infty \frac{s^j}{\Delta(s)(a_1^2+s)^{\kappa_1}(a_2^2+s)^{\kappa_2}(a_3^2+s)^{\kappa_3}}\left(1-\frac{a_1^2}{a_1^2+s}\right) ds\\ &= a_1^2 A_{j+1}(\kappa_1+1,\kappa_2,\kappa_3)\end{aligned}$$

which agrees with that equation for the chosen value of μ. Then L79 [**5**, equation (46)] follows from this by permuting μ and ν and subtracting, and equation (47) also follows from (45) in a similar way.

Any of these symbols can be expressed as a definite integral on $[0,1]$. This can be done (for example) by expressing the part of the integral on $[1,\infty)$ by the substitution $s = u^{-2}$, which is then extended over $(0,1]$. Carrying out the details results in the expression

$$\begin{aligned}A_j(a,\kappa) = a_1a_2a_3 \int_0^1 \Bigg\{ &\frac{s^j}{(a_1^2+s)^{\kappa_1+1/2}(a_2^2+s)^{\kappa_2+1/2}(a_3^2+s)^{\kappa_3+1/2}}\\ &+ \frac{2\, s^{2(|\kappa|-j)}}{(s^2a_1^2+1)^{\kappa_1+1/2}(s^2a_2^2+1)^{\kappa_2+1/2}(s^2a_3^2+1)^{\kappa_3+1/2}}\Bigg\}\, ds\end{aligned} \tag{C.4}$$

which is well-defined and has a finite integrand as long as $|\kappa| \geqslant j$. Numerical integration of this is no problem. Appendix A of L79 [**5**] shows that if we want $A_j(\kappa)$ for $j = 0, 1, \dots, N$ and for $|\kappa| = j, j+1, \dots, N$, we need only construct those

[20] See e.g., EFE [**2**].

for $j = 0$ and the others can be obtained from the basic identity (45) in L79 [**5**]. This would save some computation if we want to do that (or represent a check on the calculations).

Appendix D. Ellipsoidal Figures of Equilibrium (EFE) [2] derivation of equation (2.37)

The derivation of equation (2.37) for the potential in the interior of an ellipsoid of constant density is given in EFE [**2**].[21] It is not perfectly logical, though it leads to the correct conclusion. I present a (thoroughly logical!) version of this approach here. My derivation is, however, *very* lengthy compared with that given in EFE [**2**], or with that given above (§ 2.5).

For the solid ellipsoid D with semiaxes a_1, a_2, a_3 and density $\rho = 1$, we seek the potential

$$\Psi(\mathbf{x}) = \int_D \frac{d\mathbf{y}}{|\mathbf{x} - \mathbf{y}|} \tag{D.1}$$

at an interior point $\mathbf{x}$. We first shift the origin of coordinates to this point by the transformation $\mathbf{y} = \mathbf{x} + \eta$ so that the expression for V becomes

$$\Psi(\mathbf{x}) = \int_D \frac{d\eta}{|\eta|}. \tag{D.2}$$

Here the domain D must be re-expressed in the coordinate η:

$$D\colon \sum_{i=1}^{3} \left\{ \frac{\eta_i^2}{a_i^2} + 2\frac{x_i \eta_i}{a_i^2} + \frac{x_i^2}{a_i^2} \right\} < 1. \tag{D.3}$$

The introduction of spherical coordinates r, θ, ϕ for η will enable us to perform the r integration in a manner similar to that used above (§2.2) for the homoeoidal shell.[22] To that end put

$$\eta_i = r l_i \ (i = 1, 2, 3), \qquad (l_1, l_2, l_3) = (\sin\theta\cos\phi, \sin\theta\sin\phi, \cos\theta). \tag{D.4}$$

For a given direction $\mathbf{l} = (l_1, l_2, l_3)$ the value at the boundary $R(\theta, \phi)$ of r is obtained by solving equation (D.3) with the inequality replaced by equality, i.e., by solving the equation[23]

$$P(\mathbf{l})R(\mathbf{l})^2 + 2Q(\mathbf{l})R(\mathbf{l}) - D = 0, \tag{D.5}$$

where

$$P(\mathbf{l}) = \sum \frac{l_i^2}{a_i^2}, \ Q(\mathbf{l}) = \sum \frac{l_i X_i}{a_i^2} \quad \text{and} \quad D = 1 - \sum \frac{X_i^2}{a_i^2}. \tag{D.6}$$

It is easy to see that this quadratic equation for $R = R(\mathbf{l})$ has one (and only one) positive root for each choice of $\mathbf{l}$.

> This is contrary to the statements made in EFE [2] regarding equations (43) and (44) of Chapter 3, in the proof of Theorem 3. While the statements in question are erroneous the subsequent formulas (47) and beyond are correct. The reason for this is that the second, negative root of equation (D.6) above is the negative of the positive root of this

[21] And perhaps in earlier references? I haven't checked this as of 5/03/2020.

[22] We are exploiting the fact that, from every interior point of an ellipsoid, an arbitrary ray intersects the surface once and only once.

[23] I believe that this abuse of notation $R(\mathbf{l}) = R(\theta, \phi)$ results in no confusion in this case.

equation when $-\mathbf{l}$ is substituted for $\mathbf{l}$; since only the squares of these roots appear in the formula for the potential, the result is correct.

The integral (D.2) becomes

$$\text{(D.7)} \qquad \Psi(\mathbf{x}) = \int_0^{\pi}\int_0^{2\pi}\int_0^{R(\theta,\phi)} r\sin\theta\, dr\, d\theta\, d\phi$$

$$\text{(D.8)} \qquad = \frac{1}{2}\int_0^{\pi}\int_0^{2\pi} R(\theta,\phi)^2 \sin\theta\, d\theta\, d\phi$$

with $R(\theta,\phi) = R(\mathbf{l})$ supplied by the positive solution of equation (D.5) above. The evaluation of this integral by solving the quadratic equation (D.5) to express R as a function of $\mathbf{l}$ is difficult but is simplified (to the form of equation (D.12) below) by the following observations.

The unit vector $\mathbf{l}$ defines a ray emanating from X. The opposite ray $\mathbf{l}' = -\mathbf{l}$ is described by the values $\theta' = \pi - \theta, \phi' = \phi + \pi$ where the angle ϕ' is evaluated mod 2π. The corresponding boundary value $R(\theta', \phi') = R(-\mathbf{l})$ satisfies the equation (compare equation (D.5) above)

$$\text{(D.9)} \qquad P(\mathbf{l})R(-\mathbf{l})^2 - 2Q(\mathbf{l})R(-\mathbf{l}) - D = 0$$

where the symmetries of the functions P and Q have been used. Adding and subtracting equations (D.5) and (D.9) lead to the pair of equations

$$\text{(D.10)} \qquad P(\mathbf{l})\left(R(\mathbf{l})^2 + R(-\mathbf{l})^2\right) + 2Q(\mathbf{l})\left(R(\mathbf{l}) - R(-\mathbf{l})\right) - 2D = 0$$

and

$$P(\mathbf{l})\left(R(\mathbf{l})^2 - R(-\mathbf{l})^2\right) + 2Q(\mathbf{l})\left(R(\mathbf{l}) + R(-\mathbf{l})\right) = 0.$$

Extracting and cancelling a factor of $R(\mathbf{l}) + R(-\mathbf{l})$ (which cannot vanish!) leads to

$$R(\mathbf{l}) - R(-\mathbf{l}) = -2Q(\mathbf{l})/P(\mathbf{l})$$

and therefore we find from equation (D.10) that

$$\text{(D.11)} \qquad R(\mathbf{l})^2 + R(-\mathbf{l})^2 = \frac{4Q(\mathbf{l})^2}{P(\mathbf{l})^2} + \frac{2D}{P(\mathbf{l})}.$$

The following result is (nearly) obvious but we present it to be thorough:

LEMMA D.1. *For any function $f(\theta,\phi)$ that is 2π-periodic in ϕ,*

$$\int_0^{\pi}\int_0^{2\pi} f(\pi-\theta, \phi+\pi)\sin\theta\, d\theta\, d\phi = \int_0^{\pi}\int_0^{2\pi} f(\theta,\phi)\sin\theta\, d\theta\, d\phi.$$

Proof: Substituting $\theta' = \pi - \theta, \phi' = \phi + \pi$ converts the integral on the left to

$$-\int_{\pi}^{0}\int_{\pi}^{3\pi} f(\theta',\phi')\sin\theta'\, d\theta' d\phi' = \int_0^{\pi}\int_0^{2\pi} f(\theta,\phi)\sin\theta\, d\theta\, d\phi,$$

where we have observed that a $2\pi-$periodic function has the same integral over any interval of length 2π.

It follows from this that

$$\int_0^{\pi}\int_0^{2\pi} R(-\mathbf{l})^2\sin\theta\, d\theta\, d\phi = \int_0^{\pi}\int_0^{2\pi} R(\mathbf{l})^2\sin\theta\, d\theta\, d\phi$$

and therefore (see equation (D.8)) that

$$\text{(D.12)}\qquad \begin{aligned}\Psi(\mathbf{x}) &= \frac{1}{4}\int_0^{\pi}\int_0^{2\pi} \left(R(\theta,\phi)^2 + R(\pi-\theta,\phi+\pi)^2\right)\sin\theta\, d\theta\, d\phi \\ &= \frac{1}{4}\int_0^{\pi}\int_0^{2\pi}\left[\frac{2D}{P(\mathbf{l})} + \frac{4Q(\mathbf{l})^2}{P(\mathbf{l})^2}\right]\sin\theta\, d\theta\, d\phi.\end{aligned}$$

By virtue of the definitions in equation (D.6), this expression is quadratic in $\mathbf{x}$. That it is indeed the same as that given by equation (2.37) above is now not surprising but requires lengthy calculations that are carried out below.

Write the potential $\Psi = V_1 + V_2$ where (equation (D.12) above)

$$\text{(D.13)}\qquad V_1(\mathbf{x}) = \frac{1}{2}\int_0^{\pi}\int_0^{2\pi} P(\mathbf{l})^{-1}\left(1 - \sum_{i=1}^{3}\frac{X_i^2}{a_i^2}\right)\sin\theta\, d\theta\, d\phi.$$

The latter expression can be evaluated immediately in terms of the semiaxes on the basis of those same calculations leading to equation (2.17):

$$\text{(D.14)}\qquad V_1(\mathbf{x}) = \pi I(a)\left(1 - \sum_{i=1}^{3}\frac{X_i^2}{a_i^2}\right),$$

where $I(a) = I(a_1, a_2, a_3)$ is defined by equation (2.16).

Consider then the second term $V_2(X)$ in equation (D.12).

$$\text{(D.15)}\qquad V_2(X) = \int_0^{\pi}\int_0^{2\pi}\frac{Q(\mathbf{l})^2}{P(\mathbf{l})^2}\sin\theta\, d\theta\, d\phi.$$

Expanding the numerator (cf. equation (D.6)) we find

$$Q(\mathbf{l})^2 = \sum_{i=1}^{3}\frac{X_i^2}{a_i^4}l_i^2 + 2\sum_{cyclic}\frac{X_iX_j}{a_i^2a_j^2}l_il_j,$$

where the second sum is over terms involving cross-products l_1l_2 and their cyclic permutations. These latter terms make no contribution to the integral V_2.

To see that the terms involving cross products do not contribute, consider for example the term in l_1l_2. Its contribution to the integral is of the form

$$\int_0^{\pi}\int_0^{2\pi} P^{-2}\sin^3\theta\sin\phi\cos\phi\, d\theta\, d\phi = \int_0^{\pi}\int_{-\pi}^{\pi} P^{-2}\sin^3\theta\sin\phi\cos\phi\, d\theta\, d\phi$$

where we have again exploited our knowledge that the function P, which has appeared previously as the denominator in the integral expression (2.11), is 2π-periodic in ϕ. From this it is seen that the ϕ integration may be expressed in the form

$$\int_{-\pi}^{\pi} f(\sin^2\phi)\sin\phi\cos\phi\, d\phi$$

which vanishes since the integrand is an odd function of ϕ. Likewise, the term in l_2l_3 produces the ϕ-integral

$$\int_{-\pi}^{\pi} f(\sin^2\phi)\sin\phi\, d\phi = 0$$

since the integrand is again odd. The remaining cross term has the form

$$\int_{-\pi}^{\pi} f(\sin^2\phi)\cos\phi\, d\phi = 2\int_0^{\pi} f(\sin^2\phi)\cos\phi\, d\phi$$

since this integrand is even. But the latter integral is easily seen to vanish by splitting the integral at $\pi/2$ and, in the integral on $[\pi/2, \pi]$, substituting $\psi = \pi - \phi$ for ϕ.

The expression above for V_2 may now be written

$$V_2(\mathbf{x}) = \int_0^{\pi} \int_0^{2\pi} P(\mathbf{l})^{-2} \sum_{i=1}^{3} \frac{X_i^2}{a_i^4} l_i^2 \sin\theta \, d\theta \, d\phi. \tag{D.16}$$

Define

$$J_i = \int_0^{\pi/2} \int_0^{\pi/2} P(\mathbf{l})^{-2} l_i^2 \sin\theta \, d\theta \, d\phi, \quad i = 1, 2, 3$$

so that

$$V_2(\mathbf{x}) = 8 \sum_{i=1}^{3} J_i \frac{X_i^2}{a_i^4}. \tag{D.17}$$

For the computations below we remind ourselves of the definitions of the vector $\mathbf{l}$ (equation (D.4)) and of the function P (equation (D.6)) above. If we first perform the ϕ integrations we are left with the θ integrations

$$J_1 = \int_0^{\pi/2} (\sin\theta)^3 j_1(\theta) \, d\theta, \tag{D.18}$$

$$J_2 = \int_0^{\pi/2} (\sin\theta)^3 j_2(\theta) \, d\theta, \tag{D.19}$$

$$J_3 = \int_0^{\pi/2} (\cos\theta)^2 \sin\theta \, j_3(\theta) \, d\theta, \tag{D.20}$$

where we have further introduced the notations j_1, j_2, j_3 for the ϕ integrations. These are

$$j_1 = \int_0^{\pi/2} \frac{(\cos\phi)^2}{P^2} \, d\phi, \quad j_2 = \int_0^{\pi/2} \frac{(\sin\phi)^2}{P^2} \, d\phi, \quad j_3 = \int_0^{\pi/2} \frac{d\phi}{P^2}, \tag{D.21}$$

where

$$P = P(\theta, \phi) = \frac{(\sin\theta\cos\phi)^2}{a_1^2} + \frac{(\sin\theta\sin\phi)^2}{a_2^2} + \frac{(\cos\theta)^2}{a_3^2}.$$

With the substitution $t = \tan\phi$ as in equation (2.13) above and the same definitions of a and b as those given in equation (2.12), we find after some manipulations,

$$j_1 = \int_0^{\infty} \frac{dt}{(a + bt^2)^2}, \quad j_2 = \int_0^{\infty} \frac{t^2}{(a + bt^2)^2} \, dt, \quad j_3 = j_1 + j_2, \tag{D.22}$$

where $j_0 = \pi/(2\sqrt{ab})$ is given above in equation (2.13). Calculating j_1 and j_2 can be done for example by differentiating the formula for j_0 with respect to a and b, respectively, or by the method of residues. One finds

$$j_1 = \frac{\pi}{4}(\sqrt{a^3 b})^{-1}, \quad j_2 = \frac{\pi}{4}(\sqrt{ab^3})^{-1}, \quad j_3 = \frac{\pi}{4} \frac{a + b}{\sqrt{a^3 b^3}}.$$

We can now rewrite the integrals for J_1, J_2, J_3 with the aid of the integration variable $s = a_3^2(\tan\theta)^2$ as in the reductions leading to equation (2.17) above. After a series of manipulations we obtain

$$\begin{aligned} J_1 &= \frac{\pi}{8}\int_0^\infty \frac{s\,ds}{(a_1^2+s)^{3/2}(a_2^2+s)^{1/2}(a_3^2+s)^{1/2}} \\ &= \frac{\pi}{8}a_1^3a_2a_3\int_0^\infty \frac{s\,ds}{(a_1^2+s)\Delta(s)}, \end{aligned} \tag{D.23}$$

where Δ is defined in equation (2.15). After a similar series of manipulations we find for J_2 the similar expression

$$J_2 = \frac{\pi}{8}a_1a_2^3a_3\int_0^\infty \frac{s\,ds}{(a_2^2+s)\Delta(s)} \tag{D.24}$$

as one expects since it could be obtained by a different choice of coordinates. However, for J_3, proceeding in the same way, one finds

$$J_3' = \frac{\pi}{8}a_1a_2a_3^3\left\{a_1^2\int_0^\infty \frac{ds}{(a_1^2+s)\Delta(s)} + a_2^2\int_0^\infty \frac{ds}{(a_2^2+s)\Delta(s)}\right\}, \tag{D.25}$$

which is provisionally labeled J_3' since it fails to follow the pattern of J_1 and J_2, which should have the form

$$J_3 = \frac{\pi}{8}a_1a_2a_3^3\int_0^\infty \frac{s\,ds}{(a_3^2+s)\Delta(s)}. \tag{D.26}$$

Can the expression J_3' be right?

Yes. To see this form the difference

$$\begin{aligned} J_3' - J_3 &= \frac{\pi}{8}a_1a_2a_3^3\int_0^\infty \left\{\frac{a_1^2}{a_1^2+s} + \frac{a_2^2}{a_2^2+s} - \frac{s}{a_3^2+s}\right\}\frac{ds}{\Delta(s)} \\ &= \frac{\pi}{8}a_1a_2a_3^3\int_0^\infty \left\{\frac{a_1^2}{a_1^2+s} + \frac{a_2^2}{a_2^2+s} + \frac{a_3^2}{a_3^2+s} - 1\right\}\frac{ds}{\Delta(s)}. \end{aligned}$$

This indeed vanishes by virtue of the identity EFE [**2**, Chapter 3, equation (22)]:

$$\sum_{i=1}^3 a_i^2A_i = I, \quad \text{where } A_i = a_1a_2a_3\int_0^\infty \frac{ds}{(a_i^2+s)\Delta(s)} \tag{D.27}$$

and I is given by equation (2.16) above.

The formula (D.27) is derived in EFE [**2**, Chapter 3, Lemma 3]. We give here an alternative derivation based on the fact that the function I is homogeneous of degree two in a_1, a_2, a_3: this fact is easily checked. It follows by Euler's theorem on homogeneous functions that

$$\sum_{i=1}^3 a_i\frac{\partial I}{\partial a_i} = 2I.$$

It is easily verified from the defining formulas that

$$\frac{\partial I}{\partial a_1} = \frac{I}{a_1} - a_1^2a_2a_3\int_0^\infty \frac{ds}{(a_1^2+s)\Delta(s)} = \frac{I}{a_1} - a_1A_1,$$

and similarly for a_2 and a_3. Using these expressions in Euler's theorem now leads to the identity (D.27).

This gives for J_i the formula

$$J_i = \frac{\pi}{8}a_1a_2a_3\int_0^\infty \frac{a_i^2}{a_i^2+s}\frac{s\,ds}{\Delta(s)} \tag{D.28}$$

determining the expression V_2 (equation (D.17) above). Combining this with equation (D.14) above for V_1, we easily arrive at equation (2.37) above.

Appendix E. Axisymmetric ellipsoids

The formula for the potential coefficients EFE [**2**, Chapter 3, equation (40)] simplifies when two axes are equal in that they are then expressible in terms of elementary functions.

When $a_1 = a_2 > a_3$, the case of an oblate spheroid, the potential coefficients $A_1 = A_2$ and A_3 EFE [**2**, Chapter 3, equation (18)] are

$$A_1 = a_1^2 a_3 \int_0^\infty (a_1^2 + u)^{-2}(a_3^2 + u)^{-1/2}\, du$$

and

$$A_3 = a_1^2 a_3 \int_0^\infty (a_1^2 + u)^{-1}(a_3^2 + u)^{-3/2}\, du.$$

We record some integral formulas here:

$$g(x, y) \equiv \int_x^\infty \frac{dv}{y + v^2} = y^{-1/2}\left(\frac{\pi}{2} - \tan^{-1}\left(xy^{-1/2}\right)\right). \tag{E.1}$$

and

$$\begin{aligned} f(x, y) \equiv \int_x^\infty \frac{dv}{(y + v^2)^2} &= -\frac{\partial}{\partial y}\int_x^\infty \frac{dv}{y + v^2} = -\frac{\partial g}{\partial y} \\ &= \frac{1}{2}\left\{\left(\frac{\pi}{2} - \tan^{-1}(xy^{-1/2})\right) y^{-3/2} - \frac{x}{y(x^2 + y)}\right\}. \end{aligned} \tag{E.2}$$

The expression for A_1, under the change of variable $v = (a_3^2 + u)^{1/2}$, takes the form

$$A_1(a_1, a_1, a_3) = 2a_1^2 a_3 \int_{a_3}^\infty \frac{dv}{(a_1^2 - a_3^2 + v^2)^2}$$

which is easily expressed with the aid of the definition of eccentricity

$$e = \sqrt{1 - a_1^2/a_2^2} = 2a_1^2 a_3 f(a_3, a_1^2 - a_3^2)$$

and equation (E.2) as

$$A_1 = \left(\frac{\pi}{2} - \tan^{-1}\left(\frac{\sqrt{1 - e^2}}{e}\right)\right)\frac{\sqrt{1 - e^2}}{e^3} - \frac{1 - e^2}{e^2}.$$

From the formula $\sin(\frac{\pi}{2} - \theta) = \cos\theta$ we infer that

$$\frac{\pi}{2} - \theta = \sin^{-1}(\cos\theta),$$

and if as above $\theta = \tan^{-1}(\frac{\sqrt{1-e^2}}{e})$, then $\cos\theta = e$. The formula above for A_1 now becomes

$$A_1 = \frac{\sqrt{1 - e^2}}{e^3}\sin^{-1} e - \frac{1 - e^2}{e^2} \tag{E.3}$$

agreeing with EFE [**2**, Chapter 3, the first of equations (36)].

With the same change of variables we find for A_3 the expression

$$A_3 = 2a_1^2 a_3 \int_{a_3}^\infty \frac{dw}{(y + w^2)w^2} = \frac{2a_1^2 a_3}{y}\int_{a_3}^\infty \left(\frac{1}{w^2} - \frac{1}{y + w^2}\right) dy$$

with the same choice of y as above. With the aid of the first integral formula above, this becomes

$$A_3 = \frac{2}{e^2} - 2\frac{\sqrt{1-e^2}}{e^3}\sin^{-1} e, \tag{E.4}$$

again agreeing with the formula in EFE [**2**].

References

[1] S. Chandrasekhar, *An introduction to the study of stellar structure*, Dover Publications, Inc., New York, 1957. MR92663

[2] S. Chandrasekhar, *Ellipsoidal figures of equilibrium*, Dover Publications, New York, 1969.

[3] N. M. Ferrers, On the potentials of ellipsoids, ellipsoidal shells, elliptic laminae, and elliptic rings, of variable densities, *Quart. J. Pure Appl. Math.*, 14:1–22, 1877.

[4] O. D. Kellogg, *Foundations of potential theory*, Die Grundlehren der mathematischen Wissenschaften, Band 31, Springer-Verlag, Berlin-New York, 1967. Reprint from the first edition of 1929. MR222317

[5] N. R. Lebovitz, Ellipsoidal potentials of polynomial distributions of matter, *Astrophys. J.*, 234:619–627, December 1979.

[6] E. T. Whittaker and G. N. Watson, *A course of modern analysis*, Cambridge Mathematical Library, Cambridge University Press, Cambridge, 1996. An introduction to the general theory of infinite processes and of analytic functions; with an account of the principal transcendental functions; Reprint of the fourth (1927) edition, DOI 10.1017/CBO9780511608759. MR1424469

Dept. of Mathematics, University of Chicago, Chicago, Illinois 60637

Contemporary Mathematics
Volume **819**, 2025
https://doi.org/10.1090/conm/819/16391

Spherical distributions on the de Sitter space and their spectral singularities

Iswarya Sitiraju

ABSTRACT. A spherical distribution is an eigendistribution of the Laplace–Beltrami operator with certain invariance on the de Sitter space. Let $G' = O_{1,n}(\mathbb{R})$ be the Lorentz group and $H' = O_{1,n-1}(\mathbb{R})$ be its subgroup. The authors Ólafsson and Sitiraju have constructed the spherical distributions, which are H'-invariant, as boundary values of some sesquiholomorphic kernels. In this article we explore the connections of these kernels with reflection positivity and representations of the group $G = SO_{1,n}(\mathbb{R})_e$, which is the connected component of the Lorentz group. We also discuss the singularities of spherical distributions in terms of their wavefront sets.

1. Introduction

The n-dimensional de Sitter space dS^n, which is a one-sheeted hyperboloid, is a simple model of the universe in special relativity with constant positive curvature. There have been several studies done to understand quantum field theories on the de Sitter space including the papers [**4**–**7**] where the authors studied free fields and the related two point functions $\mathcal{W}_\lambda(x_1, x_2)$. The theory of interacting quantum fields on the de Sitter space of dimension 2 is discussed in the paper [**2**]. On the other hand, the authors in [**32**] study some aspects of algebraic quantum field theory on causal symmetric spaces of which the de Sitter space is an example. One of the tools used in [**32**] is an open complex domain called the crown domain Ξ.

In [**30**], the authors showed that the de Sitter space lies on the boundary of Ξ and its conjugate $\overline{\Xi}$, which are open subsets of the complex unit sphere. The crown Ξ is holomorphically equivalent to the Lorentzian tuboid $\mathcal{T}^+$ in the complexified de Sitter space $\mathrm{dS}^n_\mathbb{C}$ defined in [**4**–**7**]. Similarly, $\overline{\Xi}$ is equivalent to the tuboid $\mathcal{T}^-$ in $\mathrm{dS}^n_\mathbb{C}$. The de Sitter space lies on the boundary of $\mathcal{T}^+$ and $\mathcal{T}^-$.

Let $\rho = (n-1)/2$. For $\lambda \in i[0,\infty) \cup (0,\rho)$, the distribution $\mathcal{W}_\lambda(x_1, x_2)$ given in [**4**] satisfies the Klein-Gordon equation $\Delta + (\rho^2 - \lambda^2)$ in both variables. Moreover, this distribution is the boundary value of some analytic kernels in $\mathcal{T}^+$ and $\mathcal{T}^-$ called "perikernels." In [**30**, **31**], the authors have introduced the kernel Ψ_λ up to a constant which corresponds to the perikernels and showed that Ψ_λ was represented as a hypergeometric function. Analogous to $\mathcal{W}_\lambda$, the boundary values of the kernels Ψ_λ and $\widetilde{\Psi}_\lambda$ defined below, are studied in [**36**], where

$$\Psi_\lambda(z, w) = {}_2F_1\Big(\rho + \lambda, \rho - \lambda; \frac{n}{2}; \tfrac{1+[z,\bar{w}]}{2}\Big), \quad z, w \in \Xi,$$

and

$$\widetilde{\Psi}_\lambda(z,w) = {}_2F_1\Big(\rho+\lambda, \rho-\lambda; \frac{n}{2}; \tfrac{1+[z,\bar{w}]}{2}\Big), \quad z,w \in \Xi.$$

These kernels were obtained by reflection positivity on the sphere. Reflection positivity is one of the Osterwalder–Schrader axioms of constructive quantum field theory. It is a necessary and sufficient condition for a Euclidean field theory that has Euclidean symmetries to analytically continue to a relativistic field theory with Lorentzian symmetries. The initial attempt was done by E. Nelson in the paper [**29**] and the breakthrough was done in the paper [**34, 35**].

The kernels Ψ_λ and $\widetilde{\Psi}_\lambda$ are well-defined sesquiholomorphic, positive-definite, G-invariant kernels. The perikernels are holomorphic in the cut domain of $\mathrm{dS}^n \times \mathrm{dS}^n$ of the form $\mathrm{dS}^n \times \mathrm{dS}^n \setminus \Sigma$, where Σ is the set of tuples (x,y) with $[x-y, x-y] \leqslant 0$. In the paper [**36**], we showed that the boundary values of the kernels Ψ_λ and $\widetilde{\Psi}_\lambda$ are real analytic on the cut domain $\mathrm{dS}^n \times \mathrm{dS}^n \setminus \{(x,y) : [x-y, x-y] = 0\}$ and have jump discontinuities along the cut. The cut is where the distributions have singularities, and we studied these singularities in terms of analytic wavefront sets.

The wavefront set of a distribution was introduced by L. Hörmander in 1970. For a distribution Θ on a real analytic manifold, the *analytic wavefront set* $WF_A(\Theta)$ describes the set of points where Θ is not given by a real-analytic function and the direction in which the singularity occurs (see [**20**]). The wavefront set is a crucial concept in quantum field theory (QFT). One of the initial papers using wavefront sets in QFT was [**10**]. Later the wavefront set was brought into the context of Hadarmard distributions in [**37**]. It was shown that the Hadamard condition of a two point distribution of a quasi-free quantum field is equivalent to a condition on its wavefront set. In algebraic quantum field theory, the condition on the wavefront set of the states of quantum fields is related to the Reeh–Schlieder property (see [**38, 39**]). It was also conceptualized in the context of unitary representations of Lie groups in [**23**] and studied for induced representations in [**24**].

In this article we will collect the main theorems of [**36**] and their connections to representation theory ([**9**]) and reflection positivity ([**31**]). The geometrical setup is given in Section 2. The connection with reflection positivity on the sphere is established in Section 3 and Section 4 explains the representation theoretic perspective. In Sections 5 and 6, the spherical distributions and their analytic wavefront sets are reviewed.

Notations: We use the following notations throughout the article:

- $\mathrm{G} = \mathrm{SO}_{1,n}(\mathbb{R})_e$,
- $\mathrm{H} = \mathrm{SO}_{1,n-1}(\mathbb{R})_e \subset G$,
- $\mathrm{K} = \mathrm{SO}_n(\mathbb{R}) \subset G$,
- $[z,w] = -z_0w_0 + z_1w_1 + \cdots + z_nw_n$ for $z,w \in \mathbb{C}^{1+n}$,
- $\mathbb{R}^{1,n} = (\mathbb{R}^{1+n}, [\,,\,])$,
- $\mathrm{dS}^n = \{x \in \mathbb{R}^{1+n} \mid [x,x] = 1\} = G/H$,
- $\mathrm{dS}^n_{\mathbb{C}} = \{z \in \mathbb{C}^{1+n} \mid [z,z] = 1\}$,
- $\mathbb{H}^n = \{x \in i\mathbb{R}^{1+n} \mid x_0 > 0, [x,x] = 1\} \simeq G/K \subset \mathrm{dS}^n_{\mathbb{C}}$,
- $\overline{\mathbb{H}}^n = \{x \in i\mathbb{R}^{1+n} \mid x_0 < 0, [x,x] = 1\} \simeq G/K \subset \mathrm{dS}^n_{\mathbb{C}}$,
- $\mathbb{S}^n = \{(ix_0, \mathbf{x}) \mid x_0^2 + \mathbf{x}^2 = 1\}$,
- $\mathbb{S}^n_{\pm} = \{(ix_0, \mathbf{x}) \in \mathbb{S}^n \mid \pm x_0 > 0\}$,
- $\Gamma^{\pm}(x) = \{y \in \mathrm{dS}^n \mid \text{for } x \in \mathrm{dS}^n, [y-x, y-x] < 0, \pm y_0 > x_0\}$,

- $\Gamma(x) = \Gamma^+(x) \cup \Gamma^-(x)$,
- $\rho = (n-1)/2$ for $n \geqslant 2$.

2. Geometrical setup

We are going to follow the geometrical setup described in the paper [**36**]. However, the setups in [**4, 31, 36**] are all equivalent.

Let $z, w \in \mathbb{C}^{1+n}$ and let the bilinear form on $\mathbb{C}^{1+n}$ be given by

$$[z, w] = -z_0 w_0 + \sum_{j=1}^{n} z_j w_j = -z_0 w_0 + \mathbf{z} \cdot \mathbf{w}.$$

We say that a vector $v \in \mathbb{R}^{1,n}$ is time-like if $[v, v] < 0$ and space-like if $[v, v] > 0$.

2.1. The hyperboloid and the de Sitter space. The *hyperbolic spaces* $\mathbb{H}^n$ and $\overline{\mathbb{H}}^n$ are given as follows:

$$\mathbb{H}^n = \{x \in i\mathbb{R}^{1+n} \mid [x, x] = 1, x_0 > 0\} \quad \text{and} \quad \overline{\mathbb{H}}^n = \{x \in i\mathbb{R}^{1+n} \mid [x, x] = 1, x_0 < 0\}.$$

The de Sitter space dS^n is described as

$$\mathrm{dS}^n = \{x \in \mathbb{R}^{1+n} \mid [x, x] = 1\}.$$

All the spaces defined above are closed submanifolds of the complex manifold

$$\mathrm{dS}^n_{\mathbb{C}} = \{z \in \mathbb{C}^{1+n} \mid [z, z] = 1\}.$$

Let $x \in \mathrm{dS}^n$, then we denote the future (past) cone of x as $\Gamma^+(x)(\Gamma^-(x))$ where

$$\Gamma^\pm(x) := \{y \in \mathrm{dS}^n \mid [y - x, y - x] < 0, \pm(y - x)_0 > 0\}.$$

For $x \in \mathrm{dS}^n$ the set $\{y \in \mathrm{dS}^n \mid [y - x, y - x] = 0\}$ is called the light cone of x in dS^n.

Let $G = \mathrm{SO}(1, n)_e$ be the connected component of identity of the isometry group of $[\cdot, \cdot]$. We denote by $K = \mathrm{SO}(n)$ the maximal compact subgroup

$$K = \{k \in G \mid g \cdot e_0 = e_0\},$$

$$A = \left\{ a_t = \begin{pmatrix} \cosh t & 0 & \sinh t \\ 0 & I_{n-1} & 0 \\ \sinh t & 0 & \cosh t \end{pmatrix} : t \in \mathbb{R} \right\},$$

and by

$$H = \{h \in G \mid h \cdot e_n = e_n\} = \mathrm{SO}(1, n-1)_e.$$

The group G acts transitively on $\mathbb{H}^n$ and dS^n and,

$$\mathbb{H}^n = G \cdot ie_0 \simeq G/K \simeq \overline{\mathbb{H}}^n = G \cdot (-ie_0) \quad \text{and} \quad \mathrm{dS}^n = G \cdot e_n \simeq G/H.$$

The de Sitter space is a homogenous Lorentzian manifold with local co-ordinates at each point given by its tangent space. For $x \in \mathrm{dS}^n$ we have

$$\mathrm{T}_x(\mathrm{dS}^n) = \{y \in \mathbb{R}^{1+n} \mid [x, y] = 0\} \cong \mathbb{R}^{1,n-1}.$$

The following lemma has been proved in [**31**, Lemma 6.3].

LEMMA 2.1. *$G = HAK = KAH$, and*

$$G/H = KA.e_n = \mathrm{dS}^n.$$

There exists a unique up to a constant G-invariant measure on dS^n. For more discussions see [**9**, p. 159]. Let Δ be the Laplace–Beltrami operator obtained from the Lorentz metric on dS^n induced from its ambient space $(\mathbb{R}^{1+n}, [\cdot,\cdot])$. It is known from [**13**] that the algebra of polynomials in Δ is the algebra of G-invariant differential operators on dS^n.

2.2. The crown Ξ, $\overline{\Xi}$. The complex crown Ξ of a non compact Riemannian symmetric space G/K has the property that the eigenfunctions of the algebra of invariant differential operators extend to Ξ. It was introduced in [**1**]. It was studied by several authors, including in the articles [**16, 17, 27**] and it was shown in [**17**] that the non-compactly causal symmetric spaces [**25**], including the de Sitter space, can be realized as open orbits in the boundary of the crowns. We will now discuss how the kernel Ψ_λ obtained from reflection positivity is connected to representation theory.

Let $h \in \mathfrak{so}(1,n)$ be the operator

$$h(x_0, x_1, \ldots, x_{n-1}, x_n) = (x_n, 0, \ldots, 0, x_1).$$

The unit sphere is realized in $i\mathbb{R}e_0 + \mathbb{R}^n$ by $\mathbb{S}^n = \{x \in i\mathbb{R}e_0 + \mathbb{R}^n \mid [x,x] = 1\}$. Set $\mathbb{S}^n_+ = \{x \in \mathbb{S}^n \mid x_0 > 0\}$ and $\mathbb{S}^n_- = \{x \in \mathbb{S}^n \mid x_0 < 0\}$.

The crown domain of $\mathbb{H}^n$ is described as

$$\Xi = G \cdot \mathbb{S}^n_+ = G \exp(i(-\pi 2, \pi 2)h) \cdot (ie_0)$$

and that for $\overline{\mathbb{H}}^n$ as

$$\overline{\Xi} = G \cdot \mathbb{S}^n_- = G \exp i(-\pi/2, \pi/2)h \cdot (-ie_0).$$

The crown domain $\Xi \subset \mathrm{SO}_{d+1}(\mathbb{C})(ie_0) \simeq G_{\mathbb{C}}/K_{\mathbb{C}} \simeq \mathrm{dS}^n_{\mathbb{C}}$. Similarly, for $\overline{\Xi}$.

REMARK 2.2. The crown domain depends on $(\mathfrak{g}, \mathfrak{k})$ where $\mathfrak{g}, \mathfrak{k}$ are the Lie algebras of G and K, respectively. It does not depend the choice of Lie group G with Lie algebra $\mathfrak{g}$.

The crown domains $\Xi(\overline{\Xi})$ contains $\mathbb{S}^n_+(\mathbb{S}^n_-)$ and $\mathbb{H}^n(\overline{\mathbb{H}}^n)$ as submanifolds. More details on the crown domains can be found in [**31, 36**].

The following statement that the de Sitter space lies on the boundary of Ξ and $\overline{\Xi}$ was proven in [**31, 36**]. Hence, the crown domains form a bridge between the Riemannian manifolds $\mathbb{H}^n, \overline{\mathbb{H}}^n$ and the Lorentzian manifold dS^n.

LEMMA 2.3. $\mathrm{dS}^n = \partial \Xi \cap \partial \overline{\Xi}$.

REMARK 2.4. It was shown in [**4, 36**] that around each point $x \in \mathrm{dS}^n$, the crown can be represented locally as a tuboid of the form $U + i\Omega'$, where U is an open set and Ω' is a pointed cone in the tangent space of x.

3. Reflection positivity, kernels Ψ_λ and $\widetilde{\Psi}_\lambda$

We will now recall reflection positivity on the sphere [**31**](see also [**32**]), which leads to the positive definite kernel Ψ_λ (with a different normalization in [**31**]) given for $\lambda \in i[0,\infty) \cup [0, \frac{n-1}{2})$ and $\rho = (n-1)/2$ by

$$\Psi_\lambda(z,w) = {}_2F_1\left(\rho+\lambda, \rho-\lambda; \frac{n}{2}; \frac{1+[z,\overline{w}]}{2}\right), \quad z, w \in \Xi. \tag{3.1}$$

We will also consider the following kernel:

$$\widetilde{\Psi}_\lambda(z,w) = {}_2F_1\left(\rho+\lambda,\rho-\lambda;\frac{n}{2};\frac{1+[z,\overline{w}]}{2}\right),\quad z,w\in\Xi. \tag{3.2}$$

As both n and λ are fixed most of the time, we simplify our notation and write

$$ {}_2F_1(z) = {}_2F_1\left(\rho+\lambda,\rho-\lambda;\frac{n}{2};z\right).$$

Here ${}_2F_1(a,b;c;z)$ denotes the Gauss hypergeometric function

$$ {}_2F_1(a,b;c;z) = \sum_{n=0}^{\infty}\frac{(a)_n(b)_n}{(c)_n}\frac{z^n}{n!},$$

where $(d)_n = d(d+1)\cdots(d+n-1)$, $c\notin -\mathbb{N}_0$ and $|z|<1$. The hypergeometric function ${}_2F_1$ extends to a holomorphic function on $\mathbb{C}\setminus[1,\infty)$ (see [**28**]).

Consider the unit sphere $\mathbb{S}^n$ realized in $i\mathbb{R}e_0+\mathbb{R}^n$. Let $\sigma\colon\mathbb{S}^n\to\mathbb{S}^n$ be the reflection $\sigma(x_0,\mathbf{x}) = (-x_0,\mathbf{x})$. Let us denote by $\Box$ the Laplacian on $\mathbb{S}^n$. We will now consider a distribution Φ_λ on $\mathbb{S}^n\times\mathbb{S}^n$ as follows:

$$\Phi_\lambda(\phi\otimes\overline{\psi}) = \int_{\mathbb{S}^n}\overline{\phi(x)}(-\Box+(\rho^2-\lambda^2))^{-1}\psi(x)\,d\mu(x)\quad\text{for}\quad \phi,\psi\in C_c^\infty(\mathbb{S}^n),$$

where $(-\Box+(\rho^2-\lambda^2))^{-1}$ is a bounded positive operator on $L^2(\mathbb{S}^n)$ for $\lambda\in i[0,\infty)\cup(0,\rho)$.

It was proven in [**31**, Sec. 2] that the distribution Φ_λ is *reflection positive* with respect to $(\mathbb{S}^n,\mathbb{S}^n_+,\sigma)$. That is, the distribution $\Phi_\lambda^\sigma = \Phi_\lambda\circ(id,\sigma)$ is positive definite on $\mathbb{S}^n_+\times\mathbb{S}^n_+$. It was also proven that the distribution Φ_λ^σ is given by the kernel $\Psi_\lambda(x,y)$ for $x,y\in\mathbb{S}^n_+$. For a general Riemannian manifold a similar construction is done to get a reflection positive distribution (see [**11,26**]).

This kernel has a natural extension to $\Xi\times\Xi$. Part (1) of the following theorem is in [**31,32**], part (2) can be found in [**36**] and part (3) follows by part (2).

THEOREM 3.1. *Let* $\rho=\frac{n-1}{2}$. *Then*

(1) *The kernel* $\Psi_\lambda(z,w)$ *is a positive definite* G *invariant kernel on* $\Xi\times\Xi$ *which is holomorphic in the first variable and anti-holomorphic in the second variable. It is given by*

$$\Psi_\lambda(z,w) = {}_2F_1\left(\rho+\lambda,\rho-\lambda;\frac{n}{2};\tfrac{1+[z,\overline{w}]}{2}\right),\quad z,w\in\Xi.$$

(2) *Let* $z,w\in\Xi$. *Then*

$$\overline{\Psi_\lambda(z,w)} = \Psi_\lambda(\overline{z},\overline{w}).$$

(3) *The kernel* $\widetilde{\Psi}_\lambda(z,w)$ *is a* G*-invariant positive definite kernel on* $\overline{\Xi}\times\overline{\Xi}$ *holomorphic in the first variable and anti-holomorphic in the second variable, given by*

$$\widetilde{\Psi}_\lambda(z,w) = {}_2F_1\left(\rho+\lambda,\rho-\lambda;\frac{n}{2};\tfrac{1+[z,\sigma(w)]}{2}\right),\quad z,w\in\overline{\Xi}.$$

We also have that $\Psi_\lambda(z,w) = \overline{\widetilde{\Psi}_\lambda(\overline{z},\overline{w})} = \widetilde{\Psi}_\lambda(\overline{w},\overline{z})$.

The following lemma has been proved in [**31**, Lem. 6.4].

LEMMA 3.2. *We have*

$$[\mathrm{dS}^n,\Xi]\cap\mathbb{R} = [\mathrm{dS}^n,\overline{\Xi}]\cap\mathbb{R} = (-1,1).$$

From this and the properties of the hypergeometric function we get:

PROPOSITION 3.3. *The kernel* Ψ_λ *can be extended continuously to* $\Xi \times (\mathrm{dS}^n \cup \Xi)$ *and the kernel* $\widetilde{\Psi}_\lambda$ *can be extended continuously to* $\overline{\Xi} \times (\mathrm{dS}^n \cup \overline{\Xi})$.

For $y \in \mathrm{dS}^n$ and $z = e_n$, we have that $(1 + [z, y])/2 \notin [1, \infty)$ iff $y_n < 1$. In particular for $y \in \mathrm{dS}^n$ we have that $y \mapsto \Psi_\lambda(y, e_n), \Psi_\lambda(e_n, y), \widetilde{\Psi}_\lambda(y, e_n), \widetilde{\Psi}(e_n, y)$ are analytic on $\{y \in \mathrm{dS}^n \mid y_n < 1\}$. We will discuss these singularities in Sections 5 and 6.

4. Representation theory perspective

We will now discuss how the kernel Ψ_λ obtained by reflection positivity is related to representations. We have the Iwasawa decomposition $G = KAN$. We write $g = k(g)a(g)n(g)$, and G acts on S^{n-1} by $g \cdot v = k(g)v$. The principal series representation π_λ with spectral parameter λ acting on the Hilbert space $\mathrm{H}_\lambda = L^2(\mathrm{S}^{n-1})$ is given by

$$\pi_\lambda(g)f(v) = a(g^{-1}k)^{-\lambda-\rho} f(g^{-1} \cdot v),$$

where $v \in \mathrm{S}^{n-1}$, $g \in G$ and $f \in L^2(\mathrm{S}^{n-1})$. The constant function $e_\lambda(v) = 1$ is K-invariant with norm 1 and the associated spherical function is

$$\phi_\lambda(g) = \langle \pi_\lambda(g)e_\lambda, e_\lambda \rangle = \int_{\mathrm{S}^n} a(g^{-1}v)^{-\lambda-\rho}\, dv.$$

We note that $g \mapsto \pi_\lambda(g)e_\lambda$ is right K-invariant, hence $\pi_\lambda(z)e_\lambda$ is well defined for z and w in $\mathbb{H}^n$ and the kernel Ψ_λ is given by

$$\Psi_\lambda(z, w) = \langle \pi_\lambda(z)e_\lambda, \pi_\lambda(w)e_\lambda \rangle.$$

Therefore,

$$\phi_\lambda(x) = \Psi_\lambda(x, ie_0) = {}_2F_1\left(\rho + \lambda, \rho - \lambda; \frac{n}{2}; \frac{1 + ix_0}{2}\right), \quad ix \in \mathbb{H}^n$$

is an eigenfunction of the algebra of G-invariant differential operators on $\mathbb{H}^n$ (see [**9**]).

We have that $\exp(-ith)e_n$ is an element in Ξ. It was proven in [**18**], [**3**, Thm. 2.1] that for $\phi \in C_c^\infty(\mathrm{dS}^n)$ the limit

$$\Theta^\lambda(\varphi) = \lim_{t \to \pi/2^-} \int_{\mathrm{dS}^n} \overline{\varphi(y)} \Psi_\lambda(\exp(-ith)e_n, y)\, d\mu_{\mathrm{dS}^n}(y) \tag{4.1}$$

defines an H-invariant distribution (see also [**8**, Thm. 1]), using representation theoretic techniques. Another proof using Hardy space approximation was given in [**30**] and a different proof was given in [**14**]. A third proof independent of representation theory is dealt in [**36**] which we will see in the next section.

Furthermore, we also obtain that

$$\Delta\Theta^\lambda = (\rho^2 - \lambda^2)\Theta^\lambda. \tag{4.2}$$

Similar discussion holds for Ξ, $\overline{\mathbb{H}}^n$ and $\widetilde{\Psi}_\lambda$.

5. Spherical distributions as boundary values of holomorphic functions

We use the usual notation $\mathcal{D}(\mathrm{dS}^n) = C_c^\infty(\mathrm{dS}^n)$ with the standard topology. We consider the distributions as anti-linear functionals on $\mathcal{D}(\mathrm{dS}^n)$. From now onwards we will denote the elements in $\overline{\Xi}$ as $\bar{z}$, since $\overline{\bar{z}} = z$ lies in Ξ.

Let Θ be a distribution on dS^n. Then G acts on Θ by

$$\pi_{-\infty}(g)\Theta(\varphi) = \Theta(\pi_\infty(g^{-1})\varphi), \quad \varphi \in \mathcal{D}(\mathrm{dS}^n),$$

where $\pi_\infty(g)\varphi(x) = \varphi(g^{-1}\cdot x)$.

DEFINITION 5.1. Let H be a closed subgroup of G. We say that a distribution Θ is H-invariant if $\pi_{-\infty}(h)\Theta = \Theta$ for all $h \in H = G_{e_n}$.

DEFINITION 5.2. A distribution Θ is said to be a spherical distribution if it is an H-invariant eigendistribution of the Laplace–Beltrami operator Δ.

Let $G' = O_{1,n}(\mathbb{R})$ be the full Lorentz group and $H' = O_{1,n-1}(\mathbb{R})$ be its subgroup which fixes e_n. Then G'/H' is isomorphic to dS^n. It was proven in [**9**, Thm. 9.2.5] that the dimension of H'-invariant spherical distributions with eigenvalue $\rho^2 - \lambda^2$ is two. The analytic wavefront sets and the characterization of an H'-invariant spherical distribution based on its singularities have been studied in [**36**]. However, in the case of (G, H), it is a hypothesis that the dimension of H-invariant spherical distribution is four, based on [**33**]. Two of those are distributions that are constructed as boundary values of analytic functions in this section. Others will be dealt in another paper.

From (4.2), we observe that the distribution Θ^λ is a spherical distribution with eigenvalue $(\rho^2 - \lambda^2)$. Let $\mathcal{D}_\lambda^H(\mathrm{dS}^n)$ be the space of spherical distributions on the de Sitter space with $\Delta(\Theta) = (\rho^2 - \lambda^2)\Theta$.

Before understanding the limit, we will need the following theorem [**36**].

THEOREM 5.3. *The limit* $\lim_{y\to 0} {}_2F_1(\rho+\lambda, \rho-\lambda, n/2, x \pm iy)$ *for* $y > 0$ *exists in the sense of distributions where for* $\mathrm{Re}(z) > 1$ *the limit converges uniformly on compact sets. We have that for* $1 < x < 2$, *if* n *is odd,*

(5.1)
$$\begin{aligned}{}_2F_1(x \pm i0) = &\frac{\Gamma(n/2)\Gamma((2-n)/2)}{\Gamma(1/2+\lambda)\Gamma(1/2-\lambda)}\,{}_2F_1\left(\rho+\lambda, \rho-\lambda; \frac{n}{2}; 1-x\right) \\ &+ e^{\mp i\pi(\frac{2-n}{2})}(x-1)^{\frac{2-n}{2}}\frac{\Gamma(n/2)\Gamma((n-2)/2)}{\Gamma(\rho+\lambda)\Gamma(\rho-\lambda)}\,{}_2F_1\left(1/2-\lambda, 1/2+\lambda; \frac{4-n}{2}; 1-x\right),\end{aligned}$$

and if n *is even,*

$$\begin{aligned}{}_2F_1(x \pm i0) = &\frac{\Gamma(n/2)}{\Gamma(\rho+\lambda)\Gamma(\rho-\lambda)}\sum_{k=0}^{n/2-2}\frac{(-1)^k(n/2-k-2)!(1/2+\lambda)_k(1/2-\lambda)_k}{k!}(1-x)^{k+1-\frac{n}{2}} \\ &+ \frac{(-1)^{\frac{n-2}{2}}\Gamma(n/2)}{\Gamma(1/2+\lambda)\Gamma(1/2-\lambda)}\sum_{k=0}^{\infty}\frac{(\rho+\lambda)_k(\rho-\lambda)_k}{k!(n/2-1+k)!}\Big[\psi(k+1)+\psi(n/2+k) \\ &- \psi(\rho+\lambda+k) - \psi(\rho-\lambda+k) - \ln(x-1) \pm i\pi\Big](1-x)^k,\end{aligned} \tag{5.2}$$

where $\psi(z) = \Gamma'(z)/\Gamma(z)$. *Furthermore, the behaviour of the hypergeometric function near* $z = 1$ *as distributions is given as follows: for* $n = 2$,

$$ {}_2F_1(z) \approx \frac{1}{\Gamma(\rho+\lambda)\Gamma(\rho-\lambda)}(-\ln(1-z)), \tag{5.3}$$

and for $n \geqslant 3$,

$$ {}_2F_1(z) \approx \frac{\Gamma(n/2)\Gamma((n-2)/2)}{\Gamma(\rho+\lambda)\Gamma(\rho-\lambda)}(1-z)^{\frac{2-n}{2}}. \tag{5.4}$$

For $x > 2$, ${}_2F_1(x \pm i0)$ can be analytically continued.
In [**36**] we have proven that

THEOREM 5.4. *For $n \geqslant 2$ and $\lambda \in i[0,\infty) \cup (0,\rho)$,*

(1) *the limits $\lim_{t\to\pi/2} \Psi_\lambda(\exp(-ith)e_n, y)$ in Ξ, $\lim_{t\to\pi/2} \widetilde{\Psi}_\lambda(\exp(ith)e_n, y)$ in $\overline{\Xi}$, taken in the sense of (4.1), converge to distributions Ψ^λ and $\widetilde{\Psi}^\lambda$ respectively, on dS^n.*

(2) *The limits also satisfy the equation $(\Delta - m^2)\Psi^\lambda = 0 = (\Delta - m^2)\widetilde{\Psi}^\lambda$. Hence, they are spherical distributions.*

(3) *Moreover, Ψ^λ and $\widetilde{\Psi}^\lambda$ can be represented as analytic functions in the following regions:*

$$\Psi^\lambda(y) = \begin{cases} {}_2F_1\left(\frac{1+y_n}{2}\right) & \textit{if } y \notin \overline{\Gamma(e_n)}, \\ {}_2F_1\left(\frac{1+y_n}{2} - i0\right) & \textit{if } y \in \Gamma^+(e_n), \\ {}_2F_1\left(\frac{1+y_n}{2} + i0\right) & \textit{if } y \in \Gamma^-(e_n); \end{cases}$$

$$\widetilde{\Psi}^\lambda(y) = \begin{cases} {}_2F_1\left(\frac{1+y_n}{2}\right) & \textit{if } y \notin \overline{\Gamma(e_n)}, \\ {}_2F_1\left(\frac{1+y_n}{2} + i0\right) & \textit{if } y \in \Gamma^+(e_n), \\ {}_2F_1\left(\frac{1+y_n}{2} - i0\right) & \textit{if } y \in \Gamma^-(e_n). \end{cases}$$

6. Wavefront set of spherical distributions on dS^n

We will now briefly introduce the wavefront sets and study these for the distributions Ψ_λ and $\widetilde{\Psi}_\lambda$.

Let $X \subset \mathbb{R}^{1,n}$ be an open subset. Suppose Θ is a distribution with compact support; then we can define the Fourier transform of Θ at $\xi \in (\mathbb{R}^{1,n} \setminus 0)$ as follows:

$$\widehat{\Theta}(\xi) = \Theta(e^{2\pi i[x,\xi]}),$$

where $[x,\xi] = -x_0\xi_0 + x_1\xi_1 + \cdots + x_n\xi_n$.

The definition of analytic wavefront set follows from [**20**, Proposition 8.4.2].

DEFINITION 6.1. If X is an open subset of $\mathbb{R}^{1,n}$ and $\Theta \in \mathcal{D}'(X)$, we define $WF_A(\Theta)$ to be the complement in $X \times (\mathbb{R}^{1,n} \setminus 0)$ of the set (x_0, ξ_0) such that there is an open neighbourhood $U \subset X$ of x_0, a conic neighbourhood Γ of ξ_0 and a bounded sequence of $\Theta_N \in \mathcal{E}'(X)$ which is equal to Θ in U and satisfies

$$|\widehat{\Theta_N}(\xi)| \leqslant C^{N+1}(N/|\xi|)^N, \quad N = 1, 2, \ldots$$

when $\xi \in \Gamma$ and for some $C > 0$.

REMARK 6.2. The $WF_A(\Theta)$ is a closed conic set; that is, if $(x,\xi) \in WF(\Theta)$, then for $\tau > 0$, $(x, \tau\xi) \in WF(\Theta)$.

The next theorem describes the analytic wavefront sets of distributions which are boundary values of analytic functions. Let Γ be an open convex cone; then the *dual cone* Γ° is defined as

$$\Gamma^\circ = \{\eta \in \mathbb{R}^{1+n} \mid \eta_0\xi_0 + \cdots + \eta_n\xi_n \geqslant 0,\ \forall \xi \in \Gamma\}.$$

THEOREM 6.3. *Let $X \subset \mathbb{R}^{1,n}$ be an open set and Γ an open convex cone in $\mathbb{R}^{1,n}$ and for some $\gamma > 0$,*

$$Z = \{z \in \mathbb{C}^{1+n} \mid \operatorname{Re} z \in X,\ \operatorname{Im} z \in \Gamma,\ |\operatorname{Im} z| < \gamma\}.$$

If Θ is an analytic function in Z such that

$$|\Theta(z)| \leqslant C|\operatorname{Im} z|^{-N}$$

for some $N \in \mathbb{N}$ and some constant $C > 0$, then $\lim_{y\searrow 0}\Theta(.+iy) = \Theta_0$ exists in terms of distribution and is of order N. We also have that

$$WF_A(\Theta_0) \subset X \times (\Gamma^\circ \setminus 0).$$

PROOF. The proof follows from Theorem 3.1.15 and Theorem 8.4.8 in [**20**]. □

EXAMPLE 6.4. Let $\Theta = {}_2F_1(x + i0)$, the boundary value of the holomorphic function ${}_2F_1(x+iy)$ for $y > 0$. From Theorem 5.3 we have that it is a distribution which has an analytic singularity at $x = 1$. As a result of Theorem 6.3, the analytic wavefront set is

$$WF_A({}_2F_1(x + i0)) = \{(1, \tau) \mid \tau > 0\},$$

and it also follows that

$$WF_A({}_2F_1(x - i0)) = \{(1, \tau) \mid \tau < 0\}.$$ □

The following is the second main theorem of the paper [**36**].

THEOREM 6.5. *The analytic wavefront sets of the spherical distributions Ψ^λ and $\widetilde{\Psi}^\lambda$ are given by*

$$\begin{aligned} WF_A(\Psi^\lambda) &= \{(e_n, v) \mid v_0 < 0\} \cup \{(e_n + v, \tau(-v_0, v_1, \dots, v_{n-1})) \mid v_0 > 0\} \\ &\quad \cup \{(e_n + v, \tau(v_0, -v_1, \dots, -v_{n-1})) \mid v_0 < 0\}; \\ WF_A(\widetilde{\Psi}^\lambda) &= \{(e_n, v) \mid v_0 > 0\} \cup \{(e_n + v, \tau(v_0, -v_1, \dots, -v_{n-1})) \mid v_0 > 0\} \\ &\quad \cup \{(e_n + v, \tau(-v_0, v_1, \dots, v_{n-1})) \mid v_0 < 0\}, \end{aligned}$$

for $\tau > 0$ and $v \in T_{e_n}(\mathrm{dS}^n)$ such that $[v - e_n, v - e_n] = 0$.

In Figure 1 we can see the analytic wavefront set in $T_{e_n}\mathrm{dS}^n$. The proof involves calculating the analytic wavefront sets in local co-ordinates using Theorem 6.3 and propagation of singularities from [**22**].

Using the analytic wavefront sets we can prove that the distributions cannot vanish on any non-empty open set O of dS^n.

The following theorem is due to Strohmaier, Verch and Wollenberg, see [**38**, Proposition 5.3].

THEOREM 6.6. *Let X be a real analytic manifold and $\Theta \in \mathcal{D}'(X)$. If $WF_A(\Theta) \cap -WF_A(\Theta) = \varnothing$ then for an open region O in X*

$$\Theta|_O = 0 \Rightarrow \Theta = 0,$$

where $-WF_A(\Theta) = \{(x, \xi) \mid (x, -\xi) \in WF_A(\Theta)\}$.

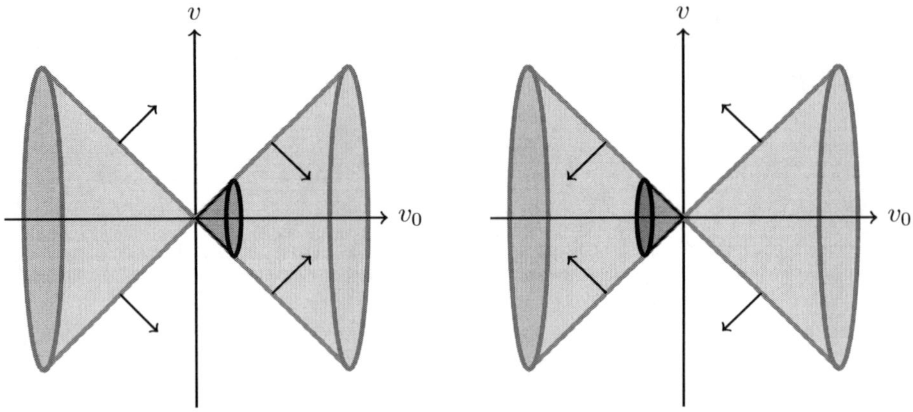

Analytic wavefront set of $\widetilde{\Psi}^\lambda$. Analytic wavefront set of Ψ^λ.

FIGURE 1. In each of (a) and (b), the tangent space at a point e_n has been identified with its cotangent space at e_n. The light grey region, which is the light cone of 0, is the analytic singular support. The arrows and the dark grey cone are the directions in which the singularity occurs at that point.

Since the wavefront sets of the distributions Ψ^λ are such that $WF_A(\Psi^\lambda) \cap -WF_A(\Psi^\lambda) = \varnothing$, which is same for $\widetilde{\Psi}^\lambda$, immediately we obtain:

COROLLARY 6.7. *The distributions Ψ^λ and $\widetilde{\Psi}^\lambda$ cannot vanish on any open regions of* dS^n.

REMARK 6.8. Some of the questions that we are going to address in the future are as follows:

- Can we construct all four spherical distributions as boundary values of some analytic kernels for the pair $(SO_{1,n}(\mathbb{R})_e, SO_{1,n-1}(\mathbb{R})_e)$?
- Can we obtain similar results to Theorem 5.4 for any $\lambda \in \mathbb{C}$?
- What are their analytic wavefront sets?
- Let $\tau\colon G \to G$ be the involution given by $\tau(g) = JgJ$, where J is the orthogonal reflection in the hyperplane $x_n = 0$. Furthermore,

$$\mathfrak{g} = \mathfrak{h} \oplus \mathfrak{q}$$

with $\mathfrak{h} = \ker(\tau - 1)$ and $\mathfrak{q} = \ker(\tau + 1)$. The wavefront sets of Ψ^λ can also be written as $WF_A(\Psi^\lambda) \subset \exp(\mathfrak{q}) \cdot e_n \times \Omega^\circ$, where $\Omega = \{v \in \mathbb{R}^n : [v, v] = 0\}$. Can we draw a similar conclusion in the context of non-compact causal symmetric spaces?

References

[1] D. N. Akhiezer and S. G. Gindikin, *On Stein extensions of real symmetric spaces*, Math. Ann. **286** (1990), no. 1-3, 1–12, DOI 10.1007/BF01453562. MR1032920

[2] J. C. A. Barata, C. D. Jäkel, and J. Mund, *The $\mathcal{P}(\varphi)_2$ model on de Sitter space*, Mem. Amer. Math. Soc. **281** (2023), no. 1389, xiv–261, DOI 10.1090/memo/1389. MR4539363

[3] E. van den Ban and P. Delorme, *Quelques propriétés des représentations sphériques pour les espaces symétriques réductifs* (French, with English summary), J. Funct. Anal. **80** (1988), no. 2, 284–307, DOI 10.1016/0022-1236(88)90002-X. MR961900

[4] J. Bros and U. Moschella, *Two-point functions and quantum fields in de Sitter universe*, Rev. Math. Phys. **8** (1996), no. 3, 327–391, DOI 10.1142/S0129055X96000123. MR1388256

[5] J. Bros and U. Moschella, *Fourier analysis and holomorphic decomposition on the one-sheeted hyperboloid*, Géométrie complexe. II. Aspects contemporains dans les mathématiques et la physique, Hermann Éd. Sci. Arts, Paris, 2004, pp. 27–58. MR2493569

[6] J. Bros and G. A. Viano, *Connection between the harmonic analysis on the sphere and the harmonic analysis on the one-sheeted hyperboloid: an analytic continuation viewpoint. I*, Forum Math. **8** (1996), no. 5, 621–658, DOI 10.1515/form.1996.8.621. MR1404805

[7] J. Bros and G. A. Viano, *Connection between the harmonic analysis on the sphere and the harmonic analysis on the one-sheeted hyperboloid: an analytic continuation viewpoint. III*, Forum Math. **9** (1997), no. 2, 165–191, DOI 10.1515/form.1997.9.165. MR1431119

[8] J.-L. Brylinski and P. Delorme, *Vecteurs distributions H-invariants pour les séries principales généralisées d'espaces symétriques réductifs et prolongement méromorphe d'intégrales d'Eisenstein* (French), Invent. Math. **109** (1992), no. 3, 619–664, DOI 10.1007/BF01232043. MR1176208

[9] G. van Dijk, *Introduction to harmonic analysis and generalized Gelfand pairs*, De Gruyter Studies in Mathematics, vol. 36, Walter de Gruyter & Co., Berlin, 2009, DOI 10.1515/9783110220209. MR2640609

[10] J. Dimock, *Scalar quantum field in an external gravitational field*, J. Math. Phys. **20** (1979), no. 12, 2549–2555, DOI 10.1063/1.524062. MR553516

[11] J. Dimock, *Markov quantum fields on a manifold*, Rev. Math. Phys. **16** (2004), no. 2, 243–255, DOI 10.1142/S0129055X04001947. MR2048318

[12] J. J. Duistermaat and L. Hörmander, *Fourier integral operators. II*, Acta Math. **128** (1972), no. 3-4, 183–269, DOI 10.1007/BF02392165. MR388464

[13] J. Faraut, *Distributions sphériques sur les espaces hyperboliques* (French), J. Math. Pures Appl. (9) **58** (1979), no. 4, 369–444. MR566654

[14] K.-H. Neeb and G. Ólafsson, *Nets of standard subspaces on Lie groups*, Adv. Math. **384** (2021), Paper No. 107715, 69, DOI 10.1016/j.aim.2021.107715. MR4242900

[15] C. Gérard, *Microlocal analysis of quantum fields on curved spacetimes*, ESI Lectures in Mathematics and Physics, EMS Publishing House, Berlin, 2019, DOI 10.4171/094. MR3972066

[16] S. Gindikin and B. Krötz, *Invariant Stein domains in Stein symmetric spaces and a nonlinear complex convexity theorem*, Int. Math. Res. Not. **18** (2002), 959–971, DOI 10.1155/S1073792802112049. MR1902298

[17] S. Gindikin and B. Krötz, *Complex crowns of Riemannian symmetric spaces and non-compactly causal symmetric spaces*, Trans. Amer. Math. Soc. **354** (2002), no. 8, 3299–3327, DOI 10.1090/S0002-9947-02-03012-X. MR1897401

[18] S. Gindikin, B. Krötz, and G. Ólafsson, *Holomorphic H-spherical distribution vectors in principal series representations*, Invent. Math. **158** (2004), no. 3, 643–682, DOI 10.1007/s00222-004-0376-1. MR2104795

[19] I. M. Gel'fand and G. E. Shilov, *Generalized functions. Vol. I: Properties and operations*, Academic Press, New York-London, 1964. Translated by Eugene Saletan. MR166596

[20] L. Hörmander, *Linear partial differential operators*, Springer-Verlag, Berlin-New York, 1976. MR404822

[21] L. Hörmander, *Linear differential operators*, Actes du Congrès International des Mathématiciens (Nice, 1970), Gauthier-Villars Éditeur, Paris, 1971, pp. 121–133. MR513000

[22] L. Hörmander, *Uniqueness theorems and wave front sets for solutions of linear differential equations with analytic coefficients*, Comm. Pure Appl. Math. **24** (1971), 671–704, DOI 10.1002/cpa.3160240505. MR294849

[23] R. Howe, *Wave front sets of representations of Lie groups*, Automorphic forms, representation theory and arithmetic (Bombay, 1979), Tata Inst. Fundam. Res. Stud. Math., vol. 10, Springer, Berlin-New York, 1981, pp. 117–140. MR633659

[24] B. Harris, H. He, and G. Ólafsson, *Wave front sets of reductive Lie group representations*, Duke Math. J. **165** (2016), no. 5, 793–846, DOI 10.1215/00127094-3167168. MR3482333

[25] J. Hilgert and G. Ólafsson, *Causal symmetric spaces: Geometry and harmonic analysis*, Perspectives in Mathematics, vol. 18, Academic Press, Inc., San Diego, CA, 1997. MR1407033

[26] A. Jaffe and G. Ritter, *Reflection positivity and monotonicity*, J. Math. Phys. **49** (2008), no. 5, 052301, 10, DOI 10.1063/1.2907660. MR2421900

[27] B. Krötz and R. J. Stanton, *Holomorphic extensions of representations. I. Automorphic functions*, Ann. of Math. (2) **159** (2004), no. 2, 641–724, DOI 10.4007/annals.2004.159.641. MR2081437

[28] N. N. Lebedev, *Special functions and their applications*, Dover Publications, Inc., New York, 1972. Revised edition, translated from the Russian and edited by Richard A. Silverman; Unabridged and corrected republication. MR350075

[29] E. Nelson, *Construction of quantum fields from Markoff fields*, J. Functional Analysis **12** (1973), 97–112, DOI 10.1016/0022-1236(73)90091-8. MR343815

[30] K.-H. Neeb and G. Ólafsson, *Reflection positivity*, SpringerBriefs in Mathematical Physics, vol. 32, Springer, Cham, 2018. A representation theoretic perspective, DOI 10.1007/978-3-319-94755-6. MR3822306

[31] K.-H. Neeb and G. Ólafsson, *Reflection positivity on spheres*, Anal. Math. Phys. **10** (2020), no. 1, Paper No. 9, 59, DOI 10.1007/s13324-019-00353-3. MR4047208

[32] K.-H. Neeb and G. Ólafsson, *Algebraic quantum field theory and causal symmetric spaces*, Geometric methods in physics XXXIX, Trends Math., Birkhäuser/Springer, Cham, 2023, pp. 207–231, DOI 10.1007/978-3-031-30284-8_20. MR4647403

[33] T. Ōshima and J. Sekiguchi, *Eigenspaces of invariant differential operators on an affine symmetric space*, Invent. Math. **57** (1980), no. 1, 1–81, DOI 10.1007/BF01389818. MR564184

[34] K. Osterwalder and R. Schrader, *Axioms for Euclidean Green's functions*, Comm. Math. Phys. **31** (1973), 83–112. MR329492

[35] K. Osterwalder and R. Schrader, *Axioms for Euclidean Green's functions. II*, Comm. Math. Phys. **42** (1975), 281–305. With an appendix by Stephen Summers. MR376002

[36] G. Ólafsson and I. Sitiraju: *Analytic wavefront sets of spherical distributions on the de Sitter space.* arXiv:2309.10685.

[37] M. J. Radzikowski, *Micro-local approach to the Hadamard condition in quantum field theory on curved space-time*, Comm. Math. Phys. **179** (1996), no. 3, 529–553. MR1400751

[38] A. Strohmaier, R. Verch, and M. Wollenberg, *Microlocal analysis of quantum fields on curved space-times: analytic wave front sets and Reeh-Schlieder theorems*, J. Math. Phys. **43** (2002), no. 11, 5514–5530, DOI 10.1063/1.1506381. MR1936535

[39] R. Verch, *Wavefront sets in algebraic quantum field theory*, Comm. Math. Phys. **205** (1999), no. 2, 337–367, DOI 10.1007/s002200050680. MR1712591

Department of Mathematics, Louisiana State University, Baton Rouge, Louisiana 70803

Email address: isitir1@lsu.edu

q-Extensions

Contemporary Mathematics
Volume **819**, 2025
https://doi.org/10.1090/conm/819/16393

The surprising first q-Appell function

George E. Andrews

ABSTRACT. Attention is drawn to the unique and surprising reduction formula for the first q-Appell function and the equivalent q-integral formula for the same function. The history of this phenomenon is presented and a combinatorial proof of a special case is given.

1. Introduction

The function to be studied in this article is the first q-Appell function

$$\Phi^{(1)}(a;b,b';c;x,y) := \sum_{m=0}^{\infty}\sum_{n=0}^{\infty}\frac{(a;q)_{m+n}(b;q)_m(b';q)_n x^m y^n}{(c;q)_{m+n}(q;q)_m(q;q)_n}, \tag{1.1}$$

where

$$(a;q)_N := \prod_{j=0}^{\infty}\frac{1-aq^j}{1-aq^{j+N}} = \prod_{m=0}^{N-1}(1-aq^m) \tag{1.2}$$

with N is a non-negative integer, and

$$(a_1,a_2,\ldots,a_r;q)_N = \prod_{i=1}^{n}(a_i;q)_N. \tag{1.3}$$

We may take the limit as $N\to\infty$, in which case we require $|q|<1$ and we write $(a;q)_\infty$.

The full domain of analyticity for $\Phi^{(1)}$ is considered extensively in [**4**]. For our purposes in this paper it suffices to require all the parameters and variables to be in absolute value $\leq 1-\varepsilon$. This guarantees absolute convergence of all series and products being considered.

F. H. Jackson in 1942, 1944 [**6**], [**7**] was the first to study the q-Appell functions. In [**6**, p. 81], he states a q-integral for the $\Phi^{(1)}$ and provides instructions for how to produce the proof. His result is as follows: Define [**5**, p. 19, eq. (1.11.1)]

$$\int_0^1 f(t,q)\,d_q(t) := (1-q)\sum_{n=0}^{\infty} f(q^n)q^n. \tag{1.4}$$

2020 *Mathematics Subject Classification.* Primary 33D70; Secondary 33C65, 05A18.

The author was partially supported by Simons Foundation Grant 633284.

Then Jackson asserted (in slightly different notation) that

$$\Phi^{(1)}[q^a; q^b, q^{b'}; q^c; x, y] = \frac{\Gamma_q(c)\Gamma_q(1)}{\Gamma_q(c-a)\Gamma_q(a)} \int_0^1 \frac{(uq;q)_{c-a-1}}{(xu;q)_b(yu;q)_{b'}} d_q u, \tag{1.5}$$

where

$$\Gamma_q(z) := \frac{(q;q)_\infty}{(q^z;q)_\infty}(1-q)^{1-z}. \tag{1.6}$$

In 1972, it was proved [**1**] that

$$\Phi^{(1)}[a; b, b'; c; x, y] = \frac{(a, bx, b'y; q)_\infty}{(c, x, y; q)_\infty} {}_3\Phi_2 \begin{bmatrix} c/a, x, y & ; q, a \\ bx, b'y & \end{bmatrix}, \tag{1.7}$$

where

$${}_3\Phi_2 \begin{bmatrix} A, B, C & ; q, z \\ D, E & \end{bmatrix} := \sum_{n=0}^{\infty} \frac{(A, B, C; q)_n z^n}{(q, D, E; q)_n}. \tag{1.8}$$

Identity (1.7) can actually be deduced from Jackson's integral (1.5). Thomas Ernst first noted this in print [**4**, Th. 10.4.1]. Indeed, (1.5) and (1.7) are effectively the same assertion.

Identity (1.7) is, at first glance, somewhat surprising because there is no reduction of the ordinary Appell function $F^{(1)}$ to a ${}_3F_2$ (see [**3**] for details). However, the $F^{(1)}$ has an integral representation [**8**, p. 215, eq. (8.2.5)] of which (1.5) is the exact q-analog. Thus, in light of this observation, (1.7) loses some of its magic.

In Scction 2 we provide a short recap of this history. In Section 3, we provide a combinatorial proof of (1.7) when $a \approx 0$.

2. History

First we recapitulate the proof of (1.7) from [**1**];

$$\begin{aligned}
\Phi^{(1)}[a; b, b'; c; x, y] &= \sum_{m=0}^{\infty}\sum_{n=0}^{\infty} \frac{(a;q)_{m+n}(b;q)_m(b';q)_n x^m y^n}{(c;q)_{m+1}(q;q)_m(q;q)_n} \\
&= \frac{(a;q)_\infty}{(c;q)_\infty} \sum_{m=0}^{\infty}\sum_{n=0}^{\infty} \frac{(b;q)_m(b';q)_n x^m y^n}{(q;q)_m(q;q)_n} \frac{(cq^{m+n};q)_\infty}{(aq^{m+n};q)_\infty} \\
&= \frac{(a;q)_\infty}{(c;q)_\infty} \sum_{m=0}^{\infty}\sum_{n=0}^{\infty} \frac{(b;q)_m(b';q)_n x^m y^n}{(q;q)_m(q;q)_n} \sum_{s=0}^{\infty} \frac{(c/a;q)_s a^s q^{s(m+n)}}{(q;q)_s}
\end{aligned}$$

(by [**5**, p. 7, eq. (1.3.2)])

$$= \frac{(a;q)_\infty}{(c;q)_\infty} \sum_{s=0}^{\infty} \frac{(c/a;q)_s a^s}{(q;q)_s} \frac{(bxq^s;q)_\infty}{(xq^s;q)_\infty} \frac{(b'yq^s;q)_\infty}{(xq^s;q)_\infty(yq^s;q)_\infty}$$

(by [**5**, p. 7, eq. (1.3.2)])

$$= \frac{(a, bx, b'y; q)_\infty}{(c, x, y; q)_\infty} \sum_{s=0}^{\infty} \frac{(c/a, x, y; q)_s a^s}{(q, bx, b'y; q)_s}$$

which proves (1.7).

We now recapitulate the proof of (1.5) as a rewriting of (1.7) following [**3**]. To begin we replace a, b, b', and c by $q^a, q^b, q^{b'}$, and q^c respectively. Hence by (1.8),

$$\begin{aligned}\Phi^{(1)}[q^a; q^b, q^{b'}, q^c; x, y] &= \frac{(q^a; xq^b, yq^{b'}; q)_\infty}{(q^c; x, y; q)_\infty} \sum_{s=0}^{\infty} \frac{(q^{c-a}; q)_s (x; q)_s (y; q)_s q^{as}}{(q; q)_s (xq^b; q)_s (yq^{b'}; q)_s} \\ &= \frac{(q^a, q^{c-a}; q)_\infty}{(q; q)_\infty (q^s; q)_\infty} \sum_{s=0}^{\infty} \frac{(q^{n+1}; q)_{c-a-1} q^{(a-1)s} q^s}{(xq^s; q)_b (yq^s; q)_{b'}} \\ &= \frac{\Gamma_q(c)\Gamma_q(1)}{\Gamma_q(c-a)\Gamma_q(a)} \int_0^1 \frac{(uq; q)_{c-a-1} u^{a-1} d_q u}{(xu; q)_b (yu; q)_{b'}}.\end{aligned}$$

3. Combinatorics

The charm of $\Phi^{(1)}$ suggested by the results recreated in Section 2 suggests that further study is merited. To provide evidence in this direction we provide a combinatorial proof of (1.7) when $a = 0$. We replace b, b', c, x, and y by $-bq, -b'q, -cq, xq$, and yq respectively:

(3.1)
$$\begin{aligned}&\sum_{m=0}^{\infty}\sum_{n=0}^{\infty} \frac{(-bq; q)_m (-b'q; q)_n (-cq^{m+n+1}; q)_\infty x^m q^m y^n q^n}{(q; q)_m (q; q)_n} \\ &\quad = \sum_{s=0}^{\infty} \frac{c^s q^{\binom{s+1}{2}} (-bxq^{s+1}; q)_\infty (-b'yq^{s+2})_\infty}{(q; q)_s (xq^{s+1}; q)_\infty (yq^{s+1}; q)_\infty}.\end{aligned}$$

We analyse the above identity piece by piece. I.e., the summand on each side is split into three parts generating a component of the tripartite partition. First we focus on

$$\frac{(-bq; q)_m x^m q^m}{(q; q)_m}.$$

This is the generating function for overpartitions (cf. [**2**]) in which the largest is m (and at least one of the appearances of m is not overlined). The exponent of x counts the largest part. The exponent on b counts the number of overlined parts. For example, consider the overpartition

$$5 + \overline{5} + 4 + 4 + 3 + 3 + \overline{3} + 1$$

whose Ferrers graph is

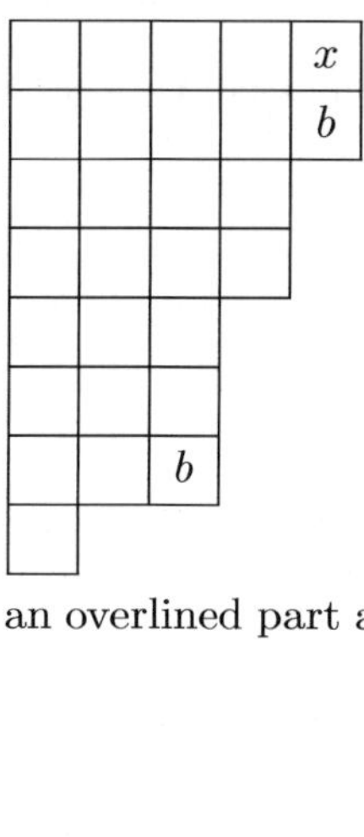

We have placed a b to designate an overlined part and an x to designate the largest part which is counted by x.

Of course, this same evaluation applies to

$$\frac{(-b'q;q)_n y^n q^n}{(q;q)_n}$$

with b, x, and m replaced by b', y, and n respectively.

The one remaining piece on the left-hand side is

$$(-cq^{m+n+1};q)_\infty,$$

since this is the generating function for ordinary partitions into distinct parts each $> m+n$.

We now move to the right-hand side. First, we focus on

$$\frac{c^s q^{\binom{s+1}{2}}}{(q;q)_s} = c^s \sum_{k_1,\ldots,k_s \geq 1} q^{k_1+2k_2+\cdots+sk_s}.$$

The term $c^s q^N$ is counting the number of partitions of N into parts $\leq s$ with each such part occurring at least once.

This is the generating function for partitions into exactly s distinct parts, where the power of c is counting the number of parts, and conjugation reveals that this is the generating function for partitions into parts $\leq s$ with each part occurring at least once with the power of c counting the largest part. Next,

$$\frac{(-bxq^{s+2};q)_\infty}{(xq^{s+1};q)_\infty}$$

is the generating function for overpartitions where all parts are $> s$ and only parts $> s+1$ may be overlined.

As before,

$$\frac{(-b'yq^{s+2};q)_\infty}{(yq^{s+1};q)_\infty}$$

is the same with b' replacing b and y replacing x.

Now we consider a Ferrers graph type representation for the collection of partitions on each side.

For the left-hand side we have the outline of a generic Ferrers graph as follows where the parts are read by rows:

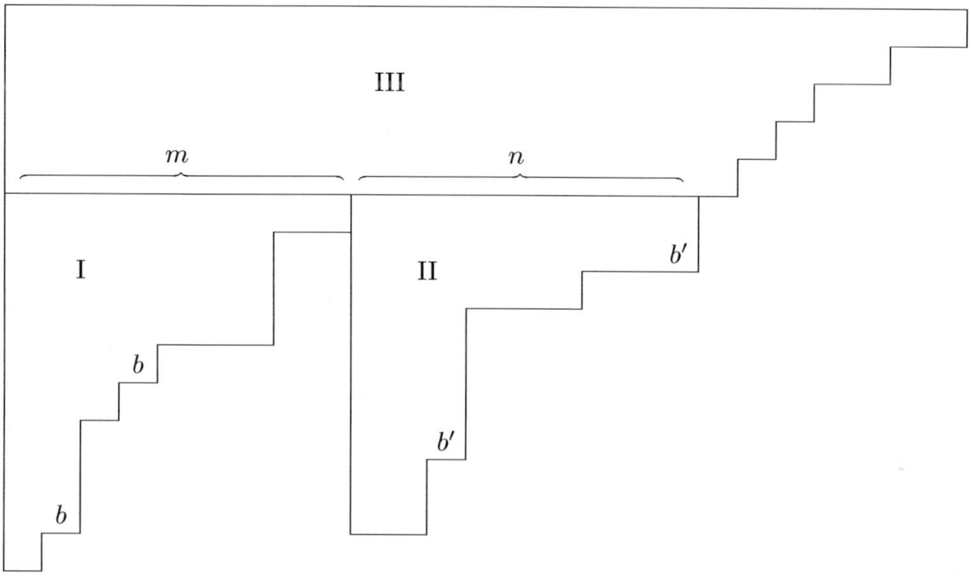

The three sets of partitions generated by the (m, n)-term on the left side appear as regions I, II, and III, respectively.

We note that the top row of both I and II does not have an x at the end (even though an instance of the largest part may be overlined if it is repeated).

Now draw the same Ferrers graph equivalently as indicated below and read it by columns:

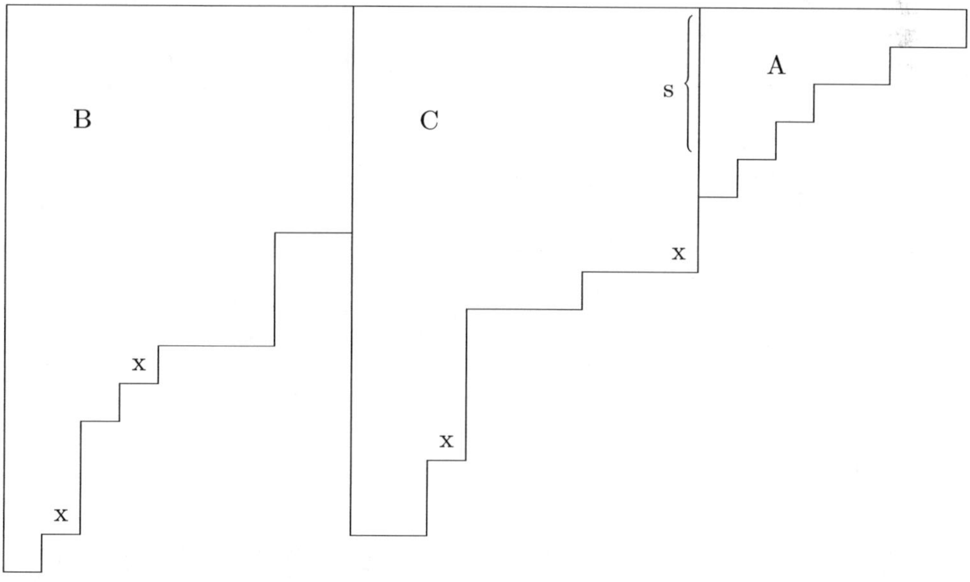

Here the three sets of partitions related to the right-hand side are given by the regions A, B, and C respectively.

We now note that we have a bijection between the partition triples generated by the left-hand side of (3.1) and those generated by the right-hand side of (3.1). To see this, we observe that the coefficient of $b^u b'^v c^s x^m y^n q^N$ in the summand on the left of (3.1) is equal to the number of ways N can be written as a sum of three (over)partitions of type I, II, III, where overpartitions of type I have largest part m, have u overlined parts, and have at least one part m which is not overlined, where overpartitions of type II have largest part n, have v overlined parts, and have at least part n which is not overlined, and where partitions of type III have partitions into s distinct parts, each $> m + n$.

Equivalently, the coefficient of $b^u b'^v c^s x^m y^n q^N$ in the summand on the left of (3.1) is equal to the number of ways that N can be written as a sum of three (over)partitions of type B, C, A, where overpartitions of type B have largest part m, have u overlined parts, and have at least $s+1$ parts m which are not overlined, where overpartitions of type C have largest part n, have v overlined parts, and have at least $s+1$ parts n which are not overlined, and where partitions of type III have partitions into s distinct parts.

Then observe that the coefficient of $b^u b'^v c^s x^m y^n q^N$ in the summand on the right of (3.1) is equal to the number of ways that N can be written as a sum of three (over)partitions of type B, C, A, where overpartitions of type B have m parts which are all $> s$ among which u overlined parts which are all $> s+1$, where overpartitions of type C have n parts which are all $> s$ among which v overlined parts which are all $> s+1$, and where partitions of type A have all parts $\leq s$ and each such part occurring at least once.

Thus the validity of (3.1), and consequently, (1.7) when $a = 0$ is established.

4. Conclusion

The purpose of this article is two-fold. First, we wish to draw attention to the surprising $\Phi^{(1)}$ whose charms are buried in the literature. Second, we hoped to suggest that this surprising function has substantial combinatorial interest. In particular, where $c = bb'$ or $a = q$, there are numerous identities associated with the ${}_3\Phi_2$ [**5**, Appendix III]. Each of these may have a combinatorial proof. Also, identities such as [**1**, Th. 2]

$$\Phi^{(1)}[b'/x; b, b'; bb'; x, y] = \frac{(bx, b', b'y/x; q)_\infty}{(bb', x, y; q)_\infty}$$

are just waiting to be explored in the world of partitions.

Acknowledgments

I thank Thomas Ernst for his extensive advice on this article.

References

[1] G. E. Andrews, *Summations and transformations for basic Appell series*, J. London Math. Soc. (2) **4** (1972), 618–622, DOI 10.1112/jlms/s2-4.4.618. MR306559

[2] S. Corteel and J. Lovejoy, *Overpartitions*, Trans. Amer. Math. Soc. **356** (2004), no. 4, 1623–1635, DOI 10.1090/S0002-9947-03-03328-2. MR2034322

[3] T. Ernst, A Comprehensive Treatment of q-Calculus, Birkhäuser, Basel (2012)

[4] T. Ernst, *Convergence aspects for q-Appell functions I*, J. Indian Math. Soc. (N.S.) **81** (2014), no. 1-2, 67–77. MR3157019

[5] G. Gasper and M. Rahman, Basic Hypergeometric Series, Cambridge Univ. Press, Cambridge, 2nd ed. (2004).

[6] F. H. Jackson, *On basic double hypergeometric functions*, Quart. J. Math. Oxford Ser. **13** (1942), 69–82, DOI 10.1093/qmath/os-13.1.69. MR7453
[7] F. H. Jackson, Basic double hypergeometric functions, Quart. J. Math. **15** (1944), 49–61.
[8] L. J. Slater, *Generalized Hypergeometric Functions.* Cambridge Univ. Press, Cambridge (1966).

Department of Mathematics, The Pennsylvania State University, University Park, Pennsylvania 16802

Email address: `geal@psu.edu`

Contemporary Mathematics
Volume **819**, 2025
https://doi.org/10.1090/conm/819/16394

Ramanujan's q-continued fractions

Gaurav Bhatnagar

Dedicated to the memory of Dick Askey

ABSTRACT. Ramanujan's q-continued fractions are a central part of Ramanujan's development of basic hypergeometric series. They appear in Chapter 16 of Part III and Chapter 32 of Part V of *Ramanujan's Notebooks* edited by Berndt, and in Volume I of Andrews and Berndt's *Ramanujan's Lost Notebook*. In these references the continued fractions are presented in the order in which they appear in Ramanujan's original notebooks. We summarize the work of several authors on this topic and re-organize Ramanujan's q-continued fractions.

1. Introduction

Ramanujan considered the q-analogue of the well-known Fibonacci continued fraction

$$\cfrac{1}{1+\cfrac{1}{1+\cfrac{1}{1+\cdots}}}$$

given by

$$\cfrac{1}{1+\cfrac{q}{1+\cfrac{q^2}{1+\cfrac{q^3}{1+\cdots}}}}, \tag{1.1a}$$

2020 *Mathematics Subject Classification.* Primary 33D15; Secondary 30B70.
Key words and phrases. q-continued fractions, Ramanujan's second notebook, Lost notebook.

and showed that for $|q| < 1$, this continued fraction can be written as a ratio of two very similar sums:

$$\frac{\displaystyle\sum_{k=0}^{\infty} \frac{q^{k^2+k}}{(1-q)(1-q^2)\cdots(1-q^k)}}{\displaystyle\sum_{k=0}^{\infty} \frac{q^{k^2}}{(1-q)(1-q^2)\cdots(1-q^k)}}. \tag{1.1b}$$

The continued fraction (1.1a) is called the Rogers–Ramanujan continued fraction. In view of the Rogers–Ramanujan identities, this ratio can be expressed as a ratio of infinite products

$$\frac{(1-q)(1-q^6)(1-q^{11})\cdots}{(1-q^2)(1-q^7)(1-q^{12})\cdots} \times \frac{(1-q^4)(1-q^9)(1-q^{14})\cdots}{(1-q^3)(1-q^8)(1-q^{13})\cdots}. \tag{1.1c}$$

It is clear that Ramanujan considered his work on this continued fraction as one of the highlights of his work. Consider the last two entries of Chapter 16 of Ramanujan's second notebook [**48**]. Entry 38 concerns the continued fraction above and the Rogers–Ramanujan identities. Entry 39 computes special cases (see Berndt [**12**, pp. 77–86]). As a corollary, Ramanujan obtains

$$\cfrac{1}{1+\cfrac{e^{-2\pi}}{1+\cfrac{e^{-4\pi}}{1+\cfrac{e^{-6\pi}}{1+\cdots}}}} = \left(\sqrt{\frac{5+\sqrt{5}}{2}} - \frac{\sqrt{5}+1}{2}\right) e^{2\pi/5} \tag{1.2a}$$

and

$$\cfrac{1}{1-\cfrac{e^{-\pi}}{1+\cfrac{e^{-2\pi}}{1-\cfrac{e^{-3\pi}}{1+\cdots}}}} = \left(\sqrt{\frac{5-\sqrt{5}}{2}} - \frac{\sqrt{5}-1}{2}\right) e^{\pi/5}. \tag{1.2b}$$

Askey [**10**] suggested that Ramanujan's q-extension (1.1a) was the primary motivation for Ramanujan to discover the Rogers–Ramanujan identities, and indeed for Ramanujan's development of basic hypergeometric series. In support of this hypothesis, we note that Ramanujan ended his development of basic hypergeometric series with (1.2a) and (1.2b). Thus q-continued fractions may be quite central to Ramanujan's thought process.

Ramanujan himself included these two continued fractions in his first letter [**15**, p. 29] to Hardy in 1913; this is additional evidence that these were results Ramanujan liked and thought likely to impress Hardy.

And impress him, they did. About these, and another formula related to this continued fraction, Hardy [**31**, p. 9] wrote

> (they) defeated me completely; I had never seen anything in the least like them before. A single look at them is enough to show that they could only be written by a mathematician of the highest class.

The objective of this article is to summarize the work of many mathematicians, whose work has made it easier to understand and organize Ramanujan's continued fractions. The three formulas (1.1a)–(1.1c) serve as a model for this organization.

As it turns out, there are only five general results. In all of these a ratio of two series is expressed as a continued fraction. The remaining continued fractions are special cases of these. These general results are given in §§2, 3 and 6. In some cases, the same ratio of two series is expressed as two or three continued fractions; thus, they are transformation formulas too. We note some special cases in §4. There are some special cases where the ratio reduces to a single series, and many where they reduce to a ratio of infinite products (see §5). When this happens, such as in the case of the Rogers–Ramanujan continued fraction, Ramanujan computes special values. In this article, we list all of Ramanujan's q-continued fractions. (Formulas for the computation of special values have not been considered here.)

Apart from this organization, we are interested in the ideas involved in the proofs. A variety of clever ideas are required to prove Ramanujan's q-continued fractions; we present a sample, giving details when possible, and references otherwise. We conclude in §7 with some remarks arising from our study, and pointers to recent work in this area.

Notation. The *q-rising factorial* is defined as $(a;q)_0 := 1$,

$$(a;q)_n := (1-a)(1-aq)\cdots(1-aq^{n-1}) \text{ for } n = 1,2,\ldots;$$

and, for $|q| < 1$,

$$(a;q)_\infty := \prod_{k=0}^{\infty}(1-aq^k).$$

In addition, we use the short-hand notation

$$(a_1, a_2, \ldots, a_r; q)_k := (a_1;q)_k(a_2;q)_k\cdots(a_r;q)_k.$$

We will require the q-binomial coefficient, defined as

$$\begin{bmatrix} n \\ k \end{bmatrix}_q := \frac{(q;q)_n}{(q;q)_k(q;q)_{n-k}},$$

where $n \geqslant k$ are non-negative integers. When $n < k$, we take $\begin{bmatrix} n \\ k \end{bmatrix}_q = 0$.

Ramanujan's entries. We record the notation used to refer to Ramanujan's entries from [**48, 49**]. An entry takes its number from either one of the five volumes of Ramanujan's Notebooks edited by Berndt, or the Lost Notebook edited by Andrews and Berndt. We have prefixed the volume number and/or the chapter to the entry. For example, Entry III.16.39 refers to Entry 39, of Chapter 16 in Part III of Berndt's five volumes on *Ramanujan's Notebooks*; that is, it refers to [**12**, Ch. 16, Entry 39]. We place an additional L when referring to the Lost Notebook. Thus, L.I.6.2.1 refers to Entry 6.2.1 in Part I of the *Ramanujan's Lost Notebook* edited by Andrews and Berndt, that is, [**6**, Entry 6.2.1].

2. A more general continued fraction

In his second letter to Hardy, Ramanujan [**15**, p. 57] mentions that Entry III.16.29 (i) is "a particular case of a theorem on the continued fraction

$$\cfrac{1}{1+\cfrac{ax}{1+\cfrac{ax^2}{1+\cfrac{ax^3}{1+\cfrac{ax^4}{1+\cfrac{ax^5}{1+\cdots}}}}}},$$

which is a particular case of the continued fraction

$$\cfrac{1}{1+\cfrac{ax}{1+bx+\cfrac{ax^2}{1+bx^2+\cfrac{ax^3}{1+bx^3+\cdots}}}},$$

which is a particular case of a general theorem on continued fractions." A theorem regarding this continued fraction involves the sum

$$g(b,\lambda) := \sum_{k=0}^{\infty} \frac{\lambda^k q^{k^2}}{(q,-bq;q)_k}. \tag{2.1}$$

For the expressions corresponding to (2.1) in the letter, Ramanujan had x in place of q, and a in place of λ.

Ramanujan records the following continued fractions in the Lost Notebook [**49**, p. 40]

L.I.6.3.1 (i), (ii), (iii). Let $|q| < 1$. Then

$$\frac{g(b,\lambda q)}{g(b,\lambda)} = \cfrac{1}{1+\cfrac{\lambda q}{1+\cfrac{bq+\lambda q^2}{1+\cfrac{\lambda q^3}{1+\cfrac{bq^2+\lambda q^4}{1+\cdots}}}}} \tag{2.2a}$$

$$= \cfrac{1}{1+\cfrac{\lambda q}{1+bq+\cfrac{\lambda q^2}{1+bq^2+\cfrac{\lambda q^3}{1+bq^3+\cfrac{\lambda q^4}{1+bq^4+\cdots}}}}}. \tag{2.2b}$$

Let $|q| < 1$ and $|b| < 1$. Then

$$\frac{g(b,\lambda q)}{g(b,\lambda)} = \cfrac{1}{1-b+\cfrac{b+\lambda q}{1-b+\cfrac{b+\lambda q^2}{1-b+\cfrac{b+\lambda q^3}{1-b+\cfrac{b+\lambda q^4}{1-b+\cdots}}}}}. \tag{2.2c}$$

REMARK 2.1. A continued fraction of Touchard [**56**] can be considered to be a special case of (2.2c); see Prodinger [**44**]. The first two continued fractions can be obtained as special cases of the general continued fractions in §3. We prove the third.

PROOF OF (2.2c). We present J. Cigler's proof (sent by email, Sept. 2012), which illustrates an idea of Askey [**10**] regarding a discovery approach to the Rogers–Ramanujan identities beginning with (1.1a) (see also [**18**]).

Let $F(b, \lambda)$ denote the continued fraction (2.2c). Then

$$F(b,\lambda) = \frac{1}{1-b+(b+\lambda q)F(b,\lambda q)}.$$

To change this non-linear equation into a linear one, we substitute

$$F(b,\lambda) = \frac{g(b,\lambda q)}{g(b,\lambda)}$$

to obtain Ramanujan's recurrence relation given in Entry L.I.6.3.1(iv):

$$g(b,\lambda) = (1-b)g(b,\lambda q) + (b+\lambda q)g(b,\lambda q^2).$$

Now assume that

$$g(b,\lambda)=\sum_{n=0}^{\infty}a_n\lambda^n,$$

substitute in Ramanujan's recurrence relation, and compare coefficients of λ^n, to obtain

$$a_n=(1-b)q^na_n+bq^{2n}a_n+q^{2n-1}a_{n-1},$$

or

$$a_n=\frac{q^{2n-1}}{(1-q^n)(1+bq^n)}a_{n-1}.$$

Iterating this we obtain

$$a_n=\frac{q^{n^2}}{(q,-bq;q)_n}a_0.$$

Now taking $a_0=1$, we obtain the expression (2.1) for $g(b,\lambda)$ and a (formal) proof of (2.2c). □

The continued fraction (2.2b) appeared earlier as Entry III.16.15. When $b=0$ this reduces to a relation for an extension of the Rogers–Ramanujan continued fraction.

Corollary to III.16.15. If $|q|<1$, then

$$\frac{\displaystyle\sum_{k=0}^{\infty}\frac{\lambda^k q^{k^2+k}}{(q;q)_k}}{\displaystyle\sum_{k=0}^{\infty}\frac{\lambda^k q^{k^2}}{(q;q)_k}}=\cfrac{1}{1+\cfrac{\lambda q}{1+\cfrac{\lambda q^2}{1+\cfrac{\lambda q^3}{1+\cfrac{\lambda q^4}{1+\cdots}}}}}.$$

When $\lambda=1$, this reduces to the Rogers–Ramanujan continued fraction. In the very next entry Ramanujan gives a formula for the convergents of (the reciprocal of) this continued fraction.

III.16.16.

$$\frac{\mu_n(0)}{\mu_n(1)}=1+\cfrac{\lambda q}{1+\cfrac{\lambda q^2}{1+\cfrac{\lambda q^3}{\ddots+\cfrac{\lambda q^n}{1}}}},\tag{2.3}$$

where $\mu_n(s)$ is defined as

$$\mu_n(s):=\sum_{k=0}^{\infty}q^{k^2+sk}\lambda^k\begin{bmatrix}n-k-s+1\\k\end{bmatrix}_q$$

for $s=0,1,2,\ldots$.

REMARKS 2.2.

(1) Hirschhorn [**32–34**] proves Ramanujan's continued fractions by finding analogous formulas for the convergents of continued fractions. See also [**12, 20**].
(2) Proofs appear in [**12, 32, 43**]. This formula was rediscovered by Menon [**43**].
(3) In our proof we illustrate a method due to Euler [**26**] (see [**17, 19**]), that works on all the general q-continued fractions mentioned in this article. Euler's approach is based on using the elementary identity
$$\frac{N}{D} = 1 + \frac{N-D}{D}$$
to 'divide' two series, both of whose first terms are 1.

PROOF OF III.16.16. Note that the sum $\mu_n(s)$ is finite; the index k goes from 0 to $\lfloor \frac{n-s+1}{2} \rfloor$. We will show

$$\frac{\mu_n(s)}{\mu_n(s+1)} = 1 + \frac{\lambda q^{s+1}}{\dfrac{\mu_n(s+1)}{\mu_n(s+2)}}, \tag{2.4}$$

for $s = 0, 1, 2, 3, \ldots, n-1$. Formula (2.3) follows by taking $s = 0, 1, \ldots, n-1$ in turn. Observe that the iteration stops when $s = n-1$, because
$$\mu_n(n) = 1 = \mu_n(n+1).$$

To prove (2.4), note that
$$\begin{aligned}
\frac{\mu_n(s)}{\mu_n(s+1)} &= 1 + \frac{\mu_n(s) - \mu_n(s+1)}{\mu_n(s+1)} \\
&= 1 + \frac{1}{\mu_n(s+1)} \\
&\quad \times \sum_{k=0}^{\infty} \frac{q^{k^2+sk}\lambda^k (q;q)_{n-k-s}}{(q;q)_k (q;q)_{n-2k-s+1}} \left[(1 - q^{n-k-s+1}) - (1 - q^{n-2k-s+1})q^k\right] \\
&= 1 + \frac{1}{\mu_n(s+1)} \sum_{k=1}^{\infty} \frac{q^{k^2+sk}\lambda^k}{(q;q)_{k-1}} \frac{(q;q)_{n-k-s}}{(q;q)_{n-2k-s+1}} \\
&= 1 + \frac{\lambda q^{s+1}}{\mu_n(s+1)} \sum_{k=0}^{\infty} \frac{q^{k^2+(s+2)k)}\lambda^k}{(q;q)_k} \frac{(q;q)_{n-k-s-1}}{(q;q)_{n-2k-s-1}},
\end{aligned}$$
on shifting the sum. From this (2.4) follows and the proof is complete. □

III.16.13 and Corollary 6.2.4 to L.I.6.2.1 (Eisenstein). If $|q| < 1$, then
$$\sum_{k=0}^{\infty} (-a)^k q^{\frac{k(k+1)}{2}} = \cfrac{1}{1 + \cfrac{aq}{1 + \cfrac{a(q^2-q)}{1 + \cfrac{aq^3}{1 + \cfrac{a(q^4-q^2)}{1 + \cdots}}}}}.$$

REMARK 2.3. Eisenstein's continued fraction appears in [**25**, pp. 35–39]. The case $a = 1$ was known to Gauss in 1797; see [**6**, p. 152]. See [**4, 34, 46**] for more proofs. See Folsom [**27**] for a history, and an account of Eisenstein's continued fractions.

PROOF OUTLINE. The special case $\lambda = a$ and $b = -a$ of (2.2a) can be written as Entry III.16.13, where to obtain the sum on the left hand side, we have to apply the transformation formula III.16.9 to the sums.

$$(aq;q)_\infty \sum_{k=0}^{\infty} \frac{q^{k^2} b^k}{(q, aq; q)_k} = \sum_{k=0}^{\infty} (-1)^k q^{\frac{k(k+1)}{2}} \frac{(b/a;q)_k}{(q;q)_k} a^k. \qquad \square$$

Ramanujan (see [**12**]) noted formulas for the denominators of the convergents of this continued fraction. They are given by

$$D_{2n} = \sum_{k=0}^{n} a^k q^{nk} \begin{bmatrix} n \\ k \end{bmatrix}_q$$

and

$$D_{2n+1} = \sum_{k=0}^{n} a^k q^{(n+1)k} \begin{bmatrix} n \\ k \end{bmatrix}_q .$$

3. Still more general

Some candidates for what Ramanujan may have meant by the phrase 'general theorem on continued fractions' in his letter, concern the sum

$$G(a,b,\lambda) := \sum_{k=0}^{\infty} \frac{(-\lambda/a;q)_k \, a^k \, q^{(k^2+k)/2}}{(q, -bq; q)_k}. \tag{3.1}$$

However, only special cases of these theorems appear in Ramanujan's notebooks, which is the only record of his discoveries before he wrote the aforementioned letter to Hardy. In the Lost Notebook, Ramanujan gave the following continued fractions and derived numerous special cases.

L.I.6.2.1 and L.I.6.4.1. Let a, b, λ, and q be complex numbers, with $|q| < 1$. Then

$$\frac{G(aq,b,\lambda q)}{G(a,b,\lambda)} = \cfrac{1}{1+\cfrac{aq+\lambda q}{1+\cfrac{bq+\lambda q^2}{1+\cfrac{aq^2+\lambda q^3}{1+\cfrac{bq^2+\lambda q^4}{1+\cdots}}}}} \tag{3.2a}$$

$$= \cfrac{1}{1+aq+\cfrac{\lambda q - abq^2}{1+aq^2+bq+\cfrac{\lambda q^2 - abq^4}{1+aq^3+bq^2+\cfrac{\lambda q^3-abq^6}{1+aq^4+bq^3+\cdots}}}}. \tag{3.2b}$$

REMARKS 3.1.

(1) The continued fraction in (3.2b), Entry L.I.6.4.1, appears a few pages after (3.2a) in the Lost Notebook [**49**], as a transformation formula given by the second equality above.

(2) Note that if we take $a = 0$ in (3.2a) we obtain (2.2a); and taking $a = 0$ in (3.2b) we obtain (2.2b). But there is no general continued fraction in Ramanujan's notebooks for $G(aq, b, \lambda q)/G(a, b, \lambda)$ which reduces to (2.2c). A candidate for such a continued fraction is [**6**, Theorem 6.4.1]

$$\frac{G(aq,b,\lambda q)}{G(a,b,\lambda)} = \cfrac{1}{1+\cfrac{aq+\lambda q}{1-aq+bq+\cfrac{aq+\lambda q^2}{1-aq+bq^2+\cfrac{aq+\lambda q^3}{1-aq+bq^3+\cdots}}}}, \tag{3.2c}$$

where $|aq| < 1$. When $b = 0$ and $a \mapsto b/q$, this reduces to (2.2c). This continued fraction appears in different forms in Hirschhorn [**34**] and Bhargava and Adiga [**16**]. We need to appeal to Entry III.16.8 in order to match the two forms.

(3) In addition to (3.2a), (3.2b), and (3.2c), there are two more continued fractions for $G(aq, b, \lambda q)/G(a, b, \lambda)$ in [**17**] (see also [**39**]).

(4) Another candidate for a 'general theorem on continued fractions' is Andrews [**3**, Theorem 6]. The most general continued fraction of this type is due to Andrews and Bowman [**9**]; see also Gupta and Masson [**30**].

ON PROOFS OF (3.2a)–(3.2c). Andrews [**4**] included a proof of (3.2a) in his introductory article reporting his discovery of the Lost Notebook. Bhargava and Adiga [**16**] proved (3.2b) and (3.2c). See [**6, 17, 34**] for more proofs. The proofs involve finding recurrence relations. □

There are several special cases of these continued fractions noted by Ramanujan. In each of the following, we assume the conditions of L.I.6.2.1 (if applicable).

L.I.6.2.3.

$$\frac{G(a,b,0)}{G(aq,b,0)} = 1 + \cfrac{aq}{1+\cfrac{bq}{1+\cfrac{aq^2}{1+\cfrac{bq^2}{1+\cdots}}}}.$$

PROOF. Set $\lambda = 0$ in (3.2a) and take reciprocals. □

Corollary 6.2.9 to L.I.6.2.1.

$$\sum_{k=0}^{\infty}(-1)^k q^{3k^2+2k}\left(1+q^{2k+1}\right) = \cfrac{1}{1+\cfrac{q^2-q}{1+\cfrac{q^4-q^2}{1+\cfrac{q^6-q^3}{1+\cfrac{q^8-q^4}{1+\cdots}}}}}.$$

PROOF. We take $q \mapsto q^2$ and $a = -1/q$, $b = -1$, and $\lambda = 1$ in (3.2a). The sum side requires an iterate of Heine's transformation formula, and L.I.9.5.1, see [**6**, p. 155]. Heine's formula is Entry III.16.6. □

Corollary 6.2.11 to Entry L.I.6.2.1.

$$1-\sum_{k=1}^{\infty} q^{k(3k-1)/2}\left(1-q^k\right) = \cfrac{2}{2+\cfrac{q+q}{1+\cfrac{q^2+q^3}{1+\cfrac{q^3+q^5}{1+\cfrac{q^4+q^7}{1+\cdots}}}}}.$$

PROOF. Take $q \mapsto q^2$, $a = 1/q$, $b = 1$, and $\lambda = 1/q$, in (3.2a). The sum side requires elementary series manipulations, a special case of the q-binomial theorem (III.16.2), the second iterate of Heine's transformation ([**28**, eq. (III.3)]) and Entry L.I.9.4.7. See [**6**, p. 156] for the details. □

There are further special cases in the next two sections.

4. Transformations of continued fractions

In §§2 and 3, we saw that a ratio of two series can be written in multiple ways as continued fractions. In this section we note two transformation formulas recorded by Ramanujan, obtained by setting the same special case of two of these continued fractions equal to one another. In addition, there is one transformation which follows by taking the odd part of a continued fraction.

L.I.6.5.1. Let $k \geqslant 0$, $\alpha = (1+\sqrt{1+4k})/2$, and $\beta = (-1+\sqrt{1+4k})/2$. Then for $|q| < 1$,

$$\cfrac{1}{1+\cfrac{k+q}{1+\cfrac{k+q^2}{1+\cfrac{k+q^3}{1+\cdots}}}} \tag{4.1a}$$

$$= \cfrac{1}{\alpha + \cfrac{q}{\alpha + \beta q + \cfrac{q^2}{\alpha + \beta q^2 + \cfrac{q^3}{\alpha + \beta q^3 + \cdots}}}}. \tag{4.1b}$$

PROOF. Observe that $\alpha\beta = k$ and $\alpha - \beta = 1$. To show this transformation formula we take a special case $\lambda = 1/\alpha^2$, $b = \beta/\alpha$ in (2.2b) and (2.2c) and equate the two resulting continued fractions. After substituting in (2.2b), we multiply and divide each fraction by α, and simplify to obtain α times (4.1b). Similarly, we obtain α times (4.1a) from (2.2c). □

L.I.6.5.2. For $|q| < 1$,

$$\cfrac{1}{1 + \cfrac{2+q}{1 + \cfrac{2+q^2}{1 + \cfrac{2+q^3}{1 + \cdots}}}} = \cfrac{1}{2 + \cfrac{q}{2 + q + \cfrac{q^2}{2 + q^2 + \cfrac{q^3}{2 + q^3 + \cdots}}}}.$$

PROOF. This is a special case of Entry L.I.6.5.1, where $k = 2$, so that $\alpha = 2$ and $\beta = 1$. □

L.I.6.4.2. If $|q| < 1$, then

$$\cfrac{1}{a + c - \cfrac{ab}{a + b + cq - \cfrac{ab}{a + b + cq^2 - \cfrac{ab}{a + b + cq^3 - \cdots}}}}$$

$$= \cfrac{1}{c - b + a + \cfrac{bc}{c - b + a/q + \cfrac{bc}{c - b + a/q^2 + \cfrac{bc}{c - b + a/q^3 + \cdots}}}}.$$

PROOF. Take $\lambda = 0$, $a \mapsto -b/aq$, $b \mapsto c/a$ in (3.2c) and (3.2b), and compare the resulting continued fractions. The transformation formula is obtained after some simplification; see [**6**, p. 161] for more details. □

L.I.6.4.3. For $q \neq 0$, n a positive integer, we have

$$
1+\cfrac{a}{1+\cfrac{b}{q+\cfrac{a}{1+\cfrac{b}{q^2+\cfrac{a}{\ddots+\cfrac{b}{q^n+\cfrac{a}{1}}}}}}}
\tag{4.2}
$$

$$
=1+a-\cfrac{ab}{a+b+q-\cfrac{ab}{a+b+q^2-\cfrac{ab}{\ddots-\cfrac{ab}{a+b+q^n}}}}.
$$

PROOF. The proof is elementary. The continued fraction on the left hand side of (4.2) is of the form $1 + N_{2n+1}/D_{2n+1}$, where

$$
\frac{N_{2n+1}}{D_{2n+1}} = \cfrac{a_1}{b_1+\cfrac{a_2}{b_2+\cfrac{a_3}{\ddots+\cfrac{a_{2n+1}}{b_{2n+1}}}}}.
$$

We consider instead the continued fraction with convergents N_k/D_k. The continued fraction on the right hand side is obtained by adding 1 to the continued fraction with convergents N_1/D_1, N_3/D_3, ..., N_{2n+1}/D_{2n+1}. This is known as the odd part of the continued fraction.

The details are as follows. An equivalent form of Ramanujan's Entry II.12.1 is that N_k and D_k are both determined by a three-term recurrence relation of the form

$$
P_k = b_k P_{k-1} + a_k P_{k-2}
$$

for $k = 3, 4, \ldots, 2n+1$, with initial values $N_1 = a_1$, $D_1 = b_1$, $N_2 = a_1 b_1$ and $D_2 = a_2 + b_1 b_2$. It is easy to see that N_{2k+1} and D_{2k+1} (for $k = 2, 3, \ldots, n$) also satisfy the same recurrence relation with $a_k \mapsto -a_{2k-1}a_{2k}b_{2k+1}/b_{2k-1}$, and $b_k \mapsto a_{2k+1} + b_{2k}b_{2k+1} + a_{2k}b_{2k+1}/b_{2k-1}$. Now we substitute the values of a_k and b_k from the left hand side of (4.2), and find the corresponding recurrence for the odd part of the continued fraction. On adding 1 to the corresponding continued fraction, we obtain the right hand side of (4.2). □

5. Continued fractions written as products

So far we have seen continued fractions written as a ratio of two series. Many of Ramanujan's q-continued fractions are ratios of infinite products. It turns out that all of these are obtained when the series in question can be written as infinite products. An example is the Rogers–Ramanujan continued fraction highlighted in the introduction, which has the three expressions given by formulas (1.1a), (1.1b) and (1.1c). In this section, we list such special cases of the general continued fractions given so far. In all the formulas in this section, we assume $|q| < 1$.

Corollary 6.2.2 to L.I.6.2.1. For any complex number $a \neq -q^{2n+1}$, $n \geqslant 1$,

$$\frac{(-aq^2;q^2)_\infty}{(-aq;q^2)_\infty^2} = \cfrac{1}{1+\cfrac{aq}{1+\cfrac{q+aq^2}{1+\cfrac{aq^3}{1+\cfrac{q^2+aq^4}{1+\cdots}}}}}. \tag{5.1}$$

PROOF. Take $a = 0$, $b = 1$ and $\lambda \mapsto a$ in (3.2a). This requires the q-binomial theorem (Entry III.16.2) to sum both the sums. □

The next two continued fractions are further special cases of this continued fraction.

V.32.21; Corollary 6.2.1 to L.I.6.2.1.

$$\frac{(-q^2;q^2)_\infty}{(-q;q^2)_\infty} = \frac{(q;q^2)_\infty}{(q^2;q^4)_\infty^2} = \cfrac{1}{1+\cfrac{q}{1+\cfrac{q+q^2}{1+\cfrac{q^3}{1+\cfrac{q^2+q^4}{1+\cdots}}}}}.$$

PROOF. Take $a = 1$ in (5.1). □

Corollary 6.2.10 to L.I.6.2.1.

$$\frac{(-q^3;q^4)_\infty}{(-q;q^4)_\infty} = \cfrac{1}{1+\cfrac{q}{1+\cfrac{q^2+q^3}{1+\cfrac{q^5}{1+\cfrac{q^4+q^7}{1+\cdots}}}}}. \tag{5.2}$$

PROOF. Take $q \mapsto q^2$ and $a = 1/q$ in (5.1). □

Another continued fraction for the LHS of (5.2) is as follows.

V.32.20.

$$\frac{(q^3;q^4)_\infty}{(q;q^4)_\infty} = \cfrac{1}{1-\cfrac{q}{1+q^2-\cfrac{q^3}{1+q^4-\cfrac{q^5}{1+q^6-\cfrac{q^7}{1+q^8-\cdots}}}}}.$$

PROOF. This is obtained from (2.2b) by taking $q \mapsto q^2$, $\lambda = -1/q$, $b = 1$. The sums are special case of the q-binomial theorem (the case $a \mapsto 0$ of Entry III.16.2, with $q \mapsto q^4$). □

Next, we have two more special cases of (3.2a).

Corollary 6.2.7 to L.I.6.2.1; V.32.18.

$$\frac{(q,q^5;q^6)_\infty}{(q^3;q^6)_\infty^2} = \frac{(q;q^2)_\infty}{(q^3;q^6)_\infty^3} = \cfrac{1}{1+\cfrac{q+q^2}{1+\cfrac{q^2+q^4}{1+\cfrac{q^3+q^6}{1+\cfrac{q^4+q^8}{1+\cdots}}}}}.$$

REMARK 5.1. This continued fraction is known as Ramanujan's cubic continued fraction. See Chan [**24**] and [**6**, Chapter 3].

PROOF. Take $q \mapsto q^2$, and take $a = 1/q$, $b = 1$, and $\lambda = 1$ in (3.2a). To obtain the product side, we require some identities of Slater [**53**, **54**]. For the details, see [**6**, p. 154]. □

V.32.22; V.32.23; Corollary 6.2.8 to L.I.6.2.1.

$$\frac{(q,q^7;q^8)_\infty}{(q^3,q^5;q^8)_\infty} = \cfrac{1}{1+\cfrac{q+q^2}{1+\cfrac{q^4}{1+\cfrac{q^3+q^6}{1+\cfrac{q^8}{1+\cdots}}}}}.$$

PROOF. Take $q \mapsto q^2$, and take $a = 1/q$, $b = 0$, and $\lambda = 1$ in (3.2a). To obtain the product side, we require some identities of Slater [**54**]. See [**6**, p. 154] for more details. This continued fraction is referred to as the Ramanujan–Göllnitz–Gordon continued fraction. □

V.32.19.

$$\frac{(q^2;q^3)_\infty}{(q;q^3)_\infty} = \cfrac{1}{1 - \cfrac{q}{1+q - \cfrac{q^3}{1+q^2 - \cfrac{q^5}{1+q^3 - \cfrac{q^7}{1+q^4 - \cdots}}}}}. \tag{5.3}$$

REMARK 5.2. The product side of (5.3) and a related continued fraction appeared in an intriguing claim of Ramanujan concerning a continued fraction that converges to three different limits. This claim and related ideas have been explored by Andrews, Berndt, Sohn, Yee and Zaharescu [**7**, **8**] and Ismail and Stanton [**36**].

PROOF. The continued fraction in this entry can be found from (3.2b). Let $\omega = e^{2\pi i/3}$. Then take $\lambda = 0$, $a = -\omega/q$, $b = -\omega^2$, to obtain the continued fraction

$$\cfrac{1}{1-\omega - \cfrac{q}{1+q - \cfrac{q^3}{1+q^2 - \cdots}}}.$$

from which (5.3) can be obtained by taking reciprocals and adding ω. A proof along these lines can be found in Berndt [**13**, p. 46]. See also [**7**]. See the recurrence relation for the function $G_2(s)$ in [**17**, p. 64] for the connection with (3.2b). □

Note that the proofs of the last three entries require results that are not explicitly stated in Ramanujan's notebooks. In fact, these are the only three entries in this article with this property.

6. Entries III.16.11 and III.16.12

The first two q-continued fractions in Chapter 16 of Ramanujan's second notebook are Entries III.16.11 and III.16.12. Entry III.16.11 is motivated by a continued fraction recorded in Chapter 12 of Ramanujan's second notebook (see also [**48**, Chapter 14, Vol. 1]). In Entry II.12.18, Ramanujan finds a continued fraction for

$$\frac{(x+1)^n-(x-1)^n}{(x+1)^n+(x-1)^n}.$$

Ramanujan notes four corollaries: continued fractions for $\tan^{-1} x$, $\log \frac{1+x}{1-x}$, $\tan x$, $\frac{e^x-1}{e^x+1}$, so it would be natural to attempt an extension of II.14.18.

We can expand $(x+1)^n$ and $(x-1)^n$ using the binomial theorem. If n is a non-negative integer, then the numerator becomes a polynomial with terms of the form $a_k x^k$, where $k = n-1, n-3, \ldots$; and the denominator has terms with $k = n, n-2, n-4, \ldots$.

Now the q-analog of the binomial theorem, given in Entry III.16.2, is

$$F(a,b) := \sum_{k=0}^{\infty} \frac{(b/a;q)_k}{(q;q)_k} a^k = \frac{(b;q)_\infty}{(a;q)_\infty}.$$

If we consider

$$\frac{F(a,b)-F(-a,-b)}{F(a,b)+F(-a,-b)}$$

we obtain

$$\frac{\displaystyle\sum_{k=0}^{\infty}\frac{(b/a;q)_{2k+1}}{(q;q)_{2k+1}}a^{2k+1}}{\displaystyle\sum_{k=0}^{\infty}\frac{(b/a;q)_{2k}}{(q;q)_{2k}}a^{2k}} = \frac{(-a,b;q)_\infty-(a,-b;q)_\infty}{(-a,b;q)_\infty+(a,-b;q)_\infty}.$$

The numerator of the left hand side contains the odd terms of the q-binomial sum, and the denominator contains the even terms. The right hand appears in the following entry.

III.16.11. Let $|q|<1$ and $|a|<1$. Then we have

$$\frac{(-a,b;q)_\infty-(a,-b;q)_\infty}{(-a,b;q)_\infty+(a,-b;q)_\infty} = \cfrac{a-b}{1-q+\cfrac{(a-bq)(aq-b)}{1-q^3+\cfrac{q(a-bq^2)(aq^2-b)}{1-q^5+\cfrac{q(a-bq^3)(aq^3-b)}{1-q^7+\cdots}}}}.$$

PROOF. For a proof, see Berndt [**12**], which relies on ideas of Jacobsen [**37**] to clarify an earlier proof by Adiga, Berndt, Bhargava and Watson [**1**]. □

Entry III.16.12 has a very similar continued fraction.

III.16.12. Let $|q|<1$, and $|ab|<1$. Then we have

$$\frac{(a^2q^3,b^2q^3;q^4)_\infty}{(a^2q,b^2q;q^4)_\infty} = \cfrac{1}{1-ab+\cfrac{(a-bq)(b-aq)}{(1-ab)(1+q^2)+\cfrac{(a-bq^3)(b-aq^3)}{(1-ab)(1+q^4)+\cfrac{(a-bq^5)(b-aq^5)}{(1-ab)(1+q^6)+\cdots}}}}.$$

REMARKS 6.1.

(1) Entry III.16.12 is a q-analogue of Entry II.12.25. To see this, one has to use III.16.1(ii), which contains the definition of the q-gamma function.

(2) The continued fraction in Entry III.16.12 converges for $|ab|>1$. It also converges for $|ab|<1$ and $|q|>1$. The expression for both of these can be found using III.16.12; see [**12,19**].

ON PROOFS OF III.16.12. The first proof of III.16.12 was given by Adiga, Berndt, Bhargava and Watson [**1**] which was further clarified by Jacobsen [**37**]; see [**12**]. See also Ramanathan [**47**]. For two inter-related proofs, see the work of Ismail and the author [**19**]. One of these proofs uses Euler's approach; and the other, a standard method from the theory of orthogonal polynomials. □

This continued fraction also arises from a ratio of two sums. From the q-binomial theorem, it follows that for $|q| < 1$ and $|a| < 1$:

$$\frac{\left(a^2q^3, b^2q^3; q^4\right)_\infty}{\left(a^2q, b^2q; q^4\right)_\infty} = \frac{\displaystyle\sum_{k=0}^{\infty} \frac{\left((bq/a)^2; q^4\right)_k}{(q^4;q^4)_k}(a^2q)^k}{\displaystyle\sum_{k=0}^{\infty} \frac{\left((b/aq)^2; q^4\right)_k}{(q^4;q^4)_k}(a^2q^3)^k}.$$

7. The proofs of Ramanujan's q-continued fractions

The first proofs of many of Ramanujan's general q-continued fractions were given by Andrews [**4, 5**] and Adiga, Berndt, Bhargava and Watson [**1**]; proofs are collated in [**6**] and [**12**]. Earlier, Selberg [**52**] obtained many Rogers–Ramanujan type continued fractions, where the continued fractions are expressed as infinite products. Reuter [**51**] has surveyed proofs of Ramanujan's hypergeometric continued fractions.

As we have seen, an eclectic collection of techniques is used to prove Ramanujan's q-continued fractions. As Berndt [**14**] wrote in a similar context:

> Methods for proving these continued fraction formulas are varied and at times ad hoc. Ramanujan evidently had a systematic procedure for proving these continued fraction formulas, but we don't know what it is.

Nevertheless, when it comes to q-continued fractions, there are several interesting systematic approaches. As we have seen, all of Ramanujan's general continued fractions (in §§2, 3, 6), can be obtained as a ratio of two similar series; the remaining results are obtained as special cases. In fact, Ramanathan [**47**] derived all of these general continued fractions formally by finding three-term recurrence relations from contiguous relations for ${}_2\phi_1$ series.

The general continued fractions make sense as formal power series. On examining the ratio of series, it is obvious what the indeterminate should be. For continued fractions involving $g(b, \lambda)$ (defined by (2.1)), it is λ; for $G(a, b, \lambda)$, and the continued fractions in §6, a is more suitable. On occasion, a suitable power series variable needs to be inserted in a continued fraction to make formal power series arguments work; in this direction; see the continued fractions considered by Sokal [**55**].

There is, however, a gap in this formal approach to continued fractions, especially when one needs to transform the series, or write the answer in terms of infinite products. In that case, one needs to know that the continued fraction converges; and, if so, whether it converges to the ratio of series in question. Jacobsen [**37**] has clarified such issues (see [**41**] for a comprehensive exposition).

Another way convergence has been addressed is by associating the continued fraction with orthogonal polynomials; see Ismail and Stanton [**36**]. However, we require the parameters to be real (rather than complex) numbers in this approach. Continued fractions are closely related to the study of orthogonal polynomials (see, for example, Askey and Ismail [**11**]). Al-Salam and Ismail [**2**] have studied orthogonal polynomials related to Ramanujan's continued fractions; see also [**20, 35**]. Again, a suitable parameter may be needed to act as the indeterminate for the polynomials. A third approach to show convergence is to use a theorem of Pincherle

[**38**, p. 164], see, for examples, Masson [**42**] and Gupta and Masson [**30**]. With regard to questions of convergence of Ramanujan's continued fractions, see also [**7, 8, 17, 21, 22, 36**].

In addition to Ramanathan [**47**], Lorentzen (née Jacobsen) [**40**] suggested another systematic approach to Ramanujan's (hypergeometric) continued fractions. Yet another approach, following Euler's elementary idea, works on all of Ramanujan's general q-continued fractions; see [**17, 19**]. Sokal [**55**] has developed Euler's idea further and given several examples. Hirschhorn [**32–34**] proves many of Ramanujan's continued fractions by finding analogous formulas for the convergents of continued fractions (see also Menon [**43**], and [**20, 23, 45**].)

Regarding transformation formulas for continued fractions, Andrews and Berndt [**6**] used the Bauer–Muir transformation [**41**, pp. 76–80] to prove Entry L.I.6.5.1. Lee, Mc Laughlin and Sohn [**39**] have applied this idea to Ramanujan's transformation formulas. By contrast, [**17**] explains the transformation of continued fractions (3.2b) by transforming the series appearing on the left-hand side of (3.2a), and then obtaining the corresponding continued fractions by Euler's method.

As we have seen, taking special cases of Ramanujan's general continued fractions involves the application of other transformation and summation formulas. In particular, in §5 we saw that the Rogers–Ramanujan type identities are very useful in expressing continued fractions as infinite products. The first paper with continued fractions of this kind was written by Selberg [**52**]. Further examples appear in Gu and Prodinger [**29**].

This prolific set of ideas surrounding Ramanujan's continued fractions is only to be expected. After all, as Hardy [**50**] famously remarked

> (Ramanujan's) mastery of continued fractions was, on the formal side at any rate, beyond that of any mathematician in the world...

Acknowledgments

A shorter version of this article is due to appear in "Ramanujan: His Life, Legacy, and Mathematical Influence," a forthcoming volume to be published by Springer. We thank Krishnaswami Alladi, George Andrews, Bruce Berndt, Peter Paule, Ole Warnaar, and Ae Ja Yee for their helpful comments. In addition, we thank Mourad Ismail, Lisa Lorentzen, Krishnan Rajkumar, and Alan Sokal, for many conversations on this subject over many years.

References

[1] C. Adiga, B. C. Berndt, S. Bhargava, and G. N. Watson, *Chapter* 16 *of Ramanujan's second notebook: theta-functions and q-series*, Mem. Amer. Math. Soc. **53** (1985), no. 315, v+85, DOI 10.1090/memo/0315. MR776347

[2] W. A. Al-Salam and M. E. H. Ismail, *Orthogonal polynomials associated with the Rogers-Ramanujan continued fraction*, Pacific J. Math. **104** (1983), no. 2, 269–283. MR684290

[3] G. E. Andrews, *On q-difference equations for certain well-poised basic hypergeometric series*, Quart. J. Math. Oxford Ser. (2) **19** (1968), 433–447, DOI 10.1093/qmath/19.1.433. MR237831

[4] G. E. Andrews, *An introduction to Ramanujan's "lost" notebook*, Amer. Math. Monthly **86** (1979), no. 2, 89–108, DOI 10.2307/2321943. MR520571

[5] G. E. Andrews, *Ramunujan's "lost" notebook. III. The Rogers-Ramanujan continued fraction*, Adv. in Math. **41** (1981), no. 2, 186–208, DOI 10.1016/0001-8708(81)90015-3. MR625893

[6] G. E. Andrews and B. C. Berndt, *Ramanujan's lost notebook. Part I*, Springer, New York, 2005. MR2135178

[7] G. E. Andrews, B. C. Berndt, J. Sohn, A. J. Yee, and A. Zaharescu, *On Ramanujan's continued fraction for* $(q^2;q^3)_\infty/(q;q^3)_\infty$, Trans. Amer. Math. Soc. **355** (2003), no. 6, 2397–2411, DOI 10.1090/S0002-9947-02-03155-0. MR1973995

[8] G. E. Andrews, B. C. Berndt, J. Sohn, A. J. Yee, and A. Zaharescu, *Continued fractions with three limit points*, Adv. Math. **192** (2005), no. 2, 231–258, DOI 10.1016/j.aim.2004.04.004. MR2128700

[9] G. E. Andrews and D. Bowman, *A full extension of the Rogers-Ramanujan continued fraction*, Proc. Amer. Math. Soc. **123** (1995), no. 11, 3343–3350, DOI 10.2307/2161075. MR1277090

[10] R. Askey, *Ramanujan and hypergeometric and basic hypergeometric series [MR1117465 (92k:01019)]*, Ramanujan: essays and surveys, Hist. Math., vol. 22, Amer. Math. Soc., Providence, RI, 2001, pp. 277–324, DOI 10.1090/hmath/022/27. MR1862756

[11] R. Askey and M. Ismail, *Recurrence relations, continued fractions, and orthogonal polynomials*, Mem. Amer. Math. Soc. **49** (1984), no. 300, iv+108, DOI 10.1090/memo/0300. MR743545

[12] B. C. Berndt, *Ramanujan's notebooks. Part III*, Springer-Verlag, New York, 1991, DOI 10.1007/978-1-4612-0965-2. MR1117903

[13] B. C. Berndt, *Ramanujan's notebooks. Part V*, Springer-Verlag, New York, 1998, DOI 10.1007/978-1-4612-1624-7. MR1486573

[14] B. C. Berndt, *What is a q-series?*, Ramanujan rediscovered, Ramanujan Math. Soc. Lect. Notes Ser., vol. 14, Ramanujan Math. Soc., Mysore, 2010, pp. 31–51. MR2856954

[15] B. C. Berndt and R. A. Rankin, *Ramanujan: Letters and commentary*, History of Mathematics, vol. 9, American Mathematical Society, Providence, RI; London Mathematical Society, London, 1995, DOI 10.1090/hmath/009. MR1353909

[16] S. Bhargava and C. Adiga, *On some continued fraction identities of Srinivasa Ramanujan*, Proc. Amer. Math. Soc. **92** (1984), no. 1, 13–18, DOI 10.2307/2045144. MR749881

[17] G. Bhatnagar, *How to prove Ramanujan's q-continued fractions*, Ramanujan 125, Contemp. Math., vol. 627, Amer. Math. Soc., Providence, RI, 2014, pp. 49–68, DOI 10.1090/conm/627/12532. MR3307491

[18] ———, *How to discover the Rogers–Ramanujan identities*, Resonance. **20** (2015), no. 5, 416–430.

[19] G. Bhatnagar and M. E. H. Ismail, *On Entry II.16.12: a continued fraction of Ramanujan*, Int. J. Number Theory **17** (2021), no. 2, 251–266, DOI 10.1142/S1793042120400072. MR4235432

[20] G. Bhatnagar and M. E. H. Ismail, *Orthogonal polynomials associated with a continued fraction of Hirschhorn*, Ramanujan J. **61** (2023), no. 2, 487–514, DOI 10.1007/s11139-021-00528-2. MR4588605

[21] D. Bowman and J. Mc Laughlin, *On the divergence of the Rogers-Ramanujan continued fraction on the unit circle*, Trans. Amer. Math. Soc. **356** (2004), no. 8, 3325–3347, DOI 10.1090/S0002-9947-03-03390-7. MR2052952

[22] D. Bowman and J. Mc Laughlin, *Continued fractions with multiple limits*, Adv. Math. **210** (2007), no. 2, 578–606, DOI 10.1016/j.aim.2006.07.004. MR2303233

[23] D. Bowman, J. Mc Laughlin, and N. J. Wyshinski, *A q-continued fraction*, Int. J. Number Theory **2** (2006), no. 4, 523–547, DOI 10.1142/S179304210600070X. MR2281862

[24] H. H. Chan, *On Ramanujan's cubic continued fraction*, Acta Arith. **73** (1995), no. 4, 343–355, DOI 10.4064/aa-73-4-343-355. MR1366039

[25] G. Eisenstein, *Mathematische Werke. Band I*, Chelsea Publishing Co., New York, 1975. MR427029

[26] L. Euler, *De transformatione seriei divergentis* $1-mx+m(m+n)x^2-m(m+n)(m+2n)x^3+m(m+n)(m+2n)(m+3n)x^4+$ *etc. in fractionem continuam*, Nova Acta Academiae Scientarum Imperialis Petropolitinae **2** (1788), 36–45, Reprinted in Opera Omnia: Series 1, Volume 16, pp. 34–46; English translation (by C. E. Sandifer) available at `http://www.math.dartmouth.edu/~euler/pages/E616.html`.

[27] A. Folsom, *Modular forms and Eisenstein's continued fractions*, J. Number Theory **117** (2006), no. 2, 279–291, DOI 10.1016/j.jnt.2005.06.001. MR2213765

[28] G. Gasper and M. Rahman, *Basic hypergeometric series*, 2nd ed., Encyclopedia of Mathematics and its Applications, vol. 96, Cambridge University Press, Cambridge, 2004. With a foreword by Richard Askey, DOI 10.1017/CBO9780511526251. MR2128719

[29] N. S. S. Gu and H. Prodinger, *On some continued fraction expansions of the Rogers-Ramanujan type*, Ramanujan J. **26** (2011), no. 3, 323–367, DOI 10.1007/s11139-011-9329-7. MR2860692

[30] D. P. Gupta and D. R. Masson, *On the Andrews-Bowman continued fraction: Dedicated to Richard A. Askey on the occasion of his 65th birthday, Part II*, Methods Appl. Anal. **6** (1999), no. 2, 175–194, DOI 10.4310/MAA.1999.v6.n2.a4. MR1803889

[31] G. H. Hardy, *Ramanujan: twelve lectures on subjects suggested by his life and work*, Chelsea Publishing Co., New York, 1959. MR0106147

[32] M. D. Hirschhorn, *Partitions and Ramanujan's continued fraction*, Duke Math. J. **39** (1972), 789–791. MR314771

[33] M. D. Hirschhorn, *A continued fraction*, Duke Math. J. **41** (1974), 27–33. MR337746

[34] M. D. Hirschhorn, *A continued fraction of Ramanujan*, J. Austral. Math. Soc. Ser. A **29** (1980), no. 1, 80–86. MR566278

[35] M. E. H. Ismail and D. Stanton, *Classical orthogonal polynomials as moments*, Canad. J. Math. **49** (1997), no. 3, 520–542, DOI 10.4153/CJM-1997-024-9. MR1451259

[36] M. E. H. Ismail and D. Stanton, *Ramanujan continued fractions via orthogonal polynomials*, Adv. Math. **203** (2006), no. 1, 170–193, DOI 10.1016/j.aim.2005.04.006. MR2231045

[37] L. Jacobsen, *Domains of validity for some of Ramanujan's continued fraction formulas*, J. Math. Anal. Appl. **143** (1989), no. 2, 412–437, DOI 10.1016/0022-247X(89)90049-8. MR1022544

[38] W. B. Jones and W. J. Thron, *Continued fractions: Analytic theory and applications*, Encyclopedia of Mathematics and its Applications, vol. 11, Addison-Wesley Publishing Co., Reading, MA, 1980. With a foreword by Felix E. Browder and an introduction by Peter Henrici. MR595864

[39] J. Lee, J. Mc Laughlin, and J. Sohn, *Applications of the Heine and Bauer-Muir transformations to Rogers-Ramanujan type continued fractions*, J. Math. Anal. Appl. **447** (2017), no. 2, 1126–1141, DOI 10.1016/j.jmaa.2016.10.052. MR3573136

[40] L. Lorentzen, *An idea on some of Ramanujan's continued fraction identities*, Ramanujan J. **17** (2008), no. 3, 369–385, DOI 10.1007/s11139-007-9057-1. MR2456840

[41] L. Lorentzen and H. Waadeland, *Continued fractions with applications*, Studies in Computational Mathematics, vol. 3, North-Holland Publishing Co., Amsterdam, 1992. MR1172520

[42] D. R. Masson, *Some continued fractions of Ramanujan and Meixner-Pollaczek polynomials*, Canad. Math. Bull. **32** (1989), no. 2, 177–181, DOI 10.4153/CMB-1989-026-1. MR1006743

[43] P. K. Menon, *On Ramanujan's continued fraction and related identities*, J. London Math. Soc. **40** (1965), 49–54, DOI 10.1112/jlms/s1-40.1.49. MR172840

[44] H. Prodinger, *On Touchard's continued fraction and extensions: combinatorics-free, self-contained proofs*, Quaest. Math. **35** (2012), no. 4, 431–445, DOI 10.2989/16073606.2012.742240. MR2999999

[45] H. Prodinger, *Finite Rogers-Ramanujan type continued fractions*, J. Algebra Comb. Discrete Struct. Appl. **5** (2018), no. 3, 137–142, DOI 10.13069/jacodesmath.451218. MR3864313

[46] K. G. Ramanathan, *Ramanujan's continued fraction*, Indian J. Pure Appl. Math. **16** (1985), no. 7, 695–724. MR801801

[47] K. G. Ramanathan, *Hypergeometric series and continued fractions*, Proc. Indian Acad. Sci. Math. Sci. **97** (1987), no. 1-3, 277–296 (1988), DOI 10.1007/BF02837830. MR983621

[48] S. Ramanujan, *Notebooks of Srinivasa Ramanujan, Volumes 1, 2*, Tata Institute of Fundamental Research, Bombay, 1957. Faithful digitized colored presentation in his own handwriting, 2012. MR0099904

[49] S. Ramanujan, *The lost notebook and other unpublished papers*, Springer-Verlag, Berlin; Narosa Publishing House, New Delhi, 1988. With an introduction by George E. Andrews. MR0947735

[50] S. Ramanujan, *Collected papers of Srinivasa Ramanujan*, AMS Chelsea Publishing, Providence, RI, 2000. Edited by G. H. Hardy, P. V. Seshu Aiyar and B. M. Wilson; Third printing of the 1927 original; With a new preface and commentary by Bruce C. Berndt. MR2280843

[51] V. J. Reuter, *Hypergeometric functions, continued fractions for products of gamma functions, and q-analogues*, ProQuest LLC, Ann Arbor, MI, 2014. Thesis (Ph.D.)–University of Illinois at Urbana-Champaign. MR3407394

[52] A. Selberg, *Über einige arithmetische identitäten*, Avhandlinger utgitt av Det Norske Videnskaps-Akademi i Oslo. I. Matematisk-Naturvidenskapelig Klasse **no. 8** (1936), 1–23.

[53] L. J. Slater, *A new proof of Rogers's transformations of infinite series*, Proc. London Math. Soc. (2) **53** (1951), 460–475, DOI 10.1112/plms/s2-53.6.460. MR43235

[54] L. J. Slater, *Further identities of the Rogers-Ramanujan type*, Proc. London Math. Soc. (2) **54** (1952), 147–167, DOI 10.1112/plms/s2-54.2.147. MR49225

[55] A. D. Sokal, *A simple algorithm for expanding a power series as a continued fraction*, Expo. Math. **41** (2023), no. 2, 245–287, DOI 10.1016/j.exmath.2022.12.001. MR4602842

[56] J. Touchard, *Sur un problème de configurations et sur les fractions continues* (French), Canad. J. Math. **4** (1952), 2–25, DOI 10.4153/cjm-1952-001-8. MR46325

Ashoka University, Sonipat, Haryana 131029, India
Email address: bhatnagarg@gmail.com
URL: https://gauravbhatnagar.org

Contemporary Mathematics
Volume **819**, 2025
https://doi.org/10.1090/conm/819/16395

Orthogonality of the big -1 Jacobi polynomials for non-standard parameters

Howard S. Cohl and Roberto S. Costas-Santos

ABSTRACT. The big -1 Jacobi polynomials $(Q_n^{(0)}(x;\alpha,\beta,c))_n$ have been classically defined for $\alpha,\beta\in(-1,\infty)$, $c\in(-1,1)$. We extend this family so that wider parameter values are allowed, i.e., the parameters may be non-standard. Assuming initial conditions $Q_0^{(0)}(x)=1$, $Q_{-1}^{(0)}(x)=0$, we consider the big -1 Jacobi polynomials as monic orthogonal polynomials which satisfy the three-term recurrence relation

$$xQ_n^{(0)}(x)=Q_{n+1}^{(0)}(x)+b_nQ_n^{(0)}(x)+u_nQ_{n-1}^{(0)}(x),\quad n=0,1,2,\ldots.$$

For standard parameters, the coefficients u_n are positive for all n. We discuss the situation when Favard's theorem cannot be directly applied, as there is some positive integer n such that $u_n=0$. We express the big -1 Jacobi polynomials for non-standard parameters as a product of two polynomials. Using this factorization, we obtain a bilinear form with respect to which these polynomials are orthogonal.

1. Introduction

In 2011, Vinet and Zhedanov in [**19**] obtained a new family of "classical" orthogonal polynomials which can be obtained from the little q-Jacobi polynomials in the limit $q\to-1$. By "classical" it is meant that these polynomials satisfy (in addition to a three-term recurrence relation) a nontrivial eigenvalue equation of the form

$$\mathcal{L}(p_n(x))=\lambda_np_n(x),$$

where $\mathcal{L}$ is a linear differential–difference operator which is of first order in the derivative ∂_x, and contains a reflection operator R which acts as $Rf(x)=f(-x)$. They referred to these polynomials as being in the continuous -1 hypergeometric orthogonal polynomial scheme. This was illustrated by using a limiting process from the Askey–Wilson polynomials to the Bannai–Ito polynomials. In [**20**] the same authors applied this limit by starting with the big q-Jacobi polynomials in order to obtain hypergeometric representations for the big -1 Jacobi polynomials.

The polynomials these authors have considered in this -1 hypergeometric orthogonal polynomial scheme are q-polynomials for standard parameters, i.e., ones for which the weight function related to these families is positive definite (see [**16**]). The main aim of this work is to investigate the property of orthogonality for the big -1 Jacobi polynomials for those values of the parameters, α and β, for which

2020 *Mathematics Subject Classification.* Primary 44A20; Secondary 42C05; 33C05;

the coefficient u_n in the recurrence relation (1.2) relative to these polynomials can be equal to zero for some $n \in \mathbb{N}_0$.

In the last three decades, some of the classical orthogonal polynomials with non-classical parameters have been provided with certain non-standard orthogonality properties. In the pioneering work by Kwon and Littlejohn in 1995 [**10**], Sobolev orthogonality for the Laguerre polynomials $L_n^{(-N)}(x)$ with $N \in \mathbb{N}$ was obtained using the inner product

$$\langle f, g\rangle = \sum_{m=0}^{N-1}\sum_{j=0}^{m} B_{m,j}(N)\left(f^{(m)}(0)g^{(j)}(0) + f^{(j)}(0)g^{(m)}(0)\right) + \int_0^\infty f^{(N)}(x)g^{(N)}(x)\,\mathrm{e}^{-x}\,\mathrm{d}x,$$

where the superscripts represent differentiation and

$$B_{m,j}(N) := \begin{cases} \displaystyle\sum_{p=0}^{j}(-1)^{m+j}\binom{N-1-p}{m-p}\binom{N-1-p}{j-p}, & 0 \leqslant j \leqslant m \leqslant N-1, \\ \displaystyle\frac{1}{2}\sum_{p=0}^{m}\binom{N-1-p}{m-p}^2, & 0 \leqslant j = m \leqslant N-1. \end{cases}$$

In 1998, Álvarez de Morales and co-authors [**3**] found, using a different technique, the orthogonality for ultraspherical polynomials $C_n^{-N+\frac{1}{2}}$ with $N \in \mathbb{N}$. Later, in [**1**], M. Alfaro et al. considered the cases of the Jacobi polynomials in which both parameters, α and β, were negative integers, proving that in such a case the Jacobi polynomials satisfy a Sobolev orthogonality. In [**2**], M. Alfaro et al. considered the situation when the inner product can be written in the form

$$\langle f, g\rangle = F^t AG + \int f^{(N)}(x)g^{(N)}(x)\,\mathrm{d}\mu(x), \tag{1.1}$$

where F and G are vectors obtained by evaluating f and g and maybe their derivatives at some points, A is a symmetric real matrix, and $\mathrm{d}\mu$ is the orthogonality measure associated with the Nth derivative of either the Laguerre, ultraspherical or Jacobi polynomials.

Note that for the parameters considered in this situation there exists an $n = N \in \mathbb{N}_0$ for which the coefficient u_n in the recurrence relation

$$xp_n = p_{n+1} + b_n p_n + u_n p_{n-1}, \qquad n = 0, 1, 2, \ldots \tag{1.2}$$

vanishes, i.e., $u_N = 0$. We say a non-standard parameter (or set of parameters) is degenerate if there exists at least one positive integer N for which $u_N = 0$.

The first term in the inner product (1.1) plays the role of the orthogonality for the set of polynomials of degree less than N, since if f, g are two polynomials with $\deg f$, $\deg g < N$, then $f^{(N)}(x) = g^{(N)}(x) = 0$, and one has

$$\langle f, g\rangle = F^t AG.$$

In this term, the points for the evaluation are the roots of p_N; this ensures that the inner product vanishes when one entry is p_n with $n \geqslant N$, since p_N is a factor of any p_n with $n \geqslant N$. The second term in (1.1) is relevant for polynomials with degree greater than N, and is needed in order to have an orthogonality characterizing the sequence $(p_n)_{n=0}^\infty$. The technique used in [**3**] is applicable to all classical orthogonal polynomials for which u_n vanishes at some $n = N \in \mathbb{N}_0$, and in fact it has been used, among others, in [**12**–**15**].

In 2009 [**5**], Costas-Santos and Sanchez-Lara studied this problem for classical discrete polynomials, i.e., discrete polynomials in the Askey scheme of generalized hypergeometric orthogonal polynomials. We consider classical discrete polynomials for which there exists N for which $u_N = 0$, i.e., for non-standard degenerate parameters. Note that this can only happen with the Racah, Hahn, dual Hahn, and Krawtchouk polynomials, which are the Wilson, continuous Hahn, continuous dual Hahn, and Meixner polynomials for which some parameters are equal to a negative integer. The corresponding three-term recurrence relation for these polynomials also presents one vanishing coefficient, i.e., $u_N = 0$, and the inner product found can be written as

$$\langle f, g\rangle = \int f(x)g(x)\,\mathrm{d}\mu_d(x) + \int f^{(N)}(x)g^{(N)}(x)\,\mathrm{d}\mu_c(x), \tag{1.3}$$

where $\mathrm{d}\mu_d$ is a discrete measure with a finite number of masses and $\mathrm{d}\mu_c$ is an absolutely continuous measure. For more details and further reading see [**5**, **6**, **9**, **17**, **18**] and references therein. The organization of the paper is as follows. In Section 2 we provide some preliminary material; in Section 3 we obtain the property of orthogonality for the big -1 Jacobi polynomials for non-standard degenerate parameters; and in Section 4 we summarize the key findings and contributions of the present study and describe potential future works in the -1 orthogonal polynomial framework.

2. Preliminary material and notations

We adopt the following notations: $\mathbb{N}_0 := \{0\} \cup \mathbb{N} = \{0, 1, 2, \ldots\}$, and we use $\mathbb{P}, \mathbb{P}'$ to represent the linear space of polynomials with complex coefficients and its algebraic dual space. We denote by $\langle \mathbf{u}, p\rangle$ the duality bracket for $\mathbf{u} \in \mathbb{P}'$ and $p \in \mathbb{P}$. The big -1 Jacobi polynomials have representations given in terms of the Gauss hypergeometric function defined as [**7**, (15.2.1)]

$$ {}_2F_1\left(\begin{matrix} a, b \\ c \end{matrix}; z\right) := \sum_{n=0}^{\infty} \frac{(a)_n (b)_n}{(c)_n} \frac{z^n}{n!}, \tag{2.1}$$

where $|z| < 1$ and the shifted factorial is defined as [**7**, (5.2.4-5)] $(a)_n := (a)(a+1)\cdots(a+n-1)$, $a \in \mathbb{C}$, $n \in \mathbb{N}_0$. The shifted factorial is related to the ratio of two gamma functions [**7**, Chapter 5] for $a \in \mathbb{C} \setminus -\mathbb{N}_0$ by

$$(a)_n = \frac{\Gamma(a+n)}{\Gamma(a)}, \tag{2.2}$$

which allows one to extend the definition to non-positive integer values of n. For a description of the properties of the gamma function, see [**7**, Chapter 5].

3. Big −1 Jacobi polynomials

The monic big −1 Jacobi polynomials can be defined in terms of the Gauss hypergeometric function by

(3.1)
$$Q_n^{(0)}(x) := \kappa_n \begin{cases} {}_2F_1\left(\begin{matrix} -\frac{n}{2}, \frac{n+\alpha+\beta+2}{2} \\ \frac{\alpha+1}{2} \end{matrix}; \frac{1-x^2}{1-c^2}\right) \\ \quad + \dfrac{n(1-x)}{(c+1)(\alpha+1)} {}_2F_1\left(\begin{matrix} 1-\frac{n}{2}, \frac{n+\alpha+\beta+2}{2} \\ \frac{\alpha+3}{2} \end{matrix}; \frac{1-x^2}{1-c^2}\right) & \text{if } n \text{ is even,} \\ {}_2F_1\left(\begin{matrix} -\frac{n-1}{2}, \frac{n+\alpha+\beta+1}{2} \\ \frac{\alpha+1}{2} \end{matrix}; \frac{1-x^2}{1-c^2}\right) \\ \quad - \dfrac{(\alpha+\beta+n+1)(1-x)}{(1+c)(\alpha+1)} {}_2F_1\left(\begin{matrix} -\frac{n-1}{2}, \frac{n+\alpha+\beta+3}{2} \\ \frac{\alpha+3}{2} \end{matrix}; \frac{1-x^2}{1-c^2}\right) & \text{if } n \text{ is odd,} \end{cases}$$

where κ_n is defined by

$$\kappa_n := \begin{cases} \dfrac{(1-c^2)^{\frac{1}{2}n}(\frac{1}{2}(\alpha+1))_{\frac{1}{2}n}}{(\frac{1}{2}(n+\alpha+\beta+2))_{\frac{1}{2}n}} & \text{if } n \text{ is even,} \\ (c+1)\dfrac{(1-c^2)^{\frac{1}{2}(n-1)}(\frac{1}{2}(\alpha+1))_{\frac{1}{2}(n-1)}}{(\frac{1}{2}(n+\alpha+\beta+1))_{\frac{1}{2}(n-1)}} & \text{if } n \text{ is odd.} \end{cases} \tag{3.2}$$

The big −1 Jacobi polynomials $Q_n^{(0)}(x;\alpha,\beta,c)$ satisfy the three-term recurrence relation [**20**, (2.20), (2.21)]

$$xQ_n^{(0)}(x) = Q_{n+1}^{(0)}(x) + b_n Q_n^{(0)}(x) + u_n Q_{n-1}^{(0)}(x), \tag{3.3}$$

where

$$b_n := \begin{cases} -c + \dfrac{(c-1)n}{\alpha+\beta+2n} + \dfrac{(c+1)(\beta+n+1)}{\alpha+\beta+2n+2} & \text{if } n \text{ is even,} \\ c - \dfrac{(c-1)(n+1)}{\alpha+\beta+2n+2} - \dfrac{(c+1)(\beta+n)}{\alpha+\beta+2n} & \text{if } n \text{ is odd,} \end{cases} \tag{3.4}$$

and

$$u_n := \begin{cases} \dfrac{(c-1)^2 n(\alpha+\beta+n)}{(\alpha+\beta+2n)^2} & \text{if } n \text{ is even,} \\ \dfrac{(c+1)^2(\alpha+n)(\beta+n)}{(\alpha+\beta+2n)^2} & \text{if } n \text{ is odd,} \end{cases} \tag{3.5}$$

with initial conditions $Q_{-1}^{(0)}(x) = 0$, $Q_0^{(0)}(x) = 1$. For any real $c \neq 1$ and real α, β satisfying the restriction $\alpha, \beta > -1$, the recurrence coefficients b_n, u_n are real and positive. Hence the big −1 Jacobi polynomials are positive definite orthogonal polynomials.

REMARK 3.1. Observe that if α (resp. β) is a negative odd integer, namely $\alpha = -2N+1$ (resp. $\beta = -2N+1$) then $u_{2N-1} = 0$, i.e., α (resp. β) is a non-standard degenerate parameter. So we can apply the degenerate version of Favard's theorem (see [**6**, Theorem 2.2]), i.e., the result that there exist moment linear functionals

$\mathscr{L}_0$ and $\mathscr{L}_N$, such that the polynomial sequence is orthogonal with respect to the bilinear form

$$\langle p,r\rangle = \mathscr{L}_0(pr) + \mathscr{L}_N(\mathscr{T}^{(2N-1)}p\mathscr{T}^{(2N-1)}r), \qquad p,r\in\mathbb{P},$$

where $\mathscr{T}$ is a certain lowering operator.

By taking the former remark into account, the Gauss hypergeometric function with a suitable normalization factorizes as follows.

LEMMA 3.2 (Factorization). *Let $n, N \in \mathbb{N}$, $a, x \in \mathbb{C}$. Then*

$$\begin{aligned}(-N+1)_{n+N}\,{}_2F_1\!\left(\begin{matrix}-n-N,a\\-N+1\end{matrix};x\right)\\ = (N+1)_n(-N+1)_N\,{}_2F_1\!\left(\begin{matrix}-N,a\\-N+1\end{matrix};x\right){}_2F_1\!\left(\begin{matrix}-n,a+N\\N+1\end{matrix};x\right).\end{aligned}\tag{3.6}$$

PROOF. By definition one has

$$\begin{aligned}(-N+1)_{n+N}\,{}_2F_1\!\left(\begin{matrix}-n-N,a\\-N+1\end{matrix};x\right) &= (-N+1)_{n+N}\sum_{k=0}^{n+N}\frac{(-n-N,a)_k}{(-N+1,1)_k}x^k\\ &= \sum_{k=0}^{n+N}\frac{(-n-N,a)_k}{(1)_k}(-N+k+1)_{n+N-k}x^k.\end{aligned}$$

Then setting $k=k+N$ one has

$$\begin{aligned}(-N+1)_{n+N}\,{}_2F_1\!\left(\begin{matrix}-n-N,a\\-N+1\end{matrix};x\right) &= \sum_{k=0}^{n}\frac{(-n-N,a)_{k+N}}{(1)_{k+N}}(k+1)_{n-k}x^{k+N}\\ &= \frac{(-n-N,a)_N n!}{N!}x^N\sum_{k=0}^{n}\frac{(-n,a+N)_k}{(N+1,1)_k}x^k\\ &= (N+1)_n(a)_N(-x)^N\,{}_2F_1\!\left(\begin{matrix}-n,a+N\\N+1\end{matrix};x\right).\end{aligned}\tag{3.7}$$

Setting $n=0$ in (3.7), one obtains

$$(-N+1)_N\,{}_2F_1\!\left(\begin{matrix}-N,a\\-N+1\end{matrix};x\right) = (a)_N(-x)^N.\tag{3.8}$$

Combining (3.7) and (3.8) produces the result. □

Now that we have obtained a factorization of the Gauss hypergeometric function, we will use it to obtain a factorization for the big -1 Jacobi polynomials when the parameter α (resp. β) is equal to $-2N-1$ for some positive integer N.

LEMMA 3.3 (Factorization of big -1 Jacobi polynomials). *Let $m, N \in \mathbb{N}_0$, $\alpha,\beta,c\in\mathbb{C}$, $c\neq\pm1$, with $\alpha=-2N-1$ or $\beta=-2N-1$. Then*

$$Q_{2N+1}^{(0)}(x;-2N-1,\beta,c) = (x^2-1)^N(x-1),\tag{3.9}$$

$$Q_{2N+1}^{(0)}(x;\alpha,-2N-1,c) = (x^2-c^2)^N(x+c),\tag{3.10}$$

and

$$Q^{(0)}_{2N+1+m}(x;-2N-1,\beta,c) = (-1)^m(x^2-1)^N(x-1)Q^{(0)}_m(-x;2N+1,\beta,-c), \tag{3.11}$$

$$Q^{(0)}_{2N+1+m}(x;\alpha,-2N-1,c) = (x^2-c^2)^N(x+c)Q^{(0)}_m(x;\alpha,2N+1,-c). \tag{3.12}$$

PROOF. Identities (3.9) and (3.10) follow from (3.12) and (3.11) by setting $m=0$. We will first prove (3.11) and then (3.12). Consider $m = 2M$, $n = 2N+1+m$. Then one has

$$Q^{(0)}_{2N+1+m}(x;\alpha,-2N-1,c) = \kappa_n\left({}_2F_1\left(\begin{matrix}-N-M, M+\widehat{\beta}\\ -N\end{matrix};\frac{1-x^2}{1-c^2}\right)\right.$$
$$\left.+\frac{(M+\widehat{\beta})(1-x)}{(c+1)N}{}_2F_1\left(\begin{matrix}-N-M, M+1+\widehat{\beta}\\ -N+1\end{matrix};\frac{1-x^2}{1-c^2}\right)\right),$$

where $\widehat{\beta} = (\beta+1)/2$. Next, by using Lemma 3.2, and after a tedious calculation comparing the final expression with the one for $Q_m(-x;2N+1,\beta,-c)$ and its corresponding normalization coefficient, the identity for the even m values holds. For $n = 2N+1+2M+1$, i.e., $m = 2M+1$, for $\alpha = -2N-1$, the proof is similar and will be omitted. Hence (3.11) follows. The identity (3.12) holds by applying the same procedure, which completes the proof. □

A different way to guess and then obtain the factorization for the big -1 Jacobi polynomials for non-standard parameters throughout is to use the following two-step procedure. The first step is to use Lemma 3.2 in order to prove the identities (3.9), (3.10). Once we have that, it is straightforward to find, e.g., for $m \geqslant 0$, that

$$Q^{(0)}_{2N+1+m}(x;-2N-1,\beta,c) = Q^{(0)}_{2N+1}(x;-2N-1,\beta,c)P_m(x),$$

where $P_m \in \mathbb{P}$ of degree m. In the second step, we obtain a recurrence relation for these new polynomials, and comparing the recurrence coefficients, we prove this new polynomial sequence is related to the original one.

LEMMA 3.4. *For any $N \in \mathbb{N}$, $\alpha, \beta, c \in \mathbb{C}$, $c \neq \pm 1$, with $\alpha = -2N-1$ (resp. $\beta = -2N-1$), the following recurrence relation holds for $n \geqslant 0$:*

$$\begin{aligned} xQ^{(0)}_n(x;2N+1,\beta,-c) &= -Q^{(0)}_{n+1}(x;2N+1,\beta,-c)\\ &\quad+\widehat{b}_nQ^{(0)}_n(x;2N+1,\beta,-c)+\widehat{u}_nQ^{(0)}_{n-1}(x;2N+1,\beta,-c),\end{aligned} \tag{3.13}$$

where

$$\widehat{b}_n = -b_{n+2N+1}(-2N-1,\beta,c) = b_n(2N+1,\beta,-c), \tag{3.14}$$

$$\widehat{u}_n = u_{n+2N+1}(-2N-1,\beta,c) = u_n(2N+1,\beta,-c). \tag{3.15}$$

Respectively, the following recurrence relation holds for $n \geqslant 0$:

$$\begin{aligned} xQ^{(0)}_n(x;\alpha,2N+1,-c) &= Q^{(0)}_{n+1}(x;\alpha,2N+1,-c)\\ &\quad+\widetilde{b}_nQ^{(0)}_n(x;\alpha,2N+1,-c)+\widetilde{u}_nQ^{(0)}_{n-1}(x;\alpha,2N+1,-c),\end{aligned} \tag{3.16}$$

where

$$\widetilde{b}_n = b_{n+2N+1}(\alpha,-2N-1,c) = b_n(\alpha,2N+1,-c), \tag{3.17}$$

$$\widetilde{u}_n = u_{n+2N+1}(\alpha,-2N-1,c) = u_n(\alpha,2N+1,-c). \tag{3.18}$$

PROOF. One starts by considering $\alpha = -2N-1$, so that $u_{2N+1} = 0$. In fact,

$$xQ_{2N+1}^{(0)}(x; 2N+1, \beta, c) = (x^2-1)^N(x-1).$$

If one considers $n \geqslant m + 2N + 1$, then

$$Q_{2N+1+m}^{(0)}(x; -2N-1, \beta, c) = Q_{2N+1}^{(0)}(x; -2N-1, \beta, c)(-1)^m Q_m^{(0)}(-x; 2N+1, \beta, -c),$$

and therefore the recurrence relation for these polynomials can be written as

$$\begin{aligned} x(-1)^m Q_m^{(0)}(-x; 2N+1, \beta, -c) = (-1)^{m+1}&Q_{m+1}^{(0)}(-x; 2N+1, \beta, -c) \\ &+ b_{m+2N+1}(-1)^m Q_m^{(0)}(-x; 2N+1, \beta, -c) \\ &+ u_{m+2N+1}(-1)^{m-1} Q_{m-1}^{(0)}(-x; 2N+1, \beta, -c). \end{aligned}$$

After some straightforward calculations the result follows. For $\beta = -2N-1$ the derivation is analogous, which completes the proof. □

Note that if $u_{2N+1} = 0$ for some $N \in \mathbb{N}$ then the orthogonality property for the big −1 Jacobi polynomials holds for polynomials of degree less or equal to $2N+1$. In the next result, we construct, thanks to the factorization of the big −1 polynomials, a new property of orthogonality that is valid for the big −1 Jacobi polynomials even when $n > 2N+1$, i.e., we extend the property of orthogonality for the big −1 polynomials to non-standard degenerate parameters.

THEOREM 3.5 (Orthogonality of big −1 Jacobi polynomials for non-standard parameters). *Let $N \in \mathbb{N}_0$, $\alpha, \beta, c \in \mathbb{C}$, $c \neq \pm 1$. Define the linear operator $\mathbf{u}$* [**20**, Section 4] *by*

$$\langle \mathbf{u}, pq \rangle := \int_{[-c,-1]\cup[1,c]} p(x)q(x)\,\frac{x}{|x|}(x+1)(c-x)(x^2-1)^{\frac{1}{2}(\alpha-1)}(c^2-x^2)^{\frac{1}{2}(\beta-1)}\,\mathrm{d}x,$$

and the norm with respect to the linear functional $\mathbf{u} \in \mathbb{P}'$ by $h_n := h_n(\alpha, \beta, c)$, where

(3.19)
$$\begin{aligned} h_n(\alpha, \beta, c) &:= \langle \mathbf{u}, Q_n^{(0)} Q_n^{(0)} \rangle \\ &= \langle \mathbf{u}, 1 \rangle \begin{cases} \dfrac{2(c^2-1)^n(\frac{1}{2}n)!(\frac{1}{2}(\alpha+1), \frac{1}{2}(\beta+1))_{\frac{1}{2}n}}{(\frac{1}{2}(\alpha+\beta)+1)_n(\frac{1}{2}(\alpha+\beta+n)+1)_{\frac{1}{2}n}} & \text{if } n \text{ is even,} \\[2ex] \dfrac{2(c-1)^{n-1}(c+1)^{n+1}(\frac{1}{2}(n-1))!(\frac{1}{2}(\alpha+1), \frac{1}{2}(\beta+1))_{\frac{1}{2}(n+1)}}{(\frac{1}{2}(\alpha+\beta)+1)_n(\frac{1}{2}(\alpha+\beta+n+1))_{\frac{1}{2}(n+1)}} & \text{if } n \text{ is odd.} \end{cases} \end{aligned}$$

Then, the polynomial sequences

$$(Q_n^{(0)}(x; -2N-1, \beta, c)), \qquad (Q_n^{(0)}(x; \alpha, -2N-1, c)) \tag{3.20}$$

are orthogonal with respect to the bilinear forms

$$\langle p, q \rangle_1 = \mathscr{L}_0(p, q) + \lambda_N(\beta)\langle \mathbf{u}_1, (\tau_\alpha^{2N+1}p)(\tau_\alpha^{2N+1}q)\rangle, \tag{3.21}$$

$$\langle p, q \rangle_2 = \mathscr{L}_1(p, q) + \lambda_N(\alpha)\langle \mathbf{u}_2, (\tau_\beta^{2N+1}p)(\tau_\beta^{2N+1}q)\rangle, \tag{3.22}$$

respectively. Here τ_α, τ_β are linear operators defined such that

$$\tau_\alpha[Q_n^{(0)}(x; \alpha, \beta, c)] = Q_{n-1}(-x; \alpha+2, \beta, -c), \tag{3.23}$$

$$\tau_\beta[Q_n^{(0)}(x; \alpha, \beta, c)] = Q_{n-1}(x; \alpha, \beta+2, -c), \tag{3.24}$$

and

$$\langle \mathbf{u}_1, f(x,\alpha,\beta,c)\rangle := \langle \mathbf{u}, f(-x,-\alpha,\beta,-c)\rangle, \tag{3.25}$$

$$\langle \mathbf{u}_2, f(x,\alpha,\beta,c)\rangle := \langle \mathbf{u}, f(x,\alpha,-\beta,-c)\rangle. \tag{3.26}$$

$\mathscr{L}_0$ and $\mathscr{L}_1$ are two moment linear functionals, and

$$\lambda_N(\alpha) = \frac{1}{2}h_{2N+1}(-2N-1,\alpha,c), \qquad \lambda_N(\beta) = \frac{1}{2}h_{2N+1}(-2N-1,\beta,c).$$

PROOF. Consider the situation corresponding to (3.21), i.e., $\alpha = -2N-1$, with $N \in \mathbb{N}$, $\beta \in \mathbb{C}$. We need to prove the properties of orthogonality $\langle Q_n^{(0)}, Q_m^{(0)}\rangle_1 = 0$ for all $m < n$, and $\langle Q_n^{(0)}, Q_n^{(0)}\rangle_1 \neq 0$ for all n, in order to prove that $(Q_n^{(0)})$ is a monic orthogonal polynomial sequence with respect to $\langle\,,\rangle_1$. There are three cases: (i) $m \leqslant n < 2N+1$; (ii) $m < 2N+1 \leqslant n$; (iii) $2N+1 \leqslant n \leqslant m$. (i): If $m \leqslant n < 2N+1$ the coefficient u_k of the recurrence relation is non-zero for $k = 0,1,\ldots,2N$, so by the Favard result (see [**4**]), there exists a moment linear functional, namely $\mathscr{L}_0$, such that the big -1 Jacobi polynomials $(Q_j^{(0)})_{j=0}^{2N}$ are orthogonal with respect to it. Due to the fact that $\tau_\alpha^{2N+1}(Q_j) = 0$ for $j = 0,1,\ldots,2N$, therefore the property of orthogonality holds in this case, i.e.,

$$\langle Q_n^{(0)}, Q_m^{(0)}\rangle_1 = \langle \mathbf{u}, Q_n^{(0)}Q_m^{(0)}\rangle = h_n(-2N-1,\beta,c)\delta_{n,m},$$

where $\delta_{n,m}$ is the Kronecker (delta) symbol. (ii): If $m < 2N+1 \leqslant n$, then $\mathscr{L}_0(Q_n^{(0)}, Q_m^{(0)}) = 0$; at the same time $\tau_\alpha^{2N+1}(Q_j) = 0$ for $j = 0,1,\ldots,2N$, therefore the property of orthogonality holds in this case. (iii): If $2N+1 \leqslant n \leqslant m$, then $\mathscr{L}_0(Q_n^{(0)}, Q_m^{(0)}) = 0$; and since

$$\tau_\alpha^{2N+1}Q_n^{(0)}(x;-2N-1,\beta,c) = Q_{n-(2N+1)}^{(0)}(-x;2N+1,\beta,-c)$$

is orthogonal with respect to the linear functional $\mathbf{u}$, one has

$$\begin{aligned}\langle Q_n^{(0)}, Q_m^{(0)}\rangle_1 &= \lambda_N(\beta)\langle \mathbf{u}, \tau_\alpha^{2N+1}Q_n^{(0)}(x;-2N-1,\beta,c)\tau_\alpha^{2N+1}Q_m^{(0)}(x;-2N-1,\beta,c)\rangle \\ &= \lambda_N(\beta)\langle \mathbf{u}, Q_{n-2N-1}^{(0)}(-x;2N+1,\beta,-c)Q_{m-2N-1}^{(0)}(-x;2N+1,\beta,-c) \\ &= h_n(-2N-1,\beta,c)\delta_{n,m},\end{aligned}$$

and the property of orthogonality holds in this situation.

Moreover, since the polynomials are monic, it is straightforward to check using the recurrence relation (3.3) that

$$h_n = \langle \mathbf{u}, xQ_n^{(0)}(x)Q_{n-1}^{(0)}(x)\rangle = u_n\langle \mathbf{u}, xQ_{n-1}^{(0)}(x)Q_{n-1}^{(0)}(x)\rangle = u_n h_{n-1}. \tag{3.27}$$

Furthermore, the norm squared (3.19) and the recurrence coefficient u_n appearing in (3.3) satisfy (3.27). The situation which corresponds to (3.22), i.e., $\beta = -2N-1$, with $N \in \mathbb{N}$, $\alpha \in \mathbb{C}$, is similar so we will omit its proof. Hence the result holds. $\square$

REMARK 3.6. Note that if $\alpha = -2N-1$ and $\beta \in \mathbb{C}$, due to (3.9) one must consider the moment linear form [**11**]

$$\mathscr{L}_0(p,q) = \sum_{j=0}^{N} \lambda_j\, p^{(j)}(1)\, q^{(j)}(1) + \sum_{j=0}^{N-1} \mu_j\, p^{(j)}(-1)\, q^{(j)}(-1),$$

where λ_j, μ_j are positive. So, taking into account (3.11), one has

$$\mathscr{L}_0(Q_j^{(0)}(x), \pi(x)) = 0, \quad j \geqslant 2N+1,$$

for any $\pi \in \mathbb{P}$. Moreover, by construction, if $0 \leqslant j < 2N+1$ one has

$$\langle Q_j^{(0)}(x), Q_j^{(0)}(x)\rangle_1 = \mathscr{L}_0\left(Q_j^{(0)}(x)p, Q_j^{(0)}(x)\right) = \langle \mathbf{u}, Q_j^{(0)}Q_j^{(0)}\rangle = h_j \neq 0.$$

If $j \geqslant 2N+1$ then taking into account Lemma 3.4 one has

$$\begin{aligned}&\langle Q_j^{(0)}(x), Q_j^{(0)}(x)\rangle_1\\ &\quad= \lambda_N(\beta)\langle \mathbf{u}, Q_{j-2N-1}^{(0)}(-x; 2N+1, \beta, -c)Q_{j-2N-1}^{(0)}(-x; 2N+1, \beta, -c)\rangle\\ &\quad= \lambda_N(\beta)h_{j-2N-1}(2N+1, \beta, -c) = h_j(-2N-1, \beta, c).\end{aligned}$$

Observe that the last identity must be considered as a formal limit:

$$\tfrac{1}{2}\lim_{\epsilon\to 0} h_{2N+1}(-2N-1+\epsilon, \beta, c)h_{j-2N-1}(2N+1+\epsilon, \beta, -c) = \lim_{\epsilon\to 0} h_j(-2N-1+\epsilon, \beta, c).$$

For $\beta = -2N-1$, the situation is analogous since in (3.10) one must consider the moment linear form [**11**]

$$\mathscr{L}_1(p,q) = \sum_{j=0}^{N} \kappa_j p^{(j)}(1)q^{(j)}(-c) + \sum_{j=0}^{N-1} \ell_j p^{(j)}(-1)q^{(j)}(c),$$

where κ_j, ℓ_j are positive. In a similar way

$$\frac{1}{2}\lim_{\epsilon\to 0} h_{2N+1}(\alpha, -2N-1+\epsilon, c)h_{j-2N-1}(\alpha, 2N+1+\epsilon, -c) = \lim_{\epsilon\to 0} h_j(\alpha, -2N-1+\epsilon, c).$$

4. Conclusion and future work

We have studied the big -1 Jacobi polynomials, which are a $q \to -1$ limit of the Bannai–Ito polynomials, when at least one of the parameters α, β is non-standard and the coefficient u_n of the three-term recurrence relation (3.3) is equal to zero for a certain index, namely N, i.e., $\alpha = -2N-1$ or $\beta = -2N-1$ where N is a positive integer. We have also obtained a Gauss hypergeometric representation, its factorization, and the property of orthogonality, using techniques similar to those used in [**5**].

In [**6**, Section 4], a similar procedure for the big q-Jacobi polynomials was applied for non-standard parameters for which the coefficient of u_n in the recurrence relation is equal to zero for some $n \in \mathbb{N}_0$. Along these lines we are investigating a method for obtaining the orthogonality property of the big -1 Jacobi polynomials for non-standard parameters, following a procedure analogous to the one followed by T. E. Pérez and M. A. Piñar in [**17**] for the generalized Laguerre polynomials. In [**6**] the authors also studied the q-Racah polynomials for non-standard parameters for which the u_n coefficient vanishes for some $n \in \mathbb{N}_0$, and since the Bannai–Ito polynomials can be obtained from the q-Racah polynomials ([**8**, §14.2]) by taking an analogous limit $q \to -1$ (see [**20**]), it makes sense to perform a similar study for the Bannai–Ito polynomials. We are also working on obtaining explicit hypergeometric representations, their factorizations, and the property of orthogonality for the Bannai–Ito polynomials and other families of the continuous -1 hypergeometric orthogonal polynomial scheme for non-standard parameters.

References

[1] M. Alfaro, M. Á. de Morales, and M. L. Rezola, *Orthogonality of the Jacobi polynomials with negative integer parameters*, J. Comput. Appl. Math. **145** (2002), no. 2, 379–386, DOI 10.1016/S0377-0427(01)00589-1. MR1917278

[2] M. Alfaro, T. E. Pérez, M. A. Piñar, and M. L. Rezola, *Sobolev orthogonal polynomials: the discrete-continuous case*, Methods Appl. Anal. **6** (1999), no. 4, 593–616, DOI 10.4310/MAA.1999.v6.n4.a10. MR1795525

[3] M. Álvarez de Morales, T. E. Pérez, and M. A. Piñar, *Sobolev orthogonality for the Gegenbauer polynomials* $\{C_n^{(-N+1/2)}\}_{n\geqslant 0}$, J. Comput. Appl. Math. **100** (1998), no. 1, 111–120, DOI 10.1016/S0377-0427(98)00124-1. MR1658734

[4] T. S. Chihara, *An introduction to orthogonal polynomials*, Mathematics and its Applications, Vol. 13, Gordon and Breach Science Publishers, New York-London-Paris, 1978. MR481884

[5] R. S. Costas-Santos and J. F. Sánchez-Lara, *Extensions of discrete classical orthogonal polynomials beyond the orthogonality*, J. Comput. Appl. Math. **225** (2009), no. 2, 440–451, DOI 10.1016/j.cam.2008.07.055. MR2494714

[6] R. S. Costas-Santos and J. F. Sánchez-Lara, *Orthogonality of q-polynomials for non-standard parameters*, J. Approx. Theory **163** (2011), no. 9, 1246–1268, DOI 10.1016/j.jat.2011.04.005. MR2832754

[7] *NIST Digital Library of Mathematical Functions.* `https://dlmf.nist.gov/`, Release 1.2.0 of 2024-03-27. F. W. J. Olver, A. B. Olde Daalhuis, D. W. Lozier, B. I. Schneider, R. F. Boisvert, C. W. Clark, B. R. Miller, B. V. Saunders, H. S. Cohl, and M. A. McClain, eds.

[8] R. Koekoek, P. A. Lesky, and R. F. Swarttouw, *Hypergeometric orthogonal polynomials and their q-analogues*, Springer Monographs in Mathematics, Springer-Verlag, Berlin, 2010. With a foreword by Tom H. Koornwinder, DOI 10.1007/978-3-642-05014-5. MR2656096

[9] A. B. J. Kuijlaars, A. Martínez-Finkelshtein, and R. Orive, *Orthogonality of Jacobi polynomials with general parameters*, Electron. Trans. Numer. Anal. **19** (2005), 1–17. MR2149265

[10] K. H. Kwon and L. L. Littlejohn, *The orthogonality of the Laguerre polynomials* $\{L_n^{-k}(x)\}$ *for positive integers* k, Ann. Numer. Math. **2** (1995), no. 1-4, 289–303. Special functions (Torino, 1993). MR1343537

[11] S. G. Moreno and E. M. García-Caballero, *Linear interpolation and Sobolev orthogonality*, J. Approx. Theory **161** (2009), no. 1, 35–48, DOI 10.1016/j.jat.2008.08.005. MR2558142

[12] S. G. Moreno and E. M. García-Caballero, *Non-standard orthogonality for the little q-Laguerre polynomials*, Appl. Math. Lett. **22** (2009), no. 11, 1745–1749, DOI 10.1016/j.aml.2009.05.017. MR2560989

[13] S. G. Moreno and E. M. García-Caballero, *New orthogonality relations for the continuous and the discrete q-ultraspherical polynomials*, J. Math. Anal. Appl. **369** (2010), no. 1, 386–399, DOI 10.1016/j.jmaa.2010.03.024. MR2643877

[14] S. G. Moreno and E. M. García-Caballero, *Non-classical orthogonality relations for big and little q-Jacobi polynomials*, J. Approx. Theory **162** (2010), no. 2, 303–322, DOI 10.1016/j.jat.2009.05.008. MR2581384

[15] S. G. Moreno and E. M. García-Caballero, *Non-classical orthogonally relations for continuous q-Jacobi polynomials*, Taiwanese J. Math. **15** (2011), no. 4, 1677–1690, DOI 10.11650/twjm/1500406372. MR2848982

[16] J. Pelletier, L. Vinet, and A. Zhedanov. Continuous −1 hypergeometric orthogonal polynomials. `https://arxiv.org/abs/2209.10727`, 47 pages, 2022.

[17] T. E. Pérez and M. A. Piñar, *On Sobolev orthogonality for the generalized Laguerre polynomials*, J. Approx. Theory **86** (1996), no. 3, 278–285, DOI 10.1006/jath.1996.0069. MR1405981

[18] J. F. Sánchez-Lara, *On the Sobolev orthogonality of classical orthogonal polynomials for non standard parameters*, Rocky Mountain J. Math. **47** (2017), no. 1, 267–288, DOI 10.1216/RMJ-2017-47-1-267. MR3619764

[19] L. Vinet and A. Zhedanov, *A 'missing' family of classical orthogonal polynomials*, J. Phys. A **44** (2011), no. 8, 085201, 16, DOI 10.1088/1751-8113/44/8/085201. MR2770369

[20] L. Vinet and A. Zhedanov, *A limit* $q = -1$ *for the big q-Jacobi polynomials*, Trans. Amer. Math. Soc. **364** (2012), no. 10, 5491–5507, DOI 10.1090/S0002-9947-2012-05539-5. MR2931336

Applied and Computational Mathematics Division, National Institute of Standards and Technology, Gaithersburg, MD 20899-8910
Email address: howard.cohl@nist.gov

Dept. of Quantitative Methods, Universidad Loyola Andalucía, Sevilla, Spain
Email address: rscosa@gmail.com

Contemporary Mathematics
Volume **819**, 2025
https://doi.org/10.1090/conm/819/16396

On an identity of Chaundy and Bullard. III. Basic and elliptic extensions

Natsuko Hoshi, Makoto Katori, Tom H. Koornwinder, and Michael J. Schlosser

This paper is dedicated to the memory of Richard A. Askey

ABSTRACT. The identity by Chaundy and Bullard expresses 1 as a sum of two truncated binomial series in one variable where the truncations depend on two different non-negative integers. We present basic and elliptic extensions of the Chaundy–Bullard identity. The most general result, the elliptic extension, involves, in addition to the nome p and the base q, four independent complex variables. Our proof uses a suitable weighted lattice path model. We also show how three of the basic extensions can be viewed as Bézout identities. Inspired by the lattice path model, we give a new elliptic extension of the binomial theorem, taking the form of an identity for elliptic commuting variables. We further present variants of the homogeneous form of the identity for q-commuting and for elliptic commuting variables.

1. Introduction

In earlier papers by two of the present authors [**9, 10**], the identity

$$1 = (1-x)^{n+1}\sum_{k=0}^{m}\binom{n+k}{k}x^k + x^{m+1}\sum_{k=0}^{n}\binom{m+k}{k}(1-x)^k, \tag{1.1}$$

where $m, n \in \mathbb{N}_0 := \{0, 1, 2, \dots\}$ and x is a variable, was analysed in great detail. The authors [**9**] originally attributed this formula to Chaundy and Bullard [**3**]. As the present paper constitutes a continuation of the papers [**9, 10**], we shall keep referring to (1.1) as the *Chaundy–Bullard identity.* (In the follow-up paper [**10**] to [**9**] the authors pointed out that an identity equivalent to (1.1) was already published by Pierre Raymond de Montmort in 1713 [**12**], in one of the very first treatises on probability theory.) This fundamental formula, expressing 1 as a sum of two truncated binomial series, was rediscovered many times over more than three hundred

---

2020 *Mathematics Subject Classification.* Primary 05A19; Secondary 05A10, 05A30, 05C22, 05C81, 11B65, 33D15, 33E05.

Key words and phrases. Chaundy–Bullard identity, q-analogue, elliptic analogue, binomial identities, theta functions, lattice paths, Bézout identity.

The first two authors' research was supported by the Grant-in-Aid for Scientific Research (C)(No.24K06888), (B)(No.23K25774), (A)(No.21H04432), and the Grand-in-Aid for Transformative Research Areas (A)(No.22H05105) of the Japan Society for the Promotion of Science.

The fourth author's research was supported by the Austrian Science Fund, grant 10.55776/P32305.

years. A lot of history and occurrences of this identity in various areas of mathematics were surveyed in [**9**, **10**], where several different proofs were compiled and connections to special functions including Gauß hypergeometric series, incomplete beta integrals and Krawtchouk polynomials were made explicit.[1] One of the topics discussed in [**9**] concerns a multivariate extension of the Chaundy–Bullard identity related to Lauricella hypergeometric functions, with an explicit description of the corresponding system of partial differential equations they satisfy.

One of the main purposes of the present paper is to present an *elliptic extension of the Chaundy–Bullard identity.* Its proof (using a suitable lattice path model) leads us even to discover a new elliptic extension of the binomial theorem.

In order to formulate our theorems, we introduce some notations. First we fix $q \in \mathbb{C}$, which we call a *base*, and for $x \in \mathbb{C}$ and $k \in \mathbb{N}_0$ define the *q-shifted factorials* $(x;q)_k$ (also known as *q-Pochhammer symbols*) by

$$(x;q)_0 := 1 \quad \text{and} \quad (x;q)_k := \prod_{\ell=0}^{k-1}(1-xq^\ell), \quad k=1,2,\dots. \tag{1.2}$$

For $|q| < 1$ we may also take $k = \infty$ in (1.2); the q-shifted factorial is then a convergent infinite product. For $x_1,\dots,x_s \in \mathbb{C}$ products of q-shifted factorials are abbreviated as $(x_1,\dots,x_s;q)_k := \prod_{i=1}^{s}(x_i;q)_k$, where $x_1,\dots,x_s \in \mathbb{C}$ and $k \in \mathbb{N}_0 \cup \{\infty\}$. Next we fix another parameter $p \in \mathbb{C}$ with $0 < |p| < 1$. The *modified Jacobi theta function* with argument $x \in \mathbb{C}^\times := \mathbb{C}\setminus\{0\}$ and *nome* p is defined by

$$\theta(x;p) := (x,p/x;p)_\infty. \tag{1.3}$$

We will frequently abbreviate products of modified Jacobi theta functions using the notation $\theta(x_1,\dots,x_s;p) = \prod_{i=1}^{s}\theta(x_i;p)$ for $x_1,\dots,x_s \in \mathbb{C}^\times$. For $x \in \mathbb{C}^\times$ and $k \in \mathbb{N}_0$, the *theta-shifted factorial* $(x;q,p)_k$ (also called the *q,p-shifted factorial*) is defined by

$$(x;q,p)_0 := 1 \quad \text{and} \quad (x;q,p)_k := \prod_{\ell=0}^{k-1}\theta(xq^\ell;p), \quad k=1,2,\dots.$$

A product of theta-shifted factorials is compactly written as $(x_1,\dots,x_s;q,p)_k := \prod_{i=1}^{s}(x_i;q,p)_k$, where $x_1,\dots,x_s \in \mathbb{C}^\times$, $k \in \mathbb{N}_0$.

Three simple identities satisfied by the modified Jacobi theta function are the symmetry

$$\theta(x;p) = \theta(p/x;p),$$

the inversion formula

$$\theta(1/x;p) = -\frac{1}{x}\theta(x;p),$$

and the quasi-periodicity

$$\theta(px;p) = -\frac{1}{x}\theta(x;p),$$

[1]After seeing [**9**, **10**], Slobodan Damjanović kindly brought to the authors' attention that almost at the same time as Chaundy and Bullard published their paper [**3**], Kesava Menon gave the identity (1.1) in [**6**, Equation (1.2)] as well. His proof uses partial fraction decomposition of $x^{-m}(1-x)^{-n}$.

which all easily follow from the definition in (1.3). In this paper we will make crucial use of the following identity, the *Weierstraß–Riemann addition formula* [**19**, p. 451, Example 5]

$$\theta(xy, x/y, uv, u/v; p) - \theta(xv, x/v, uy, u/y; p) = \frac{u}{y}\theta(yv, y/v, xu, x/u; p), \tag{1.4}$$

where $x, y, u, v \in \mathbb{C}^\times$. For a discussion of (1.4) and a comparison with similar relations, see [**8**].

Whereas we refer to the extensions of the Chaundy–Bullard identity that involve q-shifted factorials as *q-extensions* or *basic extensions* (with q being the base) of the Chaundy–Bullard identity, our *elliptic extension* of the Chaundy–Bullard identity connects series containing ratios of products of modified Jacobi theta functions (these ratios are in fact *elliptic functions*, justifying the terminology). These basic and elliptic extensions actually involve truncated basic and elliptic hypergeometric series (the textbook [**5**] contains a comprehensive treatise of the theory of these series; the elliptic cases is treated in Chapter 11). In Section 2 all our extensions of the Chaundy–Bullard identity are obtained from scratch, without requiring results from the theories of basic or elliptic hypergeometric series.

By definition, a function $g(u)$ is *elliptic*, if it is a doubly-periodic meromorphic function of the complex variable u.

As a consequence of the theory of theta functions (cf. [**13**, Theorem 1.3.3]) one may assume without loss of generality that

$$g(u) = \frac{\theta(a_1q^u, a_2q^u, \dots, a_sq^u; p)}{\theta(b_1q^u, b_2q^u, \dots, b_sq^u; p)}\, z$$

for a positive integer s, a constant z and some $a_1, a_2, \dots, a_s$, $b_1, \dots, b_s$, and p, q with $|p| < 1$, where the *elliptic balancing condition* (cf. [**17**]), namely

$$a_1a_2\cdots a_s = b_1b_2\cdots b_s,$$

holds. (So, a linear combination of two such expressions is again an expression of the same form.) If one writes $q = e^{2\pi\sqrt{-1}\sigma}$, $p = e^{2\pi\sqrt{-1}\tau}$, with complex σ and τ and $\Im\tau > 0$, then $g(u)$ is indeed periodic in u with periods σ^{-1} and $\tau\sigma^{-1}$. Keeping this notation for p and q, denote the *field of elliptic functions* over $\mathbb{C}$ of the complex variable u, meromorphic in u with the two periods σ^{-1} and $\tau\sigma^{-1}$, by $\mathbb{E}_{q^u;q,p}$.

More generally, denote the *field of totally elliptic multivariate functions* over $\mathbb{C}$ of the complex variables $u_1, \dots, u_n$, meromorphic in each variable[2] with equal periods, σ^{-1} and $\tau\sigma^{-1}$, of double periodicity, by $\mathbb{E}_{q^{u_1},\dots,q^{u_n};q,p}$. The notion of totally elliptic multivariate functions was first introduced by Spiridonov, see [**17**, p. 317, Definition 11] (where the related notion of totally elliptic hypergeometric series was defined) and [**18**, Definition 6].

We are ready to state our elliptic extension of the Chaundy–Bullard identity.

[2]Here we would like to mention Hartogs' theorem and its analogue for meromorphic functions. Hartogs' theorem says informally that a multivariate function that is separably analytic (i.e., analytic in each independent variable) is analytic. Next, a separably meromorphic function is meromorphic, see [**16**, Corollary 2].

THEOREM 1.1. *Let $a,b,c,x,q \in \mathbb{C}^\times$, $p \in \mathbb{C}$ with $0<|p|<1$ and $n,m \in \mathbb{N}_0$. Then, as an identity in $\mathbb{E}_{x,a,b,c;q,p}$,*

$$\begin{aligned}1 = &\frac{(ac,c/a,bx,b/x;q,p)_{n+1}}{(ab,b/a,cx,c/x;q,p)_{n+1}}\\ &\times\sum_{k=0}^{m}\frac{\theta(acq^{n+2k};p)(acq^n,bcq^n,c/b,q^{n+1},ax,a/x;q,p)_k}{\theta(acq^n;p)(q,aq/b,abq^{1+n},ac,cq^{n+1}/x,cxq^{n+1};q,p)_k}q^k\\ &+\frac{(bc,c/b,ax,a/x;q,p)_{m+1}}{(ab,a/b,cx,c/x;q,p)_{m+1}}\\ &\times\sum_{k=0}^{n}\frac{\theta(bcq^{m+2k};p)(bcq^m,acq^m,c/a,q^{m+1},bx,b/x;q,p)_k}{\theta(bcq^m;p)(q,bq/a,abq^{1+m},bc,cq^{m+1}/x,cxq^{m+1};q,p)_k}q^k.\end{aligned}\tag{1.5}$$

In order to demonstrate that the identity (1.5) is an extension of the original Chaundy–Bullard identity (1.1), we show how (1.5) can be reduced to (1.1) by taking suitable limits. Since our formula (1.5) involves three parameters $a,b,c \in \mathbb{C}^\times$ in addition to the variable x, the base q and the nome p, we have several intermediate identities between (1.1) and (1.5). Let

$$\begin{aligned}p_{m,n}(x;a,b,c;q,p) := &\frac{(ac,c/a,bx,b/x;q,p)_{n+1}}{(ab,b/a,cx,c/x;q,p)_{n+1}}\\ &\times\sum_{k=0}^{m}\frac{\theta(acq^{n+2k};p)(acq^n,bcq^n,c/b,q^{n+1},ax,a/x;q,p)_k}{\theta(acq^n;p)(q,aq/b,abq^{1+n},ac,cq^{n+1}/x,cxq^{n+1};q,p)_k}q^k.\end{aligned}\tag{1.6}$$

Then (1.5) can be written as

$$1 = p_{m,n}(x;a,b,c;q,p) + p_{n,m}(x;b,a,c;q,p).\tag{1.7}$$

By the definition of the modified Jacobi theta function (1.3) with (1.2), one has $\lim_{p\to 0}\theta(x;p) = 1-x$, and hence $\lim_{p\to 0}(x;q,p)_k = (x;q)_k$, $k \in \mathbb{N}_0$; that is the theta-shifted factorial is reduced to the q-shifted factorial in the limit $p \to 0$. Then the identity (1.7) with $p=0$ holds with

$$\begin{aligned}p_{m,n}(x;a,b,c;q,0) := &\frac{(ac,c/a,bx,b/x;q)_{n+1}}{(ab,b/a,cx,c/x;q)_{n+1}}\\ &\times\sum_{k=0}^{m}\frac{(1-acq^{n+2k})(acq^n,bcq^n,c/b,q^{n+1},ax,a/x;q)_k}{(1-acq^n)(q,aq/b,abq^{1+n},ac,cq^{n+1}/x,cxq^{n+1};q)_k}q^k.\end{aligned}\tag{1.8}$$

Next, if in addition to $p \to 0$ we take the further limit $c \to 0$, then we have the equality (1.7) with $p=c=0$, in which

$$p_{m,n}(x;a,b,0;q,0) := \frac{(bx,b/x;q)_{n+1}}{(ab,b/a;q)_{n+1}}\sum_{k=0}^{m}\frac{(q^{n+1},ax,a/x;q)_k}{(q,aq/b,abq^{1+n};q)_k}q^k.\tag{1.9}$$

An equivalent form of this q-extension of the Chaundy–Bullard identity was obtained by Ma in [**11**, Corollary 4.2] as a consequence of his six-variable generalization of Ramanujan's reciprocity theorem.

We can obtain another variant of a three-parameter Chaundy–Bullard identity for $x \in \mathbb{C}$ with parameters a,b,q. In (1.7) with $p=c=0$ and (1.9), we make the substitution $a \mapsto \delta a$, $b \mapsto b\delta$, $x \mapsto x/\delta$ and then take the limit $\delta \to 0$. The obtained equality is

$$1 = \widetilde{p}_{m,n}(x; a, b; q) + \widetilde{p}_{n,m}(x; b, a; q) \tag{1.10}$$

with

$$\widetilde{p}_{m,n}(x; a, b; q) := \frac{(bx; q)_{n+1}}{(b/a; q)_{n+1}} \sum_{k=0}^{m} \frac{(q^{n+1}, ax; q)_k}{(q, aq/b; q)_k} q^k. \tag{1.11}$$

In the above identity (1.10) with (1.11) the variable b is actually redundant (the substitutions $a \mapsto ab$ and $x \mapsto x/b$ applied to (1.11) eliminate the variable b), but it is useful to keep the variable b for the $a \leftrightarrow b$ symmetry.

Furthermore, if we substitute $x \mapsto x/b$ in (1.10) with (1.11) and then take the limit $b \to 0$, the equality is reduced to the identity

$$1 = (x; q)_{n+1} \sum_{k=0}^{m} \begin{bmatrix} n+k \\ k \end{bmatrix}_q x^k + x^{m+1} \sum_{k=0}^{n} \begin{bmatrix} m+k \\ k \end{bmatrix}_q q^k (x; q)_k, \tag{1.12}$$

where the *q-binomial coefficient* is defined by

$$\begin{bmatrix} n \\ k \end{bmatrix}_q := \frac{(q; q)_n}{(q; q)_k (q; q)_{n-k}}, \qquad k \in \{0, 1, \ldots, n\}. \tag{1.13}$$

Finally, if we take the limit $q \to 1$, then (1.12) is reduced to the original Chaundy–Bullard identity (1.1). We note that the $q \to 1$ limit of the equality (1.10) with (1.11) is equivalent to the original Chaundy–Bullard identity (1.1) where x is replaced by $(1 - ax)/(1 - a/b)$, in light of the easily checked relation $1 - (1 - ax)/(1 - a/b) = (1 - bx)/(1 - b/a)$. To distinguish the identities better, we say that (1.12) is the *q-extension*, (1.10) with (1.11) the *$(a, b; q)$-extension of the first kind*, (1.7) in the case $p = c = 0$ with (1.9) the *$(a, b; q)$-extension of the second kind*, (1.7) in the case $p = 0$ with (1.8) the *$(a, b, c; q)$-extension*, and (1.5) the *$(a, b; q, p)$-extension*, or simply the *elliptic extension* of the Chaundy–Bullard identity, respectively. Summarizing the above linear scheme, we have the chain of objects

$$(1.6) \to (1.8) \to (1.9) \to (1.11) \to (1.12) \to (1.1)$$

which respectively contain the following free variables:

$$(x, a, b, c; q, p) \to (x, a, b, c; q) \to (x, a, b; q) \to (x, a; q) \to (x; q) \to (x).$$

At this point we would also like to remark that the various q-extensions of the Chaundy–Bullard identity give rise to parameter-dependent Chaundy–Bullard identities after replacing the parameters a, b, and c, by q^a, q^b, and q^c, respectively, and taking the limit $q \to 1$. (We leave the details to the reader.)

From the many different proofs that are known for the original Chaundy–Bullard identity (1.1) [**9, 10**], in order to prove our elliptic extension in Theorem 1.1, we develop the fifth proof given in [**9**]; a proof by enumerating weighted lattice paths. Our construction of a suitable lattice path model with the proof of Theorem 1.1 is given in Section 2. In Section 3, we show how the q-extension of the Chaundy–Bullard identity (1.12) and its two kinds of $(a, b; q)$-extensions can be alternatively derived by making use of Bézout's identity. In Section 4 we look at variants of the q-extended Chaundy–Bullard identity and relate them to corresponding identities for q-commuting variables. Finally, in Section 5, inspired by the lattice path model, we present a new elliptic binomial theorem, taking the form of an identity for elliptic commuting variables. The elliptic binomial theorem in Theorem 5.7 is similar to a result discovered earlier by one of the authors [**15**, Theorem 2] and can be regarded as a companion result to that other one. In the same section we also present variants of the homogeneous form of the Chaundy–Bullard identity for elliptic commuting variables.

2. Proof of Theorem 1.1 by enumerating weighted lattice paths

2.1. General arguments. Let $m, n \in \mathbb{N}_0$ and consider a rectangular region of the square lattice,

$$\Lambda_{m+1,n+1} := \{(i,j) : i \in \{0,1,\dots,m+1\},\ j \in \{0,1,\dots,n+1\}\}.$$

Let $\Pi_{m+1,n+1}$ be the set of all lattice paths from $(0,0)$ to $(m+1,n+1)$ in $\Lambda_{m+1,n+1}$ using only unit east steps $(i,j) \to (i+1,j)$ and unit north steps $(i,j) \to (i,j+1)$. Such a path π consists of $m+n+2$ successive unit steps, $\pi = \{s_t(\pi), t \in \{1,2,\dots,m+n+2\}\}$. Each step s in $\pi \in \Pi_{m+1,n+1}$ is assigned a weight $w(s)$. The weight $w(\pi)$ of a path $\pi \in \Pi_{m+1,n+1}$ is defined to be a product of the weights of the respective steps of the path:

$$w(\pi) := \prod_{s\in\pi} w(s) = \prod_{t=1}^{m+n+2} w(s_t(\pi)).$$

Next we specify the weights for each of the possible steps of paths in $\Pi_{m+1,n+1}$. Let $h : \Lambda_{m,n} \to \mathbb{C}$ be a function which we will specialize later. For each $i \in \{0,1,\dots,m\}$ we define the weight of a unit east step by

$$w((i,j) \to (i+1,j)) := \begin{cases} h(i,j) & \text{if } j \in \{0,1,\dots,n\}, \\ 1 & \text{if } j = n+1, \end{cases} \tag{2.1a}$$

and for each $j \in \{0,1,\dots,n\}$ the weight of a unit north step by

$$w((i,j) \to (i,j+1)) := \begin{cases} 1 - h(i,j) & \text{if } i \in \{0,1,\dots,m\}, \\ 1 & \text{if } i = m+1. \end{cases} \tag{2.1b}$$

We assume that

$$\begin{aligned} h(i,0) &\neq 0 \quad \text{for } i \in \{0,1,\dots,m-1\}, \quad \text{and} \\ h(0,j) &\neq 1 \quad \text{for } j \in \{0,1,\dots,n-1\}. \end{aligned} \tag{2.2}$$

For $\pi \in \Pi_{m+1,n+1}$ and $\tau \in \{0,1,2,\dots,m+n+2\}$, define the truncated path π_τ by the path π terminated after τ steps. In particular, $\pi_0 = \varnothing$ and $\pi_{m+n+2} = \pi$. By the specific choices of weights in (2.1), for each $\pi_{\tau-1}$, $\tau \in \{1,2,\dots,m+n+2\}$, we have

$$\sum_{\pi_\tau : \pi_\tau \setminus s_\tau(\pi_\tau) = \pi_{\tau-1}} w(s_\tau(\pi_\tau)) = 1.$$

Hence by induction we can conclude that for each $\kappa < \tau$, $\tau \in \{1,2,\dots,m+n+2\}$, we have

$$\sum_{\pi_\tau : \pi_\tau \setminus \{s_{\kappa+1}(\pi_\tau),\dots,s_\tau(\pi_\tau)\} = \pi_\kappa} \prod_{i=\kappa+1}^{\tau} w(s_i(\pi_i)) = 1,$$

and by setting $\kappa = 0$ obtain

$$\sum_{\pi_\tau : \pi \in \Pi_{m+1,n+1}} w(\pi_\tau) = 1 \quad \text{for all } \tau \in \{1,2,\dots,m+n+2\}.$$

In the case $\tau = m+n+2$, the above gives

$$\sum_{\pi\in\Pi_{m+1,n+1}} w(\pi) = 1. \tag{2.3}$$

For any $(k,\ell)\in\Lambda_{m,n}$, the generating function $A(k,\ell)$ for all weighted lattice paths from $(0,0)$ to (k,ℓ) is defined by $A(0,0) := 1$ and

$$A(k,\ell) := \sum_{\pi:(0,0)\to(k,\ell)} w(\pi), \quad \text{for } (k,\ell)\in\Lambda_{m,n}\setminus\{(0,0)\},$$

where the sum is taken over $\{\pi : (0,0)\to(k,\ell)\}$, the set of all lattice paths from $(0,0)$ to (k,ℓ). By the specific assignment of weights in (2.1), $A(k,\ell)$ satisfies the recurrence relation

$$\begin{aligned} &h(k-1,\ell)A(k-1,\ell) + (1-h(k,\ell-1))A(k,\ell-1) = A(k,\ell), \\ &\text{for} \quad k\in\{1,2,\dots,m\} \quad \text{and} \quad \ell\in\{1,2,\dots,n\} \end{aligned} \tag{2.4a}$$

with the boundary conditions

$$\begin{aligned} A(k,0) &= \prod_{i=0}^{k-1} h(i,0) && \text{for } k\in\{1,2,\dots,m\}, \\ A(0,\ell) &= \prod_{j=0}^{\ell-1}(1-h(0,j)) && \text{for } \ell\in\{1,2,\dots,n\}. \end{aligned} \tag{2.4b}$$

Moreover, since the last step of a path in $\Pi_{m+1,n+1}$ which is not a step along the north or east boundary is either a step $(k,n)\to(k,n+1)$ $(k=0,1,\dots,m)$ or a step $(m,\ell)\to(m+1,\ell)$ $(\ell=0,\dots,n)$, the above assignment of weights implies the equality

$$\sum_{\pi\in\Pi_{m+1,n+1}} w(\pi) = \sum_{k=0}^{m}(1-h(k,n))A(k,n) + \sum_{\ell=0}^{n} h(m,\ell)A(m,\ell).$$

Hence by (2.3) we have the equality

$$1 = \sum_{k=0}^{m}(1-h(k,n))A(k,n) + \sum_{\ell=0}^{n} h(m,\ell)A(m,\ell). \tag{2.5}$$

Under assumption (2.2), we put

$$B(k,\ell) := \frac{A(k,\ell)}{A(k,0)A(0,\ell)}, \quad \text{for } (k,\ell)\in\Lambda_{m,n}.$$

Then we have from (2.4) the system of difference equations

$$\begin{aligned} &\frac{h(k-1,\ell)}{h(k-1,0)}B(k-1,\ell) + \frac{1-h(k,\ell-1)}{1-h(0,\ell-1)}B(k,\ell-1) = B(k,\ell) \\ &\text{for} \quad k\in\{1,2,\dots,m\} \quad \text{and} \quad \ell\in\{1,2,\dots,n\}, \text{ with} \\ &B(k,0) = 1 \quad \text{for } k\in\{0,1,\dots,m\}, \text{ and} \\ &B(0,\ell) = 1 \quad \text{for } \ell\in\{0,1,\dots,n\} \end{aligned} \tag{2.6}$$

which conversely uniquely determines the sequence $\big(B(k,\ell)\big)_{(k,\ell)\in\Lambda_{m,n}}$.

The above argument is summarized as follows.

PROPOSITION 2.1. *Assume that* $\big(h(i,j)\big)_{(i,j)\in\Lambda_{m,n}}$ *is given so that* (2.2) *is satisfied. Let* $\big(B(k,\ell)\big)_{(k,\ell)\in\Lambda_{m,n}}$ *be uniquely given by* (2.6). *Define* $\big(A(k,\ell)\big)_{(k,\ell)\in\Lambda_{m,n}}$ *by* $A(0,0) := 1$, (2.4b) *and*

$$\begin{aligned}A(k,\ell) &:= A(k,0)A(0,\ell)B(k,\ell),\\ &\textit{for}\quad k\in\{1,2,\dots,m\}\quad\textit{and}\quad \ell\in\{1,2,\dots,n\}.\end{aligned}\tag{2.7}$$

Then the equality (2.5) *holds.*

2.2. Proof of (1.5). We choose the weight function $h(i,j)$ as

$$h(i,j)=h_{x;a,b,c;q,p}(i,j):=\frac{\theta(bcq^{i+2j},(c/b)q^i,axq^i,(a/x)q^i;p)}{\theta(abq^{i+j},(a/b)q^{i-j},cxq^{i+j},(c/x)q^{i+j};p)},\tag{2.8}$$

where $a,b,c,x,q,p\in\mathbb{C}^\times$ and $|p|<1$. This choice is motivated by the following symmetry relation.

LEMMA 2.2. *The following equality holds,*

$$1-h_{x;a,b,c;q,p}(i,j)=h_{x;b,a,c;q,p}(j,i).\tag{2.9}$$

PROOF. By (2.8),

$$\begin{aligned}1-h_{x;a,b,c;q,p}(i,j)&=1-\frac{\theta(bcq^{i+2j},(c/b)q^i,axq^i,(a/x)q^i;p)}{\theta(abq^{i+j},(a/b)q^{i-j},cxq^{i+j},(c/x)q^{i+j};p)}\\ &=\Big(\theta(abq^{i+j},(b/a)q^{j-i},cxq^{i+j},(c/x)q^{i+j};p)\\ &\qquad+(b/a)q^{j-i}\theta(bcq^{i+2j},(c/b)q^i,axq^i,(a/x)q^i;p)\Big)\\ &\qquad\times\theta(abq^{i+j},(b/a)q^{j-i},cxq^{i+j},(c/x)q^{i+j};p)^{-1}.\end{aligned}$$

Now, with the substitution of variables $(x,y,u,v)\mapsto(cq^{i+j},aq^i,bq^j,x)$ in the Weierstraß–Riemann addition formula (1.4), specifically

$$\begin{aligned}\theta(abq^{i+j},(b/a)q^{j-i},cxq^{i+j},(c/x)q^{i+j};p)+\frac{b}{a}q^{j-i}\theta(bcq^{i+2j},(c/b)q^i,axq^i,(a/x)q^i;p)\\ =\theta(acq^{2i+j},(c/a)q^j,bxq^j,(b/x)q^j;p),\end{aligned}$$

the relation

$$1-h_{x;a,b,c;q,p}(i,j)=\frac{\theta(acq^{2i+j},(c/a)q^j,bxq^j,(b/x)q^j;p)}{\theta(abq^{i+j},(b/a)q^{j-i},cxq^{i+j},(c/x)q^{i+j};p)}$$

is established as desired and the proof is complete. □

We would like to explain the motivation for our specific choice of the weight function (2.8). Our weighted lattice model has a natural companion obtained by reflection with respect to the diagonal going northeast from the origin. Then according to (2.1) the corresponding weights $\tilde{w}$ of this companion are in terms of, say, $\tilde{h}(i,j):=1-h(j,i)$. So it would be pleasant to have a nice expression for h such that $\tilde{h}$ also has a nice expression. Now observe that (1.4) can be rewritten as $1-h_{x;a,b,c;q,p}(0,0)=h_{x;b,a,c;q,p}(0,0)$ with $h_{x;a,b,c;q,p}(0,0)$ given by (2.8) (which does not yet dependent on q). This gives a motivation for defining $h_{x;a,b,c;q,p}(i,j):=h_{x;aq^i,bq^j,cq^{i+j};q,p}(0,0)$ so that we have $\tilde{h}_{x;a,b,c;q,p}(i,j)=1-h_{x;b,a,c;q,p}(i,j)$.

With $h(i,j)$ as given as in (2.8) the system (2.6) can be explicitly written as

$$
\begin{aligned}
&\frac{\theta(bcq^{k+2\ell-1}, abq^{k-1}, (a/b)q^{k-1}, cxq^{k-1}, (c/x)q^{k-1};p)}{\theta(abq^{k+\ell-1}, (a/b)q^{k-\ell-1}, cxq^{k+\ell-1}, (c/x)q^{k+\ell-1}, bcq^{k-1};p)} B(k-1,\ell) \\
&\quad + \frac{\theta(acq^{2k+\ell-1}, abq^{\ell-1}, (b/a)q^{\ell-1}, cxq^{\ell-1}, (c/x)q^{\ell-1};p)}{\theta(abq^{k+\ell-1}, (b/a)q^{\ell-k-1}, cxq^{k+\ell-1}, (c/x)q^{k+\ell-1}, acq^{\ell-1};p)} B(k,\ell-1) \\
&= B(k,\ell)
\end{aligned}
\tag{2.10}
$$

for $\;k \in \{1,2,\dots,m\}\;$ and $\;\ell \in \{1,2,\dots,n\}$, with

$B(k,0) = 1\;$ for $k \in \{0,1,\dots,m\}$, and

$B(0,\ell) = 1\;$ for $\ell \in \{0,1,\dots,n\}$.

LEMMA 2.3. *The unique solution of* (2.10) *is given by*

$$
B(k,\ell) = \frac{\theta((a/b)q^{k-\ell}, b/a;p)(bcq^{\ell};q,p)_k(acq^k, ab, cx, c/x, q^{k+1};q,p)_\ell}{\theta((a/b)q^k, (b/a)q^{\ell};p)(bc;q,p)_k(ac, abq^k, cxq^k, (c/x)q^k, q;q,p)_\ell} q^{\ell}
\tag{2.11}
$$

for all $(k,\ell) \in \Lambda_{m,n}$.

PROOF. The proof proceeds by induction on $k+\ell$. Since $B(0,0) = 1$ by the system (2.10), the $k+\ell = 0$ case is trivial. Further, if $k = 0$ or $\ell = 0$, the values $B(k,0) = B(0,\ell) = 1$ specified in (2.10) agree with those in (2.11). Finally, let $k,\ell > 0$ and assume that the solution for $B(i,j)$ is given in (2.11) for all $(i,j) \in \Lambda_{m,n}$ with $i+j < k+\ell$. Then the left-hand side of (2.10) is, with (2.11) applied to rewrite $B(k-1,\ell)$ and $B(k,\ell-1)$ by induction,

$$
\begin{aligned}
&\frac{\theta(b/a;p)(bcq^{\ell};q,p)_k(acq^k, ab, cx, c/x, q^{k+1};q,p)_\ell q^{\ell}}{\theta((a/b)q^k, (b/a)q^{\ell};p)(bc;q,p)_k(ac, abq^k, cxq^k, (c/x)q^k, q;q,p)_\ell} \\
&\quad \times \left(\frac{\theta(bcq^{k+2\ell-1}, acq^{k-1}, q^k, (a/b)q^k;p)}{\theta(acq^{k+\ell-1}, bcq^{k+\ell-1}, q^{k+\ell};p)} \right. \\
&\qquad \left. - \frac{a}{b} q^{k-\ell} \frac{\theta(acq^{2k+\ell-1}, bcq^{\ell-1}, q^{\ell}, (b/a)q^{\ell};p)}{\theta(acq^{k+\ell-1}, bcq^{k+\ell-1}, q^{k+\ell};p)} \right) \\
&= \frac{C(k,\ell)}{\theta((a/b)q^{k-\ell}, acq^{k+\ell-1}, bcq^{k+\ell-1}, q^{k+\ell};p)} \\
&\quad \times \frac{\theta((a/b)q^{k-\ell}, b/a;p)(bcq^{\ell};q,p)_k(acq^k, ab, cx, c/x, q^{k+1};q,p)_\ell}{\theta((a/b)q^k, (b/a)q^{\ell};p)(bc;q,p)_k(ac, abq^k, cxq^k, (c/x)q^k, q;q,p)_\ell} q^{\ell}
\end{aligned}
$$

with

$$
\begin{aligned}
C(k,\ell) &= \theta(bcq^{k+2\ell-1}, acq^{k-1}, q^k, (a/b)q^k;p) \\
&\quad - \frac{a}{b} q^{k-\ell} \theta(acq^{2k+\ell-1}, bcq^{\ell-1}, q^{\ell}, (b/a)q^{\ell};p).
\end{aligned}
$$

Here we again use the Weierstraß–Riemann addition formula (1.4), now with the substitution $(x,y,u,v) \mapsto (a^{\frac12}c^{\frac12}q^{k+\ell-\frac12}, a^{-\frac12}bc^{\frac12}q^{\ell-\frac12}, a^{\frac12}c^{\frac12}q^{k-\frac12}, a^{\frac12}c^{\frac12}q^{-\frac12})$. This yields the equality

$$
C(k,\ell) = \theta((a/b)q^{k-\ell}, acq^{k+\ell-1}, bcq^{k+\ell-1}, q^{k+\ell};p)
$$

and the proof is complete. $\square$

PROOF OF (1.5). With the specific choice of $h(i,j)$ in (2.8) we obtain from (2.4b) and (2.7) the following explicit formulas for $A(k,l)$:

$$\begin{aligned}
A(k,0) &= \prod_{i=0}^{k-1} \frac{\theta(bcq^i, (c/b)q^i, axq^i, (a/x)q^i; p)}{\theta(abq^i, (a/b)q^i, cxq^i, (c/x)q^i; p)} = \frac{(bc, c/b, ax, a/x; q,p)_k}{(ab, a/b, cx, c/x; q,p)_k} \\
&\quad \text{for } k \in \{0,1,\dots,m\}, \\
A(0,\ell) &= \prod_{j=0}^{\ell-1} \frac{\theta(acq^j, (c/a)q^j, bxq^j, (b/x)q^j; p)}{\theta(abq^j, (b/a)q^j, cxq^j, (c/x)q^j; p)} = \frac{(ac, c/a, bx, b/x; q,p)_\ell}{(ab, b/a, cx, c/x; q,p)_\ell} \\
&\quad \text{for } \ell \in \{0,1,\dots,n\}, \\
A(k,\ell) &= B(k,\ell) \frac{(bc, c/b, ax, a/x; q,p)_k}{(ab, a/b, cx, c/x; q,p)_k} \frac{(ac, c/a, bx, b/x; q,p)_\ell}{(ab, b/a, cx, c/x; q,p)_\ell} \\
&\quad \text{for } k \in \{1,2,\dots,m\} \quad \text{and} \quad \ell \in \{1,2,\dots,n\}.
\end{aligned}$$

Combining the last equation with Lemma 2.3, we have

$$\begin{aligned}
A(k,\ell) &= \theta\Big(\frac{a}{b}q^{k-\ell};p\Big) \frac{(bcq^\ell, c/b, ax, a/x; q,p)_k (q^{k+1}, acq^k, c/a, bx, b/x; q,p)_\ell}{(a/b; q,p)_{k+1} (q, q(b/a); q,p)_\ell (ab, cx, c/x; q,p)_{k+\ell}} q^\ell \\
&= \theta\Big(\frac{b}{a}q^{\ell-k};p\Big) \frac{(acq^k, c/a, bx, b/x; q,p)_\ell (q^{\ell+1}, bcq^\ell, c/b, ax, a/x; q,p)_k}{(b/a; q,p)_{\ell+1} (q, q(a/b); q,p)_k (ab, cx, c/x; q,p)_{\ell+k}} q^k \\
&\quad \text{for } (k,\ell) \in \Lambda_{m,n}.
\end{aligned}$$

Inserting this explicit expression for $A(k,l)$, and those for $h(i,j)$ and $1-h(i,j)$ from (2.8) and (2.9), respectively, into (2.5) settles (1.5). □

3. Chaundy–Bullard type identities viewed as Bézout identities

As pointed out in [**9**, Remark 2.2], the Chaundy–Bullard identity (1.1) can be regarded as a *Bézout identity*,

$$1 = P^{(1)}(x)Q^{(1)}(x) + P^{(2)}(x)Q^{(2)}(x). \tag{3.1}$$

Here $P^{(1)}(x)$ and $P^{(2)}(x)$ are polynomials in $\mathbb{C}[x]$ of degrees $n+1$ and $m+1$, respectively, with no common zeros. The polynomials $Q^{(1)}(x)$ and $Q^{(2)}(x)$ in (3.1) have degree m and n, respectively, and are moreover unique. If we set $P^{(1)}(x) = (1-x)^{n+1}$ and $P^{(2)}(x) = x^{m+1}$, they are polynomials without common zeros of degree $n+1$ and $m+1$, respectively. Hence we have the equality

$$1 = (1-x)^{n+1} Q^{(1)}_{m,n}(x) + x^{m+1} Q^{(2)}_{m,n}(x), \tag{3.2}$$

where $Q^{(1)}_{m,n}(x)$ and $Q^{(2)}_{m,n}(x)$ are polynomials of degree m and n, respectively. They are uniquely determined by using the symmetry $Q^{(2)}_{m,n}(x) = Q^{(1)}_{n,m}(1-x)$ and finding $Q^{(1)}_{m,n}(x)$ by dividing both sides of the identity in (3.2) by $(1-x)^{n+1}$ and carrying out Taylor expansion in x, with the result being given by (1.1).

It is natural to ask whether this interpretation extends and can also be used to prove the various extensions of the Chaundy–Bullard identity. This can indeed be done for three of the basic extensions (namely, the q-extension and the two $(a,b;q)$-extensions), as these implicitly involve polynomial bases. We did not succeed in interpreting the $(a,b,c;q)$-extension or the elliptic extension of the Chaundy–Bullard identity as Bézout identities (as the underlying bases there are not polynomial bases but special rational function bases, respectively, elliptic function bases).

The analysis in this section strongly connects to results in basic hypergeometric series (cf. [**5**]). The *basic hypergeometric series* ${}_{r+1}\phi_r$ is defined by

$$ {}_{r+1}\phi_r\left[\begin{matrix}a_0, a_1, \ldots, a_r \\ b_1, \ldots, b_r\end{matrix}; q, z\right] := \sum_{k=0}^{\infty} \frac{(a_0, a_1, \cdots, a_r; q)_k}{(q, b_1, \cdots, b_r; q)_k} z^k, $$

where $|z| < 1$ and $|q| < 1$, if the series does not terminate, for convergence [**5**].

3.1. q-Extension of the Chaundy–Bullard identity. While the first and second term on the right-hand side of original Chaundy–Bullard identity (1.1) are evidently symmetric with respect to the simple involution $(x, n, m) \mapsto (1-x, m, n)$, the corresponding symmetry for (1.12), i.e.,

$$ 1 = (x;q)_{n+1} \sum_{k=0}^{m} \begin{bmatrix} n+k \\ k \end{bmatrix}_q x^k + x^{m+1} \sum_{k=0}^{n} \begin{bmatrix} m+k \\ k \end{bmatrix}_q q^k (x;q)_k, \tag{3.3} $$

is less evident. In order to clarify the "hidden" symmetry, we shall look at the transition matrix of the two respective polynomial bases $[x^n]_{n\in\mathbb{N}_0}$ and $[(x;q)_n]_{n\in\mathbb{N}_0}$. This will enable us to interpret (3.3) as a Bézout identity (3.1).

By the q-binomial theorem, we have (cf. [**5**, Ex. 1.2 (vi)])

$$ (x;q)_n = \sum_{k=0}^{n} \begin{bmatrix} n \\ k \end{bmatrix}_q (-1)^k q^{\binom{k}{2}} x^k. $$

Now, with the lower-triangular matrix $F = (f_{nk})_{n,k\in\mathbb{N}_0}$ defined by its entries

$$ f_{nk} = \begin{bmatrix} n \\ k \end{bmatrix}_q (-1)^k q^{\binom{k}{2}}, $$

the inverse of F is known to be the lower-triangular matrix $G = (g_{nk})_{n,k\in\mathbb{N}_0}$ with entries

$$ g_{nk} = \begin{bmatrix} n \\ k \end{bmatrix}_q (-1)^k q^{\binom{k}{2}+k(1-n)}. $$

(This matrix inversion is equivalent to the case $a \to 0$ in the matrix inversion for B in [**2**].) A simple computation reveals that

$$ g_{nk}(q) = f_{nk}(q^{-1}). $$

Therefore, the relation

$$ (x;q)_n = \sum_{k=0}^{n} f_{nk}(q) x^k \tag{3.4a} $$

is equivalent to

$$ x^n = \sum_{k=0}^{n} f_{nk}(q^{-1}) (x;q)_k. \tag{3.4b} $$

Let $\mathbb{C}(q)[x]$ be the vector space of polynomials in x with coefficients that are rational functions in q over the field $\mathbb{C}$. We define the linear operator T on $\mathbb{C}(q)[x]$ by

$$ T \sum_{k\geqslant 0} c_k(q) x^k = \sum_{k\geqslant 0} c_k(q^{-1}) (x;q)_k. $$

Note that T is an *involution* (this follows immediately from (3.4)) but not a homomorphism (unless $q = 1$). To derive (1.12) using Bézout's identity, observe that for each m,n there exist unique polynomials $Q^{(1)}_{m,n}$ and $Q^{(2)}_{m,n}$ of degree m,n, respectively, such that

$$1=(x;q)_{n+1}Q^{(1)}_{m,n}(x;q)+x^{m+1}Q^{(2)}_{m,n}(x;q).$$

This implies

$$\frac{1}{(x;q)_{n+1}}=Q^{(1)}_{m,n}(x;q)+\mathcal{O}(x^{m+1}),\qquad \text{as } x\to 0.$$

The next step is to compute $Q^{(1)}_{m,n}(x;q)$ by using

$$\frac{1}{(x;q)_{n+1}}=\sum_{k=0}^{m}\begin{bmatrix}n+k\\k\end{bmatrix}_q x^k+\mathcal{O}(x^{m+1}),\qquad \text{as } x\to 0, \tag{3.5}$$

a result which is easily deduced from the non-terminating q-binomial theorem [**5**, Equation (II.3)].

If we can show that

$$T\Big((x;q)_{n+1}\sum_{k=0}^{m}\begin{bmatrix}n+k\\k\end{bmatrix}_q x^k\Big)=x^{n+1}Q^{(2)}_{n,m}(x;q)$$

for

$$Q^{(2)}_{m,n}(x;q)=\sum_{k=0}^{n}\begin{bmatrix}m+k\\k\end{bmatrix}_q q^k(x;q)_k,$$

then we are done, as $Q^{(2)}_{n,m}(x;q)$ must have degree m and be unique. The computations are as follows:

$$\begin{aligned}
&T\Big((x;q)_{n+1}\sum_{k=0}^{m}\begin{bmatrix}n+k\\k\end{bmatrix}_q x^k\Big)\\
&=T\sum_{\ell=0}^{n+1}\sum_{k=0}^{m}\begin{bmatrix}n+1\\\ell\end{bmatrix}_q(-1)^\ell q^{\binom{\ell}{2}}\begin{bmatrix}n+k\\k\end{bmatrix}_q x^{k+\ell}\\
&=\sum_{\ell=0}^{n+1}\sum_{k=0}^{m}\begin{bmatrix}n+1\\\ell\end{bmatrix}_{q^{-1}}(-1)^\ell q^{-\binom{\ell}{2}}\begin{bmatrix}n+k\\k\end{bmatrix}_{q^{-1}}(x;q)_{k+\ell}\\
&=\sum_{k=0}^{m}\begin{bmatrix}n+k\\k\end{bmatrix}_{q^{-1}}(x;q)_k\sum_{\ell=0}^{n+1}\begin{bmatrix}n+1\\\ell\end{bmatrix}_{q^{-1}}(-1)^\ell q^{-\binom{\ell}{2}}(xq^k;q)_\ell\\
&=\sum_{k=0}^{m}\begin{bmatrix}n+k\\k\end{bmatrix}_{q^{-1}}(x;q)_k\,x^{n+1}q^{k(n+1)}\\
&=x^{n+1}\sum_{k=0}^{m}\begin{bmatrix}n+k\\k\end{bmatrix}_q q^k(x;q)_k,
\end{aligned}$$

which settles (3.3).

3.2. $(a,b;q)$-extension of the first kind of the Chaundy–Bullard identity. Next we consider the $(a,b;q)$-extensions of the first kind:

$$1=\frac{(bx;q)_{n+1}}{(b/a;q)_{n+1}}\sum_{k=0}^{m}\frac{(q^{n+1},ax;q)_k}{(q,aq/b;q)_k}q^k+\frac{(ax;q)_{m+1}}{(a/b;q)_{m+1}}\sum_{k=0}^{n}\frac{(q^{m+1},bx;q)_k}{(q,bq/a;q)_k}q^k. \tag{3.6}$$

The transition matrix $F=(f_{nk})_{n,k\in\mathbb{N}_0}$ between the polynomial bases $[(ax;q)_n]_{n\in\mathbb{N}_0}$ and $[(bx;q)_n]_{n\in\mathbb{N}_0}$ is clearly symmetric in a and b. We actually do not need its explicit form here but nonetheless note that the connection coefficients are given by

$$f_{nk}(a,b;q)=\frac{(b/a;q)_n(q^{-n};q)_k}{(q,aq^{1-n}/b;q)_k}q^k$$

(the coefficients of the inverse sequence are $g_{nk}(a,b;q)=f_{nk}(b,a;q)$); the connecting relation

$$\sum_{k=0}^{n}f_{nk}(a,b;q)(ax;q)_k=(bx;q)_n$$

is equivalent to the q-Chu–Vandermonde summation [**5**, Equation (II.6)] (which conversely uniquely determines the connection coefficients $f_{nk}(a,b;q)$).

To prove (3.6) using Bézout's identity, observe that for each m,n there exist unique polynomials $Q^{(1)}_{m,n}$ and $Q^{(2)}_{m,n}$ of degree m,n, respectively, such that

$$1=(bx;q)_{n+1}Q^{(1)}_{m,n}(a,b,x;q)+(ax;q)_{m+1}Q^{(2)}_{m,n}(a,b,x;q). \tag{3.7}$$

This implies

$$\frac{1}{(bx;q)_{n+1}}=Q^{(1)}_{m,n}(a,b,x;q)\qquad \bmod\ (ax;q)_{m+1}.$$

The next step is to compute $Q^{(1)}_{m,n}(a,b,x;q)$ by using the $(a,b,c)\mapsto(q^{n+1},ax,aq/b)$ special case of [**5**, Appendix (II.23)], which is

$$\frac{(b/a;q)_{n+1}}{(bx;q)_{n+1}}={}_2\phi_1\left[\begin{matrix}q^{n+1},ax\\ aq/b\end{matrix};q,q\right]+\frac{(b/a,q^{n+1},ax;q)_\infty}{(a/b,bq^{n+1}/a,bx;q)_\infty}\,{}_2\phi_1\left[\begin{matrix}bq^{n+1}/a,bx\\ bq/a\end{matrix};q,q\right].$$

This implies

$$\frac{1}{(bx;q)_{n+1}}=\frac{1}{(b/a;q)_{n+1}}\sum_{k=0}^{m}\frac{(q^{n+1},ax;q)_k}{(q,aq/b;q)_k}q^k\qquad \bmod\ (ax;q)_{m+1},$$

for $(ax;q)_{m+1}$ divides $(ax;q)_k$ for each $k>m$. Hence we deduce

$$Q^{(1)}_{m,n}(a,b,x;q)=\frac{1}{(b/a;q)_{n+1}}\sum_{k=0}^{m}\frac{(q^{n+1},ax;q)_k}{(q,aq/b;q)_k}q^k.$$

Since, in addition to (3.7), we also have

$$1=(bx;q)_{n+1}Q^{(2)}_{n,m}(b,a,x;q)+(ax;q)_{m+1}Q^{(1)}_{n,m}(b,a,x;q),$$

it follows by uniqueness ($Q^{(1)}_{m,n}(a,b,x;q)$ and $Q^{(2)}_{n,m}(b,a,x;q)$ have the same degree m) that we must have

$$Q^{(1)}_{m,n}(a,b,x;q)=Q^{(2)}_{n,m}(b,a,x;q),$$

which settles (3.6).

3.3. $(a,b;q)$-extension of the second kind of the Chaundy–Bullard identity. The following identity of polynomials in $\mathbb{C}(q,a,b)[x+x^{-1}]$ is the $(a,b;q)$-extension of the second kind:

$$\begin{aligned} 1 &= \frac{(bx,b/x;q)_{n+1}}{(ab,b/a;q)_{n+1}} \sum_{k=0}^{m} \frac{(q^{n+1},ax,a/x;q)_k}{(q,aq/b,abq^{1+n};q)_k} q^k \\ &\quad + \frac{(ax,a/x;q)_{m+1}}{(ab,a/b;q)_{m+1}} \sum_{k=0}^{n} \frac{(q^{m+1},bx,b/x;q)_k}{(q,bq/a,abq^{1+m};q)_k} q^k. \end{aligned} \tag{3.8}$$

The transition matrix between the two polynomial sequences $[(ax,a/x;q)_n]_{n\in\mathbb{N}_0}$ and $[(bx,b/x;q)_n]_{n\in\mathbb{N}_0}$ is clearly symmetric in a and b. We actually do not need its explicit form here but nevertheless note that the connection coefficients are given by

$$\widetilde{f}_{nk}(a,b;q) = \frac{(ab,b/a;q)_n(q^{-n};q)_k}{(q,ab,aq^{1-n}/b;q)_k} q^k$$

(the coefficients of the inverse sequence are $\widetilde{g}_{nk}(a,b;q) = f_{nk}(b,a;q)$); the connecting relation

$$\sum_{k=0}^{n} \widetilde{f}_{nk}(a,b;q)(ax,a/x;q)_k = (bx,b/x;q)_n$$

is equivalent to the q-Pfaff–Saalschütz summation [**5**, Equation (II.12)] (which conversely uniquely determines the connection coefficients $\widetilde{f}_{nk}(a,b;q)$).

To prove (3.8) using Bézout's identity, observe that for each m,n there exist unique polynomials $Q^{(1)}_{m,n}$ and $Q^{(2)}_{m,n}$ of degree m,n, respectively, such that

$$1 = (bx,b/x;q)_{n+1} Q^{(1)}_{m,n}(a,b,x+x^{-1};q) + (ax,a/x;q)_{m+1} Q^{(2)}_{m,n}(a,b,x+x^{-1};q). \tag{3.9}$$

This implies

$$\frac{1}{(bx,b/x;q)_{n+1}} = Q^{(1)}_{m,n}(a,b,x+x^{-1};q) \qquad \bmod (ax,a/x;q)_{m+1}.$$

The next step is to compute $Q^{(1)}_{m,n}(a,b,x+x^{-1};q)$ by using the $(a,b,c,e,f) \mapsto (q^{n+1},ax,a/x,aq/b,abq^{n+1})$ special case of [**5**, Appendix (II.24)], which is

$$\begin{aligned} \frac{(ab,b/a;q)_{n+1}}{(bx,b/x;q)_{n+1}} &= {}_3\phi_2\left[\begin{matrix} q^{n+1},ax,a/x \\ aq/b,abq^{n+1} \end{matrix};q,q\right] \\ &\quad + \frac{(b/a,q^{n+1},ax,a/x,b^2q^{n+1};q)_\infty}{(a/b,bq^{n+1}/a,bx,b/x,abq^{n+1};q)_\infty} {}_3\phi_2\left[\begin{matrix} bq^{n+1}/a,bx,b/x \\ bq/a,b^2q^{n+1} \end{matrix};q,q\right]. \end{aligned}$$

The last equation implies

$$\begin{aligned} &\frac{1}{(bx,b/x;q)_{n+1}} \\ &= \frac{1}{(ab,b/a;q)_{n+1}} \sum_{k=0}^{m} \frac{(q^{n+1},ax,a/x;q)_k}{(q,aq/b,abq^{n+1};q)_k} q^k \qquad \bmod (ax,a/x;q)_{m+1}, \end{aligned}$$

for $(ax,a/x;q)_{m+1}$ divides $(ax,a/x;q)_k$ for each $k>m$. Hence we deduce

$$Q^{(1)}_{m,n}(a,b,x+x^{-1};q) = \frac{1}{(ab,b/a;q)_{n+1}} \sum_{k=0}^{m} \frac{(q^{n+1},ax,a/x;q)_k}{(q,aq/b,abq^{n+1};q)_k} q^k.$$

Since, in addition to (3.9), we also have

$$1=(bx,b/x;q)_{n+1}Q^{(2)}_{n,m}(b,a,x+x^{-1};q)+(ax,a/x;q)_{m+1}Q^{(1)}_{n,m}(b,a,x+x^{-1};q),$$

it follows by uniqueness ($Q^{(1)}_{m,n}(a,b,x+x^{-1};q)$ and $Q^{(2)}_{n,m}(b,a,x+x^{-1};q)$ have the same degree m) that we must have

$$Q^{(1)}_{m,n}(a,b,x+x^{-1};q)=Q^{(2)}_{n,m}(b,a,x+x^{-1};q),$$

which settles (3.8).

4. Variants of the q-extended Chaundy–Bullard identities and q-commuting variables

In the original (1.1), if we replace x by $x/(x-1)$ and multiply the identity by $(1-x)^{m+n+1}$, a variant is obtained, namely

$$\begin{aligned}(1-x)^{m+n+1}&=\sum_{k=0}^{m}\binom{n+k}{k}(-1)^k x^k(1-x)^{m-k}\\&\quad+(-1)^{m+1}x^{m+1}\sum_{k=0}^{n}\binom{m+k}{k}(1-x)^{n-k}.\end{aligned}\tag{4.1}$$

As discussed in [**9**], (1.1) is equivalent to the identity in the form involving two variables x and y,

$$\begin{aligned}(x+y)^{m+n+1}&=y^{n+1}\sum_{k=0}^{m}\binom{n+k}{k}x^k(x+y)^{m-k}\\&\quad+x^{m+1}\sum_{k=0}^{n}\binom{m+k}{k}y^k(x+y)^{n-k},\end{aligned}\tag{4.2}$$

which is in *homogeneous form.*

In order to give a q-extension of the homogeneous Chaundy–Bullard identity (4.2), we consider the unital algebra $\mathbb{C}_q[X,Y]$ defined over $\mathbb{C}$ generated by X,Y, satisfying the relation

$$YX=qXY.$$

$\mathbb{C}_q[X,Y]$ can be regarded as a q-deformation of the commutative algebra $\mathbb{C}[x,y]$. We call X,Y forming $\mathbb{C}_q[X,Y]$ *q-commuting variables.* The following *binomial theorem for q-commuting variables* is well known (see [**7, 15**] and references therein),

$$(X+Y)^n=\sum_{k=0}^{n}\begin{bmatrix}n\\k\end{bmatrix}_q X^kY^{n-k},$$

for $X,Y\in\mathbb{C}_q[X,Y]$, where the q-binomial coefficient is defined by (1.13).

Then, from Theorem 5.4 we obtain that a q-extension of (4.2) is given by

$$\begin{aligned}(X+Y)^{m+n+1}&=Y^{n+1}\sum_{k=0}^{m}\begin{bmatrix}n+k\\k\end{bmatrix}_q q^{-(n+1)k}\,X^k(X+Y)^{m-k}\\&\quad+X^{m+1}\sum_{k=0}^{n}\begin{bmatrix}m+k\\k\end{bmatrix}_{q^{-1}} q^{(m+1)k}\,Y^k(X+Y)^{n-k}\end{aligned}$$

for $X, Y \in \mathbb{C}_q[X, Y]$. This is the homogeneous form of the q-extension of the Chaundy–Bullard identity (1.12). The left-hand and right-hand sides of this identity are both invariant under the transformation $(X, Y, q, m, n) \mapsto (Y, X, q^{-1}, n, m)$.

5. Elliptic extensions of the binomial theorem

Recall (from Section 1) that we denote by $\mathbb{E}_{a_1,a_2,\ldots,a_s;q,p}$ the field of totally elliptic functions over $\mathbb{C}$, in the complex variables $\log_q a_1, \ldots, \log_q a_s$, with equal periods σ^{-1}, $\tau\sigma^{-1}$ (where $q = e^{2\pi\sqrt{-1}\sigma}$, $p = e^{2\pi\sqrt{-1}\tau}$, $\sigma, \tau \in \mathbb{C}$, $\Im\tau > 0$), of double periodicity.

5.1. An elliptic extension of the binomial theorem. In [**15**, Theorem 2] one of the authors proved an elliptic extension of the binomial theorem which we recall for convenience and for comparison with our new analogous result in Theorem 5.7 below.

For indeterminates a, b, complex numbers q, p (with $|p| < 1$), and nonnegative integers n, k, define a variant of elliptic binomial coefficients, which here, for better distinction from (5.9), we may refer to as *W-binomial coefficients* as follows (this is exactly the expression for $w(\mathcal{P}((0,0) \to (k, n-k)))$ in [**14**, Theorem 2.1]):

$$\begin{bmatrix} n \\ k \end{bmatrix}_{a,b;q,p} := \frac{(q^{1+k}, aq^{1+k}, bq^{1+k}, aq^{1-k}/b; q, p)_{n-k}}{(q, aq, bq^{1+2k}, aq/b; q, p)_{n-k}}. \tag{5.1}$$

The W-binomial coefficient is indeed elliptic. In particular, $\left[\begin{smallmatrix} n \\ k \end{smallmatrix}\right]_{a,b;q,p} \in \mathbb{E}_{a,b,q^n,q^k;q,p}$.

It is immediate from the definition of (5.1) that (for integers n, k) there holds

$$\begin{bmatrix} n \\ 0 \end{bmatrix}_{a,b;q,p} = \begin{bmatrix} n \\ n \end{bmatrix}_{a,b;q,p} = 1, \tag{5.2a}$$

and

$$\begin{bmatrix} n \\ k \end{bmatrix}_{a,b;q,p} = 0, \qquad \text{whenever} \quad k = -1, -2, \ldots, \quad \text{or} \quad k > n. \tag{5.2b}$$

Furthermore, using the Weierstraß–Riemann addition formula in (1.4) one can verify the following recursion formula for the elliptic binomial coefficients:

$$\begin{bmatrix} n+1 \\ k \end{bmatrix}_{a,b;q,p} = \begin{bmatrix} n \\ k \end{bmatrix}_{a,b;q,p} + \begin{bmatrix} n \\ k-1 \end{bmatrix}_{a,b;q,p} W_{a,b;q,p}(k, n+1-k), \tag{5.2c}$$

for nonnegative integers n and k, where the elliptic weight function $W_{a,b;q,p}$ is defined on $\mathbb{N}_0^2$ as

$$W_{a,b;q,p}(s,t) := \frac{\theta(aq^{s+2t}, bq^{2s}, bq^{2s-1}, aq^{1-s}/b, aq^{-s}/b; p)}{\theta(aq^s, bq^{2s+t}, bq^{2s+t-1}, aq^{1+t-s}/b, aq^{t-s}/b; p)} q^t. \tag{5.3}$$

Clearly, $W_{a,b;q,p}(s, 0) = 1$, for all s. If one lets $p \to 0$, $a \to 0$, then $b \to 0$ (in this order), the weights in (5.3) reduce to the standard q-weights

$$W_q(s,t) := \lim_{b \to 0} \Big(\lim_{a \to 0} \Big(\lim_{p \to 0} W_{a,b;q,p}(s,t) \Big) \Big) = q^t,$$

and the relations in (5.2) reduce to

$$\begin{bmatrix} n \\ 0 \end{bmatrix}_q = \begin{bmatrix} n \\ n \end{bmatrix}_q = 1,$$

$$\begin{bmatrix} n+1 \\ k \end{bmatrix}_q = \begin{bmatrix} n \\ k \end{bmatrix}_q + \begin{bmatrix} n \\ k-1 \end{bmatrix}_q q^{n+1-k}$$

for positive integers n and k with $n \geqslant k$, which is a well-known recursion for the q-binomial coefficients. (If instead, one lets $p \to 0$, $b \to 0$, and then $a \to 0$ (in this order), the weight function in (5.3) reduces to q^{-t} and (5.2) reduces to the recursion for the q-binomial coefficients where q has been replaced by q^{-1}.) In [**14**] lattice paths in the integer lattice $\mathbb{N}_0^2$ were enumerated with respect to precisely this weight function. A similar weight function was subsequently used by Borodin, Gorin and Rains in [**1**, Section 10] (see in particular the expression obtained for $\frac{w(i,j+1)}{w(i,j)}$ on p. 780 of that paper) in the context of weighted lozenge tilings.

DEFINITION 5.1. For two complex numbers q and p with $|p| < 1$, let $\mathbb{C}_{q,p}[X, Y, \mathbb{E}_{a,b;q,p}]$ denote the associative unital algebra over $\mathbb{C}$, generated by X, Y and the commutative subalgebra $\mathbb{E}_{a,b;q,p}$, satisfying the relations

$$YX = W_{a,b;q,p}(1,1)\, XY \tag{5.4a}$$

and

$$Xf(a,b) = f(aq, bq^2)\, X, \tag{5.4b}$$

$$Yf(a,b) = f(aq^2, bq)\, Y \tag{5.4c}$$

for all $f \in \mathbb{E}_{a,b;q,p}$.

We refer to the variables X, Y, a, b forming $\mathbb{C}_{q,p}[X, Y, \mathbb{E}_{a,b;q,p}]$ as *elliptic commuting* variables. The algebra $\mathbb{C}_{q,p}[X, Y, \mathbb{E}_{a,b;q,p}]$ reduces to $\mathbb{C}_q[X, Y]$ if one formally lets $p \to 0$, $a \to 0$, then $b \to 0$ (in this order), while (having eliminated the nome p) relaxing the condition of ellipticity. It should be noted that the monomials X^kY^l form a basis for the algebra $\mathbb{C}_{q,p}[X, Y, \mathbb{E}_{a,b;q,p}]$ as a left module over $\mathbb{E}_{a,b;q,p}$, i.e., any element can be written uniquely as a finite sum $\sum_{k,l\geqslant 0} f_{kl}X^kY^l$ with $f_{kl} \in \mathbb{E}_{a,b;q,p}$ which we call the *normal form* of the element.

The following result from [**15**, Theorem 2] shows that the normal form of the binomial $(X + Y)^n$ is nice; each (left) coefficient of X^kY^{n-k} completely factorizes as an expession in $\mathbb{E}_{a,b;q,p}$.

THEOREM 5.2 (Binomial theorem for variables in $\mathbb{C}_{q,p}[X, Y, \mathbb{E}_{a,b;q,p}]$). *Let* $n \in \mathbb{N}_0$. *Then, as an identity in* $\mathbb{C}_{q,p}[X, Y, \mathbb{E}_{a,b;q,p}]$, *we have*

$$(X+Y)^n = \sum_{k=0}^{n} \begin{bmatrix} n \\ k \end{bmatrix}_{a,b;q,p} X^kY^{n-k}. \tag{5.5}$$

In [**15**] convolution was applied to this result (together with comparison of coefficients) to recover Frenkel and Turaev's ${}_{10}V_9$ summation [**4**] (see also [**5**, Equation (11.4.1)]), an identity which is fundamental to the theory of elliptic hypergeometric series.

PROPOSITION 5.3 (Frenkel and Turaev's ${}_{10}V_9$ summation). *Let $n \in \mathbb{N}_0$ and $a, b, c, d, e, q, p \in \mathbb{C}$ with $|p| < 1$. Then there holds the following identity:*

$$\begin{aligned}\sum_{k=0}^{n} \frac{\theta(aq^{2k};p)}{\theta(a;p)} \frac{(a,b,c,d,e,q^{-n};q,p)_k}{(q,aq/b,aq/c,aq/d,aq/e,aq^{n+1};q,p)_k} q^k \\ = \frac{(aq,aq/bc,aq/bd,aq/cd;q,p)_n}{(aq/b,aq/c,aq/d,aq/bcd;q,p)_n},\end{aligned} \tag{5.6}$$

where $a^2q^{n+1} = bcde$.

It is straightforward to use the lattice path model to derive a homogeneous Chaundy–Bullard identity for the elliptic commuting variables forming the algebra $\mathbb{C}_{q,p}[X, Y, \mathbb{E}_{a,b;q,p}]$. (We omit the details; it is also in principle possible to verify the identity by induction.) The result is as follows.

THEOREM 5.4 (Homogeneous Chaundy–Bullard identity for variables in $\mathbb{C}_{q,p}[X, Y, \mathbb{E}_{a,b;q,p}]$). *Let $n, m \in \mathbb{N}_0$. Then the following identity is valid in $\mathbb{C}_{q,p}[X, Y, \mathbb{E}_{a,b;q,p}]$:*
(5.7)

$$\begin{aligned}(X+Y)^{m+n+1} &= \sum_{k=0}^{m} \begin{bmatrix} n+k \\ k \end{bmatrix}_{a,b;q,p} X^k Y^{n+1} (X+Y)^{m-k} \\ &+ \sum_{k=0}^{n} \begin{bmatrix} m+k \\ m \end{bmatrix}_{a,b;q,p} W_{a,b;q,p}(m+1,k)\, X^{m+1} Y^k (X+Y)^{n-k}.\end{aligned}$$

5.2. A new elliptic extension of the binomial theorem. Inspired by the lattice path model in Section 2 of this paper, we present a new elliptic extension of the binomial theorem which is similar to Theorem 5.2 but different, see Theorem 5.7 below.

First define $h_{x;a,b,c;q,p}(i,j)$ as in (2.8), and let

$$H_{x;a,b,c:q,p}(i,j) := \frac{h_{x;a,b,c;q,p}(i,j)}{h_{x;a,b,c;q,p}(i,0)}. \tag{5.8}$$

With $B(k,\ell)$ given by (2.11) in Lemma 2.3, for $n \in \mathbb{N}_0$ we define a variant of elliptic binomial coefficients, which (in view of Proposition 5.5) one may refer to as *H-binomial coefficients*, by

$$\begin{bmatrix} n \\ k \end{bmatrix}_{x;a,b,c;q,p} := \begin{cases} B(k, n-k), & k \in \{0,1,\dots,n\}, \\ 0, & k \in -\mathbb{N}_0 \text{ or } k > n. \end{cases} \tag{5.9}$$

Then using the symmetry (2.9) in Lemma 2.2, the assertion of Lemma 2.3 is rewritten as follows.

PROPOSITION 5.5. *For $n \in \mathbb{N}_0$, $0 \leqslant k \leqslant n$,*

$$\begin{aligned}\begin{bmatrix} n+1 \\ k \end{bmatrix}_{x;a,b,c;q,p} &= \begin{bmatrix} n \\ k-1 \end{bmatrix}_{x;a,b,c;q,p} H_{x;a,b,c;q,p}(k-1, n+1-k) \\ &+ \begin{bmatrix} n \\ k \end{bmatrix}_{x;a,b,c;q,p} H_{x;b,a,c;q,p}(n-k, k).\end{aligned} \tag{5.10}$$

It is obvious that the H-binomial coefficients $\begin{bmatrix} n \\ k \end{bmatrix}_{x;a,b,c;q,p}$ have a nice combinatorial interpretation in terms of weighted lattice paths. The generating function

$w_{x;a,b,c;q,p}$ with respect to the weights $H_{x;a,b,c;q,p}$ of all paths from $(0,0)$ to $(k,n-k)$ is clearly

$$w_{x;a,b,c;q,p}(\mathcal{P}((0,0)\to(k,n-k))) = \begin{bmatrix} n \\ k \end{bmatrix}_{x;a,b,c;q,p}. \tag{5.11}$$

We now define a new elliptic extension of the non-commutative algebra $\mathbb{C}_q[X,Y]$:

DEFINITION 5.6. For two complex numbers q and p with $|p|<1$, let $\mathbb{C}_{q,p}[X,Y,\mathbb{E}_{x,a,b,c;q,p}]$ denote the associative unital algebra over $\mathbb{C}$, generated by X,Y and the commutative subalgebra $\mathbb{E}_{x,a,b,c;q,p}$, satisfying the relations

$$YX = H_{x;a,b,c:q,p}(0,1)\,XY \tag{5.12a}$$

and

$$Xf(x,a,b,c) = f(x,aq,b,cq)\,X, \tag{5.12b}$$

$$Yf(x,a,b,c) = f(x,a,bq,cq)\,Y \tag{5.12c}$$

for all $f\in\mathbb{E}_{x,a,b,c;q,p}$.

Just as for the variables X,Y,a,b forming $\mathbb{C}_{q,p}[X,Y,\mathbb{E}_{a,b;q,p}]$ in Definition 5.1, we refer to the variables X,Y,x,a,b,c forming $\mathbb{C}_{q,p}[X,Y,\mathbb{E}_{x,a,b,c;q,p}]$ in Definition 5.6 as *elliptic-commuting* variables.

We note that from (2.8) and (5.12) follow the useful relations

$$X\,h_{x;a,b,c;q,p}(i,j)^{\pm1} = h_{x;a,b,c;q,p}(i+1,j)^{\pm1}\,X, \tag{5.13a}$$

$$X\,h_{x;b,a,c;q,p}(j,i)^{\pm1} = h_{x;b,a,c;q,p}(j,i+1)^{\pm1}\,X, \tag{5.13b}$$

$$Y\,h_{x;a,b,c;q,p}(i,j)^{\pm1} = h_{x;a,b,c;q,p}(i,j+1)^{\pm1}\,Y, \tag{5.13c}$$

$$Y\,h_{x;b,a,c;q,p}(j,i)^{\pm1} = h_{x;b,a,c;q,p}(j+1,i)^{\pm1}\,Y \tag{5.13d}$$

for all $(i,j)\in\mathbb{N}_0^2$. Further, by induction on r and s, one has

$$Y^sX^r = \Big(\prod_{i=0}^{r-1} H_{x;a,b,c;q,p}(i,s)\Big)X^rY^s \tag{5.14}$$

for all $r,s\in\mathbb{N}_0$.

Similarly as in $\mathbb{C}_{q,p}[X,Y,\mathbb{E}_{a,b;q,p}]$, the monomials X^kY^l form a basis for the algebra $\mathbb{C}_{q,p}[X,Y,\mathbb{E}_{x,a,b,c;q,p}]$, now as a left module over $\mathbb{E}_{x,a,b,c;q,p}$. That is, any element can be written uniquely as a finite sum $\sum_{k,l\geqslant0} f_{kl}X^kY^l$ with $f_{kl}\in\mathbb{E}_{x,a,b,c;q,p}$, the normal form of the element.

The following non-commutative elliptic binomial theorem which readily follows from [**15**, Theorem 3] (and could be independently proved by induction) shows that the normal form of the binomial $\big(X+h_{x;b,a,c;q,p}(0,0)\,Y\big)^n$ is nice; each (left) coefficient of X^kY^{n-k} completely factorizes in $\mathbb{E}_{x,a,b,c;q,p}$.

THEOREM 5.7 (Binomial theorem for elliptic commuting variables in $\mathbb{C}_{q,p}[X,Y,\mathbb{E}_{x,a,b,c;q,p}]$). *Let $n\in\mathbb{N}_0$. Then, as an identity in $\mathbb{C}_{q,p}[X,Y,\mathbb{E}_{x,a,b,c;q,p}]$, we have*

$$\begin{aligned}&\big(X+h_{x;b,a,c;q,p}(0,0)\,Y\big)^n\\&=\sum_{k=0}^{n}\begin{bmatrix} n \\ k \end{bmatrix}_{x;a,b,c;q,p}\Big(\prod_{j=0}^{n-k-1} h_{x;b,a,c;q,p}(j,0)\Big)X^kY^{n-k}.\end{aligned} \tag{5.15}$$

By convolution and comparison of coefficients we obtain the following identity.

COROLLARY 5.8 (An elliptic binomial convolution formula). *Let $n, m, k \in \mathbb{N}_0$ and $x, a, b, c, q, p \in \mathbb{C}$ with $|p| < 1$. Then there holds the convolution formula*

$$\begin{aligned}&\begin{bmatrix} n+m \\ k \end{bmatrix}_{x;a,b,c;q,p} \left(\prod_{j=0}^{n+m-k-1} h_{x;b,a,c;q,p}(j,0) \right) \\ &= \sum_{j=0}^{k} \left(\begin{bmatrix} n \\ j \end{bmatrix}_{x;a,b,c;q,p} \begin{bmatrix} m \\ k-j \end{bmatrix}_{x;aq^j,bq^{n-j},cq^n;q,p} \left(\prod_{\ell=0}^{n-j-1} h_{x;b,a,c;q,p}(\ell,0) \right) \right. \\ &\quad \left. \times \left(\prod_{s=0}^{m+j-k-1} h_{x;b,a,c;q,p}(s+n-j,j) \right) \left(\prod_{i=0}^{k-j-1} H_{x;a,b,c;q,p}(i+j,n-j) \right) \right).\end{aligned} \tag{5.16}$$

PROOF. Working in $\mathbb{C}_{q,p}[X, Y, \mathbb{E}_{x,a,b,c;q,p}]$, one expands in two different ways the binomial $(X + h_{x;b,a,c;q,p}(0,0)\,Y)^{n+m}$ and suitably extracts coefficients. On the one hand,

$$\begin{aligned}&\left(X + h_{x;b,a,c;q,p}(0,0)\,Y\right)^{n+m} \\ &= \sum_{k=0}^{n+m} \begin{bmatrix} n+m \\ k \end{bmatrix}_{x;a,b,c;q,p} \left(\prod_{j=0}^{n+m-1} h_{x;b,a,c;q,p}(j,0) \right) X^k Y^{n+m-k}.\end{aligned} \tag{5.17}$$

On the other hand,

$$\begin{aligned}&\left(X + h_{x;b,a,c;q,p}(0,0)\,Y\right)^{n+m} \\ &= \left(X + h_{x;b,a,c;q,p}(0,0)\,Y\right)^{n} \left(X + h_{x;b,a,c;q,p}(0,0)\,Y\right)^{m} \\ &= \sum_{j=0}^{n} \sum_{r=0}^{m} \left(\begin{bmatrix} n \\ j \end{bmatrix}_{x;a,b,c;q,p} \left(\prod_{\ell=0}^{n-j-1} h_{x;b,a,c;q,p}(\ell,0) \right) X^j Y^{n-j} \right. \\ &\quad \left. \times \begin{bmatrix} m \\ r \end{bmatrix}_{x;a,b,c;q,p} \left(\prod_{s=0}^{m-r-1} h_{x;b,a,c;q,p}(s,0) \right) X^r Y^{m-r} \right) \\ &= \sum_{j=0}^{n} \sum_{r=0}^{m} \left(\begin{bmatrix} n \\ j \end{bmatrix}_{x;a,b,c;q,p} \begin{bmatrix} m \\ r \end{bmatrix}_{x;aq^j,bq^{n-j},cq^n;q,p} \left(\prod_{\ell=0}^{n-j-1} h_{x;b,a,c;q,p}(\ell,0) \right) \right. \\ &\quad \left. \times \left(\prod_{s=0}^{m-r-1} h_{x;b,a,c;q,p}(s+n-j,j) \right) X^j Y^{n-j} X^r Y^{m-r} \right).\end{aligned} \tag{5.18}$$

Now use (5.14) to apply

$$X^j Y^{n-j} X^r Y^{m-r} = \left(\prod_{i=0}^{r-1} H_{x;a,b,c;q,p}(i+j,n-j) \right) x^{j+r} y^{n+m-j-r} \quad \text{for } n \geqslant j,$$

and extract and equate (left) coefficients of $X^k Y^{n+m-k}$ in (5.17) and (5.18). This gives the convolution formula (5.16). □

REMARK 5.9. Corollary 5.8 is not a new result, but actually a special case of the Frenkel–Turaev sum in (5.6). Writing out all the expressions in (5.16) in explicit terms using (2.8), (2.11), (5.8), and (5.9), and applying some elementary manipulations (such as those appearing in [**5**, p. 310]) of the theta-shifted factorials, it becomes clear that, up to a multiplicative factor that can be pulled out of the sum, the sum on the right-hand side of (5.16) is indeed the Frenkel–Turaev sum in (5.6) with respect to the substitutions

$$(a, b, c, d, e, n) \mapsto (aq^{-n}/b,\ acq^{n+m},\ q^{1-n}/bc,\ aq^{k-n-m}/b,\ q^{-n}, k).$$

(In particular, the factors depending on x can all be pulled out of the sum and can be canceled with those appearing on the left-hand side. Also, the variable b in the sum is redundant.) By analytic continuation, the integers n and m can be replaced by continuous variables (say $\log_q \nu$ and $\log_q \mu$) and one recovers the full Frenkel–Tureav sum, without the restriction of having parameters that are integer powers of q.

Finally, it is again straightforward to use the lattice path model to derive a homogeneous Chaundy–Bullard identity for the elliptic commuting variables forming the algebra $\mathbb{C}_{q,p}[X, Y, \mathbb{E}_{x,a,b,c;q,p}]$. (We omit the details; it is also in principle possible to verify the identity by induction.) The result is as follows.

THEOREM 5.10 (Homogeneous Chaundy–Bullard identity for elliptic commuting variables in $\mathbb{C}_{q,p}[X, Y, \mathbb{E}_{x,a,b,c;q,p}]$). *Let $n, m \in \mathbb{N}_0$. Then the following identity is valid in $\mathbb{C}_{q,p}[X, Y, \mathbb{E}_{x,a,b,c;q,p}]$:*

$$\begin{aligned}(X + h_{x;b,a,c;q,p}(0,0)\,Y)^{n+m+1} &= \sum_{k=0}^{m} \Bigg(\begin{bmatrix} n+k \\ k \end{bmatrix}_{x;a,b,c;q,p} \Bigg(\prod_{j=0}^{n-1} h_{x;b,a,c;q,p}(j,0) \Bigg) \\ &\qquad \times h_{x;b,a,c;q,p}(n,k)\, X^k Y^{n+1} (X + h_{x;b,a,c;q,p}(0,0)\,Y)^{m-k} \Bigg) \\ &+ \sum_{k=0}^{n} \Bigg(\begin{bmatrix} m+k \\ m \end{bmatrix}_{x;a,b,c;q,p} \Bigg(\prod_{j=0}^{k-1} h_{x;b,a,c;q,p}(j,0) \Bigg) \\ &\qquad \times X^{m+1} Y^k (X + h_{x;b,a,c;q,p}(0,0)\,Y)^{n-k} \Bigg). \end{aligned} \tag{5.19}$$

Acknowledgments

The authors wish to thank the referee for the referee's helpful remarks.

References

[1] A. Borodin, V. Gorin, and E. M. Rains, *q-distributions on boxed plane partitions*, Selecta Math. (N.S.) **16** (2010), no. 4, 731–789, DOI 10.1007/s00029-010-0034-y. MR2734330

[2] D. M. Bressoud, *A matrix inverse*, Proc. Amer. Math. Soc. **88** (1983), no. 3, 446–448, DOI 10.2307/2044991. MR699411

[3] T. W. Chaundy and J. E. Bullard, *John Smith's problem*, Math. Gazette **44** (1960), 253–260. DOI: https://doi.org/10.2307/3614890

[4] I. B. Frenkel and V. G. Turaev, *Elliptic solutions of the Yang-Baxter equation and modular hypergeometric functions*, The Arnold-Gelfand mathematical seminars, Birkhäuser Boston, Boston, MA, 1997, pp. 171–204. MR1429892

[5] G. Gasper and M. Rahman, *Basic hypergeometric series*, 2nd ed., Encyclopedia of Mathematics and its Applications, vol. 96, Cambridge University Press, Cambridge, 2004, DOI 10.1017/CBO9780511526251. MR2128719

[6] P. Kesava Menon, *Summation of certain series*, J. Indian Math. Soc. (N.S.) **25** (1961), 121–128. MR145226

[7] T. H. Koornwinder, *Special functions and q-commuting variables*, Special functions, q-series and related topics (Toronto, ON, 1995), Fields Inst. Commun., vol. 14, Amer. Math. Soc., Providence, RI, 1997, pp. 131–166, DOI 10.1016/s0898-1221(96)90020-6. MR1448685

[8] T. H. Koornwinder, *On the equivalence of two fundamental theta identities*, Anal. Appl. (Singap.) **12** (2014), no. 6, 711–725, DOI 10.1142/S0219530514500559. MR3271026

[9] T. H. Koornwinder and M. J. Schlosser, *On an identity by Chaundy and Bullard. I*, Indag. Math. (N.S.) **19** (2008), no. 2, 239–261, DOI 10.1016/S0019-3577(08)80002-X. MR2489329

[10] T. H. Koornwinder and M. J. Schlosser, *On an identity by Chaundy and Bullard. II. More history*, Indag. Math. (N.S.) **24** (2013), no. 1, 174–180, DOI 10.1016/j.indag.2012.08.003. MR2997759

[11] X. R. Ma, *Six-variable generalization of Ramanujan's reciprocity theorem and its variants*, J. Math. Anal. Appl. **353** (2009), no. 1, 320–328, DOI 10.1016/j.jmaa.2008.11.039. MR2508870

[12] P. R. de Montmort, *Essay d'analyse sur les jeux de hazard* (French), 2nd ed., Chelsea Publishing Co., New York, 1980. MR605303

[13] H. Rosengren, *Elliptic hypergeometric functions*, Lectures on orthogonal polynomials and special functions, London Math. Soc. Lecture Note Ser., vol. 464, Cambridge Univ. Press, Cambridge, 2021, pp. 213–279. MR4413013

[14] M. Schlosser, *Elliptic enumeration of nonintersecting lattice paths*, J. Combin. Theory Ser. A **114** (2007), no. 3, 505–521, DOI 10.1016/j.jcta.2006.07.002. MR2310747

[15] M. J. Schlosser, *A noncommutative weight-dependent generalization of the binomial theorem*, Sém. Lothar. Combin. **81** (2020), Art. B81j, 24 pp. MR4097435

[16] B. Shiffman, *Complete characterization of holomorphic chains of codimension one*, Math. Ann. **274** (1986), no. 2, 233–256, DOI 10.1007/BF01457072. MR838467

[17] V. P. Spiridonov, *Theta hypergeometric series*, Asymptotic combinatorics with application to mathematical physics (St. Petersburg, 2001), NATO Sci. Ser. II Math. Phys. Chem., vol. 77, Kluwer Acad. Publ., Dordrecht, 2002, pp. 307–327. MR2000728

[18] V. P. Spiridonov, *Elliptic hypergeometric terms* (English, with English and French summaries), Arithmetic and Galois theories of differential equations, Sémin. Congr., vol. 23, Soc. Math. France, Paris, 2011, pp. 325–345. MR3076087

[19] E. T. Whittaker and G. N. Watson, *A course of modern analysis*, Cambridge Mathematical Library, Cambridge University Press, Cambridge, 1996. An introduction to the general theory of infinite processes and of analytic functions; with an account of the principal transcendental functions; Reprint of the fourth (1927) edition, DOI 10.1017/CBO9780511608759. MR1424469

Dept. of Physics, Faculty of Science and Engineering, Chuo University, Kasuga, Bunkyo-ku, Tokyo 112-8551, Japan

Dept. of Physics, Faculty of Science and Engineering, Chuo University, Kasuga, Bunkyo-ku, Tokyo 112-8551, Japan
Email address: makoto.katori.mathphys@gmail.com

Korteweg–de Vries Institute, University of Amsterdam, P. O. Box 94248, 1090 GE Amsterdam, The Netherlands
Email address: thkmath@xs4all.nl

Fakultät für Mathematik, Universität Wien, Oskar-Morgenstern-Platz 1, A-1090 Wien, Austria
Email address: michael.schlosser@univie.ac.at

Contemporary Mathematics
Volume **819**, 2025
https://doi.org/10.1090/conm/819/16397

Generalizations of some q-product identities of Ramanujan and others

Timothy Huber, James Mc Laughlin, and Dongxi Ye

ABSTRACT. By considering certain limiting cases of a WP-Bailey chain discovered by Andrews, and also limiting cases of certain classical summation formulae for basic hypergeometric series, we derive new expressions for certain Lambert series in terms of basic hypergeometric series. In some cases, the resulting series involve an arbitrary Bailey pair. This allows for the derivation of new basic hypergeometric expansions for some q-products and series that Ramanujan expressed in terms of Lambert series. Some of Ramanujan's identities are extended to more general relations containing one or more free parameters.

1. Introduction

In the present paper certain limiting cases of known identities with one or more free parameters are considered. These limiting cases lead to several different representations for the same function in each case, one of these representation involving Lambert series (see Theorem 3.1, Corollaries 3.4 and 4.1, and Theorems 5.1, 6.1 and 7.1). While the derivation of these limiting cases themselves is elementary, our reason for doing so is that they permit new representations (in terms of basic hypergeometric series) for various functions which also have representations in terms of Lambert series. In some cases, the new representation may involve an arbitrary Bailey- or WP-Bailey pair, thus leading to a separate identity for each such pair.

As an example of this we consider the following identity of Ramanujan, [**4**, Entry 4, Ch. 21], which states that

$$\begin{aligned}(1.1)\qquad &1+6\sum_{n=1}^{\infty}\frac{nq^n}{1-q^n}-30\sum_{n=1}^{\infty}\frac{nq^{5n}}{1-q^{5n}}\\ &=\frac{\left((q;q)_\infty^{12}+22q(q;q)_\infty^{6}(q^5;q^5)_\infty^{6}+125q^2(q^5;q^5)_\infty^{12}\right)^{1/2}}{(q;q)_\infty(q^5;q^5)_\infty}.\end{aligned}$$

As an implication of one of the results in the present paper, we show that Equation (1.1) can be extended as follows: if $(\alpha_n(1,q),\beta_n(1,q))$ is any Bailey pair

2020 *Mathematics Subject Classification.* Primary 33D15; Secondary 11B65, 05A19.
Key words and phrases. Bailey pairs, WP-Bailey Chains, WP-Bailey pairs, Lambert Series, basic hypergeometric series, q-series, theta series, q-products.
The third author was supported by the Guangdong Basic and Applied Basic Research Foundation (Grant No. 2023A1515010298).

with respect to 1 (see below for the definition of a Bailey pair), then

$$
\begin{aligned}
&1+6\sum_{n=1}^{\infty}(q,q;q)_{n-1}q^{n}\beta_n(1,q)-6\sum_{n=1}^{\infty}\frac{q^n\alpha_n(1,q)}{(1-q^n)^2}\\
&\quad-30\sum_{n=1}^{\infty}(q^5,q^5;q^5)_{n-1}q^{5n}\beta_n(1,q^5)+30\sum_{n=1}^{\infty}\frac{q^{5n}\alpha_n(1,q^5)}{(1-q^{5n})^2}\\
&=\frac{\left((q;q)_\infty^{12}+22q(q;q)_\infty^{6}(q^5;q^5)_\infty^{6}+125q^2(q^5;q^5)_\infty^{12}\right)^{1/2}}{(q;q)_\infty(q^5;q^5)_\infty}.
\end{aligned}
\tag{1.2}
$$

Ramanujan's original result may then be seen as the special case deriving from the "trivial" Bailey pair ($\alpha_0(1,q)=\beta_0(1,q)=1$, and for $n>0$, $\alpha_n(1,q)=0$ and $\beta_n(1,q)=1/(q,q;q)_n$), after employing the identity

$$
\sum_{n=1}^{\infty}\frac{q^n}{(1-q^n)^2}=\sum_{n=1}^{\infty}\frac{nq^n}{1-q^n}.
$$

Any Bailey pair $(\alpha_n(1,q),\beta_n(1,q))$, including some with free parameters, may be substituted in (1.2) to produce specific identities, and we give some examples later.

Other results in the paper are even more general, extending identities of Ramanujan to identities involving an arbitrary WP-Bailey pair (as opposed to an arbitrary Bailey pair)—see, for example, Eq. (3.4) below.

Some of the new identities lack the flexibility of containing an arbitrary Bailey pair or arbitrary WP-Bailey pair, but are still quite interesting. As an example, we have that if b and λ be non-zero complex numbers such that $|q|<\max\{1,|b/\lambda|,|\lambda/b|\}$, and we define

$$
g(b,\lambda,q):=\sum_{n=1}^{\infty}\frac{(1-\lambda q^{2n})(q;q)_{n-1}\left(b,\frac{b}{\lambda};q\right)_n\left(\frac{\lambda^2q}{b};q^2\right)_n(\frac{-q\lambda}{b})^n}{(1-\lambda q^n)\left(\frac{q\lambda}{b},\frac{\lambda^2q}{b},q;q\right)_n(qb;q^2)_n},
$$

then

$$
\begin{aligned}
g(b,\lambda,q)-g\left(\frac{1}{b},\frac{1}{\lambda},q\right)&=\frac{(1-b)\lambda\left(b^2-\lambda^3\right)}{(1-\lambda)\left(b-\lambda^2\right)\left(b^2-\lambda^2\right)}\\
&-\frac{\lambda}{b}\frac{\left(b,\frac{q}{b},-\frac{\lambda^2}{b},-\frac{qb}{\lambda^2};q\right)_\infty(q^2,q^2;q^2)_\infty}{\left(\lambda^2,\frac{q^2}{\lambda^2},\frac{\lambda^2}{b^2},\frac{q^2b^2}{\lambda^2};q^2\right)_\infty}+\frac{b}{\lambda^2}\frac{\left(b,\frac{q^2}{b},\frac{\lambda^2q}{b},\frac{qb}{\lambda^2},\frac{\lambda^2}{q},\frac{q^3}{\lambda^2},q^2,q^2;q^2\right)_\infty}{\left(\frac{b}{\lambda^2},\frac{\lambda^2q^2}{b},q,q,\frac{b}{q},\frac{q^3}{b},\lambda^2,\frac{q^2}{\lambda^2};q^2\right)_\infty}.
\end{aligned}
$$

Here and throughout, we use the standard q-hypergeometric notation

$$
(a;q)_k:=\prod_{n=1}^{k}(1-aq^{n-1})\text{ and }(a_1,\dots,a_m;q)_k:=(a_1;q)_k\dots(a_m;q)_k,
$$

valid for $k\in\mathbb{N}\cup\{\infty\}$.

2. Background

The work in the present paper continues the work initiated in two previous papers, [**8**] and [**9**], but left aside for several years. We begin by recalling some definitions. A *WP-Bailey pair* was defined by Andrews [**1**] to be a pair of sequences

$(\alpha_n(a,k,q),\ \beta_n(a,k,q))$ (if the context is clear, we occasionally suppress the dependence on some or all of a, k and q) satisfying $\alpha_0(a,k,q) = \beta_0(a,k,q) = 1$ and, for $n > 0$,

$$\beta_n(a,k,q) = \sum_{j=0}^{n} \frac{(k/a;q)_{n-j}(k;q)_{n+j}}{(q;q)_{n-j}(aq;q)_{n+j}} \alpha_j(a,k,q). \tag{2.1}$$

If $k = 0$, then the pair of sequences $(\alpha_n(a,q), \beta_n(a,q))$ is called a *Bailey pair with respect to a*. In the same paper Andrews [**1**] described two constructions whereby new WP-Bailey pairs could be derived from existing pairs. For more on WP-Bailey pairs and chains, see [**2, 7, 12, 13**].

In [**8**] and [**9**], the second author examined certain limiting cases of Andrews's two constructions to derive some general transformation and summation formulae for WP-Bailey pairs and standard Bailey pairs. These identities were applied to derive new expressions for certain functions that can be represented as certain types of Lambert series. We continue these investigations in the present paper. Firstly, we consider two other special cases of Andrews's initial construction that we overlooked in [**8**]. Secondly, we also consider limiting cases of Jackson's ${}_6\phi_5$ summation formula and q-analogues of Watson's ${}_3F_2$- and Whipple's ${}_3F_2$ summation formulae. In both cases we derive new general transformations relating certain basic hypergeometric series to various Lambert series. We then use these new transformations to derive new summation formulae for a number of q-products that have known expressions in terms of certain Lambert series, as was done in the previous papers [**8**] and [**9**].

3. A second limiting case of Andrews's first WP-Bailey chain

One of Andrews's [**1**] WP-Bailey chains implies (see Corollary 1 in [**11**], for example) that if (α_n, β_n) satisfy (2.1), then subject to suitable convergence conditions,

$$\begin{aligned}\sum_{n=0}^{\infty} &\frac{(q\sqrt{k}, -q\sqrt{k}, y, z;q)_n}{(\sqrt{k}, -\sqrt{k}, qk/y, qk/z;q)_n} \left(\frac{qa}{yz}\right)^n \beta_n \\ &= \frac{(qk, qk/yz, qa/y, qa/z;q)_\infty}{(qk/y, qk/z, qa, qa/yz;q)_\infty} \sum_{n=0}^{\infty} \frac{(y,z;q)_n}{(qa/y, qa/z;q)_n} \left(\frac{qa}{yz}\right)^n \alpha_n.\end{aligned} \tag{3.1}$$

In an earlier paper [**8**] the second author investigated the implications of letting $y \to 1$ in (3.1), namely, that if (α_n, β_n) is a WP-Bailey pair, then subject to suitable convergence conditions,

$$\begin{aligned}\sum_{n=1}^{\infty} &\frac{(1-kq^{2n})(z;q)_n(q;q)_{n-1}}{(1-k)(qk, qk/z;q)_n} \left(\frac{qa}{z}\right)^n \beta_n \\ &- \sum_{n=1}^{\infty} \frac{(z;q)_n(q;q)_{n-1}}{(qa, qa/z;q)_n} \left(\frac{qa}{z}\right)^n \alpha_n = f(a,k,z,q),\end{aligned} \tag{3.2}$$

where

$$f(a,k,z,q)=\sum_{n=1}^{\infty}\frac{(1-kq^{2n})(z,k/a;q)_n}{(1-kq^n)(qk/z,qa;q)_n(1-q^n)}\left(\frac{qa}{z}\right)^n$$

$$=-\sum_{n=1}^{\infty}\frac{(1-aq^{2n})(z,a/k;q)_n}{(1-aq^n)(qa/z,qk;q)_n(1-q^n)}\left(\frac{qk}{z}\right)^n \tag{3.3}$$

$$=\sum_{n=1}^{\infty}\frac{kq^n}{1-kq^n}+\sum_{n=1}^{\infty}\frac{q^n a/z}{1-q^n a/z}-\sum_{n=1}^{\infty}\frac{aq^n}{1-aq^n}-\sum_{n=1}^{\infty}\frac{q^n k/z}{1-q^n k/z}.$$

One special case of (3.2) that we omitted in [**8**] was the result of letting $z\to 1$, and we consider that case now.

THEOREM 3.1. *If $(\alpha_n(a,k,q),\beta_n(a,k,q))$ is a WP-Bailey pair, then subject to suitable convergence conditions,*

$$\sum_{n=1}^{\infty}\frac{(1-kq^{2n})(q,q;q)_{n-1}}{(1-k)(qk,qk;q)_n}(qa)^n\,\beta_n(a,k,q)$$
$$-\sum_{n=1}^{\infty}\frac{(q,q;q)_{n-1}}{(qa,qa;q)_n}(qa)^n\,\alpha_n(a,k,q)=f_1(a,k,q),$$

where

$$f_1(a,k,q)=\sum_{n=1}^{\infty}\frac{(1-kq^{2n})(k/a;q)_n(q;q)_{n-1}}{(1-kq^n)(qk,qa;q)_n(1-q^n)}(qa)^n$$
$$=-\sum_{n=1}^{\infty}\frac{(1-aq^{2n})(a/k;q)_n(q;q)_{n-1}}{(1-aq^n)(qa,qk;q)_n(1-q^n)}(qk)^n$$
$$=\sum_{n=1}^{\infty}\frac{q^n a}{(1-q^n a)^2}-\sum_{n=1}^{\infty}\frac{q^n k}{(1-q^n k)^2}.$$

PROOF. Divide the identity at (3.2) through by $1-z$ and then let $z\to 1$. The next-to-last equality involving Lambert series follows from the fact that if we define

$$G(z):=\sum_{n=1}^{\infty}\frac{kq^n}{1-kq^n}+\sum_{n=1}^{\infty}\frac{q^n a/z}{1-q^n a/z}-\sum_{n=1}^{\infty}\frac{aq^n}{1-aq^n}-\sum_{n=1}^{\infty}\frac{q^n k/z}{1-q^n k/z},$$

then $G(1)=0$ and

$$\lim_{z\to 1}\frac{1}{1-z}\left(\sum_{n=1}^{\infty}\frac{kq^n}{1-kq^n}+\sum_{n=1}^{\infty}\frac{q^n a/z}{1-q^n a/z}-\sum_{n=1}^{\infty}\frac{aq^n}{1-aq^n}-\sum_{n=1}^{\infty}\frac{q^n k/z}{1-q^n k/z}\right)$$
$$=\lim_{z\to 1}\frac{G(z)-G(1)}{1-z}=-G'(1)=\sum_{n=1}^{\infty}\frac{q^n a}{(1-q^n a)^2}-\sum_{n=1}^{\infty}\frac{q^n k}{(1-q^n k)^2}.\qquad\square$$

One reason we consider this special case is that certain q-products may be expressed in terms of Lambert series of the type exhibited above, and the corollary now permits new representations of such products in terms of basic hypergeometric series to be given—in fact, one such representation for each WP-Bailey pair. We give one example, which leads to a large number of new representations for the product $q(q^5;q^5)_\infty^5/(q;q)_\infty$.

COROLLARY 3.2. Let $(\alpha_n(a,k,q), \beta_n(a,k,q))$ be a WP-Bailey pair. Then, subject to suitable convergence conditions,

$$
\begin{aligned}
q\frac{(q^5;q^5)_\infty^5}{(q;q)_\infty} &= \sum_{n=1}^{\infty} \frac{(1-q^{10n-2})(q^5,q^5;q^5)_{n-1}q^n}{(1-1/q^2)(q^3,q^3;q^5)_n}\beta_n\left(\frac{1}{q^4},\frac{1}{q^2},q^5\right) \\
&+ \sum_{n=1}^{\infty} \frac{(1-q^{10n-3})(q^5,q^5;q^5)_{n-1}q^{4n}}{(1-1/q^3)(q^2,q^2;q^5)_n}\beta_n\left(\frac{1}{q},\frac{1}{q^3},q^5\right) \\
&- \sum_{n=1}^{\infty} \frac{(q^5,q^5;q^5)_{n-1}q^n}{(q,q;q^5)_n}\alpha_n\left(\frac{1}{q^4},\frac{1}{q^2},q^5\right) \\
&- \sum_{n=1}^{\infty} \frac{(q^5,q^5;q^5)_{n-1}q^{4n}}{(q^4,q^4;q^5)_n}\alpha_n\left(\frac{1}{q},\frac{1}{q^3},q^5\right)
\end{aligned}
\tag{3.4}
$$

and

$$
\begin{aligned}
q\frac{(q^5;q^5)_\infty^5}{(q;q)_\infty} &= \sum_{n=1}^{\infty} \frac{(1-q^{10n-2})(q^5;q^5)_{n-1}\left(\frac{1}{q};q^5\right)_n q^n}{(1-1/q^2)(1-q^{5n})\left(q^3,q^3;q^5\right)_n} \\
&+ \sum_{n=1}^{\infty} \frac{(1-q^{10n-3})(q^5;q^5)_{n-1}\left(\frac{1}{q^9};q^5\right)_n q^{4n}}{(1-1/q^3)(1-q^{5n})\left(q^2,q^2;q^5\right)_n} \\
&- \sum_{n=1}^{\infty} \frac{(1-q^{10n-4})(q^5;q^5)_{n-1}\left(\frac{1}{q^3};q^5\right)_n \left(\frac{1}{q};q^5\right)_{2n} q^{3n}}{(1-q^{5n-4})(1-q^{5n})\left(q,q^4;q^5\right)_n \left(\frac{1}{q^2};q^5\right)_{2n}} \\
&- \sum_{n=1}^{\infty} \frac{(1-q^{10n-1})(q^5;q^5)_{n-1}\left(\frac{1}{q^7};q^5\right)_n \left(q^6;q^5\right)_{2n} q^{2n}}{(1-q^{5n-1})(1-q^{5n})\left(q^4,q^{11};q^5\right)_n \left(\frac{1}{q^3};q^5\right)_{2n}}.
\end{aligned}
\tag{3.5}
$$

PROOF. By [**3**, Entry 18.2.23, Ch. 18]

$$
q\frac{(q^5;q^5)_\infty^5}{(q;q)_\infty} = \sum_{n=1}^{\infty}\left(\frac{n}{5}\right)\frac{q^n}{(1-q^n)^2}, \tag{3.6}
$$

where $(n/5)$ is the Legendre symbol. In Theorem 3.1, replace q with q^5 and set $(a,k) = (1/q^4, 1/q^2)$ and $(a,k) = (1/q, 1/q^3)$ respectively. Add the resulting identities: it can be seen from the last representation of $f_1(a,k,q)$ that

$$
q\frac{(q^5;q^5)_\infty^5}{(q;q)_\infty} = f_1(1/q^4, 1/q^2, q^5) + f_1(1/q, 1/q^3, q^5),
$$

and (3.4) now follows. Inserting the WP-Bailey pair (see, for example, [**2**, Eqs. (3.3) and (3.4)])

$$
\begin{aligned}
\alpha_n'(a,k) &= \frac{1-aq^{2n}}{1-a}\frac{(a,k/aq;q)_n(qa^2/k;q)_{2n}}{(q^2a^2/k,q;q)_n(k;q)_{2n}}\left(\frac{k}{a}\right)^n, \\
\beta_n'(a,k) &= \frac{(k^2/qa^2;q)_n}{(q;q)_n}
\end{aligned}
$$

into (3.4) gives (3.5). □

The substitution $a = 1$ and $k = e^{\pm i\theta}$ in Theorem 3.1 results in a WP-Bailey pair iteration for a fundamental building block of elliptic modular functions.

COROLLARY 3.3. Let $\wp(\theta)$ denote the Weierstrass $\wp$-function, normalized so that [**14**, Ex. 35, p. 460]

$$\wp(\theta)=\frac{1}{4}\csc^2\frac{\theta}{2}-\frac{1}{12}+2\sum_{n=1}^{\infty}\frac{nq^n}{1-q^n}-2\sum_{n=1}^{\infty}\frac{nq^n\cos(n\theta)}{1-q^n}.$$

Suppose $(\alpha_n(a,k,q),\beta_n(a,k,q))$ is a WP-Bailey pair. Then, subject to suitable convergence conditions,

$$\begin{aligned}\wp(\theta)=\frac{1}{4}\csc^2\frac{\theta}{2}-\frac{1}{12}&+\sum_{n=1}^{\infty}\frac{(1-q^{2n}e^{i\theta})(q;q)_{n-1}^2}{(1-e^{i\theta})(qe^{i\theta};q)_n^2}q^n\beta_n(1,e^{i\theta},q)\\&+\sum_{n=1}^{\infty}\frac{(1-q^{2n}e^{-i\theta})(q;q)_{n-1}^2}{(1-e^{-i\theta})(qe^{-i\theta};q)_n^2}q^n\beta_n(1,e^{-i\theta},q)\\&-\sum_{n=1}^{\infty}\frac{q^n\alpha_n(1,e^{i\theta},q)}{(1-q^n)^2}-\sum_{n=1}^{\infty}\frac{q^n\alpha_n(1,e^{-i\theta},q)}{(1-q^n)^2}.\end{aligned}$$

We next consider another special case of (3.2), which follows from setting $k=-a$. This special case also has some interesting applications. We also consider this special case ($k=-a$) of Theorem 3.1. The proofs are straightforward, so are omitted.

COROLLARY 3.4. If $(\alpha_n(a,k),\beta_n(a,k))=(\alpha_n(a,k,q),\beta_n(a,k,q))$ is a WP-Bailey pair, then subject to suitable convergence conditions,

$$\begin{aligned}&\sum_{n=1}^{\infty}\frac{(1+aq^{2n})(z;q)_n(q;q)_{n-1}}{(1+a)(-qa,-qa/z;q)_n}\left(\frac{qa}{z}\right)^n\beta_n(a,-a)\\&\qquad-\sum_{n=1}^{\infty}\frac{(z;q)_n(q;q)_{n-1}}{(qa,qa/z;q)_n}\left(\frac{qa}{z}\right)^n\alpha_n(a,-a)=f_2(a,z,q),\end{aligned}\tag{3.7}$$

where

$$\begin{aligned}f_2(a,z,q)&=\sum_{n=1}^{\infty}\frac{(1+aq^{2n})(z,-1;q)_n}{(1+aq^n)(-qa/z,qa;q)_n(1-q^n)}\left(\frac{qa}{z}\right)^n\\&=-\sum_{n=1}^{\infty}\frac{(1-aq^{2n})(z,-1;q)_n}{(1-aq^n)(qa/z,-qa;q)_n(1-q^n)}\left(\frac{-qa}{z}\right)^n\\&=2\sum_{n=1}^{\infty}\frac{q^na/z}{1-q^{2n}a^2/z^2}-2\sum_{n=1}^{\infty}\frac{aq^n}{1-a^2q^{2n}},\end{aligned}$$

and

$$\begin{aligned}&\sum_{n=1}^{\infty}\frac{(1+aq^{2n})(q,q;q)_{n-1}}{(1+a)(-qa,-qa;q)_n}(qa)^n\beta_n(a,-a)\\&\qquad-\sum_{n=1}^{\infty}\frac{(q,q;q)_{n-1}}{(qa,qa;q)_n}(qa)^n\alpha_n(a,-a)=f_3(a,q),\end{aligned}$$

where

$$f_3(a,q) = \sum_{n=1}^{\infty} \frac{(1+aq^{2n})(-1;q)_n(q;q)_{n-1}\,(qa)^n}{(1+aq^n)(-qa,qa;q)_n(1-q^n)}$$

$$= -\sum_{n=1}^{\infty} \frac{(1-aq^{2n})(-1;q)_n(q;q)_{n-1}\,(-qa)^n}{(1-aq^n)(qa,-qa;q)_n(1-q^n)}$$

$$= 2\sum_{n=1}^{\infty} \frac{q^n a(1+a^2q^{2n})}{(1-q^{2n}a^2)^2}.$$

To give an application of (3.7), we recall another identity of Ramanujan (see [**4**, Entry 3, Ch. 19]), namely

$$q\psi(q^2)\psi(q^6) = \sum_{n=1}^{\infty} \frac{q^{6n-5}}{1-q^{12n-10}} - \sum_{n=1}^{\infty} \frac{q^{6n-1}}{1-q^{12n-2}},$$

where $\psi(q) := \sum_{n=0}^{\infty} q^{n(n+1)/2} = (q^2;q^2)_\infty(-q;q)_\infty$ (this function also appears in other identities below). Now in (3.7), replace q with q^6, set $a=1/q$ and $z=q^4$, and divide through by 2. Next insert the unit WP-Bailey pair (see [**10**, Eq. (12.15)], for example)

$$\alpha_n(a,k) = \frac{(1-aq^{2n})(a,\frac{a}{k};q)_n}{(1-a)(q,kq;q)_n}\left(\frac{k}{a}\right)^n, \qquad \beta_n(a,k) = \delta_{n,0}$$

in the form $(\alpha_n(1/q,-1/q,q^6),\, \beta_n(1/q,-1/q,q^6))$, and the following new identity results.

Corollary 3.5. If $|q|<1$, then

$$q\psi(q^2)\psi(q^6) = -\frac{1}{2}\sum_{n=1}^{\infty} \frac{(1-q^{12n-1})\left(q^4,-1;q^6\right)_n(-q)^n}{(1-q^{6n-1})(1-q^{6n})\left(-q^5,q;q^6\right)_n}.$$

4. Identities deriving from standard Bailey pairs

Upon letting $k\to 0$ in Theorem 3.1, we get the following corollary, which implies several interesting representations for the Lambert series

$$\sum_{n=1}^{\infty} \frac{aq^n}{(1-aq^n)^2}.$$

Corollary 4.1. If $(\alpha_n(a,q),\beta_n(a,q))$ is a Bailey pair, then subject to suitable convergence conditions,

$$\sum_{n=1}^{\infty}(q,q;q)_{n-1}\,(qa)^n\,\beta_n(a,q) - \sum_{n=1}^{\infty}\frac{(q,q;q)_{n-1}}{(qa,qa;q)_n}(qa)^n\alpha_n(a,q) = f_4(a,q), \tag{4.1}$$

where

$$f_4(a,q) = \sum_{n=1}^{\infty} \frac{(q;q)_{n-1}}{(qa;q)_n(1-q^n)}\,(qa)^n$$

$$= -\sum_{n=1}^{\infty} \frac{(1-aq^{2n})(q;q)_{n-1}}{(1-aq^n)(qa;q)_n(1-q^n)}\,(-a)^n\,q^{n(n+1)/2} \tag{4.2}$$

$$= \sum_{n=1}^{\infty} \frac{q^n a}{(1-q^n a)^2}.$$

Replacing q with q^5 in (4.1), setting $a = q^{-j}$, for $1 \leqslant j \leqslant 4$, and employing (3.6) in combination with (4.2) gives the following identities.

COROLLARY 4.2. If $|q| < 1$, then

$$q\frac{(q^5;q^5)_\infty^5}{(q;q)_\infty} = \sum_{j=1}^{4}\left(\frac{j}{5}\right)\sum_{n=1}^{\infty}\frac{(q^5;q^5)_{n-1}q^{(5-j)n}}{(q^{5-j};q^5)_n(1-q^{5n})}$$
$$= -\sum_{j=1}^{4}\left(\frac{j}{5}\right)\sum_{n=1}^{\infty}\frac{(1-q^{10n-j})(q^5;q^5)_{n-1}(-1)^n q^{(5n^2+5n)/2-jn}}{(1-q^{5n-j})(1-q^{5n})(q^{5-j};q^5)_n}.$$

REMARK 4.3. The identity (1.2) is a direct consequence of using (1.1) in conjunction with Corollary 4.1.

5. A limiting case of Jackson's ${}_6\phi_5$ summation formula

We now consider limiting cases of a number of other summation formulae from the literature. The applications will not be so general, as the resulting identities do not contain WP-Bailey pairs or standard Bailey pairs, but we will exhibit some interesting consequences.

We first recall Jackson's summation formula for a very-well-poised ${}_6\phi_5$ series [**6**, p. 356, Eq. (II. 20)]:

$$\sum_{n=0}^{\infty}\frac{(1-aq^{2n})(a,b,c,d;q)_n}{(1-a)\left(\frac{aq}{b},\frac{aq}{c},\frac{aq}{d},q;q\right)_n}\left(\frac{aq}{bcd}\right)^n = \frac{(aq,aq/bc,aq/bd,aq/cd;q)_\infty}{(aq/b,aq/c,aq/d,aq/bcd;q)_\infty}. \tag{5.1}$$

THEOREM 5.1. *Let a be a complex number such that $aq^n \neq 1$ for $n \geqslant 1$. Then*

$$\sum_{n=1}^{\infty}\frac{(1-aq^{2n})(q,q;q)_{n-1}(qa)^n}{(1-aq^n)(1-q^n)(qa,qa;q)_n} = \sum_{n=1}^{\infty}\frac{aq^n(1+aq^n)}{(1-aq^n)^3} = \sum_{n=1}^{\infty}\frac{n^2a^nq^n}{1-q^n}. \tag{5.2}$$

PROOF. Rewrite (5.1) as

$$\sum_{n=1}^{\infty}\frac{(1-aq^{2n})(a;q)_n(bq,cq,dq;q)_{n-1}}{(1-a)\left(\frac{aq}{b},\frac{aq}{c},\frac{aq}{d},q;q\right)_n}\left(\frac{aq}{bcd}\right)^n$$
$$= \frac{1}{(1-b)(1-c)(1-d)}\left(\frac{(aq,aq/bc,aq/bd,aq/cd;q)_\infty}{(aq/b,aq/c,aq/d,aq/bcd;q)_\infty}-1\right).$$

Let $b, c, d \to 1$ and compute the limits on the right-hand side as derivatives. The last equality follows from expanding the $1/(1-q^n)$ terms in the last expression as geometric series, changing the order of summation, and employing a standard summation identity. □

We give one example of an application of (5.2). First, we recall two cubic theta functions from [**5**]:

$$a(q) := \sum_{m,n=-\infty}^{\infty} q^{m^2+mn+n^2}, \qquad b(q) := \sum_{m,n=-\infty}^{\infty} \omega^{m-n}q^{m^2+mn+n^2},$$

where $\omega = \exp(2\pi i/3)$.

COROLLARY 5.2. If $|q| < 1$, then

$$a(q)^3 - b(q)^3 = 27\Bigg(\sum_{n=1}^{\infty} \frac{(1-q^{6n-2})(q^3,q^3;q^3)_{n-1}q^n}{(1-q^{3n-2})(1-q^{3n})(q,q;q^3)_n} - \sum_{n=1}^{\infty} \frac{(1-q^{6n-1})(q^3,q^3;q^3)_{n-1}q^{2n}}{(1-q^{3n-1})(1-q^{3n})(q^2,q^2;q^3)_n} \Bigg).$$

PROOF. By (18.2.10) and (18.2.12) in Chapter 18 of [**3**],

$$a(q)^3 - b(q)^3 = 27\left(\sum_{n=1}^{\infty} \frac{n^2 q^n}{1-q^{3n}} - \sum_{n=1}^{\infty} \frac{n^2 q^{2n}}{1-q^{3n}} \right).$$

Now use (5.2), with q replaced with q^3 and $a = 1/q^2$ and $a = 1/q$, respectively. □

6. A limiting case of a q-analogue of Whipple's ${}_3F_2$ summation formula

We next recall Bailey's q-analogue of Whipple's ${}_3F_2$ sum (see [**6**, Formula II.18, page 355], with C replaced with $-C$ and d replaced with $-d$):

$$\begin{aligned} &{}_8\phi_7 \left[\begin{matrix} C,\ q\sqrt{C},\ -q\sqrt{C},\ a,\ q/a,\ -C,\ d,\ q/d \\ \sqrt{C},\ -\sqrt{C},\ Cq/a,\ aC,\ -q,\ Cq/d,\ Cd \end{matrix} ; q, -C \right] \\ &\qquad = \frac{(C,Cq;q)_\infty (aCd, aCq/d, Cdq/a, Cq^2/ad;q^2)_\infty}{(Cd, Cq/d, aC, Cq/a;q)_\infty}. \end{aligned}$$

If this identity is rewritten as

$$\begin{aligned} &\sum_{n=1}^{\infty} \frac{(1-Cq^{2n})(aq,dq;q)_{n-1}(C,-C,q/a,q/d;q)_n(-C)^n}{(1-C)(Cq/a,aC,Cq/d,Cd,-q,q;q)_n} \\ &\qquad = \frac{1}{(1-a)(1-d)}\left(\frac{(C,Cq;q)_\infty (aCd, aCq/d, Cdq/a, Cq^2/ad;q^2)_\infty}{(Cd, Cq/d, aC, Cq/a;q)_\infty} - 1 \right) \end{aligned}$$

and we let, in turn, $a \to 1$ and $d \to 1$, the identities in the next theorem result. We omit the proofs, as they are similar to earlier proofs.

THEOREM 6.1. *For $|C| < 1$ and non-zero $d \neq q^n$ for $n \geqslant 1$,*

$$\begin{aligned} &\sum_{n=1}^{\infty} \frac{(1-Cq^{2n})(q;q)_{n-1}(-C,d,q/d;q)_n(-C)^n}{(1-C)(Cq,Cq/d,Cd,-q;q)_n} \\ &\qquad = \frac{Cd}{1-Cd} - \frac{C}{1-C} + \sum_{n=1}^{\infty}\left(\frac{(-1)^n q^n Cd}{1-q^nCd} - \frac{(-1)^n q^n C/d}{1-q^nC/d} \right), \end{aligned} \tag{6.1}$$

and

$$\begin{aligned} &\sum_{n=1}^{\infty} \frac{(1-Cq^{2n})(q,q;q)_{n-1}(-C,q;q)_n(-C)^n}{(1-C)(Cq,Cq,C,-q;q)_n} \\ &\qquad = -\frac{C}{(1-C)^2} - 2\sum_{n=1}^{\infty} \frac{(-1)^n q^n C}{(1-q^nC)^2}. \end{aligned}$$

Before giving an example, recall once again that $\psi(q) = \sum_{n=0}^{\infty} q^{n(n+1)/2}$.

COROLLARY 6.2. If $|q| < 1$, then

$$2q\psi(q^4)^2$$
$$= \frac{q}{1-q} - \frac{iq}{1-iq} - \sum_{n=1}^{\infty} \frac{(1-iq^{4n+1})(q^2;q^2)_{n-1}(i;q^2)_n(-iq;q)_{2n}(-iq)^n}{(q^3,-q,-q^2;q^2)_n(iq;q^2)_{n+1}}.$$

PROOF. Replace q with q^2 in (6.1) and then set $C = iq$ and $d = i$. The result follows after some elementary manipulations, upon recalling the following identity of Ramanujan (see [**3**, Entry 18.2.4, p. 397]):

$$\psi^2(q^4) = \sum_{n=0}^{\infty} \frac{(-1)^n q^{2n}}{1-q^{4n+2}}.$$ □

7. A limiting case of a q-analogue of Watson's ${}_3F_2$ summation formula

The formula in question is the following identity of Bailey (see [**6**, Formula II.16, page 355]):

$$ {}_8\phi_7 \left[\begin{matrix} \lambda,\ q\sqrt{\lambda},\ -q\sqrt{\lambda},\ a,\ b,\ \lambda\sqrt{q/ab},\ -\lambda\sqrt{q/ab},\ ab/\lambda \\ \sqrt{\lambda},\ -\sqrt{\lambda},\ \lambda q/a,\ \lambda q/b,\ \lambda^2 q/ab,\ \sqrt{qab},\ -\sqrt{qab} \end{matrix} ; q, -\frac{q\lambda}{ab} \right] $$
$$= \frac{(\lambda q, \lambda q/ab; q)_\infty}{(\lambda q/a, \lambda q/b; q)_\infty} \frac{(aq, bq, q^2\lambda^2/a^2b, q^2\lambda^2/ab^2; q^2)_\infty}{(q, abq, q^2\lambda^2/ab, q^2\lambda^2/a^2b^2; q^2)_\infty}.$$

The same kind of manipulations as in the proof of Theorem 6.1 leads to the following identities.

THEOREM 7.1. *If b and λ are non-zero complex numbers such that $|q| < \max\{1, |b/\lambda|\}$ and none of the denominators below vanish, then*

$$(7.1)\quad \begin{aligned} &\sum_{n=1}^{\infty} \frac{(1-\lambda q^{2n})(q;q)_{n-1}(b, b/\lambda; q)_n(\lambda^2 q/b; q^2)_n(-q\lambda/b)^n}{(1-\lambda q^n)(q\lambda/b, \lambda^2 q/b, q; q)_n(qb;q^2)_n} \\ &= \sum_{n=1}^{\infty} \left(\frac{\lambda q^n}{1-\lambda q^n} - \frac{\frac{\lambda q^n}{b}}{1-\frac{\lambda^2 q^{2n}}{b^2}} - \frac{\frac{\lambda^2 q^{2n}}{b}}{1-\frac{\lambda^2 q^{2n}}{b}} + \frac{q^{2n-1}}{1-q^{2n-1}} - \frac{bq^{2n-1}}{1-bq^{2n-1}} \right); \end{aligned}$$

and if $|q| < \max\{1, 1/|\lambda|\}$ and none of the denominators below vanish, then

$$(7.2)\quad \begin{aligned} &\sum_{n=1}^{\infty} \frac{(1-\lambda q^{2n})(q,q;q)_{n-1}(1/\lambda;q)_n(\lambda^2 q;q^2)_n(-q\lambda)^n}{(1-\lambda q^n)(q\lambda, \lambda^2 q, q; q)_n(q;q^2)_n} \\ &= \sum_{n=1}^{\infty} \left(-\frac{\lambda q^n}{(1-\lambda q^n)^2} + \frac{\lambda^2 q^{2n}}{(1-\lambda^2 q^{2n})^2} + \frac{q^{2n-1}}{(1-q^{2n-1})^2} \right) \\ &= \sum_{n=1}^{\infty} \left(-\frac{n\lambda^n q^n}{1-q^n} + \frac{n\lambda^{2n} q^{2n}}{1-q^{2n}} + \frac{nq^n}{1-q^{2n}} \right). \end{aligned}$$

PROOF. Shift the initial "1" from the left side to the right side, divide through by $1-a$ and let $a \to 1$ to get (7.1), then divide through by $1-b$ and let $b \to 1$ to get (7.2). □

We give two applications of this theorem. Before coming to the first of these, we recall a result from [**8**], namely that if $f(a,k,z,q)$ is as defined at (3.3), then

$$f(a,k,z,q)-f\left(\frac{1}{a},\frac{1}{k},\frac{1}{z},q\right)=\frac{(a-k)(1-1/z)(1-ak/z)}{(1-a)(1-k)(1-a/z)(1-k/z)}+\frac{z}{k}\frac{(z,q/z,k/a,qa/k,ak/z,qz/ak,q,q;q)_\infty}{(z/k,qk/z,z/a,qa/z,a,q/a,k,q/k;q)_\infty}.$$

We note two special cases:

$$\begin{aligned} f(\lambda/b,\lambda,-1,q)&-f(b/\lambda,1/\lambda,-1,q)\\ &=\frac{(\lambda/b-\lambda)(1+\lambda^2/b)}{(1-\lambda^2/b^2)(1-\lambda^2)}-\frac{\lambda}{b}\frac{(b,q/b,-\lambda^2/b,-qb/\lambda^2;q)_\infty(q^2,q^2;q^2)_\infty}{(\lambda^2,q^2/\lambda^2,\lambda^2/b^2,q^2b^2/\lambda^2;q^2)_\infty}, \end{aligned}\tag{7.3}$$

and

$$\begin{aligned} f\left(\frac{b}{q},\lambda^2,b,q^2\right)&-f\left(\frac{q}{b},\frac{1}{\lambda^2},\frac{1}{b},q^2\right)\\ &=\frac{(b/q-\lambda^2)(1-1/b)(1-\lambda^2/q)}{(1-b/q)(1-\lambda^2)(1-1/q)(1-\lambda^2/b)}\\ &\quad+\frac{b}{\lambda^2}\frac{(b,q^2/b,\lambda^2q/b,qb/\lambda^2,\lambda^2/q,q^3/\lambda^2,q^2,q^2;q^2)_\infty}{(b/\lambda^2,\lambda^2q^2/b,q,q,b/q,q^3/b,\lambda^2,q^2/\lambda^2;q^2)_\infty}. \end{aligned}\tag{7.4}$$

The first consequence of Theorem 7.1 is a somewhat curious reciprocity-type result.

COROLLARY 7.2. *Let b and λ be non-zero complex numbers such that $|q|<\max\{1,|b/\lambda|,|\lambda/b|\}$, and define*

$$g(b,\lambda,q):=\sum_{n=1}^{\infty}\frac{(1-\lambda q^{2n})(q;q)_{n-1}\left(b,\frac{b}{\lambda};q\right)_n\left(\frac{\lambda^2 q}{b};q^2\right)_n\left(\frac{-q\lambda}{b}\right)^n}{(1-\lambda q^n)\left(\frac{q\lambda}{b},\frac{\lambda^2q}{b},q;q\right)_n(qb;q^2)_n}.\tag{7.5}$$

If none of the denominators below vanish, then

$$\begin{aligned} g(b,\lambda,q)-g\left(\frac{1}{b},\frac{1}{\lambda},q\right)&=\frac{(1-b)\lambda\left(b^2-\lambda^3\right)}{(1-\lambda)\left(b-\lambda^2\right)\left(b^2-\lambda^2\right)}\\ -\frac{\lambda}{b}\frac{\left(b,\frac{q}{b},-\frac{\lambda^2}{b},-\frac{qb}{\lambda^2};q\right)_\infty(q^2,q^2;q^2)_\infty}{\left(\lambda^2,\frac{q^2}{\lambda^2},\frac{\lambda^2}{b^2},\frac{q^2b^2}{\lambda^2};q^2\right)_\infty}&+\frac{b}{\lambda^2}\frac{\left(b,\frac{q^2}{b},\frac{\lambda^2q}{b},\frac{qb}{\lambda^2},\frac{\lambda^2}{q},\frac{q^3}{\lambda^2},q^2,q^2;q^2\right)_\infty}{\left(\frac{b}{\lambda^2},\frac{\lambda^2q^2}{b},q,q,\frac{b}{q},\frac{q^3}{b},\lambda^2,\frac{q^2}{\lambda^2};q^2\right)_\infty}. \end{aligned}$$

PROOF. It can be seen from (7.5) and (7.1) that

$$\begin{aligned} g(b,\lambda,q)&=\sum_{n=1}^{\infty}\left(\frac{\lambda q^n}{1-\lambda^2q^{2n}}-\frac{\frac{\lambda}{b}q^n}{1-\frac{\lambda^2}{b^2}q^{2n}}\right)\\ &\quad+\sum_{n=1}^{\infty}\left(\frac{\lambda^2q^{2n}}{1-\lambda^2q^{2n}}+\frac{\frac{1}{q}q^{2n}}{1-\frac{1}{q}q^{2n}}-\frac{\frac{b}{q}q^{2n}}{1-\frac{b}{q}q^{2n}}-\frac{\frac{\lambda^2}{b}q^{2n}}{1-\frac{\lambda^2}{b}q^{2n}}\right)\\ &=f\left(\frac{\lambda}{b},\lambda,-1,q\right)+f\left(\frac{b}{q},\lambda^2,b,q^2\right), \end{aligned}$$

after setting

$$\frac{\lambda q^n}{1-\lambda q^n}=\frac{\lambda q^n}{1-\lambda^2q^{2n}}+\frac{\lambda^2q^{2n}}{1-\lambda^2q^{2n}}.$$

By similar reasoning,

$$g\left(\frac{1}{b},\frac{1}{\lambda},q\right)=f\left(\frac{b}{\lambda},\frac{1}{\lambda},-1,q\right)+f\left(\frac{q}{b},\frac{1}{\lambda^2},\frac{1}{b},q^2\right)-\frac{q}{1-q}+\frac{q/b}{1-q/b}.$$

The result now follows upon employing (7.3) and (7.4), after setting

$$\frac{(b/q-\lambda^2)(1-1/b)(1-\lambda^2/q)}{(1-b/q)(1-\lambda^2)(1-1/q)(1-\lambda^2/b)}+\frac{(\lambda/b-\lambda)(1+\lambda^2/b)}{(1-\lambda^2/b^2)(1-\lambda^2)}-\frac{q}{1-q}+\frac{q/b}{1-q/b}$$
$$=\frac{(1-b)\lambda\left(b^2-\lambda^3\right)}{(1-\lambda)\left(b-\lambda^2\right)\left(b^2-\lambda^2\right)}.$$ □

COROLLARY 7.3. If $|q|<1$, then

$$\sum_{n=0}^{\infty}\frac{(1-q^{4n+3})(q^4;q^4)_n(-q)^n}{(1-q^{2n+1})(1-q^{2n+2})(q^2;q^4)_{n+1}}=\psi^4(q^2). \tag{7.6}$$

PROOF. Replace q with q^2 in (7.2) and set $\lambda=1/q$. The left side of (7.6) follows after re-indexing and dividing through by $-q$. The right of (7.2) with the same substitutions and then divided by $-q$ simplifies to

$$\frac{1}{q}\sum_{n=1}^{\infty}\frac{(2n-1)q^{2n-1}}{1-q^{4n-2}}.$$

The result now follows from [**4**, Example (iii), p. 139]:

$$q\psi^4(q^2)=\sum_{k=0}^{\infty}\frac{(2k+1)q^{2k+1}}{1-q^{4k+2}}.$$ □

8. Concluding remarks

One side of some of the identities has a combinatorial interpretation in terms of representations by quadratic forms. For example, the coefficient of q^{2n} on the right side of (7.6) is equal to the number of representations of n as a sum of four triangular numbers. A natural question is to ask if the left side of (7.6) also has an interesting combinatorial interpretation so that (7.6) encodes a combinatorial identity. Similar questions can be asked about some of the other identities in the paper. We leave these questions to the ingenuity of the reader.

Acknowledgments

The authors would like to thank the referees for their careful reading of the paper and their many useful suggestions which helped to improve it.

References

[1] G. E. Andrews, *Bailey's transform, lemma, chains and tree*, pp. 1–22 in Special Functions 2000: Current Perspective and Future Directions (Tempe, AZ), NATO Sci. Ser. II Math. Phys. Chem., **30**, Kluwer Acad. Publ., Dordrecht, 2001.

[2] G. E. Andrews and A. Berkovich, *The WP-Bailey tree and its implications*, J. London Math. Soc. (2) **66** (2002), no. 3, 529–549, DOI 10.1112/S0024610702003617. MR1934290

[3] G. E. Andrews and B. C. Berndt, *Ramanujan's lost notebook. Part I*, Springer, New York, 2005. MR2135178

[4] B. C. Berndt, *Ramanujan's notebooks. Part III*, Springer-Verlag, New York, 1991, DOI 10.1007/978-1-4612-0965-2. MR1117903

[5] J. M. Borwein and P. B. Borwein, *A cubic counterpart of Jacobi's identity and the AGM*, Trans. Amer. Math. Soc. **323** (1991), no. 2, 691–701, DOI 10.2307/2001551. MR1010408

[6] G. Gasper and M. Rahman, *Basic hypergeometric series*, 2nd ed., Encyclopedia of Mathematics and its Applications, vol. 96, Cambridge University Press, Cambridge, 2004. With a foreword by Richard Askey, DOI 10.1017/CBO9780511526251. MR2128719

[7] Q. Liu and X. Ma, *On a characteristic equation of well-poised Bailey chains*, Ramanujan J. **18** (2009), no. 3, 351–370, DOI 10.1007/s11139-007-9060-6. MR2495553

[8] J. Mc Laughlin, *Some new transformations for Bailey pairs and WP-Bailey pairs*, Cent. Eur. J. Math. **8** (2010), no. 3, 474–487, DOI 10.2478/s11533-010-0022-7. MR2653657

[9] J. Mc Laughlin, *A new summation formula for WP-Bailey pairs*, Appl. Anal. Discrete Math. **5** (2011), no. 1, 67–79, DOI 10.2298/AADM110203004M. MR2809035

[10] J. Mc Laughlin, *Topics and methods in q-series*, Monographs in Number Theory, vol. 8, World Scientific Publishing Co. Pte. Ltd., Hackensack, NJ, 2018. With a foreword by George E. Andrews. MR3752164

[11] J. Mc Laughlin and P. Zimmer, *Some implications of the WP-Bailey tree*, Adv. in Appl. Math. **43** (2009), no. 2, 162–175, DOI 10.1016/j.aam.2009.02.001. MR2531919

[12] V. P. Spiridonov, *An elliptic incarnation of the Bailey chain.* Int. Math. Res. Not. **37** (2002), 1945–1977.

[13] S. O. Warnaar, *Extensions of the well-poised and elliptic well-poised Bailey lemma*, Indag. Math. (N.S.) **14** (2003), no. 3-4, 571–588, DOI 10.1016/S0019-3577(03)90061-9. MR2083090

[14] E. T. Whittaker and G. N. Watson, *A course of modern analysis. An introduction to the general theory of infinite processes and of analytic functions: with an account of the principal transcendental functions*, Cambridge University Press, New York, 1962. Fourth edition. Reprinted. MR178117

School of Mathematical and Statistical Sciences, University of Texas Rio Grande Valley, Edinburg, Texas 78539

Email address: timothy.huber@utrgv.edu

Dept. of Mathematics, 25 University Avenue, West Chester University, West Chester, Pennsylvania 19383

Email address: jmclaughlin2@wcupa.edu

School of Mathematics (Zhuhai), Sun Yat-sen University, Zhuhai 519082, Guangdong, People's Republic of China

Email address: yedx3@mail.sysu.edu.cn

Contemporary Mathematics
Volume **819**, 2025
https://doi.org/10.1090/conm/819/16398

Discrete orthogonality of the polynomial sequences in the q-Askey scheme

Luis Verde-Star

ABSTRACT. We obtain weight functions associated with q-linear and q-quadratic lattices that yield discrete orthogonality with respect to a quasi-definite moment functional for the Askey–Wilson polynomials and all the polynomial sequences in the q-Askey scheme, with the exception of the continuous q-Hermite polynomials, which require the transformation $q \to q^2$.

1. Introduction

The families of basic hypergeometric orthogonal polynomial sequences included in the q-Askey scheme [**8**] have been studied for a long time. Some of those families have a well-known orthogonality determined by a weight function $w(x_k)$ defined on a finite or infinite set of points x_k that belong to a q-linear or q-quadratic lattice. In such cases the orthogonality of a polynomial sequence $\{u_n(t) : n \geqslant 0\}$ is

$$\sum_{k \in M} u_n(x_k) u_m(x_k) w(x_k) = K_n \delta_{n,m}, \qquad n, m \in M, \tag{1.1}$$

where $M = \{0, 1, 2, \ldots, N\}$, for some positive integer N, or $M = \mathbb{N}$, and K_n is a nonzero constant for $n \in M$.

The polynomial families that satisfy certain types of differential or difference equations and have a discrete orthogonality with positive weights $w(x_k)$ have been characterized and studied in great detail. The main references for discrete orthogonal polynomials are the books [**12**] and [**8**]. See also [**1**], [**4**], [**5**], and [**13**].

In this paper, we present a construction that produces for each polynomial family in the q-Askey scheme a weight function $w(x_k)$ that determines a discrete orthogonality with respect to a quasi-definite moment functional. The weights $w(x_k)$ are the values at $t = 1$ of a sequence of basic hypergeometric functions $f_k(t)$ that are of type ${}_3\Phi_2$, or type ${}_2\Phi_1$. The functions $f_k(t)$ depend on the parameters that determine the coefficients in the three-term recurrence relation of the corresponding polynomial family. The basic hypergeometric functions $f_k(t)$ are convergent at $t = 1$ and we show how they are obtained from $f_0(t)$.

For some families in the q-Askey scheme, such as the big and the little q-Jacobi and the dual q-Hahn polynomials, we obtain two discrete orthogonality weights associated with different sequences of nodes x_k. In this paper we do not deal with the problem of characterizing the cases for which the weights are positive. Such a

---

2020 *Mathematics Subject Classification.* Primary 33C45.

problem seems quite difficult because the weights $w(x_k)$ are of the form $(-1)^k c_k s_k$, where the sequence c_k depends on k and some parameters, and s_k is the sum of a basic hypergeometric series with two or three parameters. It seems that there are few results about the positivity of basic hypergeometric series. See [**6**, Ch. 8].

We obtain our results using the linear algebraic approach of our previous papers [**17**], [**18**], [**19**], and [**20**]. In [**20**] we presented a unified construction of all the hypergeometric and basic hypergeometric orthogonal polynomial sequences that uses three linearly recurrent sequences of numbers that satisfy certain difference equations of order three. The initial terms of such sequences determine the sequences of orthogonal polynomials and provide us with a uniform parametrization of all the hypergeometric and basic hypergeometric sequences. In the present paper we use several results from [**20**].

Our linear algebraic approach can be used to construct alternative descriptions of the q-Askey and the Askey schemes presented in [**8**], [**9**], and [**10**]. In a forthcoming paper we will present a classification of the families in the Askey schemes constructed by considering the partial fraction decomposition of the coefficients α_n of the three-term recurrence relation (2.13), which are rational functions of q^k.

The approach used this paper for the generic case with $q \neq \pm 1$ can be used also to find discrete orthogonality weights for a subset of the class of polynomial sequences with $q = 1$, and for a subset of the class of sequences with $q = -1$. Such classes are called the Racah-Wilson and the Bannai-Ito classes, respectively, in [**21**]. Our algebraic approach does not involve taking limits as q goes to 1 or -1.

An interesting alternative to our algebraic approach is presented in [**16**], where the case of finite sequences of discrete orthogonal polynomials was studied in detail, using Leonard pairs and linear algebra methods.

In Section 2 we present some preliminary material related with the construction of the hypergeometric q-orthogonal polynomials and their uniform parametrization from [**20**]. In Section 3 we present the construction of the weight functions obtained by solving an infinite system of linear equations and use some properties of basic hypergeometric functions. In Section 4 we find general formulas for the weight functions in the cases when some of the parameters are equal to zero. In Section 5 we present explicit expressions for the weight functions for some families of polynomials in the q-Askey scheme.

2. The class $\mathcal{H}_q$ of basic hypergeometric orthogonal polynomial sequences

In this section we present some properties of the class $\mathcal{H}_q$ of the basic hypergeometric orthogonal polynomial sequences that were obtained in [**20**].

Consider the homogeneous difference equation

$$s_{k+3} = z(s_{k+2} - s_{k+1}) + s_k, \qquad k \geqslant 0, \tag{2.1}$$

where $z = 1 + q + q^{-1}$. The roots of the characteristic polynomial of this equation are $1, q$, and q^{-1}. The sequences s_k are called z-recurrent and are discussed in [**15**] in a general framework, considering different kinds of fields and values of z. In the present paper we consider only the case with distinct roots $1, q$, and q^{-1}. Therefore the general solution of (2.1) is a linear combination of the sequences $1, q^k$ and q^{-k}. Let the sequences x_k, h_k, and d_k be solutions of (2.1). Then we can write these sequences as follows:

(2.2)

$$x_k = b_0 + b_1 q^k + b_2 q^{-k}, \qquad h_k = a_0 + a_1 q^k + a_2 q^{-k}, \qquad d_k = s_0 + s_1 q^k + s_2 q^{-k}.$$

The sequence x_k determines the Newtonian basis $\{v_n(t) : n \geqslant 0\}$ of the complex vector space of polynomials in the complex variable t, defined by

$$v_n(t) = (t - x_0)(t - x_1) \cdots (t - x_{n-1}), \qquad n \geqslant 1, \tag{2.3}$$

and $v_0(t) = 1$. We define the sequence g_k by

$$g_k = x_{k-1}(h_k - h_0) + d_k, \qquad k \geqslant 1, \tag{2.4}$$

and $g_0 = 0$. Therefore we must have $d_0 = 0$ and hence $s_0 = -s_1 - s_2$. In addition, we suppose that $h_k \neq h_j$ whenever $k \neq j$, and that $g_k \neq 0$ for $k \geqslant 1$. We can take $a_0 = 0$ because a_0 does not appear in any of the formulas that we will obtain in our development.

Let $\mathcal{D}$ be the linear operator on the space of polynomials defined by

$$\mathcal{D} v_k = h_k v_k + g_k v_{k-1}, \qquad k \geqslant 0. \tag{2.5}$$

Since $g_0 = 0$ we see that $\mathcal{D} t^n$ is equal to $h_n t^n$ plus a polynomial of degree less than n. For $n \geqslant 0$ let u_n be a monic polynomial of degree n which is an eigenfunction of $\mathcal{D}$ with eigenvalue h_n. That is,

$$\mathcal{D} u_k = h_k u_k, \qquad k \geqslant 0. \tag{2.6}$$

The operator $\mathcal{D}$ is a generalized difference operator which, in concrete examples, becomes a second order difference operator on a linear or quadratic q-lattice. In [**20**] we showed that

$$u_n(t) = \sum_{k=0}^{n} c_{n,k} v_k(t), \qquad n \geqslant 0, \tag{2.7}$$

where the coefficients $c_{n,k}$ are

$$c_{n,k} = \prod_{j=k}^{n-1} \frac{g_{j+1}}{h_n - h_j}, \qquad 0 \leqslant k \leqslant n-1, \tag{2.8}$$

and $c_{n,n} = 1$ for $n \geqslant 0$. This expression for $u_n(t)$ was also obtained by Vinet and Zhedanov in [**21**] using a different approach. The idea of representing orthogonal polynomials in terms of a Newtonian basis was introduced by Geronimus in [**7**].

The matrix of coefficients $C = [c_{n,k}]$ is an infinite lower triangular matrix where the coefficients of u_n appear in the n'th row of C. Since $c_{n,n} = 1$ for $n \geqslant 0$, the semi-infinite matrix C is invertible. Let $C^{-1} = [\hat{c}_{n,k}]$ and define the polynomials

$$w_{n,k}(t) = \prod_{j=k}^{n-1} (t - h_j), \qquad 0 \leqslant k \leqslant n. \tag{2.9}$$

Using basic properties of divided differences we obtain

$$\hat{c}_{n,k} = \frac{\prod_{j=k+1}^{n} g_j}{w'_{n+1,k}(h_k)} = \prod_{j=k+1}^{n} \frac{g_j}{h_k - h_j}, \qquad 0 \leqslant k \leqslant n-1, \tag{2.10}$$

and $\hat{c}_{n,n} = 1$ for $n \geqslant 0$.

The entries in the 0-th column of C^{-1} are

$$\hat{c}_{k,0} = \prod_{j=1}^{k} \frac{g_j}{h_0 - h_j}, \qquad k \geqslant 1, \tag{2.11}$$

and $\hat{c}_{0,0} = 1$. We denote them by $m_k = \hat{c}_{k,0}$ for $k \geqslant 0$. They satisfy $m_0 = 1$ and

$$\sum_{k=0}^{n} c_{n,k}\, m_k = 0, \qquad n \geqslant 1. \tag{2.12}$$

Note that the sequence m_n satisfies a recurrence relation of order one. We will see that the numbers m_k are the generalized moments, with respect to the Newtonian basis $\{v_k(t) : k \geqslant 0\}$, of a moment functional for which the polynomial sequence $\{u_n(t) : n \geqslant 0\}$ is orthogonal.

In [**20**] we also proved that the polynomial sequence $\{u_n(t) : n \geqslant 0\}$ satisfies a three-term recurrence relation of the form

$$u_{n+1}(t) = (t - \beta_n)u_n(t) - \alpha_n u_{n-1}(t), \qquad n \geqslant 1, \tag{2.13}$$

where the coefficients are

$$\alpha_n = \frac{g_n}{h_{n-1} - h_n}\left(\frac{g_{n-1}}{h_{n-2} - h_n} - \frac{g_n}{h_{n-1} - h_n} + \frac{g_{n+1}}{h_{n-1} - h_{n+1}} + x_n - x_{n-1}\right) \tag{2.14}$$

and

$$\beta_n = x_n + \frac{g_{n+1}}{h_n - h_{n+1}} - \frac{g_n}{h_{n-1} - h_n}. \tag{2.15}$$

Writing α_n and β_n in terms of the parameters introduced in (2.2) we obtain

$$\alpha_n = \frac{(x-1)(a_1 x - q a_2) p_1(x) p_2(x)}{(a_1 x^2 - a_2)(a_1 x^2 - q a_2)^2 (a_1 x^2 - q^2 a_2)}, \qquad n \geqslant 1, \tag{2.16}$$

where $x = q^n$,

$$p_1(x) = (a_1 x - a_2)(b_1 x^2 + b_0 q x + b_2 q^2) + q x (s_1 x - s_2),$$

and

$$p_2(x) = -\frac{a_1^2 x^3}{q^2 a_2}\, p_1\!\left(\frac{q a_2}{a_1 x}\right),$$

and

$$\beta_n = \frac{q b_0 (x^2 - 1)(a_1^2 x^2 - a_2^2) + d\, x (x-1)(a_1 x - a_2) - (q s_1 - s_2)(a_1 - q a_2) x^2}{(a_1 q x^2 - a_2)(a_1 x^2 - q a_2)}, \tag{2.17}$$

where $n \geqslant 0$, $x = q^n$ and

$$d = (q+1)(q a_1 b_2 + a_2 b_1) - q(s_1 + s_2 + (a_1 + a_2) b_0).$$

Let us note that α_n and β_n are rational functions of q^n and that they are determined by the parameters $a_1, a_2, b_0, b_1, b_2, s_1, s_2$. Note also that at least one of a_1 and a_2 must be different from zero.

It is easy to verify that α_n is invariant under the substitution $x \to q a_2/(a_1 x)$ and β_n is invariant under the substitution $x \to a_2/(a_1 x)$.

The effect of the exchange $q \leftrightarrow q^{-1}$ on the sequences $\{\alpha_n : n \geqslant 1\}$ and $\{\beta_n : n \geqslant 0\}$ is equivalent to the exchanges $a_1 \leftrightarrow a_2$, $b_1 \leftrightarrow b_2$, $s_1 \leftrightarrow s_2$.

The map

$$(a_1, a_2, b_0, b_1, b_2, s_1, s_2) \to (\{\alpha_n : n \geqslant 1\}, \{\beta_n : n \geqslant 0\})$$

is not injective. Let us consider next an example that shows that all the polynomial sequences in the q-Askey scheme correspond to a vector of parameters where at least one of b_1 and b_2 is not zero. The only exception is the sequence of continuous q-Hermite polynomials.

Let $P = (a_1, a_2, b_0, 0, 0, s_1, s_2)$. We define the vectors $P_1 = (a_1, a_2, b_0, b_1, 0, \hat{s}_1, \hat{s}_2)$, where

$$b_1 = -\frac{b_0 a_2 + s_2}{a_2}, \qquad \hat{s}_1 = \frac{q s_1 - b_0 a_2 - s_2}{q}, \qquad \hat{s}_2 = -b_0 a_2,$$

and $P_2 = (a_1, a_2, b_0, 0, b_2, \tilde{s}_1, \tilde{s}_2)$, where

$$b_2 = -\frac{b_0 a_1 + s_1}{a_1}, \qquad \tilde{s}_1 = -b_0 a_1, \qquad \tilde{s}_2 = s_2 - q a_1 b_0 - q s_1.$$

A straightforward computation shows that if $a_1 \neq 0$ and $a_2 \neq 0$ then the three vectors P, P_1, P_2 yield the same pair of coefficient sequences (α_n, β_n); if $a_1 = 0$ and $a_2 \neq 0$ then P and P_1 produce the same pair of coefficient sequences; and if $a_1 \neq 0$ and $a_2 = 0$ then P and P_2 produce the same pair (α_n, β_n).

By Favard's theorem [**3**, Thm. 4.4], [**11**, Thm. 2.14], if all the α_n are positive and the β_n are real then $\{u_n : n \geqslant 0\}$ is orthogonal with respect to a positive-definite moment functional, and if all the α_n are nonzero then $\{u_n : n \geqslant 0\}$ is orthogonal with respect to a quasi-definite moment functional.

3. Generalized moments and discrete orthogonality

Let us suppose that the parameters that determine the sequences x_k, h_k, and d_k are such that $h_k \neq h_j$ whenever $k \neq j$, and the coefficients α_n, given by (2.14), are nonzero for $n \geqslant 1$. Then, by Favard's theorem [**3**, Thm. 4.4], [**11**, Thm. 2.14], there exists a unique quasi-definite moment functional τ on the space of polynomials such that the polynomial sequence $\{u_n(t) : n \geqslant 0\}$ is orthogonal with respect to τ, that is,

$$\tau(u_n(t) u_m(t)) = K_n \delta_{n,m}, \qquad K_n \neq 0, \quad n, m \in \mathbb{N}. \tag{3.1}$$

Since $u_0(t) = 1$ we have

$$\tau(u_n(t)) = \sum_{k=0}^{n} c_{n,k} \tau(v_k(t)) = 0, \qquad n \geqslant 1, \tag{3.2}$$

and by the uniqueness of the inverse of the matrix C we must have $\tau(v_k(t)) = \hat{c}_{k,0} = m_k$ for $k \geqslant 0$. This means that the numbers m_k are the generalized moments of τ with respect to the Newtonian basis $\{v_k(t) : k \geqslant 0\}$. Note that $\tau(1) = m_0 = 1$.

THEOREM 3.1. *Suppose that a_2 and b_2 are nonzero and that $\alpha_n \neq 0$ for $n \geqslant 1$. Then there exists a unique weight function w, defined on the nodes x_k, such that for every polynomial $p(t)$ we have*

$$\tau(p(t)) = \sum_{k=0}^{\infty} p(x_k) w(x_k). \tag{3.3}$$

PROOF. From (2.2) we see that the hypotheses imply that $\{x_k : k \geqslant 0\}$ and $\{h_k : k \geqslant 0\}$ are sequences of pairwise distinct numbers, and that the polynomial sequence $\{u_n(t) : n \geqslant 0\}$ is orthogonal with respect to the quasi-definite moment functional τ.

We will write $r_j = w(x_j)$ for $j \geqslant 0$. Since $\tau(v_k(t)) = m_k$ for $k \geqslant 0$, the numbers r_j that we want to find must satisfy

$$m_k = \tau(v_k(t)) = \sum_{j=0}^{\infty} v_k(x_j)\, r_j, \qquad k \geqslant 0. \tag{3.4}$$

This is an infinite system of linear equations where the r_j are unknown. Let us denote by P the matrix of coefficients. Then $P = [v_k(x_j)]$, where j is the index for rows, k is the index for columns and j and k are non-negative integers. From the definition of the polynomials $v_k(t)$ in (2.3) we can see that P is an infinite lower triangular matrix and its entries in the main diagonal are

$$v_k(x_k) = (x_k - x_0)(x_k - x_1)\cdots(x_k - x_{k-1}),$$

and hence they are nonzero. Therefore P is invertible.

Using basic properties of divided differences with respect to the nodes x_j it is easy to show that P^{-1} is the lower triangular matrix whose (j,k) entry is $1/v'_{j+1}(x_k)$. Therefore, from the system (3.4) we obtain

$$r_k = \sum_{j=k}^{\infty} \frac{m_j}{v'_{j+1}(x_k)}, \qquad k \geqslant 0. \tag{3.5}$$

We will show that all the series in (3.5) are convergent (or terminating) basic hypergeometric series.

Let us define the power series

$$f_k(t) = \sum_{j=k}^{\infty} \frac{m_j}{v'_{j+1}(x_k)} t^j, \qquad k \geqslant 0. \tag{3.6}$$

Therefore $r_k = f_k(1)$ for $k \geqslant 0$.

From (2.11) we obtain

$$m_j = \prod_{i=1}^{j} \frac{g_i}{h_0 - h_i}, \qquad j \geqslant 1 \tag{3.7}$$

and $m_0 = 1$.

The denominators in (3.6) are

$$v'_{j+1}(x_k) = \prod_{i=0, i\neq k}^{j} (x_k - x_i), \qquad k \geqslant j \geqslant 0. \tag{3.8}$$

Therefore $f_k(t)$ can be expressed as

$$f_k(t) = \sum_{j=k}^{\infty} \left(\prod_{i=1}^{j} \frac{g_i}{h_0 - h_i} \prod_{i=0, i\neq k}^{j} \frac{1}{x_k - x_i} \right) t^j, \qquad k \geqslant 0 \tag{3.9}$$

and

$$f_0(t) = 1 + \sum_{j=1}^{\infty} \left(\prod_{i=1}^{j} \frac{g_i}{(h_0 - h_i)(x_0 - x_i)} \right) t^j. \tag{3.10}$$

The coefficient of t^j, for $j \geqslant k$, in the series $f_k(t)$ depends on the parameters $a_1, a_2, b_0, b_1, b_2, s_1, s_2$. Let us introduce new parameters y_1, y_2, y_3, z_1, z_2 as follows:

$$\begin{aligned} a_1 &= \frac{z_1 a_2}{q}, \qquad b_1 = \frac{z_2 b_2}{q}, \qquad y_3 = \frac{z_1 z_2}{q y_1 y_2}, \\ s_1 &= \frac{-a_2}{q}\left(\left(y_1 y_2 + y_1 y_3 + y_2 y_3 - \frac{z_2}{q}\right) b_2 + b_0 z_1\right), \\ s_2 &= -a_2((y_1 + y_2 + y_3 - z_1) b_2 + b_0). \end{aligned} \tag{3.11}$$

The new parameters z_1 and z_2 are well defined by the first two equations because $a_2 \neq 0$ and $b_2 \neq 0$. Note that y_3 depends on y_1, y_2, z_1, z_2. We use the redundant parameter y_3 because it simplifies the equations in (3.11) and other formulas that we will obtain next.

Let us denote by $\rho(k,j)$ the coefficient of t^j in the series $f_k(t)$. By substitution in $\rho(0,j)$ of the parameters a_1, b_1, s_1, s_2, using the expressions in (3.11), we obtain

$$\rho(0,j) = \prod_{i=1}^{j} \frac{g_i}{(h_0 - h_i)(x_0 - x_i)} = \frac{(y_1;q)_j (y_2;q)_j (y_3;q)_j \, q^j}{(q;q)_j (z_1;q)_j (z_2;q)_j}, \qquad j \geqslant 1, \tag{3.12}$$

where $(t;q)_k = (1-t)(1-qt)\cdots(1-q^{k-1}t)$ for $t \in \mathbb{C}$ and $k \in \mathbb{N}$. Note the factor q^j in the numerator. Note also that only the parameters z_1, z_2, y_1, y_2, y_3 appear in the right-hand side. Thus we have essentially four parameters, since y_3 can be written in terms of the other four parameters.

The previous equation shows that $f_0(t)$ is a ${}_3\Phi_2$ basic hypergeometric series. If $|q| < 1$ or $|q| > 1$ then, by the ratio test the series $f_0(t)$ converges at $t = 1$, since $z_1 z_2 = q y_1 y_2 y_3$. See [**6**, p. 5].

Using the change of parameters (3.11) we obtain

$$\rho(k,j) = \mu(k) \frac{(y_1;q)_j (y_2;q)_j (y_3;q)_j \, q^j}{(q;q)_{j-k} (z_1;q)_j (q^k z_2;q)_j}, \qquad j \geqslant k \geqslant 1, \tag{3.13}$$

where

$$\mu(k) = (-1)^k \frac{q^{\binom{k}{2}}}{(q;q)_k} \frac{(1 - q^{2k-1} z_2)}{(1 - q^{k-1} z_2)}, \qquad k \geqslant 1. \tag{3.14}$$

The ratio test shows that the series $f_k(t)$ converges at $t = 1$ if $|q| < 1$ or $|q| > 1$. This shows that the numbers r_k defined in (3.5) are well defined and satisfy (3.4). This completes the proof. $\square$

The coefficients of $f_0(t)$ in terms of the original parameters look like the coefficients of a ${}_3\Phi_2$ function, that is, like the right-hand side of (3.12). The formulas in (3.11) for the change of parameters are obtained by substituting $a_1 = q^{-1} z_1 a_2$ and $b_1 = q^{-1} z_2 b_2$ in $f_0(t)$ and then solving for s_1, s_2 and y_3 the system of equations obtained from (3.12) by taking several values of j.

The coefficients $\rho(k,j)$ of the series $f_k(t)$ satisfy

$$\sum_{k=0}^{j} \rho(k,j) = 0, \qquad j \geqslant 1 \tag{3.15}$$

and

$$\rho(k+1,j) = \frac{-q^k (1 - q^{j-k})(1 - q^{2k+1} z_2)(1 - q^{k-1} z_2)}{(1 - q^{k+1})(1 - q^{2k-1} z_2)(1 - q^{j+k} z_2)} \rho(k,j), \qquad j > k. \tag{3.16}$$

Let us define the sequence of functions

$$\psi_k(t) = \mu(k)(z_2;q)_k \sum_{j=k}^{\infty} \frac{(q^{j-k+1};q)_k\, t^j}{(q^j z_2;q)_k}, \qquad k \geqslant 1. \tag{3.17}$$

A simple computation shows that $f_k(t)$ is equal to the Hadamard product of the series $f_0(t)$ and $\psi_k(t)$, for $k \geqslant 1$.

The functions $f_k(t)$, for $k \geqslant 0$, are invariant under the transformation

$$(q, y_1, y_2, y_3, z_1, z_2) \to (1/q, 1/y_1, 1/y_2, 1/y_3, 1/z_1, 1/z_2).$$

The change of parameters (3.11) applied to the coefficients of the three-term recurrence relation (2.13) gives us

$$\alpha_n = \frac{b_2^2(x-1)(z_1 x - q^2)}{z_1(z_1x^2-q)(z_1x^2-q^2)^2(z_1x^2-q^3)} \prod_{j=1}^{3}((y_j x - q)(z_1 x - q y_j)), \tag{3.18}$$

where $x = q^n$, and

$$\beta_n = b_0 + \frac{b_2 x(\epsilon_2 x^2 + \epsilon_1 x + \epsilon_0)}{z_1(z_1x^2-1)(z_1x^2-q^2)}, \tag{3.19}$$

where $x = q^n$, and

$$\begin{aligned}
\epsilon_2 &= z_1(z_1^2 + (y_1+y_2+y_3)qz_1 + (y_1y_2+y_1y_3+y_2y_3)z_1 + qy_1y_2y_3),\\
\epsilon_1 &= -z_1(q+1)((q+y_1+y_2+y_3)z_1 + (y_1y_2+y_1y_3+y_2y_3)q + y_1y_2y_3),\\
\epsilon_0 &= q(z_1^2 + (y_1+y_2+y_3)qz_1 + (y_1y_2+y_1y_3+y_2y_3)z_1 + qy_1y_2y_3).
\end{aligned}$$

The entries of the matrix C, which are the coefficients of the polynomials $u_n(t)$ with respect to the Newtonian basis $\{v_k(t) : k \geqslant 0\}$, become

$$c_{n,k} = \frac{b_2^{n-k}}{q^{(n-k)k}} \frac{(q^{k+1};q)_{n-k}(q^k y_1;q)_{n-k}(q^k y_2;q)_{n-k}(q^k y_3;q)_{n-k}}{(q;q)_{n-k}(z_1 q^{n+k-1};q)_{n-k}}, \qquad n \geqslant k \geqslant 0, \tag{3.20}$$

in terms of the parameters y_1, y_2, y_3, z_1, b_2.

We consider next the family of Askey–Wilson polynomials with parameters a, b, c, d, as they are presented in [**8**, Ch. 14]. By a simple comparison of our formulas (3.18) and (3.19) for α_n and β_n with the corresponding formulas for the Askey–Wilson polynomials given in [**8**, Eq. (14.1.5)], we obtain

$$y_1 = ab, \quad y_2 = ac, \quad y_3 = ad, \quad z_1 = abcd, \quad z_2 = qa^2, \quad b_0 = 0, \quad b_2 = \frac{1}{2a}. \tag{3.21}$$

Therefore the sequence of weights for the discrete orthogonality of the Askey–Wilson polynomials is

$$r_k = w(x_k) = \mu(k) \sum_{j=k}^{\infty} \frac{(ab;q)_j (ac;q)_j (ad;q)_j q^j}{(q;q)_{j-k}(abcd;q)_j (q^{k+1}a^2;q)_j}, \qquad k \geqslant 0, \tag{3.22}$$

where

$$\mu(k) = (-1)^k \frac{q^{\binom{k}{2}}}{(q;q)_k} \frac{(1-q^{2k}a^2)}{(1-q^k a^2)}. \tag{3.23}$$

The corresponding sequences of nodes x_k and eigenvalues h_k for the Askey–Wilson polynomials are

$$x_k = \frac{1}{2}\left(a\,q^k + \frac{q^{-k}}{a}\right), \qquad h_k = a_2(abcd\,q^{k-1} + q^{-k}). \tag{3.24}$$

We will show in the next section that Theorem 3.1 can be extended to the cases when at least one of a_1 and a_2 is nonzero and at least one of b_1 and b_2 is nonzero.

4. The cases with some parameters equal to zero

In this section we consider the cases with some of the four parameters a_1, a_2, b_1, b_2 equal to zero. Since our construction of the sequence of weights requires that $h_j \neq h_k$ and $x_j \neq x_k$ whenever $j \neq k$, then at least one of a_1, a_2 must be nonzero and at least one of b_1, b_2 must be nonzero. Such restrictions give us nine cases. There are four cases with exactly one of the parameters equal to zero, four cases with two parameters equal to zero, and the case with no parameters equal to zero that was considered in the previous section. Let us recall that all the pairs of sequences (α_n, β_n) of recurrence coefficients correspond to a vector of parameters that has at least one of b_1 and b_2 different from zero. It is clear that the results in the previous section apply to the case where a_1, a_2, b_1, b_2 are nonzero numbers. We show next that they can also be applied to three other cases, just using some simple substitutions.

4.1. Case 1. $a_1 = 0$. If $a_1 = 0$ and a_2, b_1, b_2 are nonzero, then from (3.11) we obtain $z_1 = 0$ and $y_3 = 0$. Therefore the change of parameters (3.11) becomes

$$\begin{aligned} a_1 &= 0, \qquad b_1 = \frac{z_2 b_2}{q}, \qquad y_3 = 0,\\ s_1 &= \frac{a_2 b_2}{q^2}(z_2 - q y_1 y_2),\\ s_2 &= -a_2((y_1 + y_2)b_2 + b_0), \end{aligned} \tag{4.1}$$

and the coefficients $\rho(k,j)$ are

$$\rho(k,j) = \mu(k)\frac{(y_1;q)_j (y_2;q)_j q^j}{(q;q)_{j-k}(q^k z_2;q)_j}, \qquad j \geqslant k, \tag{4.2}$$

where $\mu(k)$ is defined in (3.14). Therefore $f_k(t)$ is a ${}_2\Phi_1$ basic hypergeometric function.

In this case the eigenvalues are $h_k = a_2 q^{-k}$, the nodes are $x_k = b_0 + b_2(q^{k-1}z_2 + q^{-k})$ and the coefficients in the three-term recurrence relation are

$$\alpha_n = b_2^2 q^{-3}(x-1)(xy_1y_2 - z_2)(xy_1 - q)(xy_2 - q), \tag{4.3}$$

where $x = q^n$, and

$$\beta_n = b_0 - b_2 q^{-1} x(y_1 y_2(1+q)x - z_2 - q(y_1 + y_2) - y_1 y_2). \tag{4.4}$$

4.2. Case 2. $b_1 = 0$. If $b_1 = 0$ then $z_2 = 0$ and $y_3 = 0$ and the change of parameters becomes

$$\begin{aligned} a_1 &= \frac{z_1 a_2}{q}, \qquad b_1 = 0, \qquad y_3 = 0,\\ s_1 &= -a_2 q^{-1}(b_2 y_1 y_2 + z_1 b_0),\\ s_2 &= a_2((b_2(z_1 - y_1 - y_2) - b_0), \end{aligned} \tag{4.5}$$

and the coefficients $\rho(k,j)$ are

$$\rho(k,j) = (-1)^k \frac{q^{\binom{k}{2}}}{(q;q)_k} \frac{(y_1;q)_j (y_2;q)_j q^j}{(q;q)_{j-k}(z_1;q)_j}, \qquad j \geqslant k. \tag{4.6}$$

Therefore in this case $f_k(t)$ is also a ${}_2\Phi_1$ basic hypergeometric function.

The eigenvalues are $h_k = a_2(z_1 q^{k-1} + q^{-k})$, the nodes are $x_k = b_0 + b_2 q^{-k}$ and the coefficients in the three-term recurrence relation are

$$\alpha_n = \frac{b_2^2 q x(1-x)(z_1 x - q^2)(z_1 x - q y_1)(z_1 x - q y_2)(y_1 x - q)(y_2 x - q)}{(z_1 x^2 - q)(z_1 x^2 - q^2)^2 (z_1 x^2 - q^3)}, \tag{4.7}$$

$$\beta_n = b_0 + \frac{b_2 x (z_1 \epsilon_1 x^2 + (q+1)\epsilon_2 x + q\epsilon_1)}{(z_1 x^2 - 1)(z_1 x^2 - q^2)}, \tag{4.8}$$

where $x = q^n$, and

$$\epsilon_1 = z_1 + (y_1 + y_2) q + y_1 y_2,$$
$$\epsilon_1 = -(q + y_1 + y_2) z_1 + q y_1 y_2.$$

4.3. Case 3. $a_1 = 0$ **and** $b_1 = 0$**.** If $a_1 = 0$ and $b_1 = 0$ the change of parameters becomes

$$\begin{aligned} a_1 &= 0, \qquad b_1 = 0, \qquad y_3 = 0, \\ s_1 &= -a_2 b_2 q^{-1} y_1 y_2, \\ s_2 &= -a_2((b_2(y_1 + y_2) + b_0), \end{aligned} \tag{4.9}$$

and the coefficients $\rho(k,j)$ are

$$\rho(k,j) = (-1)^k \frac{q^{\binom{k}{2}}}{(q;q)_k} \frac{(y_1;q)_j (y_2;q)_j q^j}{(q;q)_{j-k}}, \qquad j \geqslant k. \tag{4.10}$$

The eigenvalues are $h_k = a_2 q^{-k}$, the nodes are $x_k = b_0 + b_2 q^{-k}$, and the recurrence coefficients are

$$\alpha_n = b_2^2 q^{-3} y_1 y_2 x (x-1)(y_1 x - q)(y_2 x - q), \tag{4.11}$$

and

$$\beta_n = b_0 - b_2 q^{-1} x (y_1 y_2 (q+1) x - q(y_1 + y_2) - y_1 y_2). \tag{4.12}$$

4.4. Case 4. $a_2 = 0$**.** We substitute $a_2 = 0$ and $b_1 = z_2 b_2 / q$ in $f_0(t)$ and observe that the coefficient $\rho(0,j)$ of t^j is

$$\rho(0,j) = \frac{(y_1;q)_j (y_2;q)_j p^j}{(q;q)_j (z_2;q)_j} \tag{4.13}$$

for some numbers y_1, y_2, p. Then we solve for s_1, s_2, y_2 the system of equations obtained from (4.13) by taking $j = 1, 2, 3$. We obtain the change of parameters

$$\begin{aligned} a_2 &= 0, \qquad b_1 = \frac{z_2 b_2}{q}, \qquad y_2 = \frac{z_2}{p y_1}, \\ s_1 &= -\frac{a_1 b_2}{q}\left(p y_1 + \frac{z_2}{y_1}\right) - a_1 b_0, \\ s_2 &= a_1 b_2 (q - p). \end{aligned} \tag{4.14}$$

This change of parameters gives us

$$\rho(k,j) = (-1)^k \frac{q^{\binom{k}{2}}(1-q^{2k-1}z_2)}{(q;q)_k(1-q^{k-1}z_2)} \frac{(y_1;q)_j \left(\frac{z_2}{py_1};q\right)_j p^j}{(q;q)_{j-k}(q^k z_2;q)_j}, \qquad j \geqslant k. \tag{4.15}$$

In this case the recurrence coefficients are

$$\alpha_n = \frac{b_2^2}{qy_1} \frac{(x-1)(x-p)(xy_1-q)(xz_2-qpy_1)}{x^4} \tag{4.16}$$

and

$$\beta_n = b_0 + \frac{b_2}{q} \frac{((q+y_1)p+q+\frac{z_2}{y_1})x - p(1+q)}{x^2}, \tag{4.17}$$

where $x = q^n$, the eigenvalues are $h_k = a_1 q^k$, and the nodes are $x_k = b_0 + b_2(z_2 q^{k-1} + q^k)$.

4.5. Case 5. $a_2 = 0$ **and** $b_1 = 0$**.** We put $b_1 = 0$ and $z_2 = 0$ in the formulas of the previous case and obtain the change of parameters

$$\begin{aligned} &a_2 = 0, \qquad b_1 = 0, \qquad z_2 = 0, \\ &s_1 = -\frac{a_1}{q}(py_1 b_2 + qb_0), \\ &s_2 = a_1 b_2(q-p), \end{aligned} \tag{4.18}$$

which gives us

$$\rho(k,j) = (-1)^k \frac{q^{\binom{k}{2}}}{(q;q)_k} \frac{(y_1;q)_j p^j}{(q;q)_{j-k}}. \tag{4.19}$$

In this case the recurrence coefficients become

$$\alpha_n = b_2^2 p \frac{(x-1)(x-p)(q-xy_1)}{x^4}, \tag{4.20}$$

$$\beta_n = b_0 + \frac{b_2}{q} \frac{(p(q+y_1)+q)x - p(q+1)}{x^2}, \tag{4.21}$$

where, as usual, $x = q^n$. The eigenvalues are $h_k = a_1 q^k$ and the nodes are $x_k = b_0 + b_2 q^{-k}$.

4.6. Case 6. $b_2 = 0$**.** We proceed as in Case 4, but now we put $b_2 = 0$ and $a_1 = z_1 a_2/q$ in $f_0(t)$, and then we see that the coefficients $\rho(0,j)$ are

$$\rho(0,j) = \frac{(y_1;q)_j (y_2;q)_j \, p^j}{(q;q)_j (z_1;q)_j}. \tag{4.22}$$

We solve for s_1, s_2, y_2 the system of equations obtained by taking $j = 1, 2, 3$ in (4.22), and obtain the change of parameters

$$\begin{aligned} &a_1 = \frac{z_1 a_2}{q}, \qquad b_2 = 0, \qquad y_2 = \frac{z_1}{py_1}, \\ &s_1 = -\frac{a_2}{qy_1}(pb_1 y_1^2 + (b_0 z_1 - b_1)y_1 + b_1 z_1), \\ &s_2 = -a_2(b_0 + pb_1), \end{aligned} \tag{4.23}$$

which gives us

$$\rho(k,j) = \frac{(-1)^k (y_1;q)_j \left(\frac{z_1}{py_1};q\right)_j p^j}{q^{\binom{k}{2}} q^{k(j-k)} (q;q)_k (q;q)_{j-k} (z_1;q)_j}. \tag{4.24}$$

In this case the recurrence coefficients become

$$\alpha_n = -\frac{qb_1^2}{y_1^2} \frac{x(x-1)(xz_1-q^2)(xz_1-qy_1)(py_1x-q)(y_1x-q)(xz_1-pqy_1)}{(x^2z_1-q)(x^2z_1-q^2)^2(x^2z_1-q^3)}, \tag{4.25}$$

$$\beta_n = b_0 + \frac{b_1}{y_1} \frac{x(z_1\epsilon_1 x^2 - (q+1)\epsilon_2 x + q\epsilon_1)}{(x^2z_1-1)(x^2z_1-q^2)}, \tag{4.26}$$

where as usual $x = q^n$, and

$$\epsilon_1 = py_1^2 + (p+1)qy_1 + z_1, \qquad \epsilon_2 = pqy_1^2 + (p+1)y_1z_1 + qz_1.$$

The eigenvalues are $h_k = a_2(z_1q^{k-1} + q^{-k})$ and the nodes are $x_k = b_0 + b_1q^k$.

4.7. Case 7. $b_2 = 0$ and $a_1 = 0$. The formulas for this case are obtained from the corresponding formulas in Case 6, just putting $a_1 = 0$, $z_1 = 0$, and $y_2 = 0$.

The change of parameters becomes

$$\begin{aligned} a_1 = 0, \quad b_2 &= 0, \quad z_1 = 0, \quad y_2 = 0, \\ s_1 &= -\frac{a_2b_1}{q}(py_1 - 1), \\ s_2 &= -a_2(b_0 + pb_1), \end{aligned} \tag{4.27}$$

which gives us

$$\rho(k,j) = \frac{(-1)^k (y_1;q)_j p^j}{q^{\binom{k}{2}} q^{k(j-k)} (q;q)_k (q;q)_{j-k}}. \tag{4.28}$$

In this case the recurrence coefficients become

$$\alpha_n = \frac{pb_1^2}{q^3} x(x-1)(xpy_1 - q)(xy_1 - q), \tag{4.29}$$

$$\beta_n = b_0 - \frac{b_1}{q} x((q+1)py_1x - q(p+1) - py_1), \tag{4.30}$$

where, as usual, $x = q^n$. The eigenvalues are $h_k = a_2q^{-k}$ and the nodes $x_k = b_0 + b_1q^k$.

4.8. Case 8. $a_2 = 0$ and $b_2 = 0$. If $a_2 = 0$ and $b_2 = 0$ it is easy to see that the coefficients $\rho(0,j)$ of the series $f_0(t)$ are

$$\rho(0,j) = (-1)^j q^{-\binom{j}{2}} \frac{(y_1;q)_j (y_2;q)_j p^j}{(q;q)_j}. \tag{4.31}$$

Solving for s_1, s_2, y_2 the system of equations obtained from (4.33) by taking $j = 1, 2, 3$ we obtain the change of parameters

$$\begin{aligned} a_2 = 0, \qquad b_2 = 0, \qquad y_2 &= \frac{1}{py_1}, \\ s_1 &= -\frac{a_1(b_1py_1^2 + b_0y_1 + b_1)}{y_1}, \\ s_2 &= -qpa_1b_1. \end{aligned} \tag{4.32}$$

This change of parameters gives us

$$\rho(k,j) = \frac{(-1)^{k+j} q^{\binom{k}{2}} (y_1;q)_j \left(\dfrac{1}{py_1};q\right)_j p^j}{q^{\binom{j}{2}} q^{k(j-1)} (q;q)_k (q;q)_{j-k}}. \tag{4.33}$$

The recurrence coefficients are

$$\alpha_n = -\frac{qpb_1^2(x-1)(x-qpy_1)(xy_1-q)}{y_1x^4} \tag{4.34}$$

and

$$\beta_n = b_0 + \frac{b_1(py_1^2+qpy_1+1)x-(q+1)py_1}{y_1x^2}, \tag{4.35}$$

where $x=q^n$. The eigenvalues are $h_k=a_1q^k$ and the nodes are $x_k=b_0+b_1q^k$.

5. Some examples

Since the nodes x_k are defined by $x_k = b_0+b_1q^k+b_2q^{-k}$, for $k \geqslant 0$, from (2.3) it is easy to see that a change in the constant term b_0 corresponds to a translation of the polynomials $v_n(t)$ in the Newton-type basis, and therefore to the same translation of the orthogonal polynomials $u_n(t)$. We will consider that the polynomial sequences $\{u_n(t) : n \geqslant 0\}$ and $\{\tilde{u}_n(t) : n \geqslant 0\}$ are equivalent if there exists a constant λ such that $\tilde{u}_n(t) = u_n(t+\lambda)$, for $n \geqslant 0$. When the coefficients β_n of the three-term recurrence relation are expressed in terms of the parameters y_1, y_2, z_1, z_2, the parameter b_0 appears as an additive constant.

Let us recall that

$$f_k(t) = \sum_{j=k}^{\infty} \rho(k,j)\, t^j, \qquad k \geqslant 0,$$

and that the weights associated with the nodes x_k are the numbers $f_k(1)$, for $k \geqslant 0$.

We next present explicit formulas for the coefficients $\rho(k,j)$ for some families of orthogonal polynomials in the q-Askey scheme.

5.1. Continuous big q-Hermite, [8, §14.18].

For this family we have

$$\alpha_n = \frac{1-q^n}{4}, \qquad \beta_n = \frac{aq^n}{2}, \qquad x_k = \frac{aq^k+(aq^k)^{-1}}{2},$$

and the corresponding parameters are

$$a_1=0, \quad a_2=\frac{1}{a}, \quad b_0=0, \quad b_1=\frac{a}{2}, \quad b_2=\frac{1}{2a}, \quad s_1=\frac{1}{2q}, \quad s_2=0.$$

This is an example of Case 1 and therefore we can use (4.1) to obtain the parameters

$$z_1=0, \qquad z_2=a^2q, \qquad y_1=0, \qquad y_2=0, \qquad y_3=0.$$

With these values of the parameters (4.2) gives us

$$\rho(k,j) = \frac{(-1)^k q^{\binom{k}{2}} (1-q^{2k}a^2)\, q^j}{(1-q^ka^2)(q;q)_k(q;q)_{j-k}(q^{k+1}a^2;q)_j}, \qquad j \geqslant k \geqslant 0.$$

5.2. Discrete q-Hermite I, [8, §14.28]. In this case we have

$$\alpha_n = \frac{q^n(1-q^n)}{q}, \qquad \beta_n = 0, \qquad x_k = q^k,$$

and the parameters are

$$a_1 = 0, \quad b_0 = 0, \quad b_1 = 1, \quad b_2 = 0, \quad p = -1, \quad y_1 = 0, \quad s_1 = \frac{a_2}{q}, \quad s_2 = a_2,$$

where a_2 is an arbitrary nonzero number. Since $a_1 = 0$ and $b_2 = 0$ this example is in Case 7 and we obtain

$$\rho(k,j) = \frac{(-1)^{k+j} q^{\binom{k}{2}}}{q^{k(j-1)}(q;q)_k(q;q)_{j-k}}, \qquad j \geqslant k \geqslant 0.$$

5.3. Little q-Jacobi, [8, §14.12]. The normalized three-term recurrence relation for the little q-Jacobi polynomials is given in [8, Eq. (14.12.4)], with parameters a and b. The recurrence coefficients can be obtained in three different ways in terms of the parameters $b_0, b_1, b_2, a_1, a_2, s_1, s_2$. The first one is

$$a_1 = abqa_2, \quad b_0 = 0, \quad b_1 = 0, \quad b_2 = \frac{1}{qb}, \quad s_1 = 0, \quad s_2 = (qa-1)a_2,$$

where a_2 is an arbitrary nonzero number. Since $b_1 = 0$ and a_1, a_2, b_2 are nonzero, this is an example of Case 2 of the previous section. The sequence of nodes is $x_k = b^{-1}q^{-k-1}$ and the coefficients $\rho(k,j)$ of the weights are given by (4.6), taking $y_1 = bq, y_2 = 0, z_1 = q^2ab, z_2 = 0$. We obtain

$$\rho(k,j) = \frac{(-1)^k q^{\binom{k}{2}}(qb;q)_j\, q^j}{(q;q)_k(q;q)_{j-k}(q^2ab;q)_j}, \qquad j \geqslant k.$$

The second set of values of the parameters is

$$a_1 = abqa_2, \quad b_0 = 0, \quad b_1 = 1, \quad b_2 = 0, \quad s_1 = -q^{-1}(qa-1)a_2, \quad s_2 = 0,$$

where a_2 is an arbitrary nonzero number. Since $b_2 = 0$ and a_1, a_2, b_1 are nonzero, this is an example of Case 6 of the previous section. The sequence of nodes is $x_k = q^k$ and the coefficients $\rho(k,j)$ of the weights are given by (4.24), taking $y_1 = bq, z_1 = q^2ab, y_2 = 0, p = 0$. We obtain

$$\rho(k,j) = (-1)^{k+j} \frac{q^{\binom{k}{2}} q^{\binom{j}{2}} (qb;q)_j a^j q^j}{q^{k(j-1)}(q;q)_k(q;q)_{j-k}(q^2ab;q)_j}, \qquad j \geqslant k \geqslant 0.$$

T. H. Koornwinder observed that using formula [6, II.5] we can show that the weight function determined by this $\rho(k,j)$ coincides with the one in [8, Eq. (14.12.2)], and it is positive if $0 < a < 1$ and $b < q^{-1}$.

The third set of values of the parameters is

$$a_1 = -qab, \quad a_2 = -1, \quad b_0 = 0, \quad b_1 = 0, \quad b_2 = 0, \quad s_1 = a, \quad s_2 = 1.$$

Since the nodes are $x_k = 0$, for $k \geqslant 0$, we do not obtain a discrete orthogonality with this set of values of the parameters.

5.4. q-Laguerre, [8, §14.21]. This family is obtained by taking

$$a_2 = 0, \quad b_0 = 1, \quad b_1 = 0, \quad b_2 = -1, \quad s_1 = 0, \quad s_2 = a_1(q^{-\alpha} - q),$$

where a_1 is an arbitrary nonzero number. This is an example of Case 5.

The coefficients $\rho(k, j)$ are

$$\rho(k, j) = \frac{(-1)^k q^{\binom{k}{2}} q^{-j\alpha}}{(q;q)_k (q;q)_{j-k}}, \qquad j \geqslant k \geqslant 0.$$

The sequence of nodes is $x_k = -q^{-k}$, for $k \geqslant 0$. Note that $\rho(0, j)$ are the coefficients of the q-exponential function of $q^{-\alpha}$.

5.5. Dual q-Hahn, [8, §14.7]. The dual q-Hahn polynomials can be obtained with three different sets of values of the parameters. We consider first the case with

$$a_1 = 0, \quad b_0 = 0, \quad b_1 = q^{-N}, \quad b_2 = q\gamma\delta q^N, \quad s_1 = a_2 q^{-N}(q^{-1} - \gamma), \quad s_2 = -qa_2\gamma(1 + \delta),$$

where a_2 is an arbitrary nonzero number and N, γ, δ are the parameters used in [8, §14.7] to describe the q-Hahn polynomials. This example is in Case 1 of the previous section. Therefore the coefficients $\rho(k, j)$ can be obtained from (4.2) by taking

$$y_1 = q^{-N}, \qquad y_2 = \frac{q^{-N}}{\delta}, \qquad z_2 = \frac{q^{-2N}}{\gamma\delta},$$

which are the values of y_1, y_2, z_2 obtained from (4.1) with the values of the parameters given above. Since $y_1 = q^{-N}$ we obtain a discrete orthogonality on the nodes $x_k = \gamma\delta q^{N+1-k} + q^{k-N}$, for $0 \leqslant k \leqslant N$.

The second set of values of the parameters is

$$a_1 = 0, \quad b_0 = 0, \quad b_1 = q\gamma, \quad b_2 = \delta, \quad s_1 = a_2\gamma(1 - q^{-N}), \quad s_2 = -a_2(\gamma\delta q + q^{-N}),$$

where a_2 is an arbitrary nonzero number. The nodes are $x_k = \gamma q^{k+1} + \delta q^{-k}$, for $k \geqslant 0$. We obtain the coefficients $\rho(k, j)$ from (4.2) taking

$$y_1 = \gamma q, \qquad y_2 = \frac{q^{-N}}{\delta}, \qquad z_2 = \frac{\gamma q^2}{\delta}.$$

The third set of values of the parameters is

$$a_1 = 0, \quad b_0 = 0, \quad b_1 = \gamma\delta q, \quad b_2 = 1, \quad s_1 = a_2\gamma(\delta - q^{-N}), \quad s_2 = -a_2(\gamma q + q^{-N}).$$

Substitution of these values in (4.1) yields

$$y_1 = q^{-N}, \qquad y_2 = \frac{q^{-N}}{\delta}, \qquad z_2 = \frac{q^{-2N}}{\gamma\delta},$$

and then (4.2) gives us the corresponding coefficients $\rho(k, j)$. The discrete orthogonality obtained in this case is essentially equivalent to the one described in [8, Eq. (14.7.2)]. In this case the nodes are $x_k = \gamma\delta q^{k+1} + q^{-k}$, for $k \geqslant 0$.

5.6. Big q-Jacobi polynomials, [**8**, §14.5]. The big q-Jacobi polynomials are determined by the normalized three-term recurrence relation [**8**, Eq. (14.5.4)], where the recurrence coefficients are described in terms of parameters a, b, and c. In this case there are four sets of values of our parameters that yield the recurrence coefficients.

The first set of values is

$$a_1 = qaba_2, \quad b_0 = 0, \quad b_1 = 0, \quad b_2 = 1, \quad s_1 = -qaca_2, \quad s_2 = (qab - a - c)qa_2,$$

where a_2 is an arbitrary nonzero number. Since $b_1 = 0$ this is an example of Case 2 of the previous section and the coefficients $\rho(j,k)$ are given by (4.6) with $z_1 = q^2ab$, $y_1 = qa$, $y_2 = qc$, and the sequence of nodes is $x_k = q^{-k}$.

The second set of values of the parameters is

$$a_1 = qaba_1, \quad b_0 = 0, \quad b_1 = 0, \quad b_2 = \frac{c}{b}, \quad s_1 = -qaba_2, \quad s_2 = (qac - a - c)qa_2,$$

where a_2 is an arbitrary nonzero number. This is also in Case 2, but now the nodes are $x_k = (c/b)q^{-k}$ for $k \geqslant 0$, and we have $z_1 = q^2ab$, $y_1 = qb$, $y_2 = qab$. As in the previous case (4.6) gives us the coefficients $\rho(j,k)$.

The third set of values of the parameters is

$$a_1 = qaba_2, \quad b_0 = 0, \quad b_1 = qa, \quad b_2 = 0, \quad s_1 = (1 - qb - qc)aa_2, \quad s_2 = -qca_2,$$

where a_2 is any nonzero number. This is an example in Case 6 and we have $z_1 = q^2ab$, $y_1 = qa$, $y_2 = qab/c$, $p = c/a$. The coefficients $\rho(j,k)$ are given by (4.24) and the nodes are $x_k = aq^{k+1}$.

The fourth set of values is

$$a_1 = qaba_2, \quad b_0 = 0, \quad b_1 = qc, \quad b_2 = 0, \quad s_1 = (c - qab - qac)a_2, \quad s_2 = -qaa_2,$$

where a_2 is an arbitrary nonzero number. This example is also in Case 6. Now we have $z_1 = q^2ab$, $y_1 = qb$, $y_2 = qc$, $p = a/c$, and the nodes are $x_k = cq^{k+1}$. The coefficients $\rho(j,k)$ are given by (4.24). It seems that there is some sort of symmetry relating the parameters a, b, c of the big q-Jacobi polynomials.

5.7. Dual of the continuous dual q-Hahn polynomials, [**8**, §14.3]. If we replace q with $1/q$ in the continuous q-Hahn polynomials, with parameters a, b, c, we obtain the polynomial sequence with parameters

$$a_2 = 0, \quad b_0 = 0, \quad b_1 = \frac{1}{2a}, \quad b_2 = \frac{a}{2}, \quad s_1 = -\frac{a_1(b+c)}{2}, \quad s_2 = \frac{a_1qa(1-bc)}{2},$$

which corresponds to Case 4. From equations (4.14) we obtain

$$y_1 = \frac{1}{ac}, \qquad y_2 = \frac{1}{ab}, \qquad z_2 = \frac{q}{a^2}, \qquad p = qbc.$$

By substitution of these values in (4.15) we obtain the coefficients $\rho(k,j)$. In this case the nodes are $x_k = (1/2a)q^k + (a/2)q^{-k}$ and the eigenvalues are $h_k = a_1q^k$.

5.8. Big q-Laguerre polynomials, [**8**, §14.11]. The big q-Laguerre polynomials, with parameters a, b, are obtained with

$$a_1 = 0, \quad a_2 = \frac{1}{q}, b_0 = 0, \quad b_1 = 0, \quad b_2 = 1, \quad s_1 = -ab, \quad s_2 = -(a+b).$$

This is an example of Case 3 and thus we use (4.9) to obtain $y_1 = aq$ and $y_2 = bq$. Substitution of these values in (4.10) gives us the coefficients $\rho(k,j)$. The nodes are in this case $x_k = q^{-k}$ and the eigenvalues are $h_k = q^{-k-1}$.

5.9. q-Meixner polynomials, [8, §14.13]. The sequence of q-Meixner polynomials, with parameters b, c, is obtained with

$$a_2 = 0, \quad b_0 = 0, \quad b_1 = qb, \quad b_2 = 0, \quad s_1 = (bc-1)a_1, \quad s_2 = ca_1,$$

where a_1 is an arbitrary nonzero number. Since $a_2 = b_2 = 0$ this is an example of Case 8 and thus we can use (4.32) to obtain

$$y_1 = qb, \qquad y_2 = \frac{-q}{c}, \qquad p = \frac{-c}{q^2 b}.$$

Substitution of these values in (4.33) gives us the coefficients $\rho(k,j)$. The nodes are $x_k = bq^{k+1}$ and the eigenvalues are $h_k = a_1 q^k$.

5.10. Continuous q^2-Hermite polynomials. Using our uniform parametrization it is not possible to obtain the continuous q-Hermite polynomials, the recurrence coefficients of which are $\alpha_k = 1 - q^k$, for $k \geqslant 1$, and $\beta_k = 0$, for $k \geqslant 0$, but we can obtain the continuous q^2-Hermite polynomials, with recurrence coefficients $\alpha_k = 1 - q^{2k}$, for $k \geqslant 1$, and $\beta_k = 0$, for $k \geqslant 0$. We can also obtain other polynomial sequences for which the recurrence coefficients are rational functions of q^{2k}. In [10] Koornwinder considers q^2 versions of the q-Jacobi polynomials and its subfamilies, which give q-polynomials by quadratic transformation.

The continuous q^2-Hermite polynomials are obtained with the parameters

$$a_1 = 0, \quad b_0 = 0, \quad b_1 = \sqrt{q}, \quad b_2 = \frac{1}{\sqrt{q}}, \quad s_1 = \frac{a_2}{\sqrt{q}}, \quad s_2 = \sqrt{q}\, a_2,$$

where a_2 is an arbitrary nonzero number. This is an example in Case 1 and has

$$z_1 = 0, \quad z_2 = q^2, \quad y_1 = -q, \quad y_2 = 0,$$

and from (4.2) we get

$$\rho(k,j) = \frac{(-1)^k q^{\binom{k}{2}} (1 - q^{2k+1})(-q;q)_j q^j}{(1 - q^{k+1})(q;q)_k (q;q)_{j-k} (q^{k+2};q)_j}, \qquad j \geqslant k \geqslant 0.$$

The sequence of nodes is $x_k = \sqrt{q}(q^k + q^{-k-1})$ and the eigenvalues are $h_k = a_2 q^{-k}$.

6. Final remarks

We have shown that all the families of basic hypergeometric polynomial sequences in the q-Askey scheme have at least one discrete orthogonality on a sequence of nodes of the form $x_k = b_0 + b_1 q^k + b_2 q^{-k}$, where at least one of b_1 and b_2 is nonzero. The only exception is the sequence of continuous q-Hermite polynomials, which requires the transformation $q \to q^2$. Some families are orthogonal on two different sequences of nodes x_k.

In some cases the basic hypergeometric series $f_k(t)$ are obtained from $f_0(t)$ by repeated application of the q derivative or the q^{-1} derivative. In some of the simplest cases the functions $f_k(t)$ are related to the q-exponential function.

Further research is needed on the following subjects.

- The construction of weight functions for the discrete orthogonality of polynomial sequences in the Racah–Wilson ($q = 1$) and the Bannai–Ito ($q = -1$) classes.

- The construction of classifications of all the hypergeometric and basic hypergeometric orthogonal polynomial sequences that improve and refine the Askey schemes. For example, using the formulas (3.18) and (3.19) for the recurrence coefficients α_k and β_k in terms of the parameters y_1, y_2, z_1, z_2.
- The characterization of the basic hypergeometric polynomial families that have discrete orthogonality with positive weight functions.
- The generalization of our linear algebra approach to orthogonal polynomials to study orthogonal polynomials in several variables and matrix orthogonal polynomials.

References

[1] R. Álvarez-Nodarse, A. G. García, and F. Marcellán, *On the properties for modifications of classical orthogonal polynomials of discrete variables*, Proceedings of the International Conference on Orthogonality, Moment Problems and Continued Fractions (Delft, 1994), J. Comput. Appl. Math. **65** (1995), no. 1-3, 3–18, DOI 10.1016/0377-0427(95)00097-6. MR1379116

[2] G. E. Andrews, R. Askey, and R. Roy, *Special functions*, Encyclopedia of Mathematics and its Applications, vol. 71, Cambridge University Press, Cambridge, 1999, DOI 10.1017/CBO9781107325937. MR1688958

[3] T. S. Chihara, *An introduction to orthogonal polynomials*, Mathematics and its Applications, Vol. 13, Gordon and Breach Science Publishers, New York-London-Paris, 1978. MR481884

[4] D. Dominici and F. Marcellán, *Discrete semiclassical orthogonal polynomials of class 2*, Orthogonal polynomials: current trends and applications, SEMA SIMAI Springer Ser., vol. 22, Springer, Cham, 2021, pp. 103–169, DOI 10.1007/978-3-030-56190-1_5. MR4238531

[5] D. Dominici and F. Marcellán, *Discrete semiclassical orthogonal polynomials of class one*, Pacific J. Math. **268** (2014), no. 2, 389–411, DOI 10.2140/pjm.2014.268.389. MR3227440

[6] G. Gasper and M. Rahman, *Basic hypergeometric series*, 2nd ed., Encyclopedia of Mathematics and its Applications, vol. 96, Cambridge University Press, Cambridge, 2004, DOI 10.1017/CBO9780511526251. MR2128719

[7] J. Geronimus, *The orthogonality of some systems of polynomials*, Duke Math. J. **14** (1947), 503–510. MR21151

[8] R. Koekoek, P. A. Lesky, and R. F. Swarttouw, *Hypergeometric orthogonal polynomials and their q-analogues*, Springer Monographs in Mathematics, Springer-Verlag, Berlin, 2010, DOI 10.1007/978-3-642-05014-5. MR2656096

[9] T. H. Koornwinder, *Charting the q-Askey scheme*, Hypergeometry, integrability and Lie theory, Contemp. Math., vol. 780, Amer. Math. Soc., Providence, RI, 2022, pp. 79–94, DOI 10.1090/conm/780/15688. MR4476417

[10] T. H. Koornwinder, *Charting the q-Askey scheme. II. The Zhedanov scheme.* Indagationes Mathematicae, available on line from December 14, 2022. `https://doi.org/10.1016/j.indag.2022.12.003`

[11] F. Marcellán and R. Álvarez-Nodarse, *On the "Favard theorem" and its extensions*, J. Comput. Appl. Math. **127** (2001), no. 1-2, 231–254, DOI 10.1016/S0377-0427(00)00497-0. Numerical analysis 2000, Vol. V, Quadrature and orthogonal polynomials. MR1808576

[12] A. F. Nikiforov, S. K. Suslov, and V. B. Uvarov, *Classical orthogonal polynomials of a discrete variable*, Springer Series in Computational Physics, Springer-Verlag, Berlin, 1991, DOI 10.1007/978-3-642-74748-9. MR1149380

[13] M. Njinkeu Sandjon, A. Branquinho, M. Foupouagnigni, and I. Area, *Characterizations of classical orthogonal polynomials on quadratic lattices*, J. Difference Equ. Appl. **23** (2017), no. 6, 983–1002, DOI 10.1080/10236198.2017.1307972. MR3683297

[14] L. J. Slater, *Generalized hypergeometric functions*, Cambridge University Press, Cambridge, 1966. MR201688

[15] P. Terwilliger, *Two linear transformations each tridiagonal with respect to an eigenbasis of the other*, Linear Algebra Appl. **330** (2001), no. 1-3, 149–203, DOI 10.1016/S0024-3795(01)00242-7. MR1826654

[16] P. Terwilliger, *An algebraic approach to the Askey scheme of orthogonal polynomials*, Orthogonal polynomials and special functions, Lecture Notes in Math., vol. 1883, Springer, Berlin, 2006, pp. 255–330, DOI 10.1007/978-3-540-36716-1_6. MR2243532

[17] L. Verde-Star, *Characterization and construction of classical orthogonal polynomials using a matrix approach*, Linear Algebra Appl. **438** (2013), no. 9, 3635–3648, DOI 10.1016/j.laa.2013.01.014. MR3028602

[18] L. Verde-Star, *Recurrence coefficients and difference equations of classical discrete orthogonal and q-orthogonal polynomial sequences*, Linear Algebra Appl. **440** (2014), 293–306, DOI 10.1016/j.laa.2013.10.051. MR3134272

[19] L. Verde-Star, *Polynomial sequences generated by infinite Hessenberg matrices*, Spec. Matrices **5** (2017), 64–72, DOI 10.1515/spma-2017-0002. MR3602625

[20] L. Verde-Star, *A unified construction of all the hypergeometric and basic hypergeometric families of orthogonal polynomial sequences*, Linear Algebra Appl. **627** (2021), 242–274, DOI 10.1016/j.laa.2021.06.012. MR4282678

[21] L. Vinet and A. Zhedanov, *Hypergeometric orthogonal polynomials with respect to Newtonian bases*, SIGMA Symmetry Integrability Geom. Methods Appl. **12** (2016), Paper No. 048, 14 pp., DOI 10.3842/SIGMA.2016.048. MR3500200

Dept. of Mathematics, Universidad Autónoma Metropolitana, Iztapalapa, Mexico City, Mexico

Email address: verde@xanum.uam.mx

Contemporary Mathematics
Volume **819**, 2025
https://doi.org/10.1090/conm/819/16399

A new three-term relation for an indefinite sum of a very well poised basic hypergeometric series

Jianan Xu, Ying Zhang, and Xinrong Ma

Abstract. The purpose of the present paper is to study a kind of indefinite basic hypergeometric sum. By establishing a recursive formula and an explicit expression via Abel's lemma on partial sums, we prove a new three-term relation and a new limit for such indefinite hypergeometric sums.

1. Introduction

Throughout this paper we will use the standard notation and terminology for q-series, as in the book [**5**] by Gasper and Rahman. As is customary, the q-shifted factorials are

$$(a;q)_\infty := \prod_{k=0}^{\infty}(1-aq^k), \qquad (a;q)_n := \frac{(a;q)_\infty}{(aq^n;q)_\infty},$$

where $n \in \mathbf{Z}$, $\mathbf{Z}$ being the set of integers. For any parameters $a_1, a_2, \ldots, a_m$, the multivariate notation is

$$(a_1, a_2, \ldots, a_m; q)_n := \prod_{k=1}^{m}(a_k;q)_n$$

and, according to Gasper and Rahman [**5**], the basic hypergeometric series with base q and variable z is defined as

$$(1.1) \qquad {}_r\phi_{r-1}\left[\begin{matrix} a_1, a_2, \ldots, a_r \\ b_1, b_2, \ldots, b_{r-1}\end{matrix}; q, z\right] := \sum_{n=0}^{\infty}\frac{(a_1, a_2, \ldots, a_r;q)_n}{(b_1, b_2, \ldots, b_{r-1};q)_n}\frac{z^n}{(q;q)_n},$$

where the parameters $b_1, b_2, \ldots, b_{r-1}$ are such that the denominator factors in the terms of the series (1.1) are never zero. In particular, the compact notation

$${}_rW_{r-1}(a; a_4, \ldots, a_r; q, z)$$

denotes a special case of the ${}_r\phi_{r-1}$ series (1.1), called *very well poised* (in brief, VWP), in which the parameters satisfy the relations

$$(1.2) \qquad b_1a_2 = b_2a_3 = \cdots = b_{r-1}a_r = aq; \; a_1 = a, a_2 = q\sqrt{a}, a_3 = -q\sqrt{a}.$$

2020 *Mathematics Subject Classification.* Primary 33D15; Secondary 33C45.

Key words and phrases. Basic hypergeometric series; very well poised; Weierstrass's theta identity; indefinite sum; Abel's lemma; summation by parts; elliptic hypergeometric series.

This work was supported by the National Natural Science Foundation of China [Grant No. 12471315].

The main subject of this paper is the following indefinite basic hypergeometric sums.

DEFINITION 1.1. For integer $n \geqslant 0$ and parameters a, b, c, d, e such that none of the denominator parameters involved is of the form $q^{-j}, 0 \leqslant j \leqslant n-1$, we define the indefinite sum

$$\Omega_n(a;b,c,d,e) := \sum_{k=0}^{n} \frac{1-aq^{2k}}{1-a} \frac{(b,c,d,e;q)_k}{(aq/b,aq/c,aq/d,aq/e;q)_k} \left(\frac{a^2q}{bcde}\right)^k. \tag{1.3}$$

We believe that $\Omega_n(a;b,c,d,e)$ deserves a careful study, since it embraces Rogers' well-known VWP ${}_6\phi_5$ summation formula [**5**, (II.20)] as the special case $n = +\infty$ and $e = a$. However, to our knowledge, it has not received any systematic treatment in the literature, except the paper [**4**] wherein a special $\Omega_\infty(a;b,c,d,e)$ has been explored. As an anonymous referee pointed out, the indefinite sum $\Omega_n(a;b,c,d,e)$ can be reformulated as

$$\Omega_n(a;b,c,d,e) = {}_{10}W_9\left(a;b,c,d,e,q,q^{-n},aq^{n+1};q,a^2q/bcde\right). \tag{1.4}$$

Further, on assuming the balancing condition $a^2 = bcde$, we see that

$$\begin{aligned}\Omega_n(a;b,c,d,e) &= \frac{(1-a/b)(1-a/c)(1-a/d)(1-a/e)}{e(1-a)(1-bc/a)(1-bd/a)(1-cd/a)} \\ &\quad\times\left(1-\frac{(b,c,d,e;q)_{n+1}}{(a/b,a/c,a/d,a/e;q)_{n+1}}\right).\end{aligned} \tag{1.5}$$

This identity is a special case of Bailey's ${}_{10}W_9$-series transformation [**5**, (2.12.9)] and also of the bibasic summation [**5**, (3.6.15)] due to Gasper and Rahman. The main purpose of the present paper is to generalize the summation formula (1.5) to the following transformation.

THEOREM 1.2. *Let $\Omega_n(a;b,c,d,e)$ be defined by (1.3), and let $t = bcde/a^2$. Then there holds*

$$\begin{aligned}\Omega_n(a;b,c,d,e) &= \frac{t(1-a/b)(1-a/c)(1-a/d)(1-at/e)}{e(1-a)(1-bc/a)(1-bd/a)(1-cd/a)}\Bigg(\Omega_n(ta/e;b,c,d,t) \\ &\quad - \frac{(1-a/e)(1-eq^n/t)}{t^n(1-ta/e)(1-eq^n)}\frac{(b,c,d,e;q)_{n+1}}{(a/b,a/c,a/d,a/e;q)_{n+1}}\Omega_n\left(\frac{tq^{-n}}{e};\frac{q^{-n}b}{a},\frac{q^{-n}c}{a},\frac{q^{-n}d}{a},t\right)\Bigg).\end{aligned} \tag{1.6}$$

It is clear that when $t = 1$, (1.6) becomes (1.5) since both Ω_n-series on the right reduce to 1. Frankly, it is owing to the referee's insightful remarks that we are able to present our main result in such an elegant and concise form as Theorem 1.2.

Our main argument for Theorem 1.2 is built on a rearrangement technique called *summation by parts*, which is analogous to integration by parts.

LEMMA 1.3 (Abel's lemma on summation by parts: [**6**, p. 313]). *For any integer $n \geqslant 0$ and any sequences $\{U_k\}_{k\geqslant 0}$ and $\{V_k\}_{k\geqslant 0}$, there always holds*

$$\sum_{k=0}^{n} V_k(U_k - U_{k+1}) = V_0U_0 - V_nU_{n+1} + \sum_{k=1}^{n} U_k(V_k - V_{k-1}). \tag{1.7}$$

In addition, we need the following basic algebraic identity, which is almost self-evident.

LEMMA 1.4. *For any complex numbers x, z, b, c with $bcz \neq 0$, there holds*

$$\nabla\left(cx, \frac{x}{c}, bz, \frac{z}{b}\right) - \nabla\left(bx, \frac{x}{b}, cz, \frac{z}{c}\right) = \frac{z}{c}\nabla\left(bc, \frac{c}{b}, xz, \frac{x}{z}\right), \tag{1.8}$$

where the multi-parameter notation $\nabla(x) := 1-x$, $\nabla(x_1, x_2, x_3, x_4) := \nabla(x_1)\nabla(x_2)\nabla(x_3)\nabla(x_4)$ is used.

LEMMA 1.5 (Weierstrass's theta identity: [11, Ex. 20.5.6]). *For any complex numbers x, z, b, c with $bcz \neq 0$, there holds*

$$\theta\left(cx, \frac{x}{c}, bz, \frac{z}{b}\right) - \theta\left(bx, \frac{x}{b}, cz, \frac{z}{c}\right) = \frac{z}{c}\theta\left(bc, \frac{c}{b}, xz, \frac{x}{z}\right), \tag{1.9}$$

where $\theta(x)$ denotes the modified Jacobi theta function $(x, q/x; q)_\infty$ and $\theta(x_1, x_2, x_3, x_4)$ equals $\theta(x_1)\theta(x_2)\theta(x_3)\theta(x_4)$.

We remark here that (1.8) is equivalent to Weierstrass's theta identity (1.9). As for this equivalence, we refer the reader to [9] for a full proof. In addition, we would like to refer the reader to [7] for a short history of Weierstrass's theta identity.

Our paper is organized as follows. In the next section, we show Theorem 1.2 by establishing two preliminary formulas for $\boldsymbol{\Omega}_n(a; b, c, d, e)$. Section 3 is devoted to a limit of $\boldsymbol{\Omega}_{n+1}(a^2q^{1-n}/bde; aq, aq^{-n}/bd, aq^{-n}/be, aq^{-n}/de)/(aq)^n$ as n tends to infinity.

2. The proof of Theorem 1.2

To show Theorem 1.2, we first proceed to establish a recursive formula for $\boldsymbol{\Omega}_n(a; b, c, d, e)$. First of all, we show

LEMMA 2.1. *For integer $n \geqslant 0$ and parameters a, b, c, d, e such that none of the denominator parameters involved is of the form q^{-n}, there holds*

$$\begin{aligned}
&\boldsymbol{\Omega}_{n+1}(bde/aq; b, d, e, bcdeq/a^2) \\
&\quad = \frac{a^2}{bcdeq}\frac{\nabla(cq^3, b, d, e, bde/a)}{\nabla(a/cq, aq/b, aq/d, aq/e, bde/aq)}\boldsymbol{\Omega}_n(bde/a; bq, dq, eq, bcdeq/a^2) \\
&\qquad + \frac{(aq)^2}{bde}\frac{\nabla(aq, bd/aq, be/aq, de/aq)}{\nabla(bde/aq, aq/b, aq/d, aq/e)} \\
&\qquad + \frac{\nabla(b, d, e, a^2q^{1-n}/bde)}{\nabla(bde/aq, aq/b, aq/d, aq/e)}\frac{(bq, dq, eq, bcdeq/a^2; q)_{n+1}}{(a/cq, bd/a, be/a, de/a; q)_{n+1}}\left(\frac{1}{cq^2}\right)^{n+1}.
\end{aligned} \tag{2.1}$$

PROOF. To establish (2.1), we apply Abel's lemma with

$$U_n := \frac{(bq, dq, eq, bde/a^2q; q)_n}{(bd/a, be/a, de/a, aq^2; q)_n} \tag{2.2}$$

and

$$V_n := \left(\frac{1}{cq^3}\right)^n \frac{(aq^2, bcdeq/a^2; q)_{n+1}}{(a/cq, bde/a^2q^2; q)_{n+1}}, \tag{2.3}$$

where the denominator factors in (2.2) and (2.3) are never zero. As a key to using Abel's lemma, it is required to find the differences $U_n - U_{n+1}$ and $V_n - V_{n-1}$ in closed form. First, in view of U_n given by (2.2), we can evaluate directly

$$\begin{aligned} U_n - U_{n+1} &= U_n\left(1-\frac{U_{n+1}}{U_n}\right)\\ &= U_n\left(1-\frac{\nabla\left(bq^{1+n}, dq^{1+n}, eq^{1+n}, bdeq^{n-1}/a^2\right)}{\nabla\left(aq^{n+2}, bdq^n/a, beq^n/a, deq^n/a\right)}\right)\\ &= U_n\,\sigma_n\left(1-\frac{\nabla\left(bq, dq, eq, bde/a^2q\right)}{\nabla\left(aq^2, bd/a, be/a, de/a\right)}\right). \end{aligned} \tag{2.4}$$

Herein and in what follows, the notation σ_n denotes the replacement of parameters

$$\sigma_n\colon (a,b,c,d,e)\mapsto (aq^n, bq^n, c, dq^n, eq^n)$$

and $\sigma_n(F)$ means applying σ_n to the function F. In order to calculate the far-right difference of (2.4), we appeal to (1.8) and make the replacement of parameters

$$(b,c,x,z)\to\left(\frac{aq}{\sqrt{bd}},\frac{\sqrt{b}}{\sqrt{d}},\sqrt{bdq},\frac{\sqrt{bde}}{a}\right).$$

Consequently

$$\nabla\left(bq, dq, eq, \frac{bde}{a^2q}\right)-\nabla\left(aq^2,\frac{bd}{a},\frac{be}{a},\frac{de}{a}\right)=\frac{de}{a}\nabla\left(\frac{aq}{d},\frac{b}{aq},\frac{bdeq}{a},\frac{aq}{e}\right). \tag{2.5}$$

By virtue of (2.5), we can further simplify:

$$\begin{aligned} U_n - U_{n+1} &= U_n\sigma_n\left(-\frac{de}{a}\frac{\nabla\left(aq/d, b/aq, bdeq/a, aq/e\right)}{\nabla\left(aq^2, bd/a, be/a, de/a\right)}\right)\\ &= -U_n\times\frac{deq^n}{a}\frac{\nabla\left(b/aq, aq/d, aq/e, bdeq^{2n+1}/a\right)}{\nabla\left(aq^{n+2}, bdq^n/a, beq^n/a, deq^n/a\right)}\\ &= -\frac{deq^n}{a}\nabla\left(b/aq, aq/d, aq/e, bdeq^{2n+1}/a\right)\frac{(bq, dq, eq, bde/a^2q;q)_n}{(bd/a, be/a, de/a, aq^2;q)_{n+1}}. \end{aligned}$$

In exactly the same way as above, we are able to find the difference $V_n - V_{n-1}$. Referring to (2.3), we easily find

$$\begin{aligned} V_n - V_{n-1} &= -V_{n-1}\left(1-\frac{\nabla(aq^{n+2}, bcdeq^{n+1}/a^2)}{\nabla(aq^{n-1}/c, bdeq^{n-2}/a^2)}\frac{1}{cq^3}\right)\\ &= -V_{n-1}\sigma_n\left(1-\frac{\nabla(aq^2, bcdeq/a^2)}{\nabla(a/cq, bde/a^2q^2)}\frac{1}{cq^3}\right). \end{aligned}$$

Performing as above, we only need to set in (1.8)

$$(b,c,x,z)\to\left(\frac{a^{3/2}\sqrt{q}}{\sqrt{bcde}},\frac{\sqrt{bde}}{\sqrt{acq^{3/2}}},0,\frac{\sqrt{bcdeq^{3/2}}}{\sqrt{a}}\right),$$

yielding

$$\nabla\left(aq^2,\frac{bcdeq}{a^2}\right)-\nabla\left(\frac{bde}{a},cq^3\right)=cq^3\nabla\left(\frac{a}{cq},\frac{bde}{a^2q^2}\right). \tag{2.6}$$

This result allows us to compute

$$V_n - V_{n-1} = V_{n-1}\sigma_n\left(\frac{\nabla(bde/a, cq^3)}{\nabla(a/cq, bde/a^2q^2)}\frac{1}{cq^3}\right)=\frac{\nabla(bdeq^{2n}/a, cq^3)}{\nabla(aq^{n-1}/c, bdeq^{n-2}/a^2)}\frac{V_{n-1}}{cq^3}.$$

Next, on substituting these expressions for $U_n - U_{n+1}$ and $V_n - V_{n-1}$ in Abel's lemma, i.e., Lemma 1.3, it is easily found that

$$\begin{aligned}(2.7)\qquad &\frac{bde}{a^2q}\frac{\nabla(aq/b,aq/d,aq/e)}{\nabla(b,d,e,bde/a^2q^2)}\sum_{k=0}^{n}\frac{\nabla\left(bdeq^{2k+1}/a\right)}{(cq^2)^k}\frac{(b,d,e,bcdeq/a^2;q)_{k+1}}{(a/cq,bd/a,be/a,de/a;q)_{k+1}}\\ &= V_0U_0 - V_nU_{n+1}\\ &\quad+\frac{\nabla\left(cq^3\right)}{\nabla(a/cq,bde/a^2q^2)}\sum_{k=1}^{n}\frac{\nabla(bdeq^{2k}/a)}{(cq^3)^k}\frac{(bcdeq/a^2,bq,dq,eq;q)_k}{(a/c,bd/a,be/a,de/a;q)_k}.\end{aligned}$$

Upon dividing both sides of (2.7) by

$$\frac{bde}{a^2q}\frac{\nabla(aq/b,aq/d,aq/e)}{\nabla(b,d,e,bde/a^2q^2)}$$

we have

$$\begin{aligned}&\sum_{k=0}^{n}\frac{\nabla\left(bdeq^{2k+1}/a\right)}{(cq^2)^k}\frac{(b,d,e,bcdeq/a^2;q)_{k+1}}{(a/cq,bd/a,be/a,de/a;q)_{k+1}}\\ &=\frac{a^2q}{bde}\frac{\nabla(b,d,e,aq^2,bcdeq/a^2)}{\nabla(a/cq,aq/b,aq/d,aq/e)}-V_nU_{n+1}\frac{a^2q}{bde}\frac{\nabla(b,d,e,bde/a^2q^2)}{\nabla(aq/b,aq/d,aq/e)}\\ &\quad+\frac{a^2q}{bde}\frac{\nabla(cq^3,b,d,e)}{\nabla(a/cq,aq/b,aq/d,aq/e)}\sum_{k=1}^{n}\frac{\nabla(bdeq^{2k}/a)}{(cq^3)^k}\frac{(bcdeq/a^2,bq,dq,eq;q)_k}{(a/c,bd/a,be/a,de/a;q)_k}.\end{aligned}$$

Restated in terms of $\mathbf{\Omega}_n(a;b,c,d,e)$, the last identity becomes

$$\begin{aligned}&cq^2\nabla(bde/aq)\left\{\mathbf{\Omega}_{n+1}(bde/aq;b,d,e,bcdeq/a^2)-1\right\}\\ &=\frac{a^2q}{bde}\frac{\nabla(cq^3,b,d,e,bde/a)}{\nabla(a/cq,aq/b,aq/d,aq/e)}\mathbf{\Omega}_n(bde/a;bq,dq,eq,bcdeq/a^2)\\ &\quad+\frac{\nabla(b,d,e,a^2q^2/bde)}{\nabla(aq/b,aq/d,aq/e)}\frac{V_nU_{n+1}-cq^3}{q}\end{aligned}$$

which, by dividing both sides by $cq^2\nabla(bde/aq)$, becomes

$$\begin{aligned}&\mathbf{\Omega}_{n+1}(bde/aq;b,d,e,bcdeq/a^2)\\ &=\frac{a^2}{bcdeq}\frac{\nabla(cq^3,b,d,e,bde/a)}{\nabla(a/cq,aq/b,aq/d,aq/e,bde/aq)}\mathbf{\Omega}_n(bde/a;bq,dq,eq,bcdeq/a^2)\\ &\quad+\frac{\nabla(b,d,e,a^2q^2/bde)}{\nabla(aq/b,aq/d,aq/e,bde/aq)}\frac{V_nU_{n+1}}{cq^3}+\left(1-\frac{\nabla(b,d,e,a^2q^2/bde)}{\nabla(aq/b,aq/d,aq/e,bde/aq)}\right).\end{aligned}$$

It is easy to verify by (1.8) that

$$(2.8)\qquad 1-\frac{\nabla(b,d,e,a^2q^2/bde)}{\nabla(aq/b,aq/d,aq/e,bde/aq)}=\frac{(aq)^2}{bde}\frac{\nabla(aq,bd/aq,be/aq,de/aq)}{\nabla(bde/aq,aq/b,aq/d,aq/e)},$$

and in view of (2.2) and (2.3),

$$\frac{V_nU_{n+1}}{cq^3}=\frac{\nabla(bdeq^{n-1}/a^2)}{\nabla(bde/a^2q^2)}\frac{(bq,dq,eq,bcdeq/a^2;q)_{n+1}}{(a/cq,bd/a,be/a,de/a;q)_{n+1}}\left(\frac{1}{cq^3}\right)^{n+1}.$$

Substituting these into the preceding expression, we get

$$\begin{aligned}&\mathbf{\Omega}_{n+1}(bde/aq;b,d,e,bcdeq/a^2)\\&\quad=f(a;\overrightarrow{M})\mathbf{\Omega}_n(bde/a;bq,dq,eq,bcdeq/a^2)+g_{n+1}(a;\overrightarrow{M}),\end{aligned}$$

where we employ the shortened notation $\overrightarrow{M}:=(b,c,d,e)$ for any four complex numbers b,c,d,e, and

(2.9)

$$f(a;\overrightarrow{M}):=\frac{a^2}{bcdeq}\frac{\nabla(bde/a,cq^3,b,d,e)}{\nabla\left(bde/aq,a/cq,aq/b,aq/d,aq/e\right)},$$

(2.10)

$$\begin{aligned}g_n(a;\overrightarrow{M}):=&\frac{(aq)^2}{bde}\frac{\nabla(aq,bd/aq,be/aq,de/aq)}{\nabla(bde/aq,aq/b,aq/d,aq/e)}\\&+\frac{\nabla(b,d,e,a^2q^{2-n}/bde)}{\nabla(bde/aq,aq/b,aq/d,aq/e)}\frac{(bq,dq,eq,bcdeq/a^2;q)_n}{(a/cq,bd/a,be/a,de/a;q)_n}\left(\frac{1}{cq^2}\right)^n.\end{aligned}$$

This gives the complete proof of (2.1). □

Note that $\nabla(cq^3)=0$ for $cq^3=1$. This simplifies (2.1) to the following equivalent version of (1.5).

COROLLARY 2.2 (The $p=q$ case of Gasper's bibasic sum: [**8**], Eq. (17.7.22)). *For any integer $n\geqslant 0$, there holds*

$$\begin{aligned}\mathbf{\Omega}_n(bde/aq;b,d,e,bde/(aq)^2)=&\frac{(aq)^2}{bde}\frac{\nabla(aq,bd/aq,be/aq,de/aq)}{\nabla(bde/aq,aq/b,aq/d,aq/e)}\\&\times\left(1-\frac{(b,d,e,bde/a^2q^2;q)_{n+1}}{(aq,bd/aq,be/aq,de/aq;q)_{n+1}}\right).\end{aligned}\tag{2.11}$$

To show Theorem 1.2, we utilize yet another useful expression of $\mathbf{\Omega}_n(a;b,c,d,e)$.

LEMMA 2.3. *Let $f(a;\overrightarrow{M})$ and $g_n(a;\overrightarrow{M})$ be given by (2.9) and (2.10). Then there holds*

$$\mathbf{\Omega}_n(bde/aq;b,d,e,bcdeq/a^2)=\sum_{k=0}^{n}g_{n-k}(aq^{2k};q^k\overrightarrow{M})\prod_{i=0}^{k-1}f(aq^{2i};q^i\overrightarrow{M}),\tag{2.12}$$

where $q^k\overrightarrow{M}:=(q^kb,q^kc,q^kd,q^ke)$.

PROOF. For the sake of clarity of our proof, we will employ the shortened notation

$$\mathbf{S}_n(a;\overrightarrow{M}):=\mathbf{\Omega}_n\left(bde/aq;b,d,e,bcdeq/a^2\right).\tag{2.13}$$

Therefore, transformation (2.1) of Lemma 2.1 can be reformulated as

$$\mathbf{S}_{n+1}(a;\overrightarrow{M})=f(a;\overrightarrow{M})\mathbf{S}_n(aq^2;q\overrightarrow{M})+g_{n+1}(a;\overrightarrow{M}).\tag{2.14}$$

In view of (2.13), we see that to show (2.12) is to show

$$\mathbf{S}_n(a;\overrightarrow{M})=\sum_{k=0}^{n}g_{n-k}(aq^{2k};q^k\overrightarrow{M})\prod_{i=0}^{k-1}f(aq^{2i};q^i\overrightarrow{M}).\tag{2.15}$$

We proceed by induction on n. Evidently, when $n = 0$, from Definition (1.3) it follows that $\mathbf{S}_0(a; \overrightarrow{M}) = 1$ and by (2.8) and (2.10), there holds $g_0(a; \overrightarrow{M}) = 1$. Thus (2.15) is true. Next, suppose that (2.15) is valid for $n = K \geqslant 0$, that is,

$$\mathbf{S}_K(a; \overrightarrow{M}) = \sum_{k=0}^{K} g_{K-k}(aq^{2k}; q^k\overrightarrow{M}) \prod_{i=0}^{k-1} f(aq^{2i}; q^i\overrightarrow{M}). \tag{2.16}$$

Now we turn to show the conclusion holds for $n = K + 1$. First, from (2.14) it follows that

$$\mathbf{S}_{K+1}(a; \overrightarrow{M}) = f(a; \overrightarrow{M})\mathbf{S}_K(aq^2; q\overrightarrow{M}) + g_{K+1}(a; \overrightarrow{M}).$$

By applying the induction hypotheses (2.16) to $\mathbf{S}_K(aq^2; q\overrightarrow{M})$, we achieve

$$\begin{aligned}\mathbf{S}_{K+1}(a; \overrightarrow{M}) &= f(a; \overrightarrow{M})\Big(\sum_{k=0}^{K} g_{K-k}(aq^{2k+2}; q^{k+1}\overrightarrow{M}) \prod_{i=0}^{k-1} f(aq^{2i+2}; q^{i+1}\overrightarrow{M})\Big) + g_{K+1}(a; \overrightarrow{M}) \\ &\overset{k\to k-1}{\underset{i\to i-1}{=\!=}} f(a; \overrightarrow{M})\Big(\sum_{k=1}^{K+1} g_{K+1-k}(aq^{2k}; q^k\overrightarrow{M}) \prod_{i=1}^{k-1} f(aq^{2i}; q^i\overrightarrow{M})\Big) + g_{K+1}(a; \overrightarrow{M}) \\ &= g_{K+1}(a; \overrightarrow{M}) + \sum_{k=1}^{K+1} g_{K+1-k}(aq^{2k}; q^k\overrightarrow{M}) \prod_{i=0}^{k-1} f(aq^{2i}; q^i\overrightarrow{M}).\end{aligned}$$

Therefore

$$\mathbf{S}_{K+1}(a; \overrightarrow{M}) = \sum_{k=0}^{K+1} g_{K+1-k}(aq^{2k}; q^k\overrightarrow{M}) \prod_{i=0}^{k-1} f(aq^{2i}; q^i\overrightarrow{M}),$$

which asserts that (2.15) is true for $n = K + 1$. To sum up, (2.15) is valid for arbitrary non-negative n. The lemma is proved. □

Having established Lemmas 2.1 and 2.3, we are now ready to show Theorem 1.2.

PROOF. According to Lemma 2.3, we now calculate directly

$$\prod_{i=0}^{k-1} f(aq^{2i}; q^i\overrightarrow{M}) = \frac{\nabla(bdeq^{k-1}/a)}{\nabla(bde/aq)} \frac{(b, d, e, cq^3; q)_k}{(a/cq, aq/b, aq/d, aq/e; q)_k} \left(\frac{a^2}{bcdeq}\right)^k$$

and

$$\begin{aligned}g_{n-k}(aq^{2k}; q^k\overrightarrow{M}) = &-\frac{aq^{k+1}}{b} \frac{\nabla(aq^{2k+1}, bd/aq, be/aq, aq/de)}{\nabla\left(bdeq^{k-1}/a, aq^{k+1}/b, aq^{k+1}/d, aq^{k+1}/e\right)} \\ &+ \frac{\nabla(bq^k, dq^k, eq^k, a^2q^{2+k}/bde)}{\nabla\left(bdeq^{k-1}/a, aq^{k+1}/b, aq^{k+1}/d, aq^{k+1}/e\right)} \left(\frac{1}{cq^{k+3}}\right)^{n-k} \\ &\times \frac{(bq^{k+1}, dq^{k+1}, eq^{k+1}, bde/a^2q^{k+1}, bcdeq/a^2; q)_{n-k}}{(bd/a, be/a, de/a, aq^{k-1}/c, bde/a^2q^{k+2}; q)_{n-k}}.\end{aligned}$$

These results together lead us to

$$\begin{aligned}
&g_{n-k}(aq^{2k};q^k\overrightarrow{M})\prod_{i=0}^{k-1}f(aq^{2i};q^i\overrightarrow{M})\\
&\quad=-\frac{aq}{b}\frac{\nabla(aq^{2k+1},bd/aq,be/aq,aq/de)}{\nabla(bde/aq)(a/cq;q)_k}\left(\frac{a^2}{bcde}\right)^k\frac{(b,d,e,cq^3;q)_k}{(aq/b,aq/d,aq/e;q)_{k+1}}\\
&\qquad+\frac{\nabla(a^2q^{k+2}/bde)}{\nabla(bde/aq)}\frac{(bde/a^2q^{k+1},bcdeq/a^2;q)_{n-k}}{(bd/a,be/a,de/a,bde/a^2q^{k+2};q)_{n-k}}\\
&\qquad\times\frac{1}{(cq^{k+3})^{n-k}}\frac{(cq^3;q)_k(b,d,e;q)_{n+1}}{(a/cq;q)_n(aq/b,aq/d,aq/e;q)_{k+1}}\left(\frac{a^2}{bcdeq}\right)^k.
\end{aligned}$$

Substituting this back into (2.12), we get

(2.17)

$$\begin{aligned}
&\mathbf{\Omega}_n\left(bde/aq;b,d,e,bcdeq/a^2\right)\\
&\quad=\frac{(aq)^2}{bde}\frac{\nabla(aq,bd/aq,be/aq,de/aq)}{\nabla(bde/aq,aq/b,aq/d,aq/e)}\mathbf{\Omega}_n(aq;b,d,e,cq^3)-\frac{(b,d,e;q)_{n+1}}{(a/cq;q)_n}L_n,
\end{aligned}$$

involving the sum

$$\begin{aligned}
L_n=\sum_{k=0}^{n}&\frac{\nabla(bdeq^{n-2k-2}/a^2)}{\nabla(bde/aq)}\frac{(bcdeq/a^2;q)_{n-k}}{(bd/a,be/a,de/a;q)_{n-k}}\left(\frac{1}{cq^{k+3}}\right)^{n-1-k}\\
&\times\frac{(cq^3;q)_k}{(aq/b,aq/d,aq/e;q)_{k+1}}\left(\frac{a^2}{bcdeq}\right)^{k+1}.
\end{aligned}$$

On considering the relation [**5**, (I.10)]

$$(a;q)_{n-k}=\frac{(a;q)_n}{(q^{1-n}/a;q)_k}\left(-\frac{q}{a}\right)^kq^{\frac{k(k-1)}{2}-nk}\tag{2.18}$$

and further simplifying, we have

$$\begin{aligned}
L_n=&-\frac{\nabla(a^2q^{2-n}/bde)(bcdeq/a^2;q)_n}{\nabla(bde/aq,aq/b,aq/d,aq/e)(bd/a,be/a,de/a;q)_n}\left(\frac{1}{cq^2}\right)^n\\
&\times\sum_{k=0}^{n}\frac{\nabla(a^2q^{2-n+2k}/bde)}{\nabla(a^2q^{2-n}/bde)}\frac{(cq^3,aq^{1-n}/bd,aq^{1-n}/be,aq^{1-n}/de;q)_k}{(a^2q^{-n}/bcde,aq^2/b,aq^2/d,aq^2/e;q)_k}\left(\frac{aq^{n-1}}{c}\right)^k.
\end{aligned}$$

Substituting this into (2.17), we obtain

(2.19)

$$\begin{aligned}
&\mathbf{\Omega}_n\left(bde/aq;b,d,e,bcdeq/a^2\right)\\
&\quad=\frac{(aq)^2}{bde}\frac{\nabla(aq,bd/aq,be/aq,de/aq)}{\nabla(bde/aq,aq/b,aq/d,aq/e)}\\
&\qquad\times\Bigg\{\mathbf{\Omega}_n(aq;b,d,e,cq^3)\\
&\qquad\quad-\frac{\nabla(a/cq^2,bdeq^{n-2}/a^2)}{\nabla(aq,bcdeq^{n+1}/a^2)}\frac{(b,d,e,bcdeq/a^2;q)_{n+1}}{(a/cq^2,bd/aq,be/aq,de/aq;q)_{n+1}}\left(\frac{1}{cq^3}\right)^n\\
&\qquad\quad\times\mathbf{\Omega}_n\left(\frac{a^2q^{2-n}}{bde};\frac{aq^{1-n}}{bd},\frac{aq^{1-n}}{be},\frac{aq^{1-n}}{de},cq^3\right)\Bigg\}.
\end{aligned}$$

Next, we make the replacement of parameters in (2.19)

$$\sigma\colon (a,b,c,d,e)\mapsto (bcd/aq,b,bcde/a^2q^3,d,c).$$

As such, it is very easy to verify

$$\begin{aligned}\sigma(bde/aq,b,d,e,bcdeq/a^2)&=(a,b,d,c,e),\\ \sigma(aq,b,d,e,cq^3)&=(bcd/a,b,d,c,bcde/a^2),\\ \sigma\left(\frac{a^2q^{2-n}}{bde},\frac{aq^{1-n}}{bd},\frac{aq^{1-n}}{be},\frac{aq^{1-n}}{de},cq^3\right)&=\left(\frac{bcdq^{-n}}{a^2},\frac{cq^{-n}}{a},\frac{dq^{-n}}{a},\frac{bq^{-n}}{a},\frac{bcde}{a^2}\right).\end{aligned}$$

Writing $bcde/a^2$ by t, we easily find that transformation (2.19) reduces to (1.6). Hence, Theorem 1.2 is proved. □

REMARK 2.4. Observe that the replacement of parameters σ is invertible, specifically

$$\tau=\sigma^{-1}\colon (a,b,c,d,e)\mapsto (bde/aq,b,e,d,bcdeq/a^2).$$

Upon applying τ to (1.6) of Theorem 1.2, we immediately arrive at (2.19). In this sense, we say that (1.6) and (2.19) are equivalent to each other.

3. A special limit and an open problem

As a direct application of Theorem 1.2, we will consider the limit of (2.19) as n tends to infinity provided that $a=cq^2$.

THEOREM 3.1. *For* $|aq^2/bde|<1<|aq|$, *there holds*

$$\begin{aligned}&\lim_{n\to+\infty}\frac{\mathbf{\Omega}_{n+1}(a^2q^{1-n}/bde;aq,aq^{-n}/bd,aq^{-n}/be,aq^{-n}/de)}{(aq)^n}\\ &\quad=\frac{(q,bde/a^2q,a^2q^2/bde,q/b,q/d,q/e;q)_\infty}{(1/a,bde/a,aq/bde,aq^2/b,aq^2/d,aq^2/e;q)_\infty}.\end{aligned}\tag{3.1}$$

PROOF. Setting $a=cq^2$ and taking $n\to\infty$ yields the following formulas:

$$\begin{aligned}\lim_{n\to\infty}\mathbf{\Omega}_n\left(bde/aq;b,d,e,bcdeq/a^2\right)&={}_6W_5(bde/aq;b,d,e;q,1/a),\\ \lim_{n\to\infty}\mathbf{\Omega}_n(aq;cq^3,b,d,e)&={}_6W_5(aq;b,d,e;q,aq^2/bde).\end{aligned}$$

These reduce (2.19) to

$$\begin{aligned}&{}_6W_5(bde/aq;b,d,e;q,1/a)\\ &\quad-\frac{(aq)^2}{bde}\frac{\nabla(aq,bd/aq,be/aq,de/aq)}{\nabla(bde/aq,aq/b,aq/d,aq/e)}{}_6W_5(aq;b,d,e;q,aq^2/bde)\\ &\quad=\lim_{n\to\infty}\left(\frac{\nabla(a^2q^{2-n}/bde)a^{-n}}{\nabla(bde/aq,aq/b,aq/d,aq/e,bdeq^{n-1}/a)}\frac{(b,d,e,bde/aq;q)_{n+1}}{(q,bd/a,be/a,de/a;q)_n}\right.\\ &\qquad\left.\times\,\mathbf{\Omega}_n(a^2q^{2-n}/bde;aq,aq^{1-n}/bd,aq^{1-n}/be,aq^{1-n}/de)\right).\end{aligned}\tag{3.2}$$

Consequently, once [**5**, (II.20)] is applied to two ${}_6W_5$ sums, we come up with

$$\begin{aligned}\text{LHS of (3.2)} &= \frac{(bde/a, b/a, d/a, e/a; q)_\infty}{(bd/a, be/a, de/a, 1/a; q)_\infty} - \frac{(aq, aq/bd, aq/be, aq/de; q)_\infty}{(aq/b, aq/d, aq/e, aq/bde; q)_\infty} \\ &= \frac{\theta(bde/a, b/a, d/a, e/a) - \theta(bd/a, be/a, de/a, 1/a)}{(bd/a, be/a, de/a, 1/a, aq/b, aq/d, aq/e, aq/bde; q)_\infty} \\ &= \frac{-bde\ \theta(bde/a^2, 1/b, 1/d, 1/e)}{a\ (bd/a, be/a, de/a, 1/a, aq/b, aq/d, aq/e, aq/bde; q)_\infty}.\end{aligned}$$

Note that the last equality results from Weierstrass's theta identity (1.9). On the other hand, it is routine to check

$$\text{RHS of (3.2)} = \frac{-a^2q^2X}{bde\nabla(bde/aq, aq/b, aq/d, aq/e)} \frac{(b, d, e, bde/aq; q)_\infty}{(q, bd/a, be/a, de/a; q)_\infty},$$

where

$$X := \lim_{n\to+\infty} \frac{\boldsymbol{\Omega}_n(a^2q^{2-n}/bde; aq, aq^{1-n}/bd, aq^{1-n}/be, aq^{1-n}/de)}{(aq)^n}.$$

Substituting all these into (3.2) and then solving for X, we obtain

$$\begin{aligned}X = &-\frac{bde}{a} \frac{\theta(bde/a^2, 1/b, 1/d, 1/e)}{(bd/a, be/a, de/a, 1/a, aq/b, aq/d, aq/e, aq/bde; q)_\infty} \\ &\times \frac{-bde\nabla(bde/aq, aq/b, aq/d, aq/e)}{a^2q^2} \frac{(q, bd/a, be/a, de/a; q)_\infty}{(b, d, e, bde/aq; q)_\infty}.\end{aligned}$$

A bit of simplification, in which the relation $\theta(x) = -x\theta(1/x) = (x, q/x; q)_\infty$ is applied and cancellation occurs, yields

$$X = \frac{1}{aq} \frac{(q, bde/a^2q, a^2q^2/bde, q/b, q/d, q/e; q)_\infty}{(1/a, bde/a, aq/bde, aq^2/b, aq^2/d, aq^2/e; q)_\infty}.$$

That is,

$$\begin{aligned}&\lim_{n\to+\infty} \frac{\boldsymbol{\Omega}_n(a^2q^{2-n}/bde; aq, aq^{1-n}/bd, aq^{1-n}/be, aq^{1-n}/de)}{(aq)^{n-1}} \\ &\quad = \frac{(q, bde/a^2q, a^2q^2/bde, q/b, q/d, q/e; q)_\infty}{(1/a, bde/a, aq/bde, aq^2/b, aq^2/d, aq^2/e; q)_\infty}.\end{aligned}$$

Shifting n to $n+1$, we obtain (3.1). □

In the above discussions, we have confined ourselves to basic (without mentioning theta or elliptic) hypergeometric series. As far as the elliptic hypergeometric series ${}_{r+1}V_r$ (see [**5**, Eq. (11.2.19)] for its precise definition) is concerned, there naturally arises a question which, to end our paper, we propose as an open problem.

OPEN PROBLEM 3.2. *Define the elliptically balanced series*

$$\Omega_n(a; b, c, d, e, f; q, p) := {}_{14}V_{13}(a; b, c, d, e, f, a^3q/bcdef, q, q^{-n}, aq^{n+1}; q, p; q).$$

Does there exist an elliptic extension of Theorem 1.2 *to* $\Omega_n(a; b, c, d, e, f; q, p)$ *?*

The validity of this open problem is based on the referee's observation: the summation formula (1.5) is covered by Frenkel and Turaev's ${}_{12}V_{11}$ transformation [**5**, Eq. (11.5.1)] as the special case when first p and then f tend to zero. In our opinion, any positive solution to this problem may depend on whether there exists an elliptic recurrence relation similar to Lemma 2.1.

Acknowledgments

The authors are very indebted to the anonymous referees for their careful reading and extremely insightful comments on the earlier versions of this manuscript.

References

[1] G. E. Andrews, *Applications of basic hypergeometric functions*, SIAM Rev. **16** (1974), 441–484, DOI 10.1137/1016081. MR352557

[2] V. Y. B. Chen, W. Y. C. Chen, and N. S. S. Gu, *The Abel lemma and the q-Gosper algorithm*, Math. Comp. **77** (2008), no. 262, 1057–1074, DOI 10.1090/S0025-5718-07-01968-0. MR2373192

[3] W. Chu, *Bailey's very well-poised ${}_6\psi_6$-series identity*, J. Combin. Theory Ser. A **113** (2006), no. 6, 966–979, DOI 10.1016/j.jcta.2005.08.009. MR2244127

[4] W. Chu and W. Zhang, *Partial sums of Bailey's ${}_6\psi_6$-series to faster convergent series*, J. Difference Equ. Appl. **18** (2012), no. 2, 239–260, DOI 10.1080/10236190903460388. MR2888954

[5] G. Gasper and M. Rahman, *Basic hypergeometric series*, 2nd ed., Encyclopedia of Mathematics and its Applications, vol. 96, Cambridge University Press, Cambridge, 2004, DOI 10.1017/CBO9780511526251. MR2128719

[6] K. Knopp, *Theory and application of infinite series.* Dover Books on Mathematics, Dover Publications, 1990.

[7] T. H. Koornwinder, *On the equivalence of two fundamental theta identities*, Anal. Appl. (Singap.) **12** (2014), no. 6, 711–725, DOI 10.1142/S0219530514500559. MR3271026

[8] F. W. J. Olver, D. W. Lozier, R. F. Boisvert, and C. W. Clark (eds.), *NIST handbook of mathematical functions*, U.S. Department of Commerce, National Institute of Standards and Technology, Washington, DC; Cambridge University Press, Cambridge, 2010. With 1 CD-ROM (Windows, Macintosh and UNIX). MR2723248

[9] J. Wang, *A new elliptic interpolation formula via the (f,g)-inversion*, Proc. Amer. Math. Soc. **148** (2020), no. 8, 3457–3471, DOI 10.1090/proc/15043. MR4108852

[10] J. Wang and X. Ma, *A common recursive formula for Ramanujan's ${}_1\psi_1$ and Bailey's very-well-poised ${}_6\psi_6$ summation formulas*, Proc. Amer. Math. Soc. **152** (2024), no. 1, 163–175, DOI 10.1090/proc/16434. MR4661072

[11] E. T. Whittaker and G. N. Watson, *A course of modern analysis—an introduction to the general theory of infinite processes and of analytic functions with an account of the principal transcendental functions*, 4th ed. (V. H. Moll, ed.), Cambridge University Press, Cambridge, 2021. MR4286926

Department of Mathematics, Soochow University, Suzhou 215006, People's Republic of China

Email address: 20224007007@stu.suda.edu.cn

Department of Mathematics, Soochow University, Suzhou 215006, People's Republic of China

Email address: yingzhang0626@163.com

Department of Mathematics, Soochow University, Suzhou 215006, People's Republic of China

Email address: xrma@suda.edu.cn

For a complete list of titles in this series, visit the
AMS Bookstore at **www.ams.org/bookstore/conmseries/**.